Shock Wave and High Pressure Phenomena

Shock Wave and High Pressure Phenomena

M. V. Zhernokletov   B. L. Glushak (Eds.)

Material Properties under Intensive Dynamic Loading

In Collaboration with
W.W. Anderson
F.J. Cherne
M.A. Zocher

With 311 Figures

 Springer

Editors:

Mikhail V. Zhernokletov
Boris L. Glushak

The Russian Federal Nuclear Center
All-Russia Scientific Research Institute
of Experimental Physics (VNIIEF)
37, Mira Prospekt
607190, Russia
E-mail: root@gdd.vniief.ru

In Collaboration with:

William W. Anderson
Frank J. Cherne
Marvin A. Zocher

Los Alamos National Laboratory
XI MS F 663
Los Alamos, NM 87545, USA
E-mail: zocher@lanl.gov

Series Editors-in-Chief:

Lee Davison

39 Cañoncito Vista Road
Tijeras, NM 87059, USA
E-mail: leedavison@aol.com

Yasuyuki Horie

AFRL/MNME Munitions Directorate
2306 Perimeter Road
Eglin AFB, FL 32542, USA
E-mail: yasuyuki.horie@eglin.af.mil

Library of Congress Control Number: 2006931202

ISSN 8063-7200
ISBN-10   3-540-36844-2 Springer Berlin Heidelberg New York
ISBN-13   978-3-540-36844-1 Springer Berlin Heidelberg New York

Springer is a part of Springer Science+Business Media
springer.com
© Springer-Verlag Berlin Heidelberg 2006

Typesetting: by the authors and techbooks using a Springer LaTeX macro package
Cover design: WMX design GmbH, Heidelberg

Printed on acid-free paper   SPIN: 11413035   54/techbooks   5 4 3 2 1 0

Preface

This book is the result of collaboration between the Russian Federal Nuclear Center – All Russian Scientific Research Institute of Experimental Physics (RFNC-VNIIEF) located in Sarov, Russia and the University of California – Los Alamos National Laboratory (UC-LANL). The genesis of this project was the compilation of a set of lecture notes used by a number of leading VNIIEF researchers in courses taught to students of the Sarov Physical and Technical Institute (SarFTI) specializing in "Theoretical and Experimental Mechanics." A revised and significantly supplemented version of those lecture notes was ultimately published as a monograph (in Russian) by VNIIEF Press, and is used today as a textbook in courses on shock mechanics being taught at SarFTI. Recognizing the potential benefit of the manuscript to students and researchers in the field of shock mechanics in the English-speaking world, a VNIIEF/LANL collaboration was established to revise/translate/update the manuscript. This book is the result of that effort.

Understanding the physical and thermomechanical response of materials subjected to intensive dynamic loading is a challenge of great significance in engineering today. When intensive dynamic loads, such as those that result from the detonation of high explosives (HE), high-velocity impact, or rapid localized heating (such as might develop under the effects of incident laser light or relativistic electron beam), are applied to condensed matter, the result is a complex pattern of material flow involving waves of discontinuous (shock) as well as continuous (expansion) nature.

Shock compression (followed by expansion) precipitates both reversible and irreversible physical, physicochemical and mechanical processes in the material. These processes include (but are not limited to), strong compression of solids, extremely high heating rates, phase transformations, electronic structure change, work hardening, thermal softening, and the initiation and evolution of damage (such as spall). The process of detonation, manifest in the rapid release of energy that occurs as a result of the chemical transformation that takes place when a shock wave (of sufficient strength) propagates through HE occupies a special niche in the shock mechanics discipline.

Accurate diagnosis of the response of materials to intensive dynamic loading, an understanding of the fundamental mechanisms that are responsible for the observed material behavior, and the development of predictive mathematical models that are capable of faithfully forecasting material response under a variety of loading scenarios, are a necessity for the successful solution of a variety of engineering problems. Moreover, shock waves (and an understanding of shock mechanics) play a significant role in scientific studies aimed at gaining a better understanding of the behavior of materials under intensive loading. Material thermodynamic properties under high and ultrahigh pressures, and the rheological properties of materials (primarily metals) subjected to high strain rate loading, are but two examples wherein shock mechanics is the scientists fundamental tool of exploration. The many challenges in science and engineering, such as those alluded to above, serve as strong motivation for the study of shock and detonation waves in condensed media.

Purposeful studies of material dynamics in extreme states (under conditions of intensive dynamic loading) took on elevated importance both in Russia and abroad in the 1940s, and became the focus of intense research efforts in places like Los Alamos and Sarov (at the time known as Arzamas-16). Results from these studies demonstrated both the uniqueness and variety of material behavior under intensive dynamic loading. Through the intervening years, a wealth of experimental data on this problem has been acquired, made possible by significant progress in the development and use of equipment for producing very intense loads and the development of a number of discrete and continuous diagnostic methods for recording the very fast processes that are important in studies of the behavior of materials under the conditions of shock loading.

Some of the early studies in shock mechanics focused on the compressibility of materials. Notable are systematic efforts beginning in the early 1960s at VNIIEF (Arzamas-16) investigating shock wave structure, material resistance to short duration tensile stresses, plastic deformation of shock-compressed bodies, and expansion isentropes (similar efforts were taking place in research centers outside of Russia). Recently, much attention has been given to a study of structural variations brought about by intense mechanical and thermal loading, and to determining the relationships between the field variables (essential for the development of accurate predictive models).

The quantity of information being published in scientific technical journals and in conference proceedings pertaining to the behavior of materials subjected to the conditions of shock wave loading is continuously increasing. This testifies to the present-day importance of this topic in science and engineering. Several well-known books are presently available, which provide analysis of some of the issues related to the challenge of studying material behavior under the conditions of shock wave loading [1–5]. While these books provide an excellent treatment of many of the theoretical aspects of shock mechanics, and in many cases of the behavior of materials under intense dynamic loading, it is our view that insufficient attention is paid to the experimental methods and diagnostics. This book is an attempt to supplement these excellent works by focusing on just these areas: experimental and diagnostic methods.

It is our view that insignificant attention has been given to a formal presentation of the methods and diagnostics that are important in obtaining an accurate understanding of the processes that take place under shock loading. In many cases, methods have been handed down from practitioner to practitioner, with little dissemination of knowledge. This volume is an attempt to rectify this. Herein we attempt to elucidate a variety of techniques that have been used or are being used today in studies of shock mechanics. We include discussions of some methods that were used in the past but are no longer used today because we believe that the reader will be well served by placing the current state-of-the art in historical context.

The authors expect that this book will be of interest to scientists, engineers, students, and postgraduates who specialize in areas of physics/engineering wherein an understanding of the behavior of materials in conditions involving intensive dynamic loading is important. A partial, but by no means exclusive list, would include those working in the areas of high pressure and high temperature physics, shock wave physics and the phenomena of fast processes, constitutive behavior (equations of state and deviator), damage mechanics, and the optical properties of compressed materials. We believe that this volume could be adopted as a textbook or could be used as supplementary reading material for courses dealing with shock physics. We believe that practitioners who are already working in the field will find this volume to be a useful source of reference material.

Acknowledgements

We would like to thank the following people for many helpful discussions as we tried to clarify the ideas presented in this book: Jonathan Boettger, Richard L. Gustavsen, Larry G. Hill, Kevin G. Honnell, Darcie Dennis-Koller, Dean L. Preston, and Paulo Rigg. In addition, we would like to thank Lee Davison and Yasuyuki Horie for their encouragement as we struggled to make this work a reality.

References

1. Reinchart, J.A., and Pirson, J., *Metal Behavior Under Pulsed Loads*, Izdatelstvo Inostrannoy Literatury Publ., Moscow, 1958.
2. Zeldovich, Ya.B., and Raizer, Yu.P., *Physics of Shock Waves and High-Temperature Hydrodynamic Phenomena*, Nauka Publ., Moscow, 1966, [English trans., Academic Press, NY, Vol. 1 (1966), Vol. 2 (1967); Reprint in a single volume by Dover Publ., Mineola, NY, 2002].
3. Baum, F.A., Orlenko, L.P., Stanyukovich, K.P., et al., *Detonation Physics*, Nauka Publ., Moscow, 1975.

4. Selivanov, V.V., Solovyev, V.S., and Sysoyev, S.S., *Shock and Detonation Waves. Investigation Methods*, MSU Publishing House, Moscow, 1990.
5. Glushak, B.L., Kuropatenko, V.F., and Novikov, S.A., *Study of Material Strength in Dynamic Loading*, Nauka Publ., Novosibirsk, 1992.

RFNC-VNIIEF, Sarov, Russia

Boris. L. Glushak
Mikhail V. Zhernokletov

UC-LANL, Los Alamos, NM, USA

William W. Anderson
Frank J. Cherne
Marvin A. Zocher

Contents

Nomenclature

For the most part, industrial nations around the world use essentially the same engineering materials. Despite this fact, the same material may be known by several different names around the world. Since this book originated in Russia, materials in the text are referred to by their common Russian name or abbreviation. These names may not be familiar to readers from other countries. Therefore we provide as an aid to the reader this table of the abbreviations used in this text with a precise description and composition of the material.

Table. List of materials.

Abbreviation used	Composition (analogue if available)
	Steels
Steel 12Kh18N10T	69.3% Fe + 18% Cr + 10%Ni + 2.7% others (*stainless steel*)
Steel KhS38	95.3% Fe + 0.42% C + 1.4% Si + 0.6% Mn + 1.6%Cr + 0.67% others
Steel St.3	(*Armco-iron*)
Steel St.20 (low-carbon steel)	98.63% Fe + 0.17% C + 0.17% Si + 0.35% Mn + 0.68% others
Steel 45 (medium-carbon steel)	98.54% Fe + 0.48% C + 0.33% Si + 0.65% Mn
Steel 30KhGSA	93.44% Fe + 0.28% C + 0.9% Si + 0.8% Mn + 0.8% Cr + 0.35% others
Steel 40Kh	97.77% Fe + 0.4% C + 0.19% Si + 0.74% Mn + 0.9% Cr

Steels (continued)	
Steel EP712	93.44% Fe + 0.16% C + 2.4% Cr + 1.2%Ni + 1.4% W + 1.4% others
Steel 36NKhTYu	Fe + 0.36% C + (Ni+Cr+Ti+Al)
Aluminum and its alloys	
Aluminum D1	91.55% Al + 4.8% Cu + 0.8% Mg + 2.85% others (*Al 2017*)
Aluminum AD1	Pure aluminum (Al no less than 99.3%) (*Al 1230*)
Aluminum alloy V95	86% Al + 7.0% Zn + 2.8% Mg + 2.0% Cu + 2.2% others (*Al 7075*)
Aluminum D16	93.25% Al + 3.8% Cu + 1.8% Mg + 1.15% others (*Al 2024*)
Aluminum AMg6	92% Al + 6.8% Mg + 1.2% others
Other metals and alloys	
Titanium alloy VT-14	86.6% Ti + 6.3% Al + 3.8% Mo + 1.9% V + 1.4% others
Titanium alloy PT-3V	99% Ti + 0.1% C + 0.25% Fe + 0.12% Si + 0.3% Zr + 0.23% others
Brass LS59-1	60% Cu + 1.4% Pb + 38.07% Zn + 0.53% others (*C*3800)
Mg95	Pure magnesium (Mg no less than 95%)
Ceramics	
KVPT	97% Al_2O_3 + 1% TiO_2 + 0.2% Fe_2O_3 +1.8% others (High density corundum with titanium)
Chylumin	93% Al_2O_3 + 4.9% SiO_2 + 0.25% TiO_2 + 0.85% Fe_2O_3 +1% CaO
Polymers	
Plexiglas	PMMA (polymethylmethacrylate)
Polytetrafluorethylene	Fluorine plastic $[-CF_2-CF_2-]_n$ (*Teflon*)
Caprolon	$[-NH(CH_2)_5CO-]_n$

1

Basic Principles of Continuum Mechanics

B.L. Glushak

A solid body is composed of a great number of elemental components (e.g., molecules). As a consequence, the response of that body to some arbitrary loading depends, fundamentally, on the interaction of each of those components with its environment. Predicting the body's response, taking into account the action of each elemental component, however, is impractical in most engineering applications. For this purpose, we need another approach. The approach most-often adopted, is to treat the body as a continuum. In taking this approach, we consider an idealized continuous body, one devoid of microstructure, as representative in an averaged sense of the real body.

A medium is called a continuum, if its volume contains an apparent continuity of material mass over the physical scale of the problem of interest. In general this requires the domain of interest to be several orders of magnitude larger than the length scale of the elemental components (characteristically on the order of 10^{-7}–10^{-8}cm). Thus, the domain of interest can, in general, be rather small and still fit the basic description of a continuum. We also assume that all mathematical functions (e.g., velocity or displacement fields) used to describe the state of the body are continuous, except possibly at a finite number of interior surfaces separating regions of continuity. Note that there is no restriction against the body being multiply connected.

Thus, under the continuum assumption, velocity is a continuously varying function, existing at a mathematical point in a mathematical domain (or on the boundary of that domain), and represents in an averaged sense, the velocities of the elemental components of the physical body in the vicinity of that point. In a similar way, volume is representative of inter-atomic spacing and temperature is representative of atomic thermal vibration.

This chapter summarizes some of the most fundamental aspects of continuum mechanics, especially with regard to describing the behavior of continuous media subjected to intensive dynamic loading. Topics to be discussed include: constitutive behavior (both spherical and deviatoric) and elements of shock and detonation wave theory (at least to the degree of completeness needed for analysis and treatment of the results of dynamic experiments). The interested reader can find detailed discussions of continuum mechanics in a number of monographs and manuals, e.g., in [1–5].

1.1 Constitutive Relationships in a Solid Body

The state of stress at a given point in a continuum is characterized by the symmetric stress tensor σ_{ik} (symmetric under the assumption that body moments are negligible or nonexistent). The components of the stress tensor σ_{ik} are a projection of force per unit area onto a plane in the body, with its normal represented by the i axis, and acting in the direction of the k axis. Components σ_{xx}, σ_{yy}, σ_{zz} are normal stresses, components σ_{ik} $(i \neq k)$ are tangential or shear stresses. The stresses are usually taken as positive in tension and negative in compression. When studying shock-wave phenomena, however, it is convenient to consider compressive stresses as positive, which is the convention adopted herein.

Regardless of the state of stress at each point in the medium, it is always possible to find three mutually perpendicular planes, on which the tangential stresses vanish, i.e. $\sigma_{ik} = 0$ $(i \neq k)$. The normals to these planes form the principal directions, or principal axes of the stress tensor. These directions depend only on the state of stress (projection of force $\vec{F}$) and are independent of the chosen coordinate system x, y, z.

Shear stresses are denoted τ_{ik} $(\tau_{ik} = \sigma_{ik}(i \neq k))$. We may rank the principal stresses by magnitude $\sigma_1 \geq \sigma_2 \geq \sigma_3$. Having done so, the maximum shear stress acts on the area bisecting the angle between the major and minor principal directions, and is equal in magnitude to half the difference between the major and minor principal stresses: $\tau_{\max} = (\sigma_1 - \sigma_3)/2$.

The symmetric stress tensor can always be divided into isotropic (or spherical) and deviatoric parts. Denoting σ_{ik} as T_σ, let us subdivide T_σ as follows: $T_\sigma = \Gamma_\sigma + D_\sigma$, where Γ_σ is the spherical stress, and D_σ is the stress deviator. The spherical part Γ_σ is given as,

$$\Gamma_\sigma = \begin{vmatrix} \sigma_0 & 0 & 0 \\ 0 & \sigma_0 & 0 \\ 0 & 0 & \sigma_0 \end{vmatrix} ; \qquad \sigma_0 = \frac{\sigma_{ii}}{3}$$

and the deviator D_σ as,

$$D_\sigma = \begin{vmatrix} \sigma_{xx} - \sigma_0 & \sigma_{xy} & \sigma_{xz} \\ \sigma_{yx} & \sigma_{yy} - \sigma_0 & \sigma_{yz} \\ \sigma_{zx} & \sigma_{zy} & \sigma_{zz} - \sigma_0 \end{vmatrix} ; \qquad D_{ik} = \sigma_{ik} - \sigma_0 \delta_{ik}.$$

The term δ_{ik} represents the Kroneker delta. Note that we are in this instance using D_{ik} to denote the stress deviator which commonly appears as

S_{ij} in western literature. The σ_0 component of the isotropic stress is equivalent to hydrostatic pressure, and consequently relates to changes in volume V, or volumetric strain at a point. On inspection, the principal stresses of the isotropic tensor are equal to the algebraic mean of the normal stresses.

The stress deviator is responsible for shape change at a given point, and characterizes the degree of deviation of a given stress state from uniform compression or expansion.

In a manner somewhat analogous to stress, the state of strain is characterized by a symmetric linear Eulerian tensor, ε_{ik}, where each component represents the strain perpendicular to axis i in the direction of axis k, and $\varepsilon_{ik} = (\partial S_i/\partial x_k + \partial S_k/\partial x_i)/2$, with S_i being the displacement. Components ε_{xx}, ε_{yy}, ε_{zz} are normal strains describing relative elongations, shear or tangential strains are described with ε_{xy}, ε_{xz}, ε_{yx}, ε_{yz}, ε_{zx}, ε_{zy}. As for stresses, we consider strains as positive in compression and negative in tension. At each point in the body, it is possible to find three mutually perpendicular directions, referred to as principal strain axes, whereupon shear strains are zero and the body experiences a change in length only. We may rank the principal strains by magnitude, $\varepsilon_1 \geq \varepsilon_2 \geq \varepsilon_3$. In an isotropic elastic body, the principal axes of stress and strain coincide. The maximum value of the shear strain, γ_{max}, is given as: $\gamma_{max} = \varepsilon_1 - \varepsilon_3$.

The strain tensor, denoted here as T_ε ($T_\varepsilon = \varepsilon_{ik}$), may be split into isotropic and deviatoric parts, $T_\varepsilon = \Gamma_\varepsilon + D_\varepsilon$ with components:

$$\Gamma_\varepsilon = \begin{vmatrix} \varepsilon_0 & 0 & 0 \\ 0 & \varepsilon_0 & 0 \\ 0 & 0 & \varepsilon_0 \end{vmatrix} ; \qquad D_\varepsilon = \begin{vmatrix} \varepsilon_{xx} - \varepsilon_0 & \varepsilon_{xy} & \varepsilon_{xz} \\ \varepsilon_{yx} & \varepsilon_{yy} - \varepsilon_0 & \varepsilon_{yz} \\ \varepsilon_{zx} & \varepsilon_{zy} & \varepsilon_{zz} - \varepsilon_0 \end{vmatrix} ,$$

where $\varepsilon_0 = \varepsilon_{ii}/3$.

The spherical part of the strain tensor Γ_ε is related to volume change only, tensor D_ε relates to shape change only, and provides a measure of the degree to which the strained state differs from uniform compression (or expansion).

The relative change in specific volume (ε), is equal to the sum of the relative elongations along the three orthogonal directions drawn through a given point, i.e., $\varepsilon = \varepsilon_{xx} + \varepsilon_{yy} + \varepsilon_{zz}$.

Strain rate components are related to the velocity components v_i, as follows:

$$\xi_{xx} = \frac{\partial v_x}{\partial x} ; \qquad \xi_{yy} = \frac{\partial v_y}{\partial y} ; \qquad \xi_{zz} = \frac{\partial v_z}{\partial z} ;$$

$$h_{xy} = \frac{1}{2}\left(\frac{\partial v_x}{\partial y} + \frac{\partial v_y}{\partial x}\right) ; \quad h_{yz} = \frac{1}{2}\left(\frac{\partial v_y}{\partial z} + \frac{\partial v_z}{\partial y}\right) ; \quad h_{zx} = \frac{1}{2}\left(\frac{\partial v_z}{\partial x} + \frac{\partial v_x}{\partial z}\right) ,$$

which forms a symmetric tensor of strain rates.

ξ_{xx}, ξ_{yy}, ξ_{zz} represent rates of relative elongations in the direction of the coordinate axes, and h_{xy}, h_{yz}, h_{xz} represent rates of angular skewing of originally right angles.

The rate of relative volume change is $\xi = \xi_{xx} + \xi_{yy} + \xi_{zz} = \text{div } \vec{v}$. The above relations for the strain rates are correct under the assumption that in an isotropic medium, the principal axes of the strain rate tensor coincide with those of the strain tensor.

Given principal stresses σ_1, σ_2, σ_3, the so-called effective or equivalent stress, σ_i (characterizing the resistance of the elementary volume to irreversible shape change) is:

$$\sigma_i = \frac{\sqrt{2}}{2}\sqrt{(\sigma_1 - \sigma_2)^2 + (\sigma_2 - \sigma_3)^2 + (\sigma_3 - \sigma_1)^2} \; .$$

Analogously, an effective or equivalent strain, ε_i (characterizing the change in shape of the elementary volume), is given as:

$$\varepsilon_i = \frac{\sqrt{2}}{3}\sqrt{(\varepsilon_1 - \varepsilon_2)^2 + (\varepsilon_2 - \varepsilon_3)^2 + (\varepsilon_3 - \varepsilon_1)^2}$$

Substituting ξ_1, ξ_2, ξ_3 for ε_1, ε_2, ε_3 in the previous expression, an effective or equivalent strain rate, ξ_i may be written as:

$$\xi_i = \frac{\sqrt{2}}{3}\sqrt{(\xi_1 - \xi_2)^2 + (\xi_2 - \xi_3)^2 + (\xi_3 - \xi_1)^2} \; .$$

For any body undergoing deformation, the first law of thermodynamics must hold. For an adiabatic process (extensive thermodynamic properties), the first law may be stated as:

$$dE = TdS - \frac{1}{\rho}\sigma_{ik}\,d\varepsilon_{ik} \; .$$

1.2 Equation of State of a Solid Body

The physical domain of interest is modeled as a continuum having certain physical and mechanical properties. Stress and strain tensors may be split into isotropic and deviatoric parts, allowing us to predict the behavior of the medium under both static and dynamic loading with separate descriptions of material constitutive behavior under hydrostatic and non-hydrostatic circumstances.

Hydrostatic constitutive behavior is characterized by a so-called equation of state (not an equation of state in the classical sense since history or path dependence may enter), establishing the relationship between pressure P (average stress σ_0), density ρ or specific volume V (average strain ε_0) and temperature T or specific internal energy E.

The most common equations of state, are simple statistical-mechanical models based on a quasi-harmonic approximation (such as in the Einstein and Debye models), in which a crystal is represented as a set of independent harmonic oscillators. Representative of this class of models is the Mie-Grüneisen equation of state (EOS) [6]:

$$P = P_x + \frac{\Gamma(V)}{V}\frac{3R}{A}TD\left(\frac{\theta}{T}\right), \qquad E = E_x + \frac{3R}{A}TD\left(\frac{\theta}{T}\right), \qquad (1.1)$$

where $P_x = -(dE_x/dV)$ is the cold compression curve ($T = 0$ K isotherm), $D(\theta/T)$ is the Debye function (equal to unity when $T \gg \theta$), θ is the Debye temperature, R is the gas constant, A is the atomic weight, and Γ is the Grüneisen factor of the lattice. One can eliminate temperature from Eq. (1.1) to obtain the caloric form of the Mie-Grüneisen equation of state:

$$P = P_x + \frac{\Gamma(V)}{V}(E - E_x). \qquad (1.2)$$

The Grüneisen gamma, $\Gamma(V)$, is estimated from experimental results involving shock compression of crystalline and porous samples, or is calculated in accordance with some approximate model that attempts to account for the relationship between average frequency and volume [6]. The first method should be preferred. In a region of small compressions, Γ is typically taken as a constant equal to $\Gamma(V = V_0)$. At higher compressions, $\Gamma(V)$ is frequently taken as $\Gamma = \Gamma_0(V/V_0)$. A useful generalized form for $\Gamma(V)$ is given by [7]:

$$\Gamma(V) = \Gamma_\infty + (\Gamma_0 - \Gamma_\infty)\left(\frac{V}{V_0}\right)^M, \qquad (1.3)$$

where Γ_∞, Γ_0, and M are constants, $\Gamma_0 = \Gamma(V = V_0)$, $\Gamma_\infty = \Gamma(V \ll V_0)$.

It is necessary to account for oscillation anharmonicity in the interatomic (ionic) potential in order to accurately predict "high-temperature" effects. Examples of models which account for these effects can be found in [8,9].

In addition, at temperatures on the order of several tens of thousands of degrees K, it becomes necessary to account for contributions from thermally excited electrons. At temperatures small compared to the Fermi energy, the contribution of thermally excited electrons to thermal pressure and thermal energy can be approximated by [2,8]:

$$P_{el} = \frac{\gamma_{el}}{2V}\beta T^2 ; \qquad E_{el} = \beta\frac{T^2}{2}, \qquad (1.4)$$

where $\beta = \beta_0(V/V_0)^{\gamma_{el}}$ is an electronic heat capacity factor, $\beta_0 = \beta(V = V_0)$, and γ_{el} is an analog of the Grüneisen factor for electrons ($\gamma_{el} = 0.5 - 0.7$).

The expanded form of the equation of state (1.1) which accounts for electronic effects is:

$$P = P_x + \frac{\Gamma(V)}{V}\frac{3R}{A}TD\left(\frac{\theta}{T}\right) + \frac{\gamma_{el}}{2V}\beta T^2,$$

$$E = E_x + \frac{3R}{A}TD\left(\frac{\theta}{T}\right) + \beta\frac{T^2}{2}. \tag{1.5}$$

If $T \gg \theta$, the Debye function is taken equal to one.

At present, it is difficult to construct accurate $P_x(V)$ and $E_x(V)$ from first principles only; experimental data and phenomenology are required. For example, the functions $P_x(V)$ and $\Gamma(V)$ may be empirically determined from experimental Hugoniots of continuous and porous materials by simultaneous solution of Eqs. (1.3), (1.5) and the Hugoniot equation.

Let us also note that it is often convenient to express the density in the form of a dimensionless compression (or extension), $\delta = V_0/V = \rho/\rho_0$, and the cold curve pressure as a function of this dimensionless quantity, that is, $P_x(\delta)$ instead of the relation $P_x(V)$.

A number of proposed analytical equations of state for the cold curve, $P_x(\delta)$ are presented in Table 1.1.

Table 1.1. Analytical dependencies $P_x(\delta)$ [5, 6, 8, 9, 10]

$\frac{\rho_0 c_0^2}{n}\left(\delta^n - 1\right)$	$n > 0$
$\frac{\rho_0 c_0^2}{n}\left(\delta^n - \delta^k\right)$	$n > 0, k > 0, n > k$
$A_i\delta^{n_i} + B_i\delta^{m_i}$	$A_i > 0,\ B_i < 0,\ n_i > 0,\ m_i > 0$
$\sum\limits_{n=1}^{7} a_n\delta^{i/3+1}$	a_n is an alternating-sign coefficient
$\frac{3}{2}B_0\left(\delta^{7/3} - \delta^{5/3}\right)\left[1 - \xi\left(\delta^{2/3} - 1\right)\right]$	$\xi = 0.75(4 - B_0'),\ B_0' = \left(\frac{\partial B}{\partial P}\right)_{T=0},$ $B = \rho_0 c_0^2$
$Q\left\{\delta^{2/3}\exp\left[q\left(1 - \delta^{-1/3}\right)\right] - \delta^{4/3}\right\}$	Q, q – constants
$B\rho^{5/3}\left(1 + \sum\limits_{n=0}^{5} \beta_n\rho^{-n/3}\right)$	β_n is an alternating-sign coefficient

Table 1.1 demonstrates a great variety of the analytical forms for the dependency of pressure on the degree of compression, $P_x(\delta)$. All the formulas are constructed so that they contain both positive terms describing repulsive interatomic forces and negative terms describing attractive forces, as well as coefficients and exponents determined from experimental data.

The compressible co-volume model [11] is a modification of the Van der Waals model, removing its limitation that the material specific volume can not be less than the natural volume of molecules, the co-volume, V_n, which the Van der Waals model treats as a constant. In this model the equation of state is:

$$P = \frac{RT}{V - V_n} + P_{attr}(V) \tag{1.6}$$

where the co-volume is a function of kinetic pressure only,

$$V_n = V_n(P - P_{attr}(V)) . \tag{1.7}$$

In Eqs. (1.6) and (1.7), R is the gas constant, P is total pressure, and $P_{attr}(V)$ is the attractive pressure. The second term on the right-hand side of Eq. (1.6) describes the interparticle attraction in the mean-field approximation.

An expression for internal energy that asymptotically approaches ideal gas behavior for large V is:

$$E = E_{id}(T) - \int_{V_O}^{V_n} P_{rep}(V_n)dV - \int_{V_n}^{V} P_{attr}(V)dV . \tag{1.8}$$

The empirical functions $P_{rep}(V)$ and $P_{attr}(V)$ are represented with simple functions with a small number of constants selected using Hugoniot data and cohesive energies.

The compressible co-volume model is discussed in detail in [11] and its extensions to ionization and molecular mixtures are discussed in [12].

1.3 Models of Plastic Deformation of Solid Bodies

The deviatoric response of a continuum subjected to some arbitrary thermo-mechanical loading, as manifested in a change in shape at a material element (or point), is determined in large part by the material's constitutive properties (modeled in terms of elastic, plastic, and viscous characteristics). Deviatoric constitutive behavior is described with the generic equation

$$\sigma_i = \sigma_i(\varepsilon_i, \dot{\varepsilon}_i, \sigma_0, T, \ldots) \tag{1.9}$$

relating the material equivalent flow stress σ_i to hydrostatic stress σ_0 (pressure P), equivalent strain ε_i, equivalent strain rate $\dot{\varepsilon}_i$, temperature T, and possibly other parameters. This relation is found computationally and experimentally for each individual material, taking into account it's hydrostatic and deviatoric properties. Material equivalent flow stress, σ_i is equivalent to yield strength in the uniaxial stress case.

Over the years, a number of finite deformation plasticity models have been developed in an attempt to describe the response of metals to shock and expansion waves. These models have differed in terms of the details inherent in modeling the stress deviator, and in terms of the deformation mechanisms considered. As determined from experiments, the response of metals to dynamic loading varies widely. Consequently, a number of different plasticity models may be shown to have applicability with regard to predicting finite deformation plasticity, depending on the circumstances of the loading. In practice, the choice of the plastic deformation model best suited to the problem to be solved is made on a case-by-case basis. By virtue of the complex nature of the constitutive properties of metals subjected to shock loading, it is hardly possible to describe in full measure the constitutive behavior throughout the wide range of states up to the state of melting at the shock front with a single model. This is all the more true in cases wherein the experimental data is limited.

The simplest model is an ideal elastic-plastic body. For this body, in a state higher than the Hugoniot elastic limit, σ_{HEL}, the flow stress (dynamic yield stress Y_d) remains constant.

Better agreement with experimental results may be achieved with the introduction of plastic strain or average stress as independent variables to the elastic-plastic yield strength model. Frequently a linear relation of Y_d with the above variables is used in this approach.

A more complete (accounting for more of the physics) model used at high deformation rates is discussed in [13, 14]. In [13] the flow stress model is:

$$Y_d = Y_0 \left[1 + \beta \left(\varepsilon_i^p - \varepsilon_{i0}^p \right)^n \right] \left[1 + \left(\frac{\partial Y}{\partial P} \right)_T \frac{P}{Y_0 \delta^{1/3}} + \left(\frac{\partial Y}{\partial T} \right)_P \frac{\Delta T}{Y_0} \right] \quad (1.10)$$

and in [14]:

$$Y_d = Y_0 \left[(1 + \beta \varepsilon_i^p)^m + \alpha P \right] \left[1 - \frac{\varepsilon_T}{\varepsilon_T^p} \right]. \quad (1.11)$$

In Eqs. (1.10) and (1.11) Y_0, β, n, $(\partial Y/\partial P)_T$, $(\partial Y/\partial T)_P$, m, and α are constants determined from experimental data, ε_i^p is the equivalent plastic strain, ε_{i0}^p is the initial equivalent plastic strain (in general equal to zero), ε_T is the current internal energy, ε_T^{melt} is the internal energy required for melt in the case of homogeneous heating at a given density.

The thermal term in the right-hand side of Eq. (1.11) was found using the Lindemann melt law [15].

Physical interpretation of Eqs. (1.10) and (1.11) reveals that either model includes work hardening and thermal softening $((\partial Y/\partial T)_P < 0)$. The work hardening value is limited: for example in Eq. (1.11), $Y_0(1 + \beta \varepsilon_i^p)^m \leq Y_{\mathrm{max}}$ (a similar limit is imposed upon Eq. (1.10)). The term Y_0 is related to the elastic precursor, and Y_{max} corresponds to maximum experimentally observed yield strength.

The simplified model of an ideal elastic-plastic body fails to describe the variety of deformation features observed for various material classes. The models represented by Eqs. (1.10) and (1.11) have been shown to satisfactorily describe basic features of metal deformation for a number of materials in a number of loading cases, and thanks to their simplicity, have been widely adopted for practical purposes.

To describe material behavior during shock loading more completely, various relaxation (viscoplastic) models of plastic deformation have been constructed using the concept of the stress deviator component as a function of plastic shear deformation rate. Relaxation models are reviewed in [5, 15].

In general, commonly used relaxation models can be divided into two classes. In the first of them, the plastic deformation rate is related to dislocation dynamics. The second class includes macroscopic phenomenological models, in which the tangential stress rate $d\tau/dt$ is represented by a function of effective tangential stresses and some relaxation parameters.

For 1D deformation of an isotropic solid body in a plane shock wave, the stress-strain relation is given as:

$$\frac{d\sigma_{xx}}{dt} = (\lambda + 2G)\frac{d\varepsilon_{xx}}{dt} - \frac{8}{3}G\frac{d\gamma}{dt} , \qquad (1.12)$$

where λ and G are Lamé parameters and σ_{xx} is the stress in the direction of shock wave propagation.

From Eq. (1.12) it may be observed that stresses grow due to total strain and relax due to plastic strain.

According to the well-known Orowan relation, plastic shear strain γ depends on dislocation density N and average dislocation displacement L_{disp}

$$\gamma = dNL_{disp} , \qquad (1.13)$$

where b is the Burgers' vector.

Plastic strain rate $d\gamma/dt$ is

$$\frac{d\gamma}{dt} = b\frac{dN}{dt}L_{disp} + bN_m\frac{dL_{disp}}{dt} , \qquad (1.14)$$

where dN/dt is the dislocation multiplication rate, dL_{disp}/dt is the average movable dislocation velocity, and N_m is the movable dislocation density.

In Eq. (1.14); dN/dt, N_m, and dL_{disp}/dt are complex functions of many variables characterizing the constitutive state and include a great number of constants that are selected by rules of thumb [5, 15].

The complex dislocation models of plastic deformation provide a detailed description of the evolution of the shock load momentum as it propagates across the solid. However, the presence of numerous material constants in the models and the fact that their determination with independent experimental methods is problematic, restricts their potential for use in practice. The relations describing dislocation dynamics do not include potential temperature effect, so the models can only be applied in cases where the stresses are low enough that significant heating is not a factor.

For practical use, Eq. (1.12) may be transformed to a more convenient form

$$\frac{d\sigma_{xx}}{dt} = \frac{dP}{dt} + \frac{4}{3}\frac{d\tau}{dt} . \qquad (1.15)$$

To describe the deviatoric component in Eq. (1.15), various analytical expressions [5, 15] have been proposed:

$$\frac{d\tau}{dt} = G\left[\frac{d\varepsilon_{xx}}{dt} - \frac{1}{t_r}\exp\left(-\frac{D}{\tau - \tau^*}\right)\right];$$ (1.16)

$$\frac{dS}{dt} = 2G\frac{d\varepsilon_{xx}}{dt} - \frac{S - S_{eq}}{t_r},$$ (1.17)

$$\frac{d\tau}{dt} = G\left(\frac{d\varepsilon_{xx}}{dt} - \frac{\tau - \tau_0'}{t_r}\right).$$ (1.18)

In Eq. (1.16), t_r and D are relaxation parameters and τ^* is a critical stress. In Eq. (1.17), $S = \frac{4}{3}\tau$ is the stress deviator, S_{eq} is the equilibrium or deformation-rate-independent stress deviator, and t_r is a characteristic relaxation time. In Eq. (1.18), $\tau_0' = \tau_0 + f(\varepsilon^p)$ is the strained material yield strength including work hardening, and t_r is as specified above. The reader can become familiar with other relaxation models by studying [15].

1.4 Equations of 1D Shock Compression of Compressible Media

We call media ideal if pressure is the only surface force characterizing macroparticle interactions (by this definition inviscid fluids are ideal).

When there is no energy release from extraneous sources in the medium, its continuous motion is described by a set of governing differential equations representing the general laws of conservation of mass, momentum and energy, complemented by an equation of state for the material. In the Eulerian coordinate system, the equations are [2]:

$$\frac{d\rho}{dt} + \operatorname{div}\rho\vec{u} = 0, \qquad \rho\frac{d\vec{u}}{dt} = -\operatorname{grad}P;$$

(1.19)

$$\frac{dE}{dt} + P\frac{dV}{dt} = 0, \qquad E = E(P, V).$$

In the general case, to find the velocity u, density ρ, and pressure P as functions of space and time, the system of equations given as Eqs. (1.19) is solved numerically. However, for a number of simple equations of state (for example the EOS of an ideal gas), the solution can be found using the method of characteristics. Despite the limited number of cases, which afford an exact solution using the method of characteristics, it is useful to examine this approach for the qualitative insights to material behavior that such an examination provides.

For 1D adiabatic (but possibly spatially nonhomogeneous in entropy) flow of a compressible medium, there are three families of characteristics described by the following equations [2]:

$$C_+ : \frac{dx}{dt} = u + c \,, \qquad C_- : \frac{dx}{dt} = u - c \,, \qquad C_0 : \frac{dx}{dt} = u$$

where c is sound speed.

Riemann invariants are retained along these characteristics:

$$du + \frac{dP}{\rho c} = 0 \text{ along } C_+ \,, \qquad du - \frac{dP}{\rho c} = 0 \text{ along } C_- \,, \qquad dS = 0 \text{ along } C_0.$$

Characteristics C_+ and C_- describe trajectories of weak perturbation propagation in the $x - t$ plane; characteristics C_0 are streamlines. The Riemann invariants describe the material states along the characteristics.

In experimental studies of shock-wave processes in condensed media, observations are recorded, as a rule, at fixed specimen cross sections. In this circumstance, it is convenient to view the wave processes in terms of Lagrangian coordinates connected to the moving medium.

In the Lagrangian frame of reference, Eqs. (1.19) are transformed to the following [2]:

$$\rho_0 \frac{\partial V}{\partial t} - \frac{\partial u}{\partial h} = 0 \,, \qquad \rho_0 \frac{\partial u}{\partial t} - \frac{\partial P}{\partial h} = 0 \,,$$

$$\frac{\partial E}{\partial t} + P \frac{\partial V}{\partial t} = 0 \,, \qquad E = E\left(P, V\right) \,,$$

(1.20)

where $h = \int_0^x (\rho dx / \rho_0)$ is the Lagrangian coordinate; ρ and ρ_0 are current and initial material densities, respectively.

In Lagrangian coordinates, the families of characteristics, C_+, C_-, and C_0 for 1D adiabatic (but possibly spatially nonhomogeneous in entropy) flow of a compressible medium, become:

$$C_+ : \frac{dh}{dt} = a \,, \qquad C_- : \frac{dh}{dt} = -a \,, \qquad C_0 : \text{streamline } \frac{dh}{dt} = 0 \,,$$

where $a = \frac{\rho}{\rho_0} c$ is Lagrangian sound speed.

For 1D isentropic flow, a change in the material state along the characteristics is described by the following Riemann integrals:

$$u = u_0 + \int_{\rho_0}^{\rho} \frac{dP}{\rho_0 a} \text{ along } C_+ \,, \qquad u = u_0 - \int_{\rho_0}^{\rho} \frac{dP}{\rho_0 a} \text{ along } C_- \,,$$

The equations of motion for a 1D elastic-plastic medium are derived from Eqs. (1.19) and (1.20) by substituting the normal stress, σ_{xx} acting in the direction of the x-axis for pressure. For most media, where sound speed increases with increasing pressure, continuous flow in the form of a compression wave will evolve in such a way as to form a "discontinuity." This is caused by the intersection of the characteristics of one family. These equations must be complemented by equations for the constitutive relationship of the stress deviator.

The aforementioned discontinuity, on the surface of which thermodynamic and kinematic values experience a jump, and which moves across the material at some velocity, is called a strong or shock discontinuity. Assume that the medium behaves in the manner that is normal for most materials, i.e. the Hugoniot of the medium is convex downward $(d^2P/dV^2 > 0)$.

In the laboratory coordinate system, the fundamental laws of conservation of mass, momentum, and energy across the shock wave are expressed by the following system of algebraic equations:

$$\rho_1(D - u_1) = \rho_0(D - u_0) \, ,$$
$$P_1 + \rho_1(D - u_1)^2 = P_0 + \rho_0(D - u_0)^2 \, , \tag{1.21}$$
$$\rho_1(D - u_1)[H_1 + 0.5(D - u_1)^2] = \rho_0(D - u_0)[H_0 + 0.5(D - u_0)^2] \, .$$

In Eqs. (1.21), D is the shock wave velocity, u_0 is the velocity of the medium ahead of the shock front, u_1 is the velocity of the medium behind the shock front, H_0 is the material enthalpy ahead of the shock front, and H_1 is the material enthalpy behind the shock front.

Equations (1.21) relate the material state on either side of the shock front and form a system of three equations in five unknowns: D, u_1, P_1, ρ_1, and H_1. The values of u_0, P_0, ρ_0, and H_0, characterizing the state of the material ahead of the shock front, are assumed known. Equations (1.21) are complemented by an equation of state, which relates three of the thermodynamic variables

$$P = P(\rho, H) \quad \text{or} \quad P = P(\rho, E) \, . \tag{1.22}$$

We have a system of four equations and five unknowns. Hence, to find all the values behind the shock front, one of them has to be given; in this case the system of Eqs. (1.21) and (1.22) becomes definite. For the simplest equations of state, the system of Eqs. (1.21), (1.22) may be solved analytically, for more complex equations of state, the solution, as a rule, must be found numerically.

Minor modifications to the system of equations expressed in Eqs. (1.21) results in the following relations, which will prove to be convenient for practical purposes:

$$\frac{\rho_1}{\rho_0} = \frac{D - u_0}{D - u_1} = \eta_1 \, , \qquad P_1 - P_0 = \rho_0(D - u_0)(u_1 - u_0) \, ,$$
$$(D - u_0)^2 = V_0^2 \frac{P_1 - P_0}{V_0 - V_1} \, , \qquad (D - u_1)^2 = V_1^2 \frac{P_1 - P_0}{V_0 - V_1} \, , \tag{1.23}$$
$$u_1 - u_0 = \sqrt{(P_1 - P_0)(V_0 - V_1)} \, .$$

Note that η denotes substance compression on the Hugoniot whereas δ in Sect. 1.2 denotes compression on the isentrope.

One possible relationship between thermodynamic values on either side of the shock front is provided by Eqs. (1.21):

$$H_1 - H_0 = 0.5\,(P_1 - P_0)(V_0 + V_1) \qquad (1.24)$$

Since $H = E + PV$, Eq. (1.24) may be transformed to

$$E_1 - E_0 = 0.5\,(P_1 + P_0)(V_0 - V_1) \qquad (1.25)$$

Relationships between the five thermodynamic variables are, in general, referred to as Rankine-Hugoniot relationships, or simply as Hugoniots. A Hugoniot, such as the one represented by Eq. (1.25), in conjunction with an equation of state $P = P(V, E)$, is sometimes referred to as the shock Hugoniot.

Of practical interest is the situation, wherein a shock wave propagates through an initially quiescent medium with a shock pressure considerably greater than the initial pressure, i.e. $u_0 = 0$ and $P_1 \gg P_0$. In this case Eqs. (1.23) simplify to:

$$\rho_0 D = \rho_1(D - u_1)\,; \qquad P_1 = \rho_0 D u_1\,; \qquad \eta_1 = \frac{D}{D - u_1}\,;$$
$$(1.26)$$

$$D^2 = V_0^2\,\frac{P_1}{V_0 - V_1}\,, \qquad (D - u_1)^2 = V_1^2\,\frac{P_1}{V_0 - V_1}\,, \qquad u_1 = \sqrt{P_1(V_0 - V_1)}\,.$$

In the remainder of this section we shall highlight some features of shock compression that are true, in general.

If we exclude the internal energy from Eqs. (1.25) and (1.26), we can consider a Hugoniot equation of the form $P = P(V, V_0, P_0)$. In doing so, we are in essence considering a 2-dimensional slice through the 5-dimensional hyperspace of the complete Rankine-Hugoniot relations. The result is a two-parameter family of curves in the $P - V$ plane, where V_0 and P_0 act as parameters. Note that it is impossible to arrive at the same final state in the $P - V$ plane by means of multiple shock compression as would be reached in a single shock compression (note the difference in final density of state 3 verses state A in Fig. 1.1). By contrast, in an isentropic process the final density will always be the same regardless the number of stages in the loading. The mutual position of the Poisson adiabat and Hugoniot of single and multiple compressive stages is depicted in Fig. 1.1.

For strong shock waves ($P_1 \gg P_0$) P_0 and E_0 are negligible compared to P_1 and E_1. Hence,

$$E_1 = 0.5 P_1 (V_0 - V_1)\,; \qquad E_k = \frac{u_1^2}{2} = 0.5 P_1 (V_0 - V_1)\,,$$

i.e., internal energy and kinetic energy are essentially the same.

In the region of moderate enough pressures the shock wave velocity is

$$D = [c_0 + c(P) + u_1]/2$$

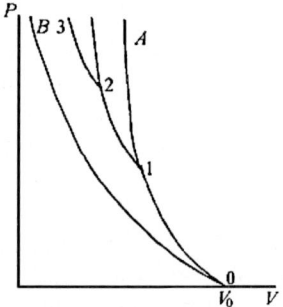

Fig. 1.1. Relative position of Poisson adiabat (isentrope) ($0B$), Hugoniot of single ($0A$), double (012) and triple (0123) continuous material compression

where c_0 is sound speed ahead of the shock front and $c(P)$ is sound speed behind the shock.

Inequalities $D > c_0$ and $u_1 + c_1 > D$, where c_1 is the isentropic sound speed in the compressed state, are true in general for shock wave processes. Thus, the shock wave propagates at supersonic speed relative to the state before the shock front, and at subsonic speed relative to the state behind the shock front. Hence: a) an expansion wave, whose head propagates at velocity $u_1 + c_1$, will overtake the shock front with time and a "stationary" shock wave will cease to exist; b) a shock front will overtake the front of a preceding shock wave.

In the shock wave, all kinematic and thermodynamic values undergo a jump-like change. The direction of the change in these values is:

$$P_1 > P_0, \ \rho_1 > \rho_0, \ c_1 > c_0, \ T_1 > T_0, \ S_1 > S_0 \ .$$

Mass velocity u_1 (or velocity increment $u_1 - u_0$) points in the shock wave propagation direction.

1.5 Wave Interaction. Arbitrary Discontinuity

Thermodynamic and kinematic values on opposite sides of a shock wave propagating through a homogeneous compressible material are related through the expressions in Eqs. (1.21). The situation becomes somewhat more complicated in cases involving arbitrary discontinuities. Consider a problem, set up in such a way that at the initial time, there is a discontinuity surface in the initial boundary value problem, thermodynamic and kinematic values on opposite sides of which are independent of one another. Such discontinuities are called *"arbitrary."* Typical examples of problems involving an arbitrary discontinuity include: the arrival of a shock wave at an interface between two materials possessing different properties, high-speed impactor deceleration on a target, and the arrival of a detonation wave at an HE-target interface. Arbitrary discontinuities are unstable. The Riemann (shock transmission/reflection) problem results in a complex flow pattern, possibly involving expansion waves, constant flow regions, and shock waves. Fortunately, the set of possible solutions is limited by the physical requirement that not more than one wave propagate in a given direction from the arbitrary discontinuity surface. The desired solution is constructed as a combination of two waves propagating in opposite directions from the initial discontinuity, and separated by regions of constant flow (in the general case these regions are two in number). The interface of the regions is the arbitrary discontinuity surface. This interface is also called a "*contact discontinuity*." The contact discontinuity resulting from the Riemann problem has a peculiarity: on opposite sides of the discontinuity, pressures and mass (or particle) velocities must be equal, whereas density, temperature, entropy, and other thermodynamic values (excluding pressure), are as a rule different.

The Riemann problem for materials with normal thermodynamic properties can result in the appearance of the following possible flow conditions at an arbitrary discontinuity:

1. shock waves propagate outward on both sides of the discontinuity surface;
2. a shock wave propagates outward on one side of the discontinuity surface and an expansion wave propagates in the opposite direction;
3. expansion waves propagate outward on both sides of the discontinuity surface.

The response of the material on either side of an arbitrary discontinuity may be characterized in terms of the function $P(u)$, which is a shock Hugoniot for pressures higher than the initial pressure, and an isentrope for pressures lower than the initial pressure. The magnitude of the slope of this function, $|dP/du|$, is a measure of material hardness. The values of P and u at the contact discontinuity may be determined through the solution of a system of two equations representing the dependencies of $P(u)$ within the two interacting materials.

One may gain a more thorough understanding of the Riemann problem by considering (in addition to the $P(u)$ relationships), $x - t$ diagrams, which depict the trajectories of shock waves, expansion waves, and interfaces.

For illustrative purposes, let us consider several characteristic examples of the Riemann problem. Let a freely flying thick plate be decelerated on a thick target. The impact plane $x = 0$ serves as an arbitrary discontinuity surface. $P_0 = 0$, ρ_0, $u_0 = 0$ and $P_1 = 0$, ρ_1, $u_1 = W_{imp}$ characterize the initial states in the target and the impactor, respectively (points 0 and 1 in Fig. 1.2a).

In the given Riemann problem, shock waves propagate through the impactor and the target. The shock Hugoniots intersect at point 2 (see Fig. 1.2a). Pressure P_2 and mass velocity u_2 characterize the state at the contact discontinuity. During the transition through the shock discontinuity, the impactor particle velocity changes by $W_{imp} - u_2$ in a jump. If the target ma-

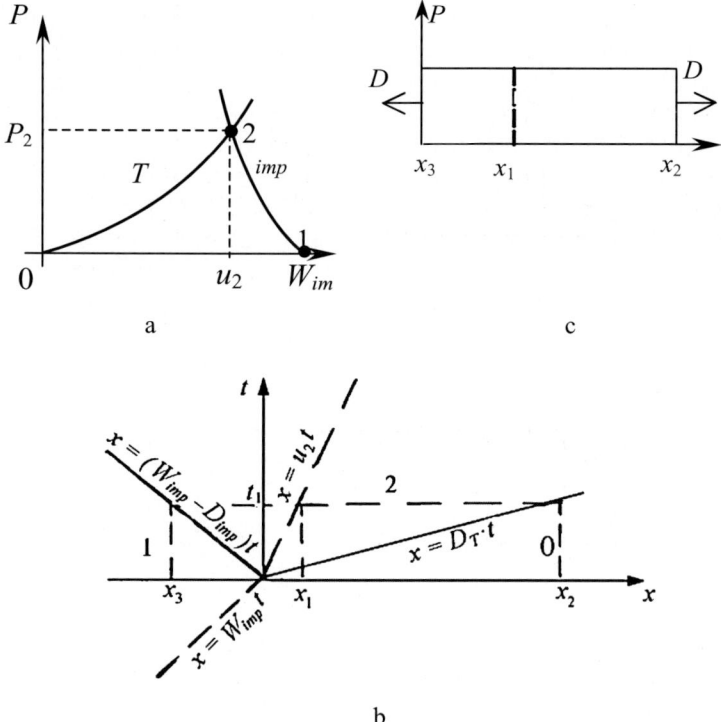

Fig. 1.2. Impact of a thick impactor on a thick plate: **a** – $P - u$ diagram; **b** – $x - t$ diagram; **c** – pressure profile at time t_1, x_1 is the interface position

terial hardness and W_{imp} are constant, the shock amplitude increases with increasing impactor hardness. For a given impactor material hardness, the pressure at the shock front increases with increasing impact velocity, W_{imp} and target material hardness. The pressure profile at time t_1 is plotted in Fig. 1.2c. The contact discontinuity moves at velocity u_2.

Next, consider the case of a normal collision between two shock waves (Fig. 1.3). Let the shock waves interact at the plane $x = 0$. This plane will be an arbitrary discontinuity surface. The resultant flow is illustrated in Fig. 1.3.

The interaction of the two shock waves results in the production of two secondary shock waves propagating in opposite directions from the discontinuity surface. The initial states, P_1, u_1 and P_2, u_2, following the fronts of the impacting waves, serve as progenitors for the $P - u$ dependencies of the secondary compression Hugoniots. A nonlinear system of algebraic equations for determination of the values of the pressure and particle velocity at the contact discontinuity is

$$P_3 - P_1 + \rho_1(D_{13} - u_1)(u_3 - u_1) = 0 \; ;$$

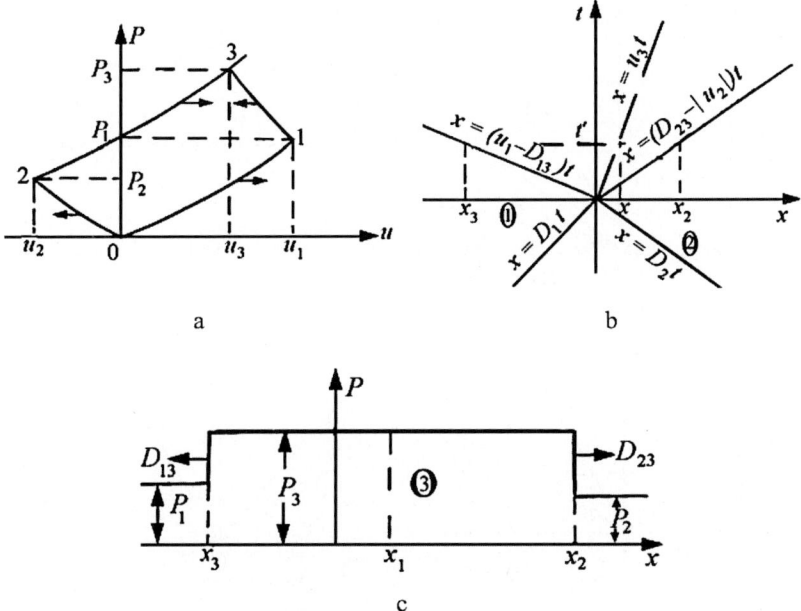

Fig. 1.3. Interaction of two contradirectional shock waves: **a** – $P - u$ diagram; **b** – $x - t$ diagram; **c** – pressure profile at time t_1, x_1 is the interface position

$$P_3 - P_2 + \rho_2(D_{23} - u_2)(u_3 - u_2) = 0 \,,$$

where D_{13} and D_{23} are secondary shock wave propagation velocities. The contact discontinuity moves at velocity u_3. Region $x_2 - x_3$ represents two regions of constant flow divided by the contact discontinuity. The pressure profile at time t_1 is plotted in Fig. 1.3c.

Let us now direct our attention to a case of immense practical significance: a shock wave impinging on an interface between two media (Fig. 1.4). Let a shock wave of known strength propagate through material 1, which has an interface with material 2. Material 2 is initially not loaded. The interface between materials 1 and 2 (at $x = 0$), will serve as an arbitrary discontinuity surface. The resultant flow that takes place in the given Riemann problem is illustrated in Fig. 1.4.

One of two possible types of flow will occur, depending on the media hardness ratio. Let us consider the first possibility (material 2 "harder" than material 1). In this case, the arrival of the shock wave at the interface with a "harder" material (curve $0B$ in Fig. 1.4a) results in the production of two shock waves traveling in opposite directions from the arbitrary discontinuity. The second possibility, the case where the shock wave travels from a "harder" medium to a "softer" one (curve $0C$), results in the production of both an expansion wave and a shock wave. The expansion wave will propagate into the "hard" medium while the shock wave propagates into the "soft" medium.

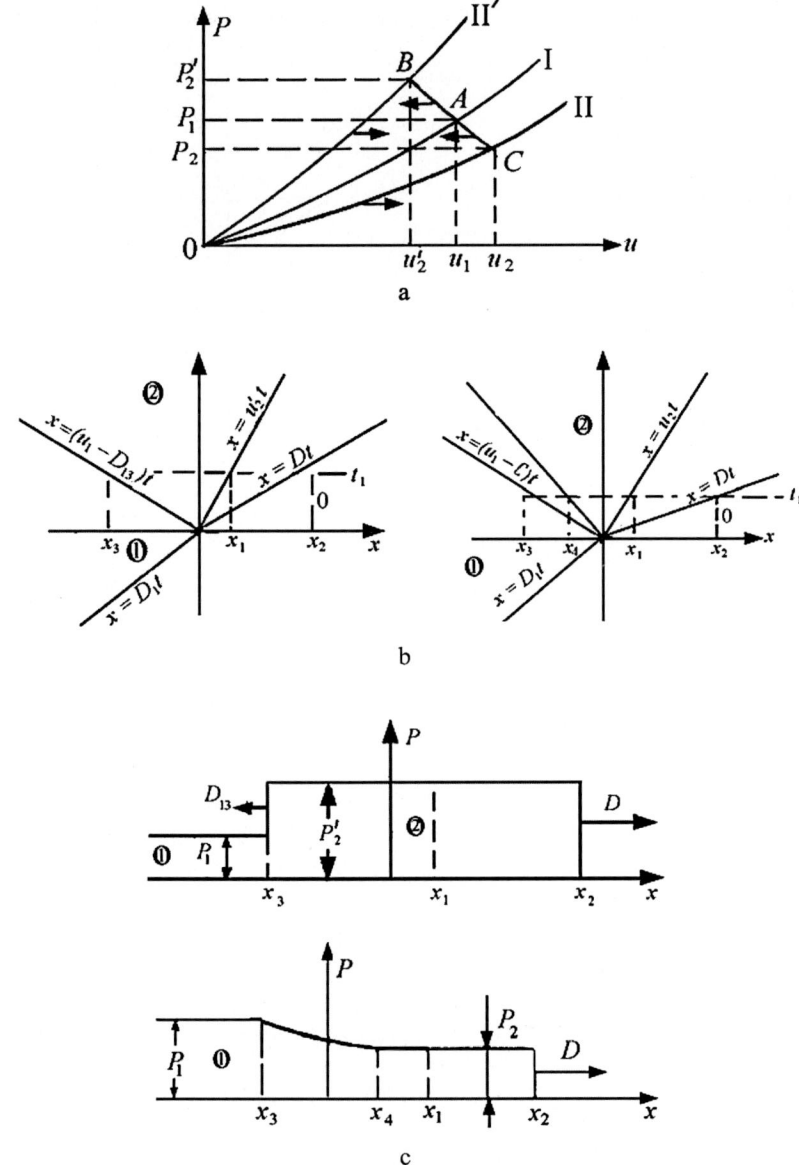

Fig. 1.4. Shock wave impinging on an interface of two media: **a** – $P - u$ diagram; **b** – $x - t$ diagram; **c** – pressure profile at time t_1, x_1 is the interface position

The contact discontinuity will move at velocity u_2. Curve AB is the shock Hugoniot for second compression of the first material produced in the first scenario, curve AC is the first material expansion isentrope produced in the

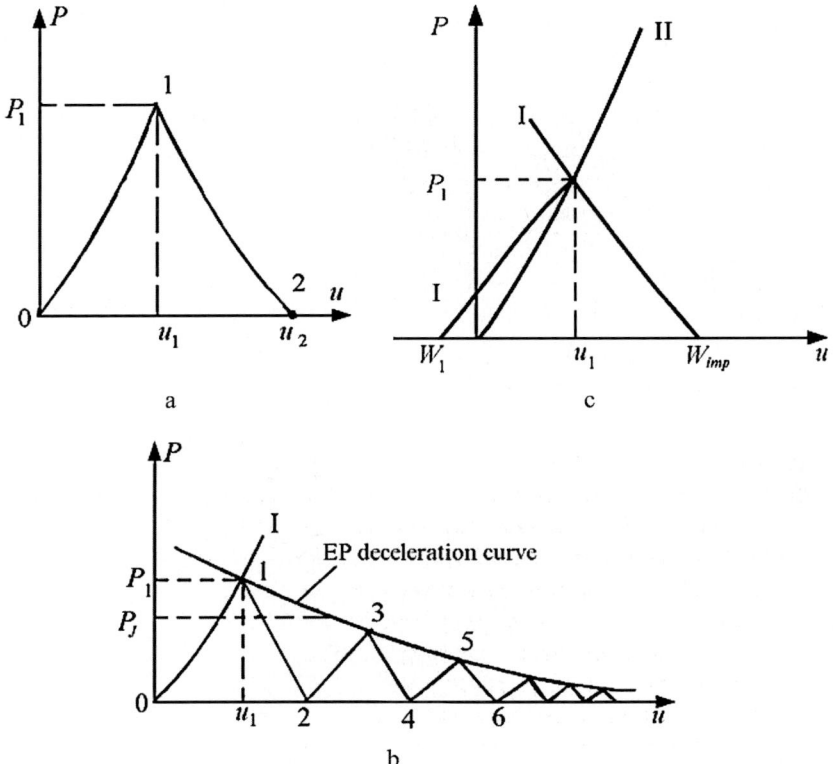

Fig. 1.5. Solution to the Riemann problem in $P - u$ plane: **a** – shock wave arrival at the free surface; **b** – compressible plate acceleration by the detonation wave; **c** – deceleration of thin soft impactor (I) on thick hard barrier (II)

second scenario. The Hugoniot in $P - u$ space for the expansion wave may be determined from the relations appearing in Sect. 1.6.

We give two additional examples of the Riemann problem solution in $P - u$ coordinates in Fig. 1.5. In the first case (Fig. 1.5a), we consider the arrival of a shock wave at a free surface possessing a zero pressure (vacuum) boundary condition. The free surface-vacuum interface is an arbitrary discontinuity surface. In the $P - u$ plane, the vacuum (as medium) is presented as axis u ($P = 0$). Curve 01 is the material shock Hugoniot, curve 12 is the isentrope of expansion from state P_1, u_1. For fairly weak shock waves (those producing small strains), the expansion isentrope may be approximated by the shock Hugoniot, hence, $u_2 - u_1 \approx u_1$ and $u_2 \approx 2u_1$, i.e. the rule of doubling applies.

In the second case (Fig. 1.5b), states 1, 3, 5... occur sequentially at the interface of explosion products (EP) and a plate. The plate is accelerated in a jump-like manner by shock waves of amplitude P_1, P_3, P_5... generated by expansion waves running from the free surface of the plate and impinging on the loaded EP-plate interface. Curves 01, 23, 45 are shock Hugoniots, 12,

34, 56 are expansion isentropes. In the limit, plate velocity tends toward the velocity of EP expansion into a vacuum.

In the third case (Fig. 1.5c), we consider the collision of two plates. The collision results in two P_1-amplitude shock waves being generated. When the shock wave propagating through the impactor arrives at the impactor's free surface, the impactor is released to $P = 0$, $u = W_1 < 0$. When the expansion wave has impinged on the interface, the target material becomes unloaded to $P = 0$, $u = 0$ (we approximate the unloading trajectory as coinciding with the loading trajectory). If the interface is not strong, the impactor will fly away from the target in the direction opposite to the original at velocity W_1.

1.6 Shock and Expansion Waves in Solid Bodies

For simple forms of the equation of state of normal media, an analytical solution for the Hugoniot can be obtained through the simultaneous solution of two equations: the Hugoniot equation and the equation of state.

If the Mie-Grüneisen form of the equation of state with $P_x = \rho_0 c_0^2/n\,(\delta^n - 1)$ is used, the shock Hugoniot equation becomes

$$P_1 = \frac{\rho_0 c_0^2 \left[\left(h - \dfrac{n+1}{n-1} \right) \eta_1^n + \dfrac{2n}{n-1}\eta_1 - (h+1) \right]}{n(h - \eta_1)}, \tag{1.27}$$

where $h = 1 + 2/\Gamma$, $\eta_1 = \rho_1/\rho_0$, and P_1 is the pressure at the shock front. Note that h can be a function of specific volume, and E_0 and P_0 are neglected.

If $\eta_1 \to h$, pressure $P_1 \to \infty$. This is an indication that at very high pressures on the Hugoniot, the dominant effect is that of thermal pressure and thermal energy. As a rule, c_0 is taken close or equal to the volumetric sound speed at standard conditions, and the values of n and h are chosen by determining the values, that afford the best fit to experimental data (for both shock and uniform isothermal compression).

For the value of h to be selected correctly, experimental results from studies of shock compressibility of materials with density ρ_{00} less than continuous material density are widely used [2,6]. In view of relation $E_1 = 0.5\,P_1(V_{00} - V_1)$, the shock Hugoniot for these materials is represented as

$$P_1 = \frac{\rho_0 c_0^2 \left[\left(h - \dfrac{n+1}{n-1} \right) \eta_1^n + \dfrac{2n}{n-1}\eta_1 - (h+1) \right]}{n(h - k\eta_1)}, \tag{1.28}$$

where $k = \rho_0/\rho_{00}$ is the porosity factor and $\eta_1 = \rho_1/\rho_0$.

The shock-compressed temperature of continuous materials $(T > T_0)$ is evaluated from the expression

$$T_1 = \frac{0.5u_1^2 - E_x}{C_V}. \tag{1.29}$$

For porous materials, the form of relation (1.29) remains unchanged, but in (1.28) $\delta_1 = \eta_1 = D/k\,(D - u_1)$ should be taken.

As the material is shock compressed from the initial state V_0, T_0 to state V_1, T_1 the change in entropy is given by:

$$S_1 - S_0 = \int_{T_0}^{T_1} C_V \frac{dT}{T} + \int_{V_0}^{V_1} \Gamma(V)\,C_V \frac{dV}{V}\,. \tag{1.30}$$

If Γ and C_V are constant, then

$$S_1 - S_0 = C_V \ln \frac{T_1}{T_0} \left(\frac{V_1}{V_0}\right)^{\Gamma}\,. \tag{1.31}$$

The isentropic sound speed, c in the shock-compressed state is given as:

$$c_1 = \sqrt{c_0 \eta_1^{n-1} + \Gamma\,(\Gamma + 1)\,C_V T_1}\,. \tag{1.32}$$

For the special case of shock waves of moderate amplitude, $c_1 = c_0 \eta_1^{(n-1)/2}$.

For strong shock waves, $c_1 = \sqrt{\Gamma\,(\Gamma + 1)\,C_V T_1}$.

In the case of successive material compressions by shock waves of amplitudes P_1, P_2, $\ldots P_n$, the expression for the shock Hugoniot numbered n is found through the solution of an equation system composed of the equation of state and the following:

$$E_1 = 0.5 P_1\,(V_0 - V_1) + 0.5\,(P_2 + P_1)\,(V_1 - V_2)$$
$$+ \ldots + 0.5\,(P_n + P_{n-1})\,(V_{n-1} - V_n)\,.$$

The equation system describing variations in the thermodynamic state along the isentrope includes the equation for the isentrope, $dE = -\,PdV$, and the equation of state.

The equation for the isentrope passing through point P_1, P_{1T}, η_1 on the shock Hugoniot is

$$P = P_x\,(\delta) + P_{1T} \left(\frac{\delta}{\eta_1}\right)^{\Gamma+1}\,. \tag{1.33}$$

On isentropic expansion from the initial state P_1, η_1 to pressure P, the increment in particle velocity, Δu is given as:

$$\Delta u = \pm \int_{P_0}^{P} \sqrt{-dPdV}\,, \tag{1.34}$$

where the function $P(V)$ is taken along the isentrope by formula (1.33).

The full velocity of shock-preheated material particles is:

$$u = u_1 \pm \Delta u\,.$$

The plus sign corresponds to the case wherein the direction of propagation of the shock front and expansion wave head is opposite, and the minus sign corresponds to the case wherein the direction of propagation is the same direction.

Consider shock wave arrival at a free surface. The expanded material velocity is determined as:

$$W = \sqrt{(P_1 - P_0)(V_0 - V_1)} + \int_{P_1}^{0} \sqrt{-dPdV} \,. \qquad (1.35)$$

For moderate-amplitude shock waves $\Delta u \approx u_1$ and, hence, $W = 2u_1$, i.e. during unloading the material returns to the original thermodynamic state that existed before the arrival of the shock front. For strong shock waves, $W > 2u_1$.

Density δ_k in the unloaded state follows from (1.33),

$$P_x(\delta_k) + P_{1T}\left(\frac{\delta_k}{\eta_1}\right)^{\Gamma+1} = 0 \,. \qquad (1.36)$$

Temperature T_k in the unloaded state is related to the shock compression temperature T_1 as:

$$T_k = T_1\left(\frac{\delta_k}{\eta_1}\right)^{\Gamma} \,.$$

The change in the internal energy on unloading to vacuum is:

$$E_k - E_0 = 0.5(P_1 + P_0)(V_0 - V_1) - \int_{V_1}^{V_k} PdV = C_P(T_k - T_0) \,. \qquad (1.37)$$

For complex equations of state, for example, equations of state involving electron components, the relevant equation system is solved numerically.

Note that, having experimentally fixed the position of the isentrope in the $P - u$ plane, V and the other thermodynamic values along the isentrope can be estimated using formulas such as the following, without resorting to an equation of state:

$$V = V_1 - \int_{P_1}^{P}\left(\frac{du}{dP}\right)^2 dP \,; \qquad E = E_1 + \int_{P_1}^{P} P\left(\frac{du}{dP}\right)^2 dP \,, \qquad (1.38)$$

where V_1, E_1, P_1 is the state on the shock Hugoniot.

Strength properties in metals affect their behavior in dynamic processes. The material behaves in an elastic manner under high-speed deformation in shock waves so long as the principal stress difference is no greater than the dynamic yield stress. With increasing shock-wave amplitude, plastic strains will begin to develop in the material, and as shock strength increases further the material melts.

In the case of 1D deformation in a shock wave, principal stresses, pressure, and dynamic yield stress are related as follows:

$$\sigma_1 = P + \frac{2}{3}Y_d \; ; \qquad \sigma_2 = \sigma_3 = P - \frac{1}{3}Y_d \; , \qquad (1.39)$$

(σ_1 is the principal stress along the normal to the shock front plane). The principal stresses are related as

$$\sigma_1 - \sigma_2 = \sigma_1 - \sigma_3 = \frac{1 - 2\nu}{1 - \nu}\sigma_1 \; ; \qquad \sigma_2 = \sigma_3 = \frac{\nu}{1 - \nu}\sigma_1 \; , \qquad (1.40)$$

where ν is the Poisson ratio.

Figure 1.6 illustrates the characteristic response of an elastic-plastic medium subjected to a 1D shock wave with subsequent 1D unloading. When the stresses are low, $\sigma_1 < \sigma_{\mathrm{HEL}}$, the deformation is elastic. As the shock amplitude increases above σ_{HEL}, the material undergoes plastic deformation, up to the melting point σ_{melt}. If $\sigma_1 = P > \sigma_{melt}$, the metal behaves as a fluid. σ_{HEL} is called the Hugoniot yield stress (or Hugoniot elastic limit). It is related to Y_d as follows:

$$\sigma_{\mathrm{HEL}} = \frac{1 - \nu}{1 - 2\nu}Y_d \; .$$

In stress range $\sigma_{\mathrm{HEL}} < \sigma_1 < \sigma_{melt}$, the difference between principal stresses and pressure at fixed V on the shock Hugoniot is determined using Eqs. (1.39), where Y_d is a function of constitutive characteristics. Upon a

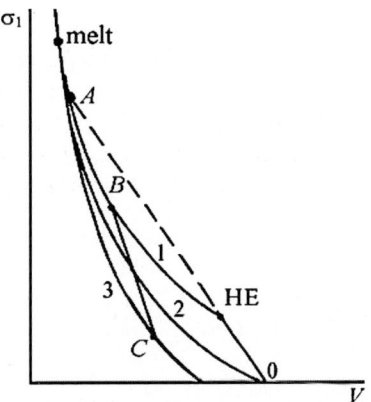

Fig. 1.6. Compression and unloading of elastic-plastic medium in variables σ_1, V: 1 – plastic loading line; 2 – uniform compression curve; 3 – plastic unloading line

change in deformation direction, compressed solids will unload elastically, as illustrated by line BC in Fig. 1.6 The elastic unloading amplitude is:

$$\Delta\sigma_{el} = \sigma_B - \sigma_C = \frac{2(1-\nu)}{1-2\nu}Y_d \ .$$

The complexity inherent in the loading and unloading shock adiabat in the $\sigma_{HEL} < \sigma_1 < \sigma_A$ stress range, results in a rather complex wave pattern (see Fig. 1.7).

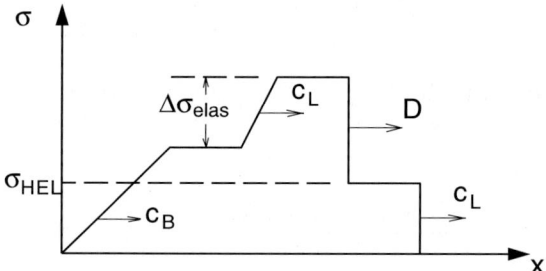

Fig. 1.7. Compression and subsequent unloading impulse profile in elastic-plastic region

In this range, both compression and expansion waves are split into two parts (one elastic, the other plastic). In the elastic-plastic region, metal properties are characterized by values of the yield strength Y_d, Young's modulus E, bulk modulus K, shear modulus G, and Poisson's ratio ν. The last four parameters are not measured directly in shock wave experiments, but are related to recorded sound speeds: elastic longitudinal c_L and volumetric c_B.

$$K = \rho c_B^2 \ ; \qquad E = \rho c_L^2 \frac{(1-2\nu)(1+\nu)}{1-\nu} \ ;$$

$$G = \frac{E}{2(1+\nu)} \ ; \qquad \nu = \frac{3-(c_L/c_B)^2}{3+(c_L/c_B)^2} \ . \tag{1.41}$$

Within the elastic deformation region, the relations on the shock-wave discontinuity are characterized by the following:

$$\rho_1(D-u_1) = \rho_0 D \ ; \qquad \sigma_1 = \rho_0 D u_1 \ ; \qquad E_1 - E_0 = 0.5\sigma_1(V_0 - V_1) \ . \tag{1.42}$$

The equations are also valid when $\sigma_A < \sigma_1 < \sigma_{melt}$. If $\sigma_1 \geq \sigma_{melt}$, the relations on the discontinuity are described by ideal medium relations (i.e. the relationships for a fluid). In the range $\sigma_{HEL} < \sigma_1 < \sigma_A$, the equations for the plastic wave are given as:

$$\rho_1 \left(D - u_1 \right) = \rho_{\mathrm{HEL}} \left(D - u_{\mathrm{HEL}} \right);$$

$$\sigma_1 - \sigma_{\mathrm{HEL}} = \rho_{\mathrm{HEL}} \left(D - u_{\mathrm{HEL}} \right) \left(u_1 - u_{\mathrm{HEL}} \right); \qquad (1.43)$$

$$E_1 - E_{\mathrm{HEL}} = 0.5 \left(\sigma_1 + \sigma_{\mathrm{HEL}} \right) \left(V_{\mathrm{HEL}} - V_1 \right).$$

From Eqs. (1.42) and (1.43), we can derive expressions for the elastic-plastic medium shock Hugoniot:

$$\sigma_1 = \frac{2E_1}{V_0 - V_1}, \quad \sigma_1 \geq \sigma_A; \qquad (1.44)$$

$$\sigma_1 = \frac{2E_1 - \sigma_{\mathrm{HEL}} \left(V_0 - V_1 \right)}{V_{\mathrm{HEL}} - V_1}, \quad \sigma_1 < \sigma_A. \qquad (1.45)$$

Since $\sigma_1 = P + 2Y_d/3$, $E_1 = E \left(V_1, T_1 \right) + Y_d^2/6\rho_1 G_1$, Eqs. (1.44) and (1.45) may be transformed to the following form, describing the elastic-plastic shock Hugoniot:

$$\sigma_1 = \frac{2E \left(V_1, T_1 \right) + Y_d^2/3\rho_1 G_1}{V_0 - V_1}, \quad \sigma_1 \geq \sigma_A; \qquad (1.46)$$

$$\sigma_1 = \frac{2E \left(V_1, T_1 \right) - \sigma_{\mathrm{HEL}} \left(V_0 - V_1 \right) + Y_d^2/3\rho_1 G_1}{V_{\mathrm{HEL}} - V_1}, \quad \sigma_1 < \sigma_A. \qquad (1.47)$$

1.7 Plane Stationary Detonation Waves

For shock waves propagating in inert media, the increment in internal energy depends only on the compression. Not all media are inert, however. Processes occur in nature, wherein chemical transformations accompanied by the release of energy proceed under compression and elevated temperature conditions. One such process occurs with the detonation of high explosive (HE). Hydrodynamic detonation theory is discussed in detail in a number of places, one resource being [16].

The material chemical transformation that occurs during detonation produces a sharp rise in pressure, a considerable rise in the density of the explosion products (compared to the density of the original HE), and medium motion. Thus detonation involves a combination of chemical and gas-dynamical processes. The laws of motion are therefore responsible for many aspects of the detonation process. In the general case, it is therefore necessary to solve the equations of motion and of chemical kinetics, simultaneously, in order to obtain a comprehensive understanding of the thermomechanical processes involved.

Many basic and practically important aspects of the detonation process can be inferred from an analysis involving the laws of conservation of mass, momentum and energy only, ignoring the equations of chemical kinetics. A detonation theory based on the conservation laws only, is usually referred to as a hydrodynamic theory of detonation. The theory implies that the detonation wave front is a shock-wave discontinuity that compresses the HE and heats it to a higher temperature. Under these conditions, the chemical exothermal reaction proceeds very rapidly, releasing energy in some zone behind the wave front. Behind this zone are explosion products (EP), which expand. The process may be thought of as one involving the supersonic propagation of chemical transformation through material. Experimental data suggests that, given sufficient HE-sample cross-sectional dimensions, the detonation wave (unlike a shock wave), can propagate at constant velocity to as great a distance as one desires. Thus, the detonation regime is stationary, and assuming that the composition of HE is homogeneous, the number of stationary detonation regimes is one.

Failing to account for the precise details of the detonation front structure, we assume that the chemical transformation takes place in a negligibly short time period in a narrow zone adjacent to the wave front. We use the algebraic form of the conservation laws to derive relationships relating the kinematic and thermodynamic values before the detonation front to the state at the Jouguet point. The state before the front is considered to be unperturbed. The first two laws of conservation – of mass and momentum – will be exactly like those for a shock wave. If the energy quantity released by HE per unit mass in the chemical reaction zone is Q, then the shock Hugoniot equation becomes $E_J - E_0 = [(P_J + P_0)/2] (V_0 - V_J) + Q$. The equations of motion are complimented by the equation of state and the Chapman-Jouguet conditions to obtain a set of five algebraic equations:

$$\rho_J (D - u_J) = \rho_0 D ; \qquad P_J + \rho_J (D - u_J)^2 = P_0 + \rho_0 D^2 ;$$

$$E_J - E_0 = \left(\frac{P_J + P_0}{2} \right) (V_0 - V_J) + Q ; \qquad (1.48)$$

$$E = E(P, V) ; \qquad u_J + c_J = D ,$$

where subscript "J" denotes the state at the Jouguet point; and D is the detonation wave velocity.

Equation system (1.48) represents a complete set of equations for the phenomenological description of the normal detonation process as described. The last of the system equations is an analytical formulation of the Chapman-Jouguet hypothesis, which states that the motion of EP at the Jouguet point is sonic relative to the state behind the detonation wave front, which just accounts for the constancy in the detonation wave velocity. The equation for specific internal energy implies that the chemical reaction products are in the thermodynamic equilibrium state. The third equation of system (1.48), in conjunction with the equation of state, describe the Hugoniot in the $P - V$ plane corresponding to total release of chemical transformation energy.

Figure 1.8 illustrates principal concepts of the hydrodynamic theory. The shock wave propagating through HE as though it were inert material, compresses it to state 1 on Hugoniot 01 of "cold" HE possessing all properties

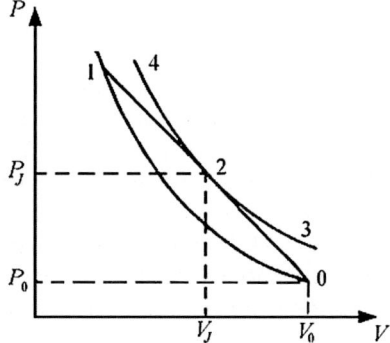

Fig. 1.8. Relative position of HE Hugoniot (01) and adiabat (324) of total release of chemical transformation energy

of the continuum Hugoniot. Due to high pressure and temperature at point 1, chemical reaction starts. Energy release causes chemical transformation product expansion. The states occurring during the chemical reactions continuously vary along the Michelson-Rayleigh straight line, 021, whose equation is $D^2 = V_0^2(P - P_0)/(V_0 - V)$. Since the reaction proceeds irreversibly, along the Michelson-Rayleigh straight line in the direction from point 1 to point 2 the EP entropy increases and reaches its maximum at point 2. At point 2, where the Michelson-Rayleigh straight line touches adiabat 324 of total chemical transformation energy release, the reactions terminate. The point of tangency between the Michelson-Rayleigh straight line and the total energy release adiabat is the Jouguet point. It has a remarkable feature: the detonation wave propagation velocity corresponding to the state at this point is the minimum possible value. Such a detonation wave is called a normal detonation wave.

Within the detonation front there is an elevated pressure zone 12 usually referred to as the chemical peak. The spatial pressure distribution at the chemical peak depends on specific chemical reaction conditions which we do not consider here.

The explosion product expansion process can be described with differential equations of motion, such as those given in Eqs. (1.19) and (1.20). This is because the process is adiabatic everywhere below the Jouguet point. At the Jouguet point the entropy is constant for all time. We choose to neglect the chemical reaction zone and consider the states at the Jouguet point and subsequent isentropic expansion (or compression) of explosion products. As at the Jouguet point all values are constant, the expansion region in the vicinity of the Jouguet point will be a simple expansion wave.

Many versions of the equation of state of explosive EPs have been developed for the description of their behavior [5, 16]. A number of applications have been found for which what is perhaps the simplest equation of state is adequate:

$$P = A\rho^n \tag{1.49}$$

or its extension

$$P = (\gamma - 1)\rho E + A\rho^n \tag{1.50}$$

with constant $A, n,$ and γ.

The principal relationships at the Jouguet point are found using the system of equations given in (1.48) and the expansion isentrope, assumed to be given in terms of the polytrope $P = A\rho^n$, where A is the entropy function. For an isentropic process, $A = \text{const.}$ at any ρ.

Sound speed c_J at the Jouguet point is:

$$c_J^2 = \left(\frac{\partial P}{\partial \rho}\right)_S = n\frac{P}{\rho} . \tag{1.51}$$

From (1.48) and (1.51), it follows that

$$c_J^2 = \frac{n\rho_0 D^2}{\rho_J}\left(1 - \frac{\rho_0}{\rho_J}\right), \qquad c_J = \frac{\rho_0}{\rho_J}D . \tag{1.52}$$

We can eliminate c_J from the equations to obtain:

$$\frac{\rho_J}{\rho_0} = \frac{n+1}{n} \tag{1.53}$$

and:

$$c_J = \frac{n}{n+1}D ; \qquad u_J = \frac{1}{n+1}D . \tag{1.54}$$

Pressure at the Jouguet point is expressed through D as follows:

$$P_J = \rho_0 D u_J = \frac{\rho_0}{n+1}D^2. \tag{1.55}$$

Let a semi-infinite layer of HE be initiated in the plane $x = 0$ at time $t = 0$. In this case, the system of differential equations can be accurately solved using the method of characteristics in order to arrive at a description of the flow in terms of a simple expansion wave:

$$u = \frac{2}{n+1}\left(\frac{x}{t} - \frac{D}{2}\right) ; \qquad c = \frac{n-1}{n+1}\frac{x}{t} + \frac{D}{n+1} ; \tag{1.56}$$

$$\frac{\rho}{\rho_J} = \frac{\left(\dfrac{n-1}{n+1}\dfrac{x}{t} + \dfrac{D}{n+1}\right)}{c_J} ; \qquad \frac{P}{P_J} = \left(\frac{\rho}{\rho_J}\right)^{\frac{2n}{n-1}} .$$

From (1.56), it is seen that $u, c, \rho,$ and P are functions of the self-similar variable x/t. Thus, the EP flow is self-similar in its nature, i.e. the wave will behave as if stretched in a self-similar manner: at equal magnitudes of x/t all values describing the flow will be invariable.

Let a planar (normal) detonation wave impinge non-obliquely on the surface of a barrier. The precise nature of the barrier media is unimportant; it may be condensed matter or a gas. To solve the Riemann problem (see Sect. 1.5), it is necessary to determine the pressure P vs. mass velocity u for the EP, the so-called explosion product deceleration curve. According to [16], the EP deceleration curve is given as:

$$P = P_J \left(\frac{3n-1}{2n} - \frac{n^2-1}{2n} \frac{u}{D} \right)^{\frac{2n}{n-1}}. \tag{1.57}$$

In expansion to vacuum ($P = 0$), the maximum EP expansion velocity is

$$u_{\max} = \frac{3n-1}{n^2-1} D.$$

If the detonation wave decelerates on a sufficiently rigid wall, the EP shock wave is reflected ($P > P_J$) and strictly speaking, it is necessary to calculate the EP Hugoniot in order to determine the flow parameters. However, recognizing that the shock amplitude and density variations are small compared to the states at the Jouguet point, the shock wave may be treated as a compression wave to a good approximation. Therefore, formula (1.57) may be extended to account for the case of the deceleration of a detonation wave on a rigid barrier ($u < u_J$).

For constant polytropic index n, the pressure at the deceleration on an absolutely rigid wall is given by:

$$P = P_J \left(\frac{3n-1}{2n} \right)^{\frac{2n}{n-1}}. \tag{1.58}$$

When solving application problems, n is often taken to be 3. This considerably simplifies the mathematical problem solution, especially when using the method of characteristics. In this case, characteristics of opposite families do not interact with one another and are straight lines throughout the flow.

It should be kept in mind, however, that when the degree of expansion of the EP is high, the exponent in (1.49) must decrease by a factor of about two. Such a decrease gives it a value roughly corresponding to that of a rarefied gas. This fact limits the regime of applicability for relation (1.49) to problems involving relatively high EP densities. In cases where the degree of expansion relative to the state at the Jouguet point is significant, [17] recommends that problem solution accuracy may be increased by using an equation of state of the following form:

$$P = Ae^{k/\rho} + B\rho^\gamma, \tag{1.59}$$

where A, k, B, and γ are constants. Equation (1.59) removes the limitations of Eq. (1.49) mentioned above, and may be employed in numerical solutions of a variety of problems for which a description of the EP motion over a wide range of specific volume is required.

References

1. Landau, L.D., and Lifshits, E.M., *Continuum Mechanics*, Gostekhizdat Publ., Moscow, 1954.
2. Zeldovich, Ya.B., and Raizer, Yu.P., *Physics of Shock Waves and High-Temperature Hydrodynamic Phenomena*, Nauka Publ., Moscow, 1966, [English trans., Academic Press, NY, Vol. 1 (1966), Vol. 2 (1967); Reprinted in a single volume by Dover Publ., Mineola, NY 2002].
3. Sedov, L.I., *Continuum Mechanics*, Nauka Publ., Moscow, 1970, [English trans., Mechanics of Continuous Media, World Scientific Publ., Singapore, 1997].
4. Godunov, S.K., *Elements of Continuum Mechanics,* Nauka Publ., Moscow, 1978. [*see also* Godunov, S.K., and Romenskii, E.I., Elements of Continuum Mechanics and Conservation Laws, Kluwer Academic/Plenum Publ., New York, NY, 2003 (in English)].
5. Glushak, B.L., Kuropatenko, V.F., and Novikov, S.A., *Study of Material Strength at Dynamic Loads*, Nauka Publ., Siberian Branch, Novosibirsk, Russia, 1992.
6. Altshuler, L.V., "Use of Shock Waves in High-Pressure Physics," *Uspekhi Fizicheskikh Nauk*, Vol. 85, No. 2, 1965, pp. 197–258, [English trans., *Soviet Physics Uspekhi*, Vol. 8, No. 1, 1965, pp. 52–91].
7. Altshuler, L.V., and Brusnikin, S.E., "Equations of State of Compressed and Heated Metals," *Teplofizika Vysokikh Temperatur*, Vol. 27, No. 1, 1989, pp. 42–51, [English trans., *High Temperature*, Vol. 27, No. 1, 1989, pp. 39–47].
8. Kormer, S.B., Funtikov, A.I., Urlin, V.D., and Kolesnikova, A.N., "Dynamic Compression of Porous Metals and the Equation of State with Variable Specific Heat at High Temperatures," *Zhurnal Eksperimentalnoi i Teoreticheskoi Fiziki*, Vol. 42, No. 3, 1962, pp. 686–702, [English trans., *Soviet Physics JETP*, Vol. 15, No. 3, 1962, pp. 477–488].
9. Glushak, B.L., Gudarenko, L.F., and Styazhkin, Yu.M., "Semi-Empirical Equation of State of Metals with Variable Heat Capacity of Nuclei and Electrons," *Voprosy Atomnoi Nauki i Tekhniki. Seriya: Matematicheskoe Modelirovanie Fizicheskikh Protsessov*, 1991, No. 2, pp. 57–62.
10. Glushak, B.L., Gudarenko, L.F., Styazhkin, Yu.M., and Zherebtsov, V.A., "Semi-Emperical Equation of State of Metals with Variable Electron Heat Capacity," *Voprosy Atomnoi Nauki i Tekhniki. Seriya: Matematicheskoe Modelirovanie Fizicheskikh Protsessov*, 1991, No. 1, pp. 32–37.
11. Medvedev, A.B, "Model of Equation of State Including Vaporization," *Voprosy Atomnoi Nauki i Tekhniki. Seriya: Teoreticheskaya i Prikladnaya Fizika,* 1990, No. 1, pp. 23–29.
12. Medvedev, A.B., "Model of Equation of State Including Vaporization, Ionization, and Melting," *Voprosy Atomnoi Nauki i Tekhniki. Seriya: Teoreticheskaya i Prikladnaya Fizika,* 1992, No. 1, pp. 12–19.
13. Steinberg, D.J., Cochran, S.G., and Guinan, M.W., "A Constitutive Model for Metals Applicable at High-Strain Rate," *Journal of Applied Physics*, Vol. 51, No. 3, 1980, pp. 1498–1504.
14. Glushak, B.L., Novikov, S.A., and Batkov, Yu.V., "Constitutive Equation for Describing High Strain Rates of Al and Mg in a Shock Wave," *Fizika Goreniya i Vzryva*, Vol. 28, No. 1, 1992, pp. 84–89 [English trans., *Combustion, Explosion, and Shock Waves*, Vol. 28, No. 1, 1992, pp. 79–83].

15. Batkov, Yu.V., Glushak, B.L., and Novikov, S.A., "Plastic-Strain Material Resistance During High-Rate Deformation in Shock Waves (Review)," TsNIIAtominform, Moscow, 1990.
16. Baum, F.A., Stanyukovich, K.B., and Shekhter, B.N., *Explosion Physics*, Fizmatgiz Publ., Moscow, 1959.
17. Zubarev, V.N., and Evstigneev, A.A., "Equations of State of the Products of Condensed-Explosive Explosions," *Fizika Goreniya i Vzryva*, Vol. 20, No. 6, 1984, pp. 114–126, [English trans., *Combustion, Explosion, and Shock Waves*, Vol. 20, No. 6, 1984, pp. 699–710].

Methods and Devices
for Producing Intense Shock Loads

M.V. Zhernokletov

It is convenient to classify experimental devices according the energy source used to drive shock loads. In this approach, they can be subdivided into a number of principal groups:

- gun type launchers, which include powder guns, light-gas guns (LGG), and ballistic shock tubes (BST);
- explosive systems based on powerful condensed high explosives;
- electric and electromagnetic guns;
- combined gun types using several energy sources;
- devices using radiation sources.

It is important to recognize that none of the above gun groups can cover the entire range of pressures, velocities, masses of launched bodies, and duration of steady-state flow needed to study the broad range of material behavior (i.e. experiments for Hugoniot estimation, or short-term pulses in studies of spallation or other failure). Therefore, the choice of device type depends on the specific problem facing the investigator.

Presently, in worldwide practice, powder guns, LGG, BST, and explosive launchers have gained the widest acceptance in studying dynamic compressibility, strength characteristics, and spallation phenomena in the laboratory. Here we consider these and other devices for producing shocks in more detail.

2.1 Gun Type Launchers

In this type of facility, the launched body (a projectile), as is common in artillery systems, moves in a tube under the force of a compressed gas. The facilities differ depending on the gas used and its method of compression. Consider the simple schematic of the gun presented in Fig. 2.1. The projectile of mass m is located in a tube of cross section S. To its right, in the barrel channel of length l_b, a vacuum is produced. The chamber of length l_c to the left of the projectile is filled with gas under pressure P_0. Denote sound speed

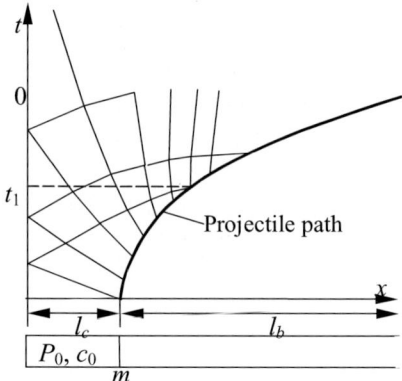

Fig. 2.1. Schematic of the gun and rarefaction wave and projectile paths

in the gas at P_0 by c_0 and assume that the polytropic exponent, γ, which is the ratio of specific heat at constant pressure (C_P) to that at constant volume (C_V), is a constant.

At the time the projectile starts moving, a rarefaction wave running to the left appears in the chamber. The wave front, i.e. the initial perturbation from the projectile, propagates to the left at a sound speed of a characteristic C_-, $x = -c_0 t$, and arrives at the rear wall of the chamber in time, l_c/c_0. The wave front is reflected from the rear wall and returns to the projectile. When the characteristics emerging from the projectile motion line interact with the reflected wave front, their paths and the path of the front itself bend, as shown in the $x - t$ diagram in Fig. 2.1.

The 1D process under discussion is described by the equations following from Eq. (1.19):

$$\frac{\partial \rho}{\partial t} + u\frac{\partial \rho}{\partial x} + \rho\frac{\partial u}{\partial x} = 0 \qquad \text{(conservation of mass)};$$

$$\frac{\partial u}{\partial t} + u\frac{\partial u}{\partial x} + \frac{1}{\partial \rho}\frac{\partial P}{\partial x} = 0 \qquad \text{(conservation of momentum)};$$

$$\frac{\partial P}{P} + \gamma\frac{\partial \rho}{\rho} = 0 \qquad \text{(isentropic process)},$$

where ρ is density, u is flow rate, and x is coordinate. The equations should be supplemented with the equation of state of gas, $P = \rho RT$. Sound speed is determined from relations $c^2 = \gamma P/\rho = \gamma RT$.

This equation system has a classic solution [1, 2]. For a rarefaction wave running to the left, u and c along the characteristics C_- that have slope $dx/dt = u - c$ will be constant. The Riemann invariant for the opposite family of the characteristics is constant in the entire flow. Hence, the following relation is valid near the rear surface of the projectile:

$$\frac{2c}{\gamma - 1} + u = \text{const} = \frac{2c_0}{\gamma - 1} .$$

As force $PS = m\,(du/dt)$ acts on the projectile and $c/c_0 = (P/P_0)^{\gamma - 1/2\gamma}$ for the isentropic process, we arrive at the following relation for the projectile velocity [1, 2]:

$$u_{pr} = \frac{2c_0}{\gamma - 1} \left\{ 1 - \left[\frac{SP_0 t}{mc_0} \frac{\gamma + 1}{2} + 1 \right]^{-\frac{\gamma - 1}{\gamma + 1}} \right\}, \qquad (2.1)$$

which is valid until the time, when the rarefaction wave front returns to the projectile rear surface (time t_1 in the $x - t$ diagram). The longer the chamber with the working gas is, the better the agreement is between the solution and the actual projectile velocity. In this case the rarefaction wave that has reflected from the chambers rear surface does not have time to overtake the projectile until it has left the barrel. After time t_1, the above scheme for computing projectile velocity is not applicable. Then the computation should be performed either graphically or numerically.

The analysis of the obtained expression suggests that to increase velocities of projectiles of equal mass, it is reasonable to select a gas, in which γ is small and the sound speed is large. This gas would "follow" the projectile better and impart to it a higher velocity.

Here we consider the types of devices in which compressed gases are used to drive impactors.

2.1.1 Powder Guns

The simplest design of the powder gun [3] is shown in Fig. 2.2. In guns of this type, the projectiles are driven by expanding hot gases produced from the combustion of a powder charge, ignited by a nichrome helix. The temperature of the powder gases, composed mainly of CO, CO_2, N_2, H_2O, and H_2, is less than or equal to 3000 K. Only in some very hot powders does the gaseous product temperature reach 3600 K [4]. At such temperatures the sound speeds (relatively low because of the high molecular mass of combustion products) are insufficient to achieve launch velocities higher than 3 km/s. The factor limiting the launch velocity is the wave processes in the space behind the projectile. The powder chamber length is much shorter than the barrel acceleration segment, and the rarefaction waves reflected from the rear of the powder chamber have time to overtake the projectile before it has left the barrel. Absence of a vacuum in the barrel also leads to a loss of velocity.

Figure 2.3 shows the VNIIEF powder gun. It has a barrel with an inner diameter of 85-mm. The total length of the gun is ~15 m. It is able to accelerate ~1-kg projectiles up to ~2 km/s velocity.

Difficulties arising from handling powder and the need for frequently cleaning of scale from the barrel limit the number of times these devices can be

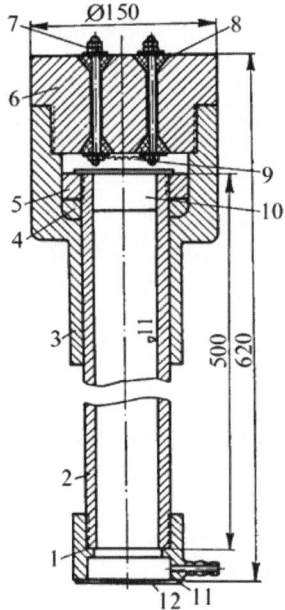

Fig. 2.2. Powder gun: 1 – muzzle nozzle, 2 – barrel, 3 – chamber housing, 4 and 5 – rings, 6 – cover, 7 – electrodes, 8 – insulator, 9 – powder weight and nichrome helix, 10 – impactor, 11 – vacuum seal, 12 – target. (Sizes are given in mm)

Fig. 2.3. VNIIEF powder rail-gun

used in the laboratory. In addition to these demerits, one should recognize the poor reproducibility of experimental results in short-barrel guns. Even with the use of modern high-energy colloid powders, which burn uniformly in parallel layers, there is no assurance of the stability of shots that is required in many cases, nor does their use preclude a wide variation in velocity from one shot to another, which makes experimental results considerably harder to compare. A very essential problem relating to powder gun viability, which is considerably inferior to that of other barrel guns, is the specific conditions of their operation (i.e. pressures up to 200–300 MPa and powder gas temperatures between 2000–3000 K for hundredths or thousandths of a second). Studies performed on 175-mm guns show [4] that lifetimes of powder gun barrels are no higher than 300 to 900 shots, if no special measures are taken.

2.1.2 Light-gas Guns (LGG)

Among LGG, the most widespread are the two-staged guns. LGG should be categorized in the combined accelerator group, as various sources are used to rapidly compress and simultaneously increase temperature of a light gas (hydrogen or helium) confined in a special chamber; the sources may be either chemical, with combustion of a powder charge or a special oxygen-hydrogen-helium mixture, or electrical with high-voltage discharge.

Figure 2.4 shows a schematic of a two-staged light-gas gun. The principle of its operation is as follows. First powder is fired in the chamber. The powder combustion results in high-molecular-mass gas and a high pressure (the first stage), under which a massive piston begins to move along the smooth channel of the compression chamber. The compression chamber, hermetically sealed at the initial time, contains gas with a relatively low molecular mass, i.e. light gas. As the massive piston moves along the compression chamber channel, the piston compresses the light gas to very high pressures (the second stage). Initially the pressure of the light gas is typically 10–100 atm yet can reach up to 10000 atm during a shot. As the piston compresses the light gas, there is a point when the pressure in the acceleration chamber is greater than the yield stresses of the rupturing diaphragm. When the diaphragm ruptures, the high-pressure gas begins to accelerate the projectile along the ballistic barrel. Note that the projectile usually begins to move before the pressure in the compression chamber has reached a maximum value. When the piston reaches the narrowing channel, its front part is squeezed farther into

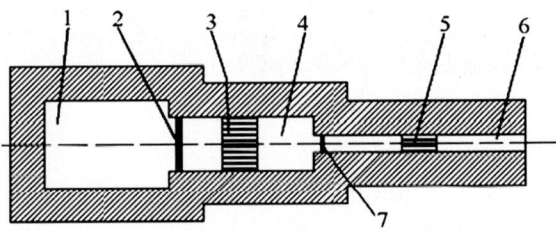

Fig. 2.4. Scheme of the two-staged light-gas gun: 1 – powder chamber (first stage); 2,7 – diaphragms; 3 – piston; 4 – pressure chamber filled with light gas (second stage); 5 – projectile; 6 – acceleration pipe

the acceleration chamber and the average piston velocity decreases. As the piston deforms, its front part moves into the acceleration tube and maintains high pressures behind the projectile. The projectile velocity, W_{pr}, depends on the sound speed in the pushing gas. The actual launch velocity is not higher than $W_{pr} \sim 1.8$–$2.5 \ c_0$ [2, 4].

The typical cycle of the two-staged LGG is affected by a fairly large number of factors. These include powder type and mass (charge mass), piston mass, boosting pressure, i.e. pressure in the powder chamber sufficient for cutting off the expanded section of the piston tail or its squeezing into the compression chamber, initial pressure and relative molecular mass of gas in the compression chamber, opening pressure of rupturing diaphragm, and projectile mass. By way of example, consider the characteristics of the US Lawrence Livermore National Laboratory two-staged light-gas gun [5], reckoned to be one of the best. It is able to accelerate projectiles of 20 grams up to 8 km/s. At its impact onto a target the projectile produces a shock wave, whose maximum pressure depends on specimen shock impedance, but ranges from 20 GPa for light substances to 500 GPa for heavy metals. Before the shot, the gun barrel (28 mm diameter) and the target chamber are evacuated down to 1–50 µm Hg, so that there is minimal projectile heating due to friction from the residual gas. The total increase in projectile temperature during the acceleration is about 1 degree. In the second gun stage, the pushing gas is hydrogen. Figure 2.5 depicts the projectile design [4]. It consists of a metal impacting plate 25 mm in diameter and 1 to 3 mm in thickness, a cylindrical shoe made of polycarbonate plastic, into which the impactor is pressed while the plastic is hot, and a polyethylene cap tightly put on the shoe on its rear thus providing a seal for the pushing gas.

The projectile velocity is experimentally measured by using a pulsed X-ray technique which measures the distance traveled by the projectile through the measured time interval (the time between X-ray pulses). The projectile positions at the beginning and end of the interval are found by

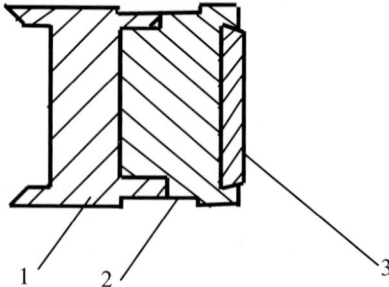

Fig. 2.5. Longitudinal section through the projectile with the plane metal impacting plate: 1 – polyethylene cap, 2 – shoe made of polycarbonate plastic, 3 – metal impacting plate

recording two pulsed X-ray images on Polaroid film with a luminescent screen made of zinc sulfide converting X-ray radiation to visible wavelengths.

In addition, diagnostic devices, including detectors (sensors) and electronic circuits, developed for the gun allow measuring shock wave velocities with nanosecond resolution.

A gas-dynamical study of operation of the two-staged LGG shows that the increase in their efficiency is due to higher light gas temperatures. In the two-staged piston facility the temperature increases solely due to a one-time gas compression. The gas temperature at the end of compression can be obtained from the equation of state of gas and the known expression for entropy, $S = P/\rho^{\gamma}$, using the initial gas parameters [2]:

$$\left(\frac{T_{\max}}{T_0}\right) = \left(\frac{P_{\max}}{P_0}\right)^{\frac{\gamma-1}{\gamma}} \left(\frac{S_{\max}}{S_0}\right)^{1/\gamma} = \left(\frac{c_{\max}}{c_0}\right)^2. \tag{2.2}$$

In the above relation the subscript *"max"* refers to the maximum temperature, pressure, entropy, and sound speed in the gas, respectively, at the end of the compression and the subscript "0" to initial state of the gas for the same variables. From Eq. (2.2) it follows that in piston facilities it is beneficial to have a high degree of compression $(P_{\max}/P_0)$ and to conduct the process with maximum possible entropy growth $(S_{\max}/S_0)$. In this case, at a given $P_{\max}$ primarily determined by facility strength, γ, and initial gas temperature T_0, the temperature of the gas and thus sound speed will be the greatest. On the whole, two-staged LGG's are quite complex and expensive apparatuses requiring a lengthy cycle of preparing and conducting experiments and recovery for subsequent experiments.

The LGG launching velocities can be also increased using various methods; preheating the working gas, modification of the projectile design, as well as, using a plastic cushion layer placed on the rear of the plate to be launched. Refinement in the latter two areas has been extensively explored by Sandia National Laboratories (located in the US), and has resulted in the development of ultrahigh-velocity launchers [6]. Below we briefly describe the principle of operation of the Sandia impactors.

Very high dynamic pressures of tens or hundreds of GPa are required to accelerate the plates up to hypervelocities. To avoid melting or vaporization of the launched plate, the loading pressure pulse must be time-dependent. For this purpose layered impactors are used, a package of layers of thin plates fabricated from materials of various dynamic hardness is used.

Figure 2.6 presents two schemes for launching ultrahigh-velocity projectiles. In the modified two-staged light-gas gun, as shown in Fig. 2.6, the plate to be launched is accelerated gradually using a buffer and a lexan impactor with a package of thin plates, whose dynamic impedance increases from left to right: plastic – magnesium – aluminum – titanium – copper – tantalum. The characteristic thickness of each layer ranges from 0.5 to 1 mm. This impactor upgrade produces a flyer plate velocity up to 10.4 km/s in flyer plates 1 mm

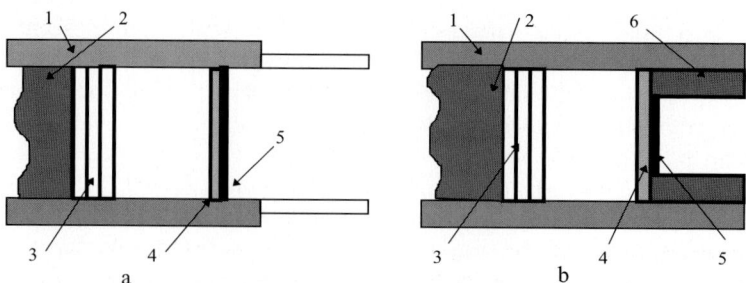

Fig. 2.6. Schematics of Sandia ultrahigh-velocity launchers: a – ultrahigh-velocity accelerator for launching plates at a velocity up to 12 km/s; b – improved ultrahigh-velocity launcher accelerating the plate in the tungsten barrel up to 16 km/s. 1 – barrel; 2 – lexan impactor; 3 – package of thin plates; 4 – buffer made of 0.8 g/cm^3 density plastic; 5 – plate to be launched; 6 – tungsten barrel

thick made of aluminum, magnesium or titanium and up to 12 km/s for plates 0.5 mm thick made of the same materials.

Further improvement of the launch characteristics can be achieved by making the launched plate diameter shorter relative to the diameter of the barrel the impactor moves along. The experimental device shown in Fig. 2.6b operates as a dynamic booster reservoir. The plate to be launched is accelerated in a tungsten barrel located inside the main LGG barrel. For example, plates with a thickness of 1 mm were accelerated to velocities of 14 km/s, and plates 0.5 mm thick were accelerated up to velocities of 15.8 km/s [6].

2.1.3 Ballistic Shock Tubes

The BST operation is analyzed in [2, 4, 7, and 8]. The key design elements of the BST are presented in Fig. 2.7. BST have much in common with LGG, but are operated under less rigid conditions: pressure no higher than a few tens of MPa, normal working gas temperatures, absence of any explosive compositions, i.e. powder or gas mixtures. The maximum launch velocity of a BST is

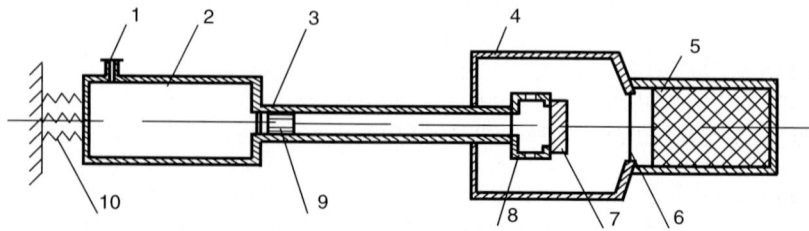

Fig. 2.7. Schematic of BST: 1 – gate; 2 – high-pressure chamber (HPC); 3 – barrel; 4 – vacuum chamber (VC); 5 – deceleration compartment; 6 – diaphragm; 7 – target; 8 – target holder; 9 – projectile; 10 – shock-absorber

about 2000 m/s. Many principles of the developed theory for two-staged LGG can be applied in BST gas-dynamic computations. The general issues include those pertaining to selection of high-pressure chamber (HPC) form, optimal volume, as well as, barrel gauge and length.

Despite differences in design of various ballistic shock tubes, essentially all of them have the same set of key elements presented in Fig. 2.7. There is a distinct relationship between the parameters of the BST key design elements. For example, it is found [8] that for a given gas working volume at a given equilibrium temperature, the HPC projectile velocity W_{pr} is a function of dimensional parameter Pd^2l/m, where P is gas pressure in the HPC, d is barrel gauge, l is barrel length, and m is projectile mass. It turns out that $Pd^2l/m = 1.5 \cdot 10^7$ m^2/s^2, if the projectile is to be accelerated to a velocity $W_{pr} = 1500$ m/s for a working gas (helium) temperature of 310 K in the HPC. For given values of d and m, using a long enough barrel enables a reduction in the working pressure in the HPC.

A clear advantage that BST have over other guns is servicing simplicity and operational safety, which facilitates the use of these devices in an ordinary research laboratory environment.

In 1974 VNIIEF introduced a BST [9] with the following characteristics: 76.2-mm barrel gauge; 10.8-m barrel length. Other characteristic sizes of VNIIEF's BST are presented in Fig. 2.8. The maximum working pressure in the HPC is 40 MPa. To avoid preloading of the target with an air cushion between the projectile and the target, the barrel is evacuated using a special vessel (vacuum chamber, VC). The residual air in front of the flying impactor and the working gas from the barrel after impact are bled into the vacuum chamber.

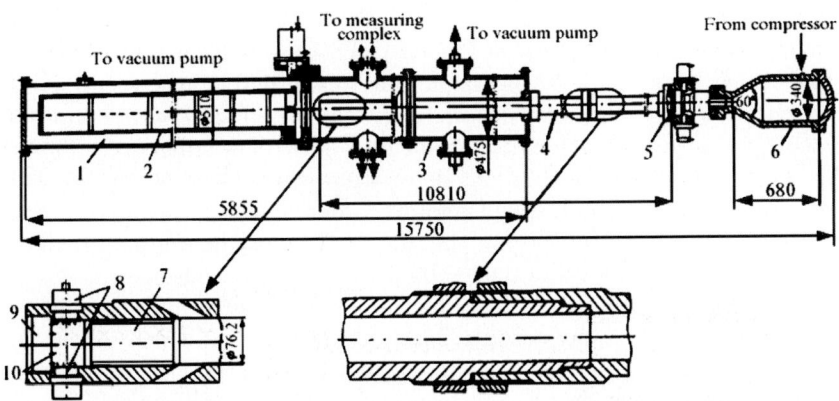

Fig. 2.8. Design of the VNIIEF BST [9]: 1 – deceleration chamber; 2 – block with buffer; 3 – residual gas release chamber; 4 – barrel; 5 – diaphragm; 6 – HPC; 7 – projectile, 8 – a set of contacts to measure projectile velocity; 9 – target; 10 – contacts to measure projectile-target impact plane skewness

The mass of the projectile, whose housing is fabricated from strong aluminum alloy AMg6, can range up to ~700 grams. When developing and manufacturing the impactors, necessary requirements taken into account include: strength and rigidity, minimum skewness at the impact and maximum reduction in the friction force during motion of the impactor. A typical design of an impactor is presented in Fig. 2.9 [9]. Experimental recorded velocities for a projectile of $0.72 + 0.02$ kg in mass are: $W_{pr} \approx 0.3$ km/s at $P = 2$ MPa pressure in the chamber; $W_{pr} \approx 0.42$ km/s at $P = 4$ MPa: and $W_{pr} \approx 0.5$ km/s at $P = 10$ MPa.

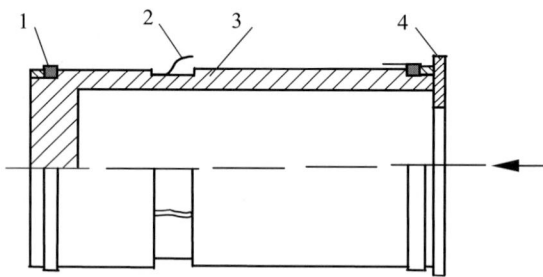

Fig. 2.9. Projectile: 1 – guiding ring; 2 – grounding contact; 3 – housing; 4 – retaining ring

2.2 Explosive Launchers

In these devices, cylindrical charges are typically used, along whose axis plane detonation waves propagate (see Fig. 1b). The waves impinge on metal impactors which cause their acceleration. The plane detonation waves are produced either with plane wave lenses, a version of which [10, 11] appears in Fig. 2.10, or with special devices ensuring multi-point initiation. The time difference in the release of different detonation waves is detected using high-speed streak-cameras (high-speed photorecorders). The lenses used by VNI-IEF [10, 11] are characterized by the time difference in release of different detonation waves at a level of 0.02–0.03 µs for a 40-mm diameter.

The charges are fabricated from high explosives with differing shattering effect (TNT, hexogen and octogen base compositions), which allows for a variation in impactor launch velocity with the same size of impactors. In the lower dynamic pressure range "contact" explosive devices [11] are used. Although the term "contact" is conventional, it is also somewhat misleading. An important element of the devices, one of which is presented in Fig. 2.11, is a small (5 mm) air gap between the charge and a thick (10 mm) plate. The gap ensures generation of a flat top shock profile in the plate. When the shock wave approaches the free surface, the plate is moving at a constant

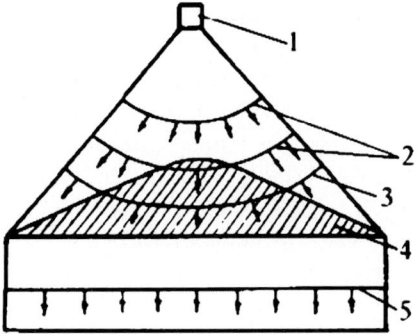

Fig. 2.10. Plane lens detonation wave initiator: 1 – detonator; 2 – diverging detonation wave front; 3 – diverging shock wave front; 4 – inert straightening lens; 5 – plane detonation wave front

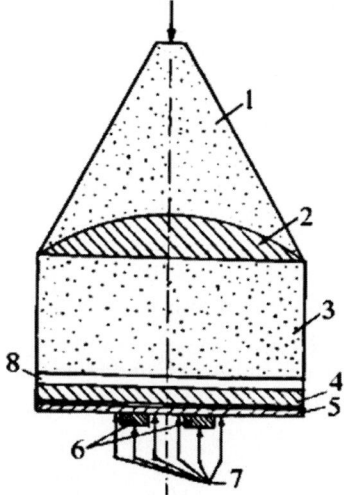

Fig. 2.11. Scheme of contact shock wave generator: 1 – HE lens charge; 2 – straightening inert lens; 3 – main HE charge; 4 – impactor; 5 – base plate made of reference material; 6 – specimens of interest; 7 – electrocontact sensors; 8 – air gap

velocity for the "first-step" of its motion. The velocity of the 10-mm-plate is measured through a second gap separating the impactor from the base plate containing the specimens of interest that are positioned on it. Aluminum and copper are used as base plate materials in the contact devices. Launchers of this type allow pressure variation from a few tenths of a GPa to 30–40 GPa with a change in the charge length, impactor material, and some modifications involving introduction of additional soft cushions between the thick plate and the specimen.

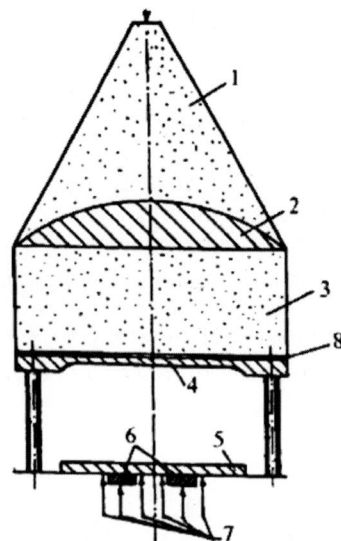

Fig. 2.12. Schematic of launching shock wave generator. 1 – HE lens charge; 2 – straightening inert lens; 3 – main HE charge; 4 – impactor; 5 – base plate made of reference material; 6 – specimens of interest; 7 – electrocontact sensors; 8 – plexiglass cushion 1 mm thick

Higher-intensity shock waves are generated by devices [11] using a long acceleration of thin discs of expanding explosion products. A schematic of one of the devices is shown in Fig. 2.12. To reduce heating, the accelerated discs are separated from the charge by plexiglass or polyethylene cushions. The impact on the base plate with the specimens of interest positioned on it occurs at a distance of 25–90 mm from the charge end when the impactor has gained constant velocity W. The velocity of the impactor depends on the HE composition and the ratio of impactor mass m to charge mass M per unit area. The final velocity is estimated with the following expression to an acceptable accuracy:

$$W = D \left[(1 + \beta) - \sqrt{(1 + \beta)^2 - 1} \right] , \qquad (2.3)$$

where D is charge detonation rate; $\beta = 27m/16M$.

This expression was obtained by E.I. Zababakhin [12] from solving the incompressible plate acceleration problem assuming that the explosion products followed a cubic equation of state.

The plane accelerating systems are capable of producing impact velocities of 4–6 km/s and specimen pressures of up to $\sim$200 GPa for metals with densities close to that of iron. VNIIEF has developed about 30 plane-wave launchers with charge diameters of 90 and 120 mm and length up to 220 mm, which are widely used in shock-wave studies.

Much higher shock-wave amplitudes are obtained with explosive systems using converging supra-compressed detonation waves. In the spherical geometry, pressures at the detonation wave front continuously increase as the center is approached. Therefore, in the stock of the dynamic methods, the hemispherical charge simultaneously initiated across the outer surface is a good tool for studying material properties at ultrahigh pressures.

Key elements of the measuring device [13] designed for these purposes are depicted in Fig. 2.13. The device is a hemispherical charge with an internal hollow cavity, into which a thin-walled steel shell is inserted. The impacting shell is accelerated and converges on the device center as a result of the converging detonation wave products. By positioning the target plate composed of specimens shielded by a base plate at various radii, pressures of 1–2 TPa can be obtained in the specimens.

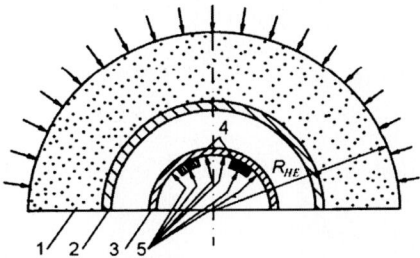

Fig. 2.13. Key elements of the hemispherical shock wave generator. 1 – hemispherical HE charge; 2 – iron shell; 3 – iron base plate; 4 – specimens under investigation; 5 – electrocontact sensors for measuring shock wave velocity

A relatively small air gap between the shell and HE is added to some hemispherical charge designs to reduce heating of the shells during the transmission of shock waves through them. The measuring device with an air gap is schematically shown in Fig. 2.14. Several types of such charges were developed and are referred to as SC ("soft charge") devices. In most of them the shell thickness is about 3 mm and the air gap is ~12 mm. These devices differ from each other in radial target position and, as a result, the measuring radii. In accordance with pressures produced at the radii, the charges are referred to as SC-3, SC-3.5, SC-4, and SC-8 (where 3, 3.5, 4, and 8 represent the iron pressures in Mbar). To obtain pressures close to 10 or more Mbar, cascade hemispherical device types are used.

The principle of 1-D multi-cascade acceleration of plates was considered and justified by E.I. Zababakhin in 1948. As depicted in Fig. 2.15, a "thick" plate is accelerated to velocity W_1 by the first charge explosion products. The thick plate impacts charge 2 which produces a supracompressed detonation wave within it. The explosion products then expand and accelerate a thinner

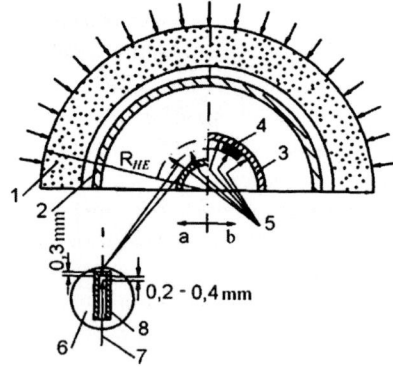

Fig. 2.14. Schematic [13] of SC type measuring device for measuring shell flight velocity (a) and shock wave velocity in specimens (b): 1 – hemispherical HE charge; 2 – steel shell; 3 – steel base plate; 5 and 6 – electrocontact sensors for measuring shock wave velocity D and shell flight velocity W; 7 – copper pin; 8 – steel cap

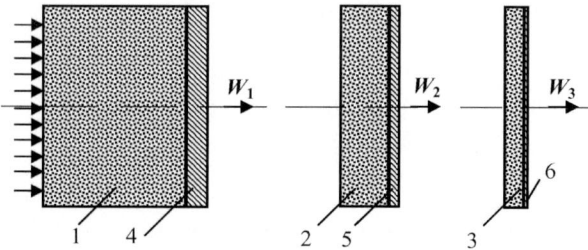

Fig. 2.15. Schematic [13] of 1-D multi-cascade plate acceleration: 1–3 – first, second and third cascade HE charges; 4–6 – first, second and third cascade plates

plate attached to charge 2 to a velocity $W_2 > W_1$. The progressing acceleration process can continue through three or more cascades.

Figure 2.16 presents the diagram of flow in a layered launch system in $x - t$ (distance-time) coordinates. The diagram illustrates the complex wave process, which arises during the deceleration of a massive, slow flying impactor against a readily compressible interlayer, followed by acceleration of the flyer of a lower mass.

This method was used experimentally [14] for high-velocity acceleration of molybdenum foils and thin copper plates. With thicknesses of 0.5 and 0.1 mm, velocities of 9.1 and 11.3 km/s were achieved, respectively. Thus, with an increasing number of cascades, an increase in the velocity of the driven plate is possible with a commensurate decrease in the plate thickness.

However, with increasing velocity, the shock heating of the driven plate also increases, which leads to maximum achievable velocities when the plate is still moving in the solid state. This limitation is not only due to the possibility of the plate heating to the melting point, but also to the reduction in plate

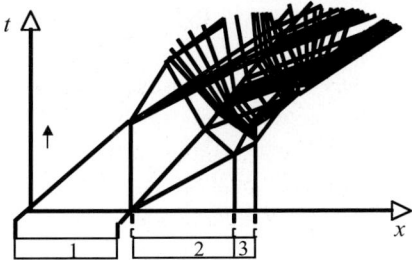

Fig. 2.16. $x - t$ diagram of flow in a layered launching system: 1 – impactor; 2 – cushion; 3 – flyer

material strength during heating. A reduction in material strength in the plate promotes Rayleigh-Taylor instabilities that can lead to destruction of the plate during the acceleration process. Estimates show [15] that given the fabrication precision (ensuring a plate thickness difference of ∼0.1%), the maximum achievable velocity of a hard plate is ∼18 km/s for titanium and ∼15 km/s for steel.

Figure 2.17 illustrates a three-cascade explosive accelerating device [16] developed by VNIIEF. In the device, between the plate that has been accelerated in the two-cascade system and the plate to be accelerated to ultrahigh velocity, there is a damper of low-density material, which ensures a reduction in the shock wave intensity down to a permissible level. The first cascade, involving an HE charge of 240 mm diameter and 600 mm length, with a mass ∼50 kg, is initiated by a plane wave generator. The steel plate of the first cascade is attached to charge 2 through an organic glass pad. At a distance of 85 mm from the glass pad, there is the second cascade. It consists of an HE charge of 120 mm diameter and 10 mm length with a second cascade steel plate (thickness of 1 mm and diameter of 90 mm) immediately adjacent to it, followed by a damper (thickness of 2 mm and diameter of 90 mm). The

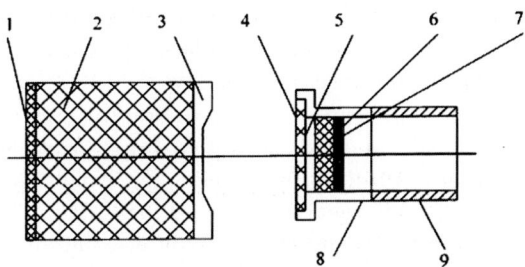

Fig. 2.17. Schematic of the three-cascade plane-geometry shock wave generator. 1 – plane detonation wave generator; 2, 4 – HE charges of the first and second cascade, respectively; 3, 5, 7 – plates of the first, second and third cascades; 6 – buffer; 8, 9 – steel and organic glass cylinders

damper is made of microporous polymethylmethacrylate, 0.05 g/cm^3 in density; and the driven plate (0.5 mm thick and 90 mm diameter) is made of titanium enclosed in a steel cylinder. The driven plate moves in an evacuated cylinder made of organic glass, which allows pulsed X-rays through the walls to determine its shape and velocity. The experiments show that the average velocity of the titanium plate is 10.2 ± 0.2 km/s at the central part and 10.9 ± 0.3 km/s at the periphery. The average velocity of the second cascade plate is 6.8 ± 0.3 km/s. The titanium plate remains intact and is in the solid state. An illustration of the cascade launching in the spherical geometry is presented in Fig. 2.18.

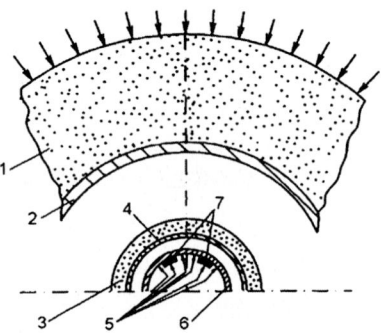

Fig. 2.18. Scheme of two-cascade hemispherical shock wave generator [13]: 1, 3 – first and second cascade HE charges; 2, 4 – first and second cascade shells; 5 – electrocontact sensors for measuring shock wave velocity; 6 – steel base plate; 7 – specimens under study

Obtaining reproducible results using explosive launchers requires that high-quality manufacturing processes be applied to the production of the charges. This high standard limits the acceptable ranges of charge composition, density, and geometry. As practice shows, when the tolerance requirements for HE charge fabrication and system components are met, the dynamic explosive system parameters are reproducible to within 1–1.5% in successive experiments.

Launch velocities are determined through special experiments and then specified more exactly while studying special materials with wave measurements. When using explosive shock wave generators, the gauge lengths run into millimeters, and characteristic recording times in microseconds or fractions of microseconds. Measurement errors are mainly due to asymmetry (skewness and curvature) of the impactor and the shock wave motion. According to experiments performed with a streak-camera, the time difference (asymmetry) of the shell flight to the target in the region not covered by rarefaction waves propagating from the HE charge base perimeter is $\Delta t \leq 6 \cdot 10^{-8}$ s. In plane-wave launchers, the time difference in the shock front release from the base plates is no more than 0.05 μs in a 40-mm-diameter circle, which just determines the working boundaries of the base plates.

Energy concentration greater than that available from planar devices can be obtained [14] with explosive conical generators using the effect of geometric convergence at an irregular Mach reflection of conically converging shock waves. The layout of the generator is shown in Fig. 2.19. The converging wave is generated in the cone either by direct action of detonation products or by an impact of the metal (aluminum, copper) impactor on its side surface. When using a conical shock wave generator, pressures produced in bismuth were up to 670 GPa, which is 3 times higher than in plane-wave devices.

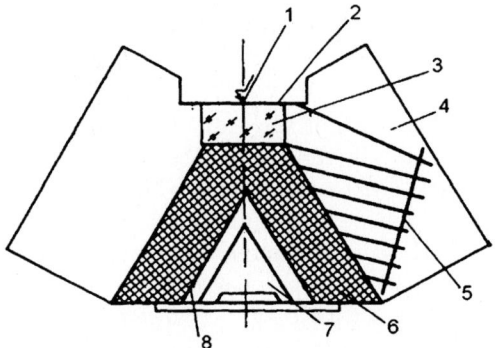

Fig. 2.19. Generator of conically converging shock waves [14]: 1 – firing initiation point; 2 – HE layer driving the detonation from the initiation point along the circumference; 3 – inert gasket; 4 – plane detonation wave generator, an element of a discrete multi-point HE conical charge initiation system; 5 – HE in plane detonation wave generator channels; 6 – conical HE charge; 7 – a conical specimen, where the converging shock wave is generated; 8 – copper conical flyer launched by charge detonation products

Studies of the HE initiation by shock waves and spallation strength for structural materials widely use explosion products for impactor acceleration which consists of a thin HE layer applied onto the impactor surface under conditions of detonation wave tangential incidence (the sliding detonation method). The impactors' motion is stable under the above loading conditions which are illustrated in Fig. 2.20.

The impactor is initially inclined at angle θ to the surface of the specimen. The angle is chosen in accordance with the following relationship $\sin\theta = W_n/D$, where D is detonation rate, $W_n = W_{imp}$ is the velocity component normal to the specimen surface. Correct choice of angle θ ensures simultaneous impact on the entire surface. As in the case of the normal detonation wave incidence, the maximum velocity of the driven plates is weakly dependent on the material compressibility, which allows the barriers to be considered incompressible to a first approximation. According to the computation, at

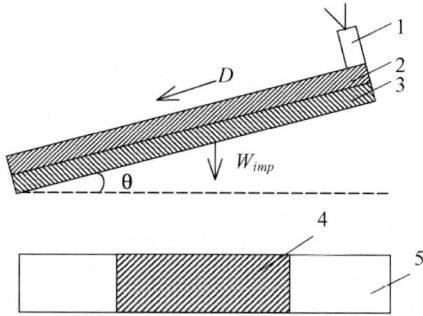

Fig. 2.20. The device for impactor acceleration by detonation wave in the tangential incidence regime. 1 – plane detonation wave initiator; 2 – HE layer; 3 – impactor; 4 – target; 5 – collar

normal and tangential detonation wave incidence using a cubic equation of state for EP on an absolutely rigid wall, the latter imparts essentially the same mechanical momentum to the barrier [17]: $i_0 = 0.296 \; mD$ and $i_0 = 0.3 \; mD$, respectively, where m is HE mass per unit surface. By time, $t_0 = 3\delta_{HE}/D$, where δ_{HE} is HE layer thickness, more than 90% of total mechanical momentum has been imparted to the barrier. Experimental data agree well with simulations for barriers with $\rho_0 c_0^2 = (0.5\text{--}1.7) \cdot 10^2$ GPa, $i_0 = 0.29 \; mD$ [17]. Thus, for barriers, whose thickness meets condition $t_0 = 3\delta_{HE}/D < 2\delta_{barr}/c_0$, the final velocity of the impactor is:

$$W_{imp} = \frac{i}{m_{imp}} = 0.29\alpha D \; , \qquad (2.4)$$

where $\alpha = m/m_{imp}$, m_{imp} is impactor specific mass per unit area. The relation presented for W_{imp} is applicable by virtue of the above time inequality for systems with small α. Reference [18] proposes the following relationship for computing barrier velocity for the geometry under discussion given by (see Fig. 2.20):

$$\frac{W_{imp}^2}{\varepsilon_0} = \frac{6\alpha^2}{\alpha^2 + 5\alpha + 4} \; , \qquad (2.5)$$

where ε_0 is EP specific energy equal to $\varepsilon_0 = D^2/16$ for the cubic equation of state of EP. After transformations, we arrive at the expression similar to the Harnie relation:

$$W_{imp} = \frac{\alpha D \sqrt{3}}{\sqrt{8 \left(\alpha^2 + 5\alpha + 4 \right)}} \; . \qquad (2.6)$$

For small α, $W_{imp} \approx 0.306\alpha D$, this reasonably agrees with Eq. (2.4). Other empirical relations for W_{imp} are proposed in [19,20]. In [19] for $0.3 \le \alpha \le 3.5$,

$$W_{imp} = 0.65 D \frac{\alpha}{\alpha + 2} \; . \qquad (2.7)$$

The authors of [20] recommend the expression

$$W_{imp} = 1.2D\frac{(1 + 1.18\alpha)^{0.5} - 1}{(1 + 1.18\alpha)^{0.5} + 1}. \tag{2.8}$$

In the range $0.3 \leq \alpha \leq 3.5$, Eqs. (2.7) and (2.8) yield similar results. For $\alpha > 1$, Eq. (2.8) leads to a lower W_{imp} compared to Eqs. (2.6) and (2.7). For $\alpha > 3.5$, W_{imp} calculated by Eq. (2.5) proves to be lower than W_{imp} calculated by Eq. (2.4). To calculate W_{imp} for small α ($\alpha < 0.3$), Eq. (2.4) is preferred, for $0.3 \leq \alpha \leq 3.5$ Eq. (2.7) is more accurate, and for $\alpha > 3.5$ Eq. (2.8) provides the best fit. The above-considered method can be used to vary principal characteristics of the pressure pulse (amplitude and duration) introduced to the barrier over a wide range by changing impactor thicknesses (mass) and HE mass. Motion stability of thin plates (of $\sim$0.1 mm thickness) [15] makes the method suitable for obtaining ultrashort pressure pulses.

In explosive launching, high velocities are also achieved with use of explosive tubular guns with cylindrical channels. RAS Siberian Branch Institute of Hydrodynamics has developed [21] a device for launching through a shaped explosion. This allows solid bodies 0.1–10 mm in size to be accelerated up to velocities of 8–14 km/s. The gas-shaped charge is a hollow HE cylinder initiated on one end with an auxiliary blasting cartridge. The body to be launched is placed on the channel axis at some distance from the exit hole. The hollow cylindrical charge explosion results in a shaped jet of detonation products moving at ultra-detonation velocity. When the jet is released, its velocity can be 13 or more km/s. A solid body put in the path of the jet can thus be accelerated to very high velocities. It was possible to successfully launch bodies made of metal, ceramics or glass using these systems. The charge-to-body mass ratio was 10^5–10^6. The highest velocities (about 14 km/s) occur by the acceleration of metal spheres $\approx$0.1 mm in size. The highest rate of detonation product outflow occurs at an inner-to-outer diameter ratio of $\approx$0.3, in this case the rate of flow is 10–11 km/s at product density 0.1–0.15 g/cm^3 in the jet. The disadvantage of such devices is the severe effect of the gas-shaped jet on the launched body. The gas-shaped jet can lead to considerable deformation of the projectiles and possible destruction during the acceleration. This makes it impossible to accelerate plane elements for the generation of plane shock waves and to measure shock compression variables in materials of interest.

Another type of explosive tubular gun (ETG) is known as the Fast Shock Tube. It has been developed [22] in the United States for studies in physics at high dynamic pressures and processes occurring during a high-velocity impact. The simplest ETG design, a version of which appears in Fig. 2.21, is composed of an initiation system, a prolonged hollow cylinder made of powerful HE and a cylindrical core made of a low-density polymer foam material. On initiation, as the detonation propagates through the HE charge, a "Mach" disc is produced in the core due to convergence of the conical shock wave. Compression and severe heating result in the decomposition of the foam material resulting on

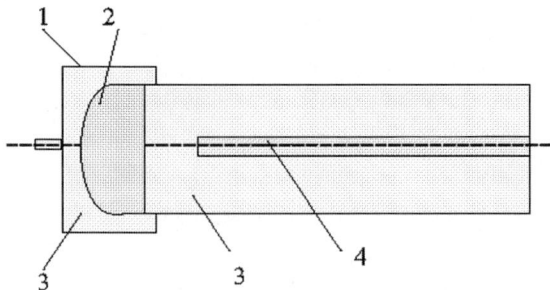

Fig. 2.21. A version of FST (Fast Shock Tubes): 1 – initiation system; 2 – porous polystyrene insertion, $\rho = 0.7$ g/cm^3; 3 – HE; 4 – polystyrene rod, $\rho = 0.52$ g/cm^3

the generation of a working body (i.e. compressed gas of a high temperature and density). The expanding gas accelerates the body to be launched which is in contact with the charge or located at some distance from it in the barrel. Devices of this kind enable the acceleration of metal discs up to velocities of 8–9 km/s with charge/launched body weight ratios of 10^3–10^4. Various polymer materials are used for fillers. The best results in velocity and preservation of the elements during the acceleration are obtained using materials with highly uniform foam cellular structure with small pore sizes and densities of 0.3–0.4 g/cm^3. A higher foam density results in a lower shock wave velocity in the inner channel, whereas a lower density leads to inner channel collapse and outflow of part of the products in the opposite direction. The relative content of hydrogen in the polymer molecular structure is of high importance. The quality of the working body and the rate of outflow of the products during the foam material destruction (due to the compression in the conically converging shock wave) become higher with increasing hydrogen quantity. It is also important to have a planar shock wave in the foam transformation products during the acceleration of the element. Violation of the planar nature leads to non-uniform effects on the element and its rapid destruction within the first several millimeters of flight. It should be noted that in such devices quite heavy demands are imposed on the quality of the foam filler, namely, on a high homogeneity of its porous structure.

In an ETG developed [23] at the N.N. Semenov Institute of Chemical Physics, the body to be launched is accelerated in another manner. The layout of the ETG is shown in Fig. 2.22.

The ETG is composed of a hollow cylindrical charge of powerful HE with a core composite charge, composed of light HE and heavy HE charges. A blasting cap through an additional HE lens initiates the device, as a whole. The body to be launched, a metal disc of diameter d, is accelerated down a short steel barrel. The steel barrel is designed to ensure the disc inserted into the light HE is at a depth where the Mach disc in the light HE has not yet been destroyed due to the rarefaction waves from the free surface of the main tubular charge. To reduce damage effects, the body to be launched can be put

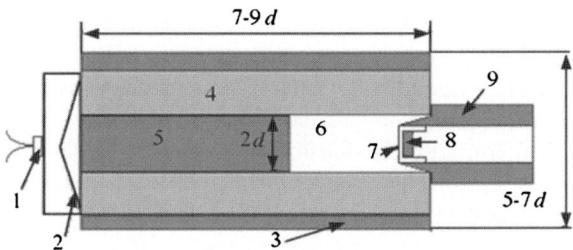

Fig. 2.22. Explosive tubular gun with explosive filler. 1 – detonator; 2 – HE lens; 3 – metal shell; 4 – main tubular charge; 5 – heavy HE with metal powder filler; 6 – light HE, 7 – plastic pan; 8 – metal impactor; 9 – steel barrel

into a plastic pan. It is possible to ensure a launch velocity gain by encasing the whole charge in a metal shell.

The gun operates as follows. Upon the firing of the detonator, a plane detonation wave is produced by the HE lens into the ETG. As it propagates across the tubular charge, the detonation front outstrips the detonation in the core ensuring the creation within it of a stationary conically converging detonation wave with the Mach disc at the center propagating at a velocity equal to the surrounding HE detonation rate. Light HE detonation products, which are light gas compressed to a high density in the converging wave and additionally compressed by heavy HE explosion products, expand and accelerate the metal disc located in the barrel.

Experiments [23] measuring flyer plate velocities showed that duralumin discs of 8 mm diameter and 2 mm thickness reached a velocity of 8 km/s and steel discs of the same sizes reached a velocity of 7 km/s. The main tubular charge was fabricated from mixed octogen and trinitrotoluene (64/36). An octogen-tungsten powder mixture was used for heavy HE and nitromethane for light HE. The choice of HE was made based on the following. Nitromethane is one of the highest-hydrogen-content HE materials and is accessible enough for studies. Compositions for the tubular charge and heavy HE were selected so that a stable supracompressed Mach-configuration detonation with a large diameter Mach disc at the center would be created in nitromethane (i.e. that the tubular HE detonation rate be higher than the nitromethane detonation rate and the heavy HE detonation rate be close to the nitromethane detonation rate). Thus, with a relatively small ratio of HE mass to flyer plate mass $(5 \cdot 10^2)$ the steel disc is accelerated to a velocity of 7 km/s. The velocity can be increased further using materials in the light HE charge with a higher hydrogen content and more powerful compositions for the tubular charge and heavy HE.

2.3 Electric and Electromagnetic Guns

In an electric gun, the impactor is driven by a dense plasma produced from capacitor bank discharge on a thin conductor [24] (for example, aluminum or copper foil). Varying the capacitor bank energy, impactor thickness, foil thickness and surface area, a wide range of impactor velocities are possible, from a few hundred meters per second to tens of kilometers per second. These devices are suitable for various studies: generation of unsteady shock waves [25], X-ray diffraction recording of shock-compressed states [26], high-intensity shock wave generation [27].

The electric gun can be considered as a gas gun prototype, where the exploding metal foil, insulator, and metal flyer play the roles of gas or powder compressed to high pressure and a projectile with an impacting plate located in its front plane. When the flyer is accelerated to high velocities, the exploding foil in the electric gun behaves like an ultrahigh-power HE. Simulations using a simple model for energy and momentum balance suggest that the final flyer velocity, W_{imp}, is proportional to the square root of characteristic specific energy, E_g, and the coefficient which is a function of the ratios between flyer and exploding foil or HE masses per unit area. E_g is energy converted from foil internal energy or HE chemical energy to flyer kinetic energy. The experiments show that W_{imp} in the electric gun can reach 3–10 W_{imp} in a system where HE of the same mass ratio per unit area drives the same flyer. This means that in the electric gun, E_g is 10–100 times as high as that in an equivalent HE.

In experiments conducted according to the layout in Fig. 2.23 [27], the plasma expansion from the aluminum foil explosion accelerated a tantalum impactor 12.7 μm in thickness and 5.08 mm in diameter to a velocity of 16 km/s. Symmetric motion of the impactor was ensured by a special system restricting expansion on the plasma side and forming a planar leading edge. A tantalum flyer moving at that velocity on the tantalum target produces a shock wave of 2 TPa. To achieve even higher pressures, it is suggested [28]

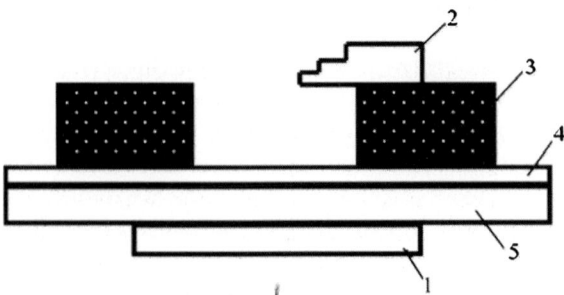

Fig. 2.23. Scheme of shock-wave measurements using foil electric explosion [27]: 1 – exploding metal foil; 2 – stepped target; 3 – distance ring; 4 – metal impactor; 5 – insulating pad

that the flyer velocity be increased. Most realistic methods to increase the flyer velocity include either injecting higher energy into the exploding foil for a shorter period of time or decreasing the flyer's total mass. However, the thinner the metal flyer, the shorter the duration of shock-compressed steady state.

A railgun [29], which is composed of two parallel metal buses that the impactor moves between when the electric circuit is closed (Fig. 2.24), can be used to achieve even higher velocities (up to 40 km/s) with thicker (1 mm) impactors.

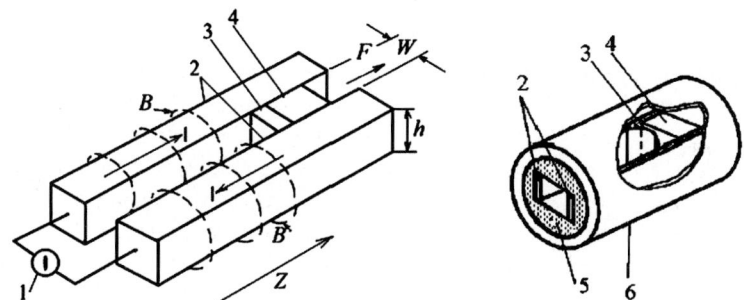

Fig. 2.24. Schematic view of the railgun: 1 – power source; 2 – conducting rails; 3 – insulating shell of the unit; 4 – impactor; 5 – closing switch; 6 – barrel

When a powerful electric energy source, for which a capacitor bank, homopolar or explosive magnetic generator can be used, is discharged to the system, ponderomotive force produced by flowing megaampere current begins to act on the impactor. Thermal insulation of the impactor from electric arc appearing at the rear of the unit is provided for the system, and measures are undertaken to keep the impactor intact.

2.4 Combined Guns

All of the LGG types can be attributed to this gun group, however, in view of the specific importance of high-velocity light-gas devices; they have been discussed previously in detail. Here we dwell on some of the most promising methods for driving solid bodies using energy released from HE detonation as one of the sources. These include a method for acceleration up to ultrahigh velocities using a shaped charge surrounding a metal shell filled with hydrogen or helium (convergence-shaping guns) [30]. The principle of operation of the convergence-shaping gun and sequential phases of the process evolution are elucidated in Figs. 2.25 and 2.26. A powerful HE charge is initiated from one end with the angle of the conic collapsing shell being chosen so as to avoid the formation of a metal shaped jet. Gas contained in the shell produces a

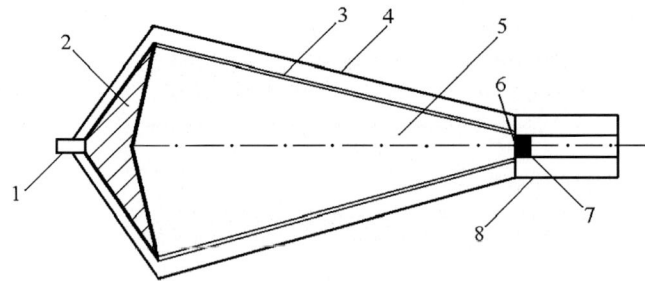

Fig. 2.25. Schematic view of the gun with a shaped charge and collapsing metal shell [30]: 1 – detonator; 2 – shock wave attenuating material; 3 – metal shell; 4 – powerful HE; 5 – gas (typically hydrogen or helium); 6 – diaphragm; 7 – impactor; 8 – metal tube

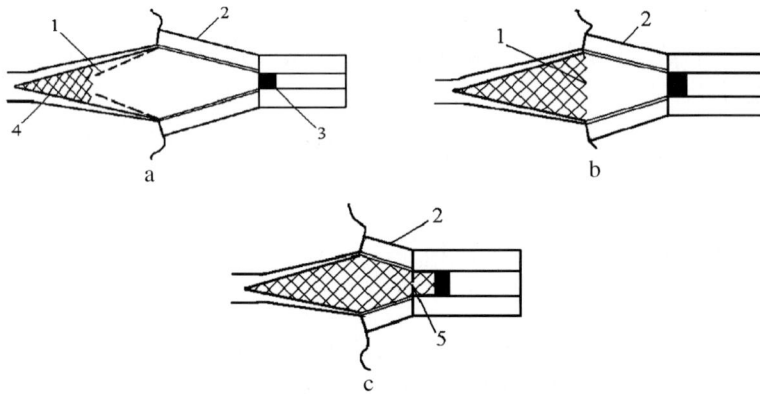

Fig. 2.26. Shell positions in times t_1 **(a)**, t_2 **(b)**, and t_3 **(c)** after the firing: 1 – shock front; 2 – detonation front; 3 – impactor; 4 – high-density gas; 5 – ruptured diaphragm

high-velocity jet achieving velocities up to 140 km/s, which then drives the impactor.

Ultrahigh magnetic fields generated in magnetocumulative generators are also used for high-velocity acceleration of solid bodies [31]. In the magnetocumulative generator, whose schematic view appears in Fig. 2.27, the magnetic field in the closed volume, (6), is almost promptly compressed as the detonation wave propagates down the cylindrical HE charge. Under action of the magnetic field, a 2-kg annular aluminum impactor, (5), can be accelerated to very high velocities. Theoretically, the magnetocumulative generator base gun can acheive velocities up to 100 km/s for impactors of several grams in mass.

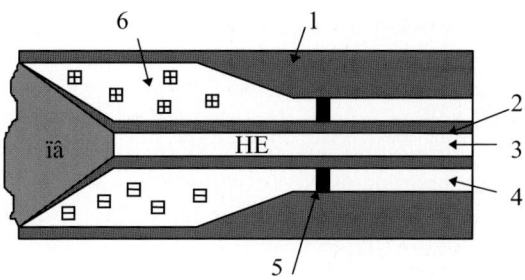

Fig. 2.27. Schematic view of the device for high-velocity launching of aluminum impactor by compressive magnetic field energy: 1 – magnetocumulative generator body; 2 – central metal tube; 3 – cylindrical HE charge; 4 – annular gap (acceleration path); 5 – annular impactor; 6 – compressed magnetic field

2.5 Using Radiation Sources for Shock Wave Generation in Solids

Shock waves can be generated in solids not only through the application of an external pulsed load, but also through rapid material heating by external or internal energy sources. Materials can be heated rapidly with several methods: exposure to laser or X-ray radiation, or through the use of electron or ion beams, etc. Macroscopically, the stress formation mechanism is the same for all radiation types, yet depends on absorbed energy concentration. A threshold separating one mechanism from another is bond energy, Q_{bond}. Consider mechanisms of stress wave and shock wave generation in solids at different absorbed energy concentrations.

When energy release is rapid, the body does not have time to expand and the absorbing medium density does not have time to change according to increasing temperature, so its pressure increases. When energy release in a substance retaining its initial density is prompt, pressures $P = \Gamma \rho_0 e_T$ develop, where e_T is the concentration of absorbed radiation energy.

Because of the medium's sluggishness and the existing pressure gradient, the medium is set in motion, compression and expansion waves then propagate through it. The wave interaction leads to the appearance of both compressive and tensile stresses (i.e. positive and negative pressures) in the solid. If the absorbed energy concentration and, hence, the arising pressure level are such that no spallation phenomena occur, the continuum motion is vibrational in nature due to wave circulation. This phenomenon is referred to as heat shock. The heat shock phenomenon occurs when a solid or liquid medium is exposed to laser radiation, pulsed flows of charged particles, or X-ray pulses several nanoseconds in duration. The heat shock method is of the highest priority when studying phenomena responsible for medium behavior at pressures greater than 10 GPa, temperatures up to 10^4 K, times between 10^{-6}–10^{-10}s, with the initial temperature range, $T_0 \sim 4$ K to $0.8\,T_{melt}$.

The above processes pertain to the case where the absorbed energy concentration is no higher than the bond energy, Q_{bond}. Another shock wave generation mechanism takes place when the absorbed energy concentration is higher than the bond energy, Q_{bond}. In this case, the interaction of powerful radiation fluxes with condensed matter is characterized by complex physical processes relating to radiation absorption and scattering, energy transfer by electron heat conduction, hydrodynamic instability of compressed plasma, radiation pulse duration, etc. Therefore, we restrict our discussion to the qualitative treatment of the motion of plasma absorbing high-density light flux. Detailed information about laser radiation – material interaction physics can be found in the review by Anisimov et al. [32] and references therein.

Let the radiation flux of density $q(t)$ fall onto a plane solid surface. Vaporization and ionization near the surface results in the generation of a plasma layer. Optical thicknesses of the layer will increase until a noticeable portion of the light flux begins to be absorbed leading to a reduced vaporization rate. The plasma layer will begin to expand, its density will decrease, and its external part will become transparent to incident radiation. At the same time, a shock wave appears in the solid body as a result of the force attributed to the recoil of momentum on the side of the expanding plasma. This results in the formation of three regions [33] (shown in Fig. 2.28) that are considerably different in their physical properties: 1) the shock-compressed solid body; 2) a strongly absorbing layer with high density and temperature gradients; 3) a rarefied plasma layer that absorbs essentially no radiation.

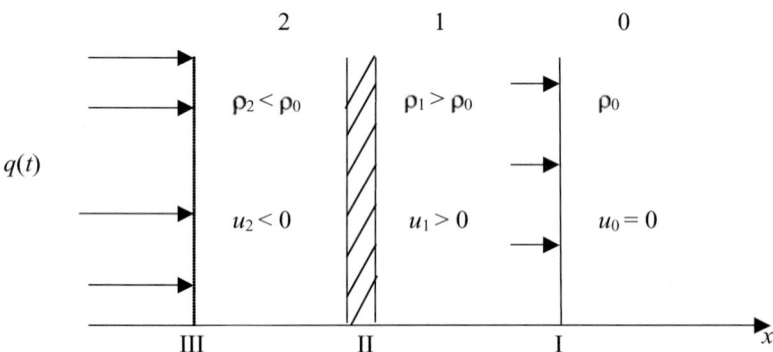

Fig. 2.28. Qualitative schematic of light-absorbing medium motion: I – shock front; II – absorbing layer; III – rarefied plasma

The values pertaining to the unperturbed solid, the region following the shock front, and the rarefied plasma jet are denoted with subscripts 0, 1, and 2, respectively. Density and temperature of the medium in region 1 depend on its equation of state and shock wave intensity.

Among the above radiation sources used to generate shock waves in the laboratory, the most advantageous are pulsed lasers of various power and heavy-current electron beam accelerators. Therefore, we will dwell in more detail on their abilities for the generation of powerful shock waves and general requirements for targets and diagnostic tools.

The major purpose in the development of powerful pulsed lasers, able to generate light pulses up to 10^4 J in energy and tens of terawatts in power was their employment for controlled fusion. Focusing light pulses onto small surfaces ($\sim 10^{-4}$cm^2) results in a high energy concentration. Specific powers supplied to targets recently achieved are $\sim 10^{14}$–10^{17} W/cm^2 and, can be brought to 10^{21} W/cm^2 in the near future [34]. Analysis of hydrodynamic computations for shock waves resulting from such laser systems on various materials suggests that it is feasible to obtain pressures in the terapascal range in metals. Experiments on the powerful shock wave laser generation system confirmed the computational results and illustrated the significant potential in using lasers to study high pressure and temperature behavior of materials. However, measurement of parameters for materials compressed by laser radiation in the above pressure range proves to be an extremely complicated technical task.

To date a majority of the laser experiments have been performed using a plane quasi-stationary wave in a setup that imposes certain limitations on target and laser pulse parameters. High dynamic pressures are produced by radiation pulses of nano- or subnanosecond duration; therefore, all measurements should be taken with very thin targets, whose thickness is not larger than several tens of micrometers. In experiments evaluating the equation of state with dynamic methods (for more details, see Chap. 4), one of the parameters to be measured is shock wave velocity in the specimen. Estimations performed in [32] indicate that at specimen thickness of 20–50 μm and shock wave velocity 20–50 km/s, the characteristic time of the experiment is about 1 ns. Therefore, optical method based devices of ~ 10 ps time resolution are used for the recording of physical processes. Figure 2.29 presents the typical layout [35, 36] of an experiment for the measurement of shock wave velocity in a laser-irradiated specimen. This layout typically includes a laser radiation source, a focusing system, a stepped specimen, a high-speed slit camera, and photographic film.

The target sizes are taken such that several criteria are met: The laser focal spot diameter has to be sufficiently small to ensure the required high energy density. On the other hand, it has to be much larger than target thickness, l, to ensure a planar shock wave. As with the pulsed effect, a rarefaction wave, which overtakes and attenuates the shock wave, propagates into the target. Target thickness, l, is taken so that condition $l \leq D\tau$ can be met, where τ is the pulse duration. However, l has to be larger than the non-thermal electron range to ensure that the opposite side of the target is not severely heated. Typically, in experiments involving strong shock waves and 0.1–2.0 TPa pressures at the front, the focal spot diameter is 300–700 μm (small

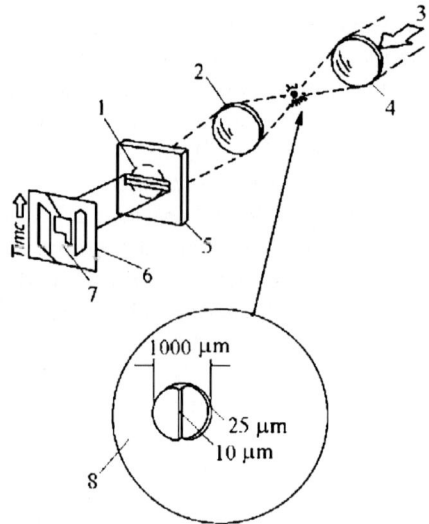

Fig. 2.29. Scheme of the experiment on laser generation of shock waves [35, 36].
1 – slit image; 2 – high-speed framing camera; 3 – laser radiation; 4 – focusing
system; 5 – camera slit; 6 – film; 7 – image to be recorded; 8 – stepped target

focal spot sizes are undesirable, as in this case the shock waves rapidly decay
and degenerate to spherical ones), the target diameter is ∼1 mm at a total
thickness of 20–40 μm.

To ensure that the accuracy in the measurement of shock wave velocity is on the order of 2–3% (this accuracy has not yet been achieved), the
required target base fabrication and measurement accuracy has to be ∼0.1 μm.
Thus, the laser-generated shock wave experiments require a more advanced
technology for the manufacture and quality control of planar microtargets.
The targets are fabricated using methods of ion bombardment and metal deposition from the gaseous phase, which enables the production of steps with
∼0.1-μm surface non-flatness. The target sizes are measured using optical,
electron and interference microscopy methods.

In shock waves generated by laser facilities of various power, the pressure
range is quite wide, from a few GPa to more than a thousand GPa [32],
which is higher than the values for light-gas guns and conventional HE based
facilities.

Now consider the response of metals to rapid heating by a relativistic electron beam (REB). Electrons with energies of a few megaelectronvolts incident
on the metal targets are absorbed in a region of macroscopic size. The characteristic absorption zone thickness is a few tenths of a millimeter. For this
reason, the interaction of short-duration electron beams with the substance
under study results in rapid material heating in the region of energy absorption. The continuum motion, which therewith arises, is qualitatively similar

to that from the material exposure to laser light. Power density in the focal spot on the anode, which can be achieved with beam self-focusing using the technology involving modern pulsed heavy-current REB accelerators, is higher than 10^{12} W/cm^2 [37]. At the above power density, energy releases into the thin anode layer, which is determined by the depth of electron penetration into the anode material, results in a thermal explosion of the anode surface for pulse time ($\tau \leq 10^{-7}$s). The expansion of material vapors at high velocities produces considerable momentum, leading to the appearance, as previously noted, of strong shock waves in the material under study. In turn, these lead to the appearance of a crater and structural changes in the anode material.

By way of example, Fig. 2.30 presents the layout of an experiment [37] for the purpose of studying the decay of shock waves actuated with heavy-current REB in metals. The heavy-current REB is emitted by the cathode and penetrates the metal anode. The REB-anode interaction generates a shock wave, which propagates inside the anode. When the shock wave arrives at the external side of the anode, the surface of the latter begins to move at double the shock wave mass velocity, and closes the gap between the anode and the metal plate. In so doing, a voltage pulse appears across the oscillograph's input resistance due to capacitance discharge through the resistance. We will not go into the details of the experiment, only noting that the REB can be used to simulate many phenomena that occur during an ultrahigh-velocity impact

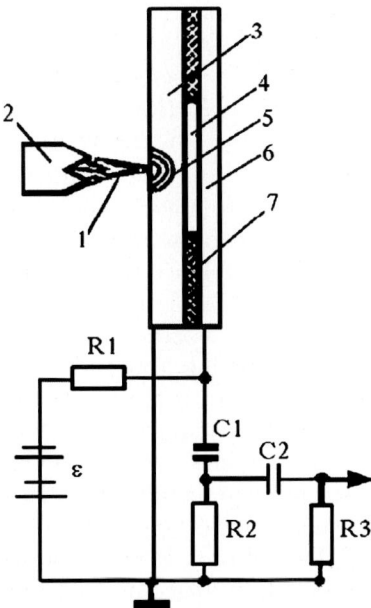

Fig. 2.30. Schematic of experiment [37] measuring shock-wave velocities in metals actuated with REB: 1 – REB; 2 – cathode; 3 – anode; 4 – gap; 5 – shock front; 6 – metal plate, 7 – dielectric pad

and near-surface explosion, in particular, crater formation. The pulsed REB is successfully used for studying metal failure kinetics under pulsed volume heating conditions [38].

2.6 Devices for Shock-Compressed Material Preservation

Experimental data suggests that various chemical and physical-chemical transformations occur in many condensed materials under shock wave conditions. The appearance of new phases violates the monotonic nature of shock compression curves. Phase changes during shock transitions are evidenced by kinks in experimental curves in coordinates $P - V$ or $D - u$ (see Chaps. 4 and 5). Besides polymorphic transformations, Hugoniot kinks can be caused by phenomena such as melting, which are most pronounced for transparent materials in $T - P$ coordinates, and plastic flows during transition from uniaxial elastic compression to bulk compression.

Detected kinks in Hugoniots allow estimation of the shock wave parameters where the transformation occurs, however, this is insufficient to answer questions concerning what transformation is experienced by the material, whether the transformation is reversible or irreversible, what fraction of the material has experienced the transformation, etc. The questions can be answered only by recovering the sample affected by the shock waves so that it could be studied with conventional physical and chemical methods. The principal point of shock-compression chemistry is analyzing recovered materials placed in metal containers for the physical and chemical transformations which may have taken place under shock loading [39–41]. This area of science has gained wide development by virtue of new material synthesis techniques, including production of ultrahard materials, diamond and cubic boron nitride specimens, metal strengthening, powder compaction, metal explosive welding, etc.

A set of devices have been developed which preserve specimens under various loading types during physical and chemical transformations under dynamic compression of condensed substances which allows for their subsequent analysis. After shock compression, especially under high pressures (100 or more GPa), specimens are hard to preserve because of rarefaction waves generated in any device and the resulting tensile stresses, which leads to the rupture of metal containers or ampules and destruction of the specimens under study. Therefore, when designing explosive devices, the main task is to place the specimens to be preserved in the working volume regions where tensile forces appearing as rarefaction wave interactions would be minimal. Here we consider the most typical methods for preserving ampule loading with HE drive.

2.6.1 Plane Dynamic Loading

Various design options exist for the preservation of materials after plane-wave shock compression loading [41–43]. The basic concepts involved in achieving this end are illustrated schematically in the two layouts of Fig. 2.31.

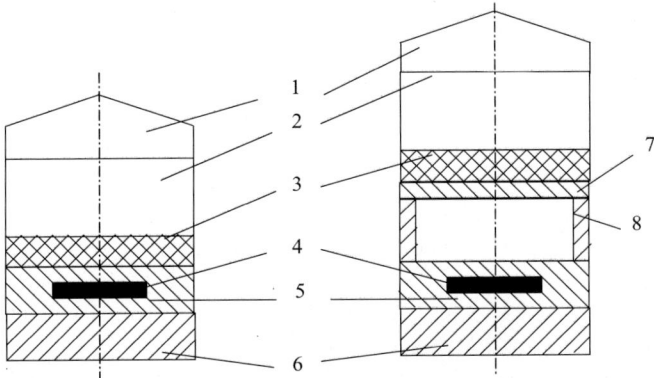

Fig. 2.31. Schematics of devices [43] for planar shock-wave loading of specimens in the preserving ampules: 1 – plane wave generator; 2 – HE charge; 3 – damping pad; 4 – ampule; 5 – material under study; 6 – metal substrate; 7 – impactor; 8 – ring

In the first scheme, the shock waves in the ampule and the specimen are generated by a plane wave detonation of an HE charge contacting the ampule. In the other layout, the shock waves are generated by an impact of a metal impactor driven by detonation products. The second scheme is preferable, as the plate impact generates a shock wave with a rectangular profile. A Change in the impactor thickness allows for variation in duration of the studied material under compression. The amplitudes of the shock wave are varied by changing thickness of HE charges, impactors, and damping pads and can range from 1 GPa to 150 GPa [42, 43]. Pressures in the studied material and the ampule are equalized through multiple shock wave reflections from the ampule bottom and cover, which are usually fabricated from a strong material of high dynamic hardness.

Increase in shock front pressure is interrelated with increase in temperature making it extremely difficult to obtain fairly high pressures without concurrent severe material heating. Shock compression of pre-cooled material does not lead to a noticeable temperature reduction. Therefore, the $P - T$ diagram region immediately adjacent to axis P is basically inaccessible in the shock-wave experiments. However, for states in the pressure range up to 100 GPa, small heating is important for studying metallization of dielectrics, including condensed gases, avoidance of annealing of high-pressure phases which form, and decomposition of the new chemical compounds produced. To ensure given thermodynamic conditions during condensed matter dynamic loading in plane geometry, isentropic shock compression methods are being developed [42] that allow a considerable reduction in thermal components of pressure and energy.

According to the shock wave theory (see Chap. 1), material heating during shock compression can be attenuated through a stepwise compression of the specimen by a series of sequentially acting shock waves of lower amplitudes. At identical final pressure, the stepwise compression leads to a decrease in final temperature and an increase in material compression degree so that the states corresponding to multiple shock compressions become close to those on the cold compression curve $P_x(V)$.

Figure 2.32 presents the experimental designs simplest to implement for specimen compression by: a) a shock wave passing through layers of different hardness.

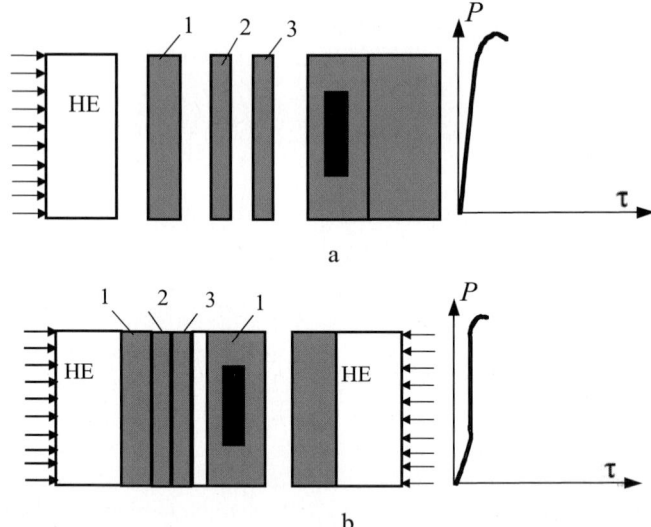

Fig. 2.32. Schemes of devices [42]: (**a**) for dynamic and quasi-isentropic compression and (**b**) sequential material compression by isentropic and then shock wave; the three-layered impactor is composed of plates of plexiglass (3), aluminum (2), and copper (1)

On the right (a) are pressure profiles in the specimens versus time during loading and (b) sequential impacts on the target with the specimen by plane impactors made of materials with increasing dynamic hardness that are apart during their travel to the target. The material compression implemented in such loading systems is attributed in [42] to dynamic isentropic compression (DIC). The actual compression pattern during material compression, according to the scheme of Fig. 2.32b, becomes more complex due to interaction of the impactors after reflection from the target and possible

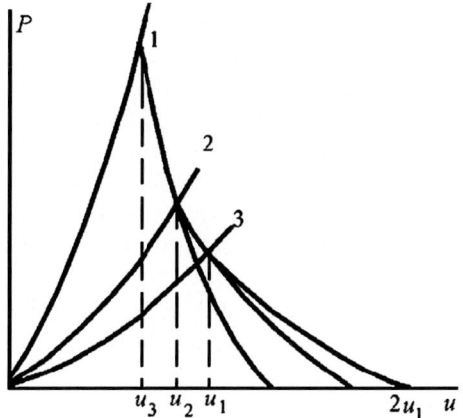

Fig. 2.33. Hugoniots and expansion isentropes of copper (1), aluminum (2), and plexiglass (3)

spallation phenomena. Thus, in order to obtain the actual loading pattern, it is reasonable to record compression wave profiles by manganin or other pressure sensors.

The $P - u$ diagram presented in Fig. 2.33 elucidates the principle of operation of the scheme in Fig. 2.32b. After the HE charge detonation, a shock wave propagates through the set of impactors (assume that they are fabricated from copper, aluminum, and plexiglass). When the shock wave arrives at the free surface of the plexiglass, the plate will fly at a velocity close to double the mass velocity at the shock front as a result of the Riemann problem. The aluminum and copper impactors will have lower velocities. The target with the specimen will be further loaded through the sequential impacts of the plates, with the final pressure being determined by the impact of the copper plate, whose thickness should be larger than the total thickness of the aluminum and plexiglass plates.

2.6.2 Cylindrical Scheme of Loading

Cylindrical loading device options with minor modifications are discussed in many papers [39, 41–43]. The material under study is loaded, in accordance with the cylindrical scheme presented in Fig. 2.34, using a sliding detonation wave produced from the explosion of HE, inside which a cylindrical ampule with the material is placed. The detonation wave sliding on the cylinder produces a converging cone-shaped shock wave in it and the material under study.

It has been experimentally shown [44] that the wave transforms into a three-impact configuration and propagates at a velocity equal to the detonation wave velocity for the HE used. The central part of the three-impact configuration is occupied by the head wave called the Mach wave. There is a linear dependence [45] between the Mach wave diameter and the specimen

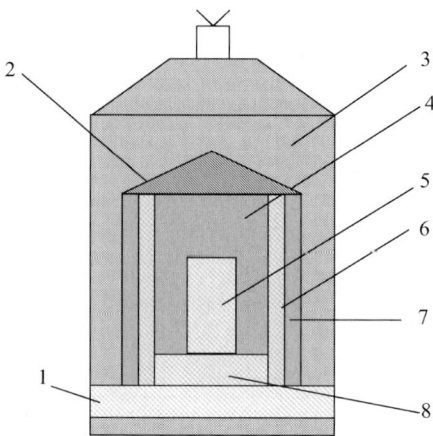

Fig. 2.34. Scheme of the material compression in cylindrical ampule [45]: 1 – substrate; 2 – lens for oblique shock wave formation; 3 – HE charge initiated by plane detonation wave generator; 4 – sleeve; 5 – sample under study; 6, 7 – cylindrical pads; 8 – plug

diameter. It should be noted that the material experiences a single shock compression by the Mach wave in the central part of the ampule and a double compression due to the incident and reflected shock waves near the ampule walls.

Gas-dynamic computations [46] for convergence of oblique conic shock waves for the loading device, presented in Fig. 2.34, corroborate the steady three-impact configuration wave formation. The three-impact wave intensities can be varied using different compositions of HE as well as HE charges of different density.

Studies of specimens loaded in the cylindrical preserving ampules detected [46] a characteristic feature that a melted-through axial channel forms in the specimen beginning with some intensity of the three-impact wave, whereas no channel forms at a lower pressure. In the specimen cross section a trace is clearly seen corresponding to the three-impact wave propagation zone cross-sectional size and accompanied with material structural changes.

The diameter of the melted-through hole is somewhat less than that of the three-impact wave trace, which is due to non-uniform pressure distribution in the conic Mach wave.

The experimentally obtained relationship between three-impact configuration wave propagation velocity and rate of detonation of HE charge used allows evaluation of pressures developed in the material under study by its known Hugoniot without any special direct measurements. Thus, when using HE whose detonation rate is ∼8000 m/s, one can obtain ∼500 GPa pressure in metals close in density to tungsten and ∼35 GPa pressures in organic materials, such as plexiglass.

The possibility of obtaining very high pressures, temperatures, and macroplastic strains in samples employing the cylindrical preserving ampules is beneficial for the realization of many physical and chemical processes. However, these conditions and the complex pattern of wave processes in specimens during dynamic compression limit the studies of material behavior laws to metals and high-melting point compounds [42].

2.6.3 Spherical Implosion of Preserving Ampules

Reference [47] presents results for experimental studies of the structural changes in samples of several metals and alloys preserved after their loading by spherically converging shock waves. Specimens which are cylinders 10 mm in diameter and 6 mm in height, were positioned diametrically in a copper core 30 mm in radius at 15–18 mm depth from its external surface on either side. A gradual increase in core pressure amplitude was ensured through a system of interleaving light (plexiglass, polystyrene foam) and heavy (lead, steel) layers. One of the layers was composed of HE, which was initiated across its surface at a number of points by spark detonators. The ampule containing the samples, the role of which was played by the copper core, was kept intact thanks to the external steel and internal lead shells available in the system.

The imploded ampule expanded during unloading, which was decelerated by the internal shell slowly flying away, yet remained intact in the experiments. The maximum computed pressure was 100 GPa on the core surface and 400 GPa at the sample positions. It should be noted that the pressure pulse duration can be considerably longer in systems with converging shock waves as compared to other systems.

References

1. Stanyukovich, K.P., *Unsteady Motion of Continuum*, Nauka Publ., Moscow, 197.1
2. Zlatin, N.A., and Mishin, G.I., *Ballistic Devices and their use in Experimental Studies*, Nauka Publ., Moscow, 1974.
3. Nabatov, S.S., and Yakushev, V.V., "A facility for Shock-Wave Experiments," *Problemy Prochnosti*, 1975, No. 3, pp. 101–102, [English trans., *Strength of Materials*, Vol. 7, No. 3, 1975, pp. 367–368].
4. Lekornte, C.L., "High-Velocity Launching," *in* High-Speed Physics, Zlatin, N.A., ed., Mir Publ., Moscow, 1971, Vol. 2, pp. 247–275 (in Russian) [see also High-Speed Physics, Vollrath, K., and Thomer, G., eds., Springer-Verlag, Wien, 1967 (in German)].
5. Mitchell, A.C., and Nellis, W.J., "Diagnostic System of the Lawrence Livermore National Laboratory Two-Stage Light-Gas Gun," *Review of Scientific Instruments*, Vol. 52, No. 3, 1981, pp. 347–359.

6. Chhabildas, L.C., Kmetyk, L.N., Reinhart, W.D., and Hall, C.A., "Launch Capabilities to 16 km/s," Shock Compression of Condensed Matter – 1995, Schmidt, S.C., and Tao, W.C., eds., AIP Press, Woodbury, NY, 1996, pp. 1197–1200.

7. Fowles, G.R., Duval, G.E., Asay, J.R., Bellamy, P., Feistmann, F, Grady D., Michaels, T, and Mitchell, R. "Gas Gun for Impact Studies," *Review of Scientific Instruments*, Vol. 41, No. 7, 1970, pp. 984–996 [Russian trans., *Pribory dlya Nauchnykh Issledovaniya*, 1970, No. 7, pp. 78–90].

8. Thunborg, S., Jr., Ingram, G.E., and Graham, R.A., "Compressed Gas Gun for Controlled Planar Impacts Over a Wide Velocity Range," *Review of Scientific Instruments*, Vol. 35, No. 1, 1964, pp. 11–14, [Russian trans., *Pribory dlya Nauchnykh Issledovaniya,* 1964, No. 1, pp. 13–17].

9. Mineev, V.N., Pogorelov, V.P., Ivanov, A.G., Svidinskii, V.A., Rusak, V.N., Bukreev, Yu.V., Tkachenko, I.A., Shitov, A.T., and Krivov, S.A., "Unit for Investigation of the Behavior of Materials and Constructions with Dynamic Loads," *Fizika Goreniya i Vzryva*, Vol. 14, No. 3, 1978, pp. 129–133, [English trans., *Combustion, Explosion, and Shock Waves*, Vol. 14, No. 3, 1978, pp. 377–380].

10. Altshuler, L.V., "Use of Shock Waves in High-Pressure Physics," *Uspekhi Fizicheskikh Nauk*, Vol. 85, No. 2, 1965, pp. 197–258, [English trans. *Soviet Physics Uspekhi*, Vol. 8, No. 1, 1965, pp. 52–91].

11. Trunin, R.F., (ed.), Properties of Condensed Materials at High Pressures and Temperatures, RFNC-VNIIEF, Sarov, Russia, 1992.

12. Zababakhin, E.I., *Some Problems of the Gasdynamics of Explosions,* RFNC-VNIITF, Snezhinsk, Russia, 1997, [English trans. RFNC-VNIITF, Snezhinsk, Russia, 2001].

13. Altshuler, L.V., Trunin, R.G., Krupnikov, K.K., and Panov, N.V., "Explosive Laboratory Devices for Shock Wave Compression Studies," *Uspekhi Fizicheskikh Nauk*, Vol. 166, No. 5, 1996, pp. 575–581, [English trans., *Physics – Uspekhi*, Vol. 39, No. 5, 1996, pp. 539–544].

14. Glushak, B.L., Zharkov, A.P., Zhernokletov, M.V., Ternovoi, V.Ya., Filimov, A.S., and Fortov, V.E., "Experimental Investigation of the Thermodynamics of Dense Plasmas Formed from Metals at High Energy Concentrations," *Zhurnal Eksperimentalnoi i Teoreticheskoi Fiziki*, Vol. 96, No. 4, 1989, pp. 1301–1318, [English trans., *Soviet Physics – JETP*, Vol. 69, No. 4, 1989, pp. 739–749].

15. Raevsky, V.A., "Effect of Rayleigh-Taylor Instability on Solid Plane Layer Acceleration*," Voprosy Atomnoi Naukii i Tekhniki. Seriya: Teoreticheskaya i Prikladnaya Fizika*, 1994, No. 1, pp. 55–58.

16. Batkov, Yu.V., Kovalev, N.P., Kovtun, A.D., Kuropatkin, V.G., Lebedev, A.I., Makarov, Yu.M., Manachkin, S.F., Novikov, S.A., Raevskii, V.A., and Styazhkin, Yu.M., "Acceleration of Metal Plates up to Velocities Higher than 10 km/s," *Doklady Akademii Nauk*, Vol. 357, No. 6, 1997, pp. 765–767, [English trans., *Physics – Doklady*, Vol. 42, No. 12, 1997, pp. 680–682].

17. Pogorelov, A.P., Glushak, B.L., Novikov, S.A., Sinitsyn, V.A., and Chernov, A.V., "Dependence of Recoil Impulse from a Rigid Barrier Under Sliding Conditions of Detonation of an Explosive Layer," *Fizika Goreniya i Vzryva*, Vol. 13, No. 5, 1977, pp. 771–776, [English trans., *Combustion, Explosion, and Shock Waves*, Vol. 13, No. 5, 1977, pp. 654–656].

18. Hoskin, N.E., Allan, J.W.S, Bailey, W.A., Lethaby, J.W., and Skidmore, I.C., "The Motion of Plates and Cylinders Driven by Detonation Waves at Tangential Incidence," *proc, 4th International Symposium on Detonation*, White Oak, MD, 1965, pp. 14–26.
19. Glushak, B.L., Novikov, S.A., Pogorelov, A.P., and Sinitsyn, V.A., "Investigation of TNT and TH 50/50 Initiation by Short-Duration Shocks," *Fizika Goreniya i Vzryva*, Vol. 17, No. 6, 1981, pp. 90–95, [English trans., *Combustion, Explosion, and Shock Waves*, Vol. 17, No. 6, 1981, pp. 660–665].
20. Deribas, A.A., *Strengthening and Welding Physics*, Nauka Publ., Siberian Branch, Novosibirsk, 1980.
21. Titov, V.M., Fadeenko, Yu.I., and Titova, N.S., "Acceleration of Solid Particles by Cumulative Explosion," *Doklady Akademii Nauk SSSR*, Vol. 180, No. 5, 1968, pp. 1051–1052, [English trans., *Soviet Physics – Doklady*, Vol. 13, No. 6, 1968, pp. 549–550].
22. Meier, J.K., and Kerrisk, J.F., "An Introduction to the Fast Shock Tupe (FST)," *Shock Compression of Condensed Matter – 1991*, Schmidt, S.C., Dick, R.D., Forbes, J.W., and Tasker, D.G., eds., Elsivier, Amsterdam, 1992, pp. 1045–1048.
23. Dolgoborodov, A.Yu., "Explosive Tubular Gun for High-Velocity Launching," *Khimicheskaya Fizika*, Vol. 14, No. 1, 1995, pp. 27–32.
24. Keller, D.V., and Penning, J.R., Jr., Exploding Foils – The Production of Plane Shock Waves and the Acceleration of Thin Plates," *in* Exploding Wires, Vol. 2, Chace, W.G., and Moore, H.K., eds., Plenum Press, NY, 1962, pp. 263-277, [Russian trans., *in* Conductor Electric Explosion, Vol. 2, Mir Publ., Moscow, 1965, pp. 299–316].
25. Pavlovskii, A.I., Kashintsov, V.I., Glushak, B.L., and Novikov, S.A., "Generation of a Mechanical Impulse by Electrical Explosion of a Conductor," *Fizika Goreniya i Vzryva*, Vol. 19, No. 3, 1983, pp. 124–126, [English trans., *Combustion, Explosion and Shock Waves*, Vol. 19, No. 3, 1983, pp. 361–369].
26. Barenboim, A.I., Egorov, L.A., Kalinin, V.G., Makeev, N.G., Mokhova, V.V., and Rumyantsev, V.G., "Laboratory Complex for X-Ray Diffraction Studies of Shock-Compressed Materials with Exposure Time of ~50 nsec," *Pribory i Tekhnika Eksperimenta*, 1992, No. 1, pp. 189–192, [English trans., *Instruments and Experimental Techniques*, Vol. 35, No. 1, Part 2, 1992, pp. 145–148].
27. Chau, H.H., Dittbenner, G., Hofer, W.W., Honodel, C.A., Steinberg, D.J., Stroud, J.R., and Weingart, R.C., "Electric Gun: a Versatile Tool for High-Pressure Shock-Wave Research," *Review of Scientific Instruments*, Vol. 51, No. 12, 1980, pp. 1676–1681.
28. Froeschner, K.E., Chau, H., Dittbenner, G., Lee, R.S., Mikkelson, K., Steinberg, D., and Weingart, R.C., "Shock Hugoniot Experiments Using an Electric Gun," Shock Waves in Condensed Matter – 1981, Nellis, W.J., Seaman, L., and Graham, R.A., eds., AIP Press, New York, 1982, pp. 174–178.
29. Hawke, R.S., Brooks, A.L., Mitchell, A.C., Fowler, C.M., Peterson, D.R., and Shaner, J.W., "Railguns for Ewuation-of-State Research," Shock Waves in Condensed Matter – 1981, Nellis, W.J., Seaman, L., and Graham, R.A., eds., AIP Press, New York, 1982, pp. 179–183.
30. Nikolayevsky, V.N., (ed.), *High-Velocity Shock Phenomena*, (trans. from English), Mir Publ., Moscow, 1973.
31. Sakharov, A.D., "Magnetoimplosive Generators," *Uspekhi Fizicheskikh Nauk*, Vol. 88, No. 4, 1966, pp. 725–734, [English trans., *Soviet Physics Uspekhi*, Vol. 9, No. 2, 1966, pp. 294–299].

32. Anisimov, S.I., Prokhorov, A.M., and Fortov, V.E., "Application of High-Power Lasers to Study Matter at Ultrahigh Pressures," *Uspekhi Fizicheskikh Nauk*, Vol. 142, No. 3, 1984, pp. 395–434, [English trans., *Soviet Physics Uspekhi*, Vol. 27, No. 3, 1984, pp. 181–205.

33. Ansimov, S.I., Imas, Ya.A., Romanov, G.S., and Khodyko, Yu.V., *Action of High-Power Radiation on Metals,* Nauka Publ., Moscow, 1970.

34. Kanel, G.I., Razorenov, S.V., Utkin, A.V., and Fortov, V.E., Shock-Wave Phenomena in Condensed Matter, Yanum-K Publ., Moscow, 1996 [*see also* Kanel, G.I., Razorenov, S.V., and Fortov, V.E., Shock-Wave Phenomena and the Properties of Condensed Matter, Springer-Verlag, New York, 2004].

35. Trainor, R.J., Shaner, J.W., Auerbach, J.M., and Holmes, N.C., "Ultrahigh-Pressure Laser-Driven Shock-Wave Experiments in Aluminum," *Physical Review Letters*, Vol. 42, No. 17, 1979, pp. 1154–1157.

36. Veeser, L.R., Solem, J.C., and Lieber, A.J., "Impedance-Match Experiments using Laser-Driven Shock Waves," *Applied Physics Letters*, Vol. 35, No. 10, 1979, pp. 761–763.

37. Demidov, B.A., and Martynov, A.I., "Experimental Investigation of Shock Waves Excited in Metals by an Intense Relativistic Electron Beam," *Zhurnal Eksperimentalnoi i Teoreticheskoi Fiziki*, Vol. 80, No. 3, 1981, pp. 738–744, [English trans., *Soviet Physics JETP*, Vol. 53, No. 2, 1981, pp. 374–377].

38. Bonyushkin, E.K., Zavada, N.I., Novikov, S.A., and Uchayev, A.Ya., *Metal Dynamic Failure Kinetics under Pulsed Volume Heating Conditions*, RFNC-VNIIEF, Sarov, Russia, 1998.

39. Dremin, A.N., and Breusov, O.N., "Processes Occurring in Solids Under the Action of Powerful Shock Waves," *Uspekhi Khimii*, Vol. 37, No. 5, 1968, pp. 898–916, [English trans., *Russian Chemical Reviews*, Vol. 37, No. 5, 1968, pp. 392–402].

40. Adadurov, G.A., Experimental Study of Chemical Processes under Dynamic Compression Conditions," *Uspekhi Khimii*, Vol. 55, No. 4, 1986, pp. 555–578, [English trans., *Russian Chemical Reviews*, Vol. 55, No. 4, 1986, pp. 282–296].

41. Batsanov, S.S., "Inorganic Chemistry of High Dynamic Pressures," *Uspekhi Khimii*, Vol. 55, No. 4, 1986, pp. 579–607, [English trans., *Russian Chemical Reviews*, Vol. 55, No. 4, 1986, pp. 297–315].

42. Adadurov, G.A., and Goldanskii, V.I., "Transformations of Condensed Substances under Shock-Wave Compression in Controlled Thermodynamic Conditions," *Uspekhi Khimii*, Vol. 50, No. 10, 1981, pp. 1810–1827, [English trans., *Russian Chemical Reviews*, Vol. 50, No. 10, 1981, pp. 948–957].

43. Osipov, R.S., Funtikov, A.I., and Tsyganov, V.A., "Determination of the Thermodynamic Parameters of Shock Compression of Lead, Tin, Copper, and Nickel by Their Melting in Conservation Ampoules," *Teplofizika Vysokikh Temperatur*, Vol. 36, No. 4, 1998, pp. 590–595, [English trans., *High Temperature*, Vol. 36, No. 4, 1998, pp. 566–571].

44. Adadurov, G.A., Dremin, A.N., Kanel, G.I., and Pershin, S.V., "Determination of the Shock Wave Parameters in Materials Preserved in Cylindrical Bombs," *Fizika Goreniya i Vzryva*, Vol. 3, No. 2, 1967, pp. 281–285, [English trans., *Combustion, Explosion, and Shock Waves*, Vol. 3, No. 2, 1967, pp. 175–177].

45. Adadurov, G.A., Dremin, A.N., and Kanel, G.I., "Mach Reflection Parameters for Plexiglas Cylinders," *Zhurnal Prikladnoi Mekhaniki i Tekhnicheskoi Fiziki*, Vol. 10, No. 2, 1969, pp. 126–128, [English trans., *Journal of Applied Mechanics and Technical Physics*, Vol. 10, No. 2, 1969, pp. 302–305].

46. Funtikov, A.I., Osipov, R.S., and Tsyganov, V.A., "Isentropes of Relief of Iron and Austenitic Steel from the State of Shock Compression at a Pressure of 150 GPa," *Teplofizika Vysokikh Temperatur*, Vol. 37, No. 6, 1999, pp. 887–894, [English trans., *High Temperature*, Vol. 37, No. 6, 1999, pp. 857–864].
47. Voinov, B.A., Nadykto, B.A., Novikov, S.A., Sinitsyna, L.M., Tkachenko, I.A., and Yukina, N.A., "Study of Structural Changes in Specimens of Different Materials Preserved After the Action of Pulsed High Pressures," *Fizika Goreniya i Vzryva*, Vol. 27, No. 4, 1991, pp. 109–116, [English trans., *Combustion, Explosion, and Shock Waves*, Vol. 27, No. 4, 1991, pp. 490–496].

Recording Fast Processes in Dynamic Studies

Yu.V. Batkov, V.A. Borisenok, S.I. Gerasimov, V.A. Komrachkov,
A.D. Kovtun, and M.V. Zhernokletov

Diagnostic methods for obtaining quantification of the field variables that
are important in understanding the physics of fast time-dependent processes,
such as the thermomechanical processes associated with shock waves, have
their own particular features. These features grow out of the demands of
the challenging environment within which the measurements must be taken.
The data must be taken within a very short time period; the diagnostic device
should be remote since destruction is inevitable in an explosion or impact; and
the measurements should be as complete as possible since it is impossible to
return a system (assembly, sample) to its original state in order to check the
results.

Because a measurement of the field variables that are important in understanding explosions and impacts poses particular challenges, a variety of
methods have been developed for supplying the needed information at the
appropriate level [1–3]. In general, the name of an experimental method corresponds to the physical principle that is exploited in order to obtain the measurements. For electrical measurements: the electrocontact method which involves the closing of an electrical circuit; the capacitive method which involves
a measurement of capacitance; the electromagnetic method which makes use
of changes in magnetic flux that result in the development of electromotive
forces (E.M.F.) in the circuit, etc. Similarly, optical methods are named in
accordance with the physical principle that is exploited. For example, the
method using the phenomenon of complete internal reflection, the flash gap
method, or by known analogies: the aquarium method, the foil method, and
the optical lever method, etc.

The experimental methods can be divided into two principal types [2]:
(1) discrete methods, in which the signal corresponding to a certain event
in space is recorded; and (2) continuous methods, in which either motion of
a given surface, or the mass velocity profile, or the stress profile is recorded
continuously in time. In turn, the above methods can be subdivided into
internal and external categories depending on the positioning of the sensor in

the experimental device. Methods that combine sensors operated on several physical principles are referred to as combined methods.

A most important component of any method, any measuring system, is the sensor (transducer, detector) that transforms the quantity being observed into a signal convenient for the measurement. The sensors and their relevant measuring schemes are typically categorized according to the functional thermophysical measurement that can be obtained from the experiment. Depending on the aims of the experiment, sensors may be used to record time t, path in time $x - t$ (both discretely and continuously), velocity (wave velocity D, mass velocity u, sound velocity c) in time (continuously), pressure P (stress σ) in time (continuously), density ρ or volume $V = 1/\rho$ in time (both discretely and continuously).

The physical principles used as a basis for the sensors during studies of fast processes can be subdivided into three groups: mechanical, electrical (electromagnetic), and optical (X-ray, electro-optical).

3.1 Discrete Methods for Wave and Mass Velocity Measurements

When studying the behavior of shock-compressed materials, the measurements most often taken are of the kinematic field variables, i.e. wave velocity, D, and mass velocity, u. These quantities may then be used to determine pressure P, density ρ, and shock compression energy, E, in accordance with the laws of conservation of mass, momentum, and energy. Impactor and target free surface velocities, W, are typically measured in the same experiment [1–3]. The velocities are measured by fixing time intervals, within which shock discontinuities arrive at given points in space. The time of wave arrival is measured using sensors designed to provide signals that can be recorded by (depending upon the nature of the signal) high-speed cathode-ray oscilloscopes (or digitizers), analog-to-digital converters, film cameras, photoreceivers, or electrooptical converters.

Most widespread are the electrocontact sensors, actuated by the breakdown of insulation in contacts under electric tension located at different points along the path of the shock wave or flying plate propagation. Their actuation time spread can be from 1 to 10 ns depending on the sensor design. In an attempt to minimize errors in the measurement of shock arrival time, and to account for potential shock wave decay, in each experimental device 10 to 15 electrocontact sensors, as a rule, are positioned at several points of reference and through the specimen depth. Efficiency of the electrocontact sensors becomes low when the shock wave strength is small or the free surface velocities are low. For low pressures, the best results are achieved using piezoelectric transducers, signals from which are formed due to the piezoelectric effect appearing as the compression waves pass through piezoelectric cells [3, 4].

Time interval measurements with photorecording devices – mechanical photorecorders, photodiodes, photoelectric cells, and electrooptical converters – use gas luminosity in gaps as the shock waves pass through them, a change in luminous intensity occurs when the shock wave arrives at a free surface or when a flying body is decelerated by impact onto a transparent barrier [2, 3]. It is possible to control deviations from one-dimensional hydrodynamic flow along with the velocity measurements using this method. Recently fiber optic lines have been widely used to record optical signals with measuring devices [5]. A variation of transparent barriers is a package of thin plates separated by small gaps that collapse under force of the moving shock wave [6]. A similar idea is used to measure wave velocities in packages of thin metal plates [7]. When the shock wave moves through the package, contact resistance between the plates change, and this is recorded by oscilloscopes. This method is in essence a discrete rheostat method.

The discrete methods include a pulsed X-ray method, in which X-ray tubes positioned at known distances and actuated at given times produce a pulsed X-ray image, for example, of a flying projectile, or to determine material displacement by recording interface or reference point position. In view of the significant progress achieved recently in flash radiography, which is largely due to multi-frame recording system development, a separate section will be devoted to this method.

3.1.1 Electrocontact Sensors

The most widespread discrete technique for measuring wave and mass velocities of shock waves employs shorting electrocontact sensors. Electrocontact sensors record times of the shock wave or surface passage through gauge length reference points (D or W). A typical design of the coaxial electrocontact sensor is presented in Fig. 3.1 [3]. The gap between the central conductor and the sensor casing provides a break in the electrical circuit. Upon shock wave arrival, the sensor gap closes and the electrical circuit produces a current pulse that can be recorded by a cathode-ray oscilloscope. The obtained oscillograms are used to determine the time intervals between the times of actuation of several sensors positioned along the shock wave path in the specimen or along the path of the free surface.

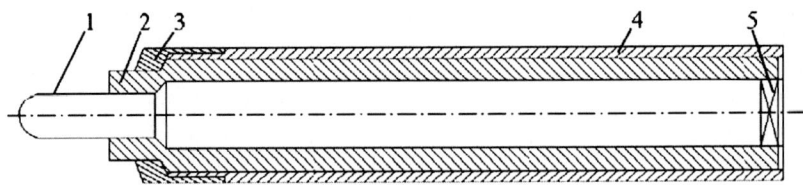

Fig. 3.1. Coaxial electrocontact sensor [3]: 1 – brass rod; 2 – electrical insulation; 3 – tip; 4 – sensor casing; 5 – mica washer

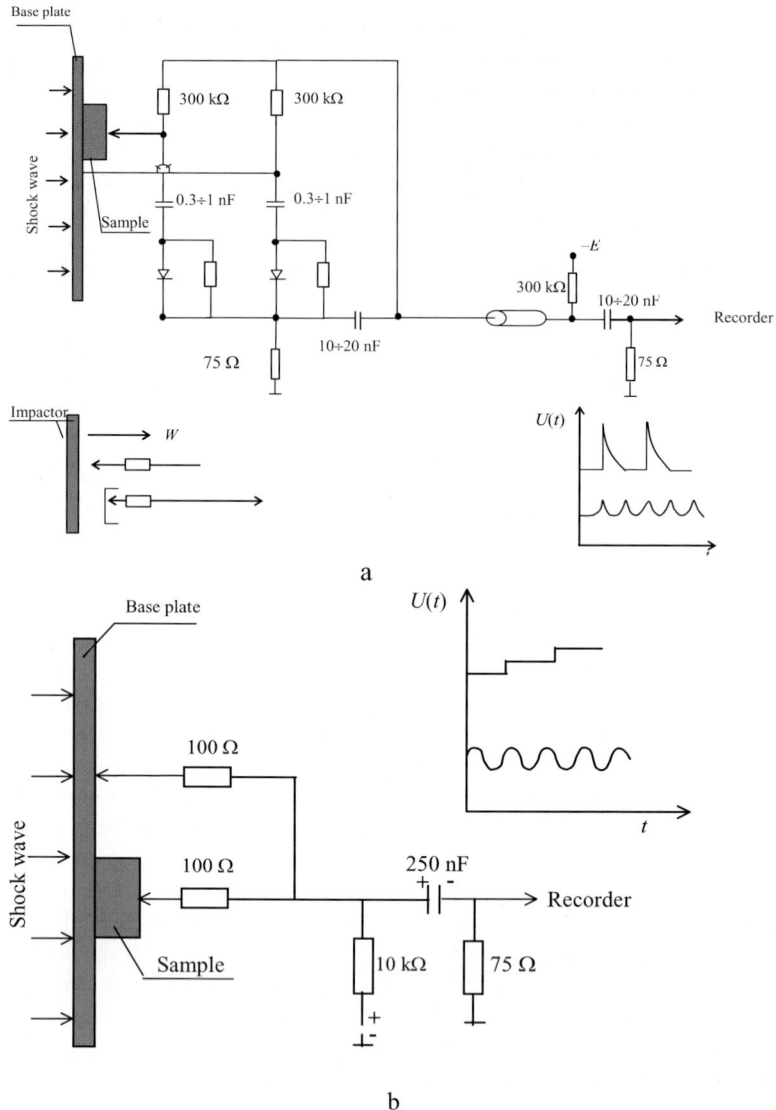

Fig. 3.2. Electrical circuit, which electrocontact sensor [3] is included in: **a** – circuit for measuring D and W; **b** – circuit for measuring D

Typical layouts of sensors used in experiments measuring D and W, and two variations of electrocontact sensor power supply circuits appear in Figs. 3.2a and 3.2b. The circuit depicted in Fig. 3.2a produces electrical signals, whose duration is much longer than the time interval to be measured. In the circuit shown in Fig. 3.2b, the contact closing results in signals, whose duration is several times shorter than the time interval to be measured. In

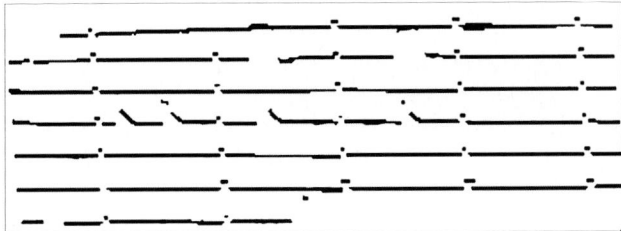

Fig. 3.3. Characteristic oscillogram of output pulses from electrocontact sensors

either circuit, standard frequency signals determining the time sweep scale are fed onto the recorder. A typical oscillogram of the signals supplied by the electrocontact sensors is presented in Fig. 3.3.

Figure 3.4 depicts other electrocontact sensor types (twisted and pinned), gives examples of their employment in experimental devices and their layout. The simplest sensor for measuring shock wave velocities at impactor velocities of ≥ 1.0 km/s is the sensor fabricated from PEV-2 twisted wire of 0.14 mm diameter and 0.01 mm insulation thickness (see Fig. 3.4a). When shock wave amplitudes are small, the sensor contacts are manufactured in the form of thin bare metal rods separated from base plates and samples by a 0.03-mm gap (Fig. 3.4b) [8]. When measuring the impactor velocity W_{imp}, the wire and rod sensor contacts are placed in special tubes and covered with caps (Fig. 3.4c) to avoid premature sensor actuation from airborne shock waves [8].

The reproducibility of the electrocontact sensor actuation time depends on sensor design features: surface uniformity, sensor position precision, part finish, and the accuracy of sensor location in each individual experiment. Characteristic time delays of the best electrocontact sensors is at a level of about 1–2 ns. With a 5-mm gauge length and velocities of 5–10 km/s, recorded time intervals are 0.5–1.0 μs. Modern high-speed oscilloscopes record these intervals within 1% accuracy. Features of the electrical discharge and the hydrodynamics of the gap closing upon shock wave arrival are such that, at pressures below 7–5 GPa, the sensors are not typically used.

A classic example of the use of simple electrocontact sensors to locate the free surface position as a function of time and determine complex shock wave profiles in iron is provided by the data obtained by Minshal [9] and presented in Fig. 3.5. In the figure, the time of the shock wave arrival at the probe is laid off on the axis t and the distance from the probe to the surface on the axis x.

The electrocontact technique has been recently improved in order to enhance the accuracy and reliability of measured free-surface velocity [10]. For this purpose, a miniature electrocontact sensor has been developed (Fig. 3.6); this is a cylinder of capacitor paper, along the surface of which copper wires (contacts) made of winding wire are attached. Each wire is cut together with the paper at a certain height (depending on the gauge length), so that the

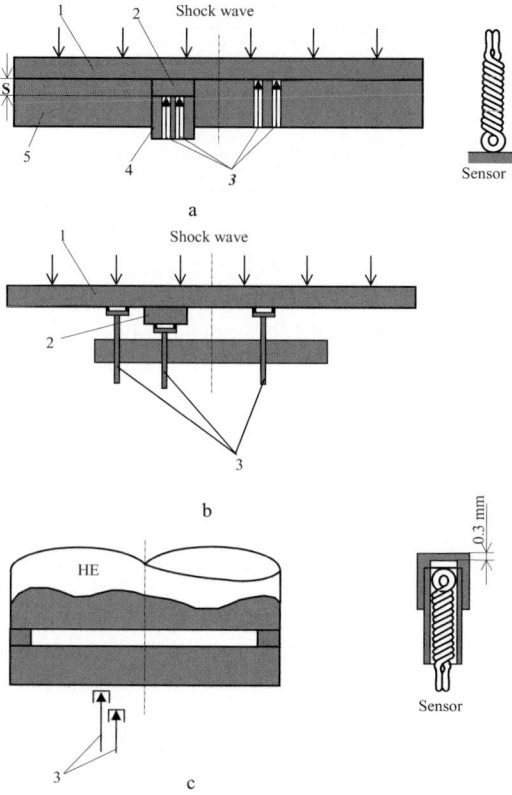

Fig. 3.4. Examples of electrocontact sensor designs and schemes of their application for recording wave and plate velocities: **a** – twisted contact; **b** – pin contact; **c** – contact covered with a cap. 1 – base plate; 2 – sample; 3 – electric contacts; 4 – clamping insertion; 5 – disc with holes for samples and contacts

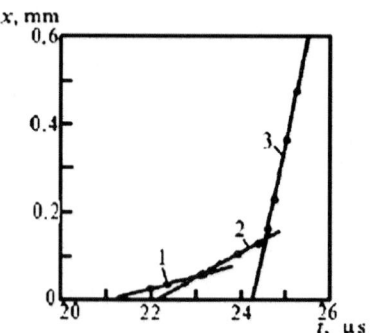

Fig. 3.5. Data from electrocontact measurements [9] illustrating existence of three waves in iron: 1 – elastic wave; 2 – first plastic wave; 3 – main wave

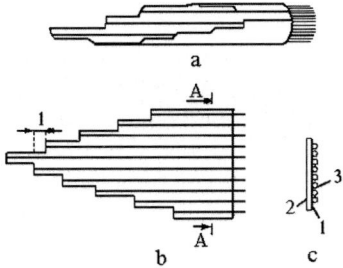

Fig. 3.6. Multi-contact sensor: **a** – helical; **b** – plane; **c** – section through AA. 1 – gauge length; 2 – capacitor paper 0.01 mm thick; 3 – 0.06-mm-diameter wire PEV-2

wire ends are arranged helically (the name "helical sensor" comes from this). The diameter of the sensor with ten contacts is 2.3 mm. The wires are cut by a template, which allows fabrication of the sensors with gauge lengths ranging within ± 0.015 mm. The total thickness of the paper and glue in the sensor is no greater than 0.04 mm, however sensor stiffness is quite high because of its tubular shape.

When refining the sensors, much attention was paid to distances between the sensor wires, so that there would be no interference between contacts. For the helical sensor with 0.06-mm-diameter wire contacts, the optimal inter-contact distance proved to be 0.6 mm. Grouping contacts in a single small-diameter helical sensor reduces measurement error due to surface asymmetry (each sensor actually records the $x - t$ diagram of motion of one surface point), facilitates measuring unit mounting, and ensures a tighter location of the contacts.

Since the contact ends are not insulated (for more reliable closing with a moving body), sensor operation depends on the degree of ionization of the gas medium within which the measurements are taken. If the measurements are taken in air with open contacts, it is desirable that the velocity of the body be no greater than 2.5 km/s. When the velocity is greater, the shock front before the moving body ionizes the air, which upsets normal operation of the contacts (they are closed by the air shock wave rather than by the body surface). For measurements at high velocities, either the air is evacuated or helical sensor gas protection is used, which consists of replacing the air with another gas of a lower ionization degree behind the shock front. Helium having a high ionization potential or many-atom gases (carbonic acid gas, methane, propane, etc.), which have a higher heat capacity and are heated less in shock waves than air, can be used.

3.1.2 Piezoelectric and Ferroelectric Time Markers

The wave and mass velocities in shock loading experiments are also measured using piezoelectric and ferroelectric time markers. The principle of operation of the piezoelectric markers is based on the straightforward piezoelectric effect: polarization in non-centrosymmetrical dielectrics under the action of mechanical stress [11]. The piezoelectrics are typified by single-crystal quartz.

A feature of ferroelectrics is spontaneous (residual) polarization with no external effects. Under the effects of mechanical loading, as pressure increases materials of this class first exhibit the direct piezoelectric effect (within the elastic region) and then partial and full depolarization [12]. Ferroelectrics used most frequently in shock experiments include single-crystal lithium niobate, piezoceramics of lead zirconate-titanate system (PZT), and polyvinylidene fluoride (PVDF) polymer. The sensor in the time marker type under discussion is a capacitor filled with piezoelectric or ferroelectric. Under the effects of mechanical loading, the appearance of polarization or a change in polarization is attended by a current pulse in the sensor's external circuit which is recorded as a time mark.

The piezoelectric and ferroelectric time markers are used both in experiments at low pressures (below 5.0 GPa), where electrocontact sensors operate unreliably, and at higher pressures [13, 14]. For example, [14], records PVDF electric response at ~100 GPa pressure. This marker type requires no external power supply and can be designed either as a sectional structure with a clamping spring or as a monolithic device as shown in Fig. 3.7. The above-discussed effects are essentially inertia less. The absolute error of the mechanical effect arrival time recording depends in the main on the marker design and is typically a few nanoseconds.

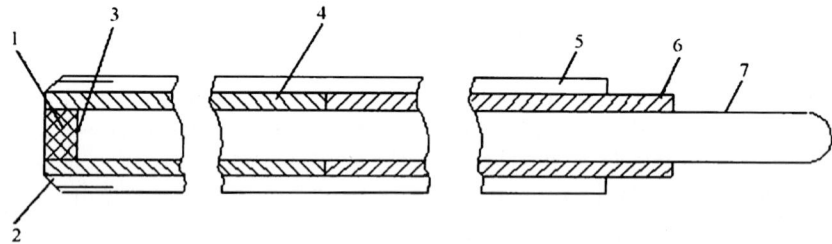

Fig. 3.7. Coaxial piezoelectric transducer: 1 – copper substrate; 2 – piezoelectric crystal; 3 – silver-epoxy paste; 4 – epoxy resin; 5 – brass tube; 6 – teflon tube; 7 – brass wire

3.1.3 Method of Flash Gaps

The method of flash gaps uses thin (50 μm) gas-filled gaps between a transparent medium (through which the observation proceeds) and the surface of the sample [2,3]. The gas in the gap is compressed and heated in an adiabatic manner by the shock wave as it leaves the sample. The gas flashes occurring in the gaps are noted by a high-speed photorecorder (streak camera) with a rotating-mirror operated in the slit scanning mode [15–17].

The basic diagram of operation of the camera as a photo recorder appears in Fig. 3.8. Objective (10) constructs an image of the process under study in the plane of vertical adjustable slit (9), which cuts a thin strip from the image. Objective (4), with the help of rotating mirror (6), constructs a strip image on the focal surface, upon which photographic film 1 lies. When the mirror rotates, the light strip cut from the process image is moved across the focal surface thereby exposing the film. In this way the time history of the phenomenon is recorded.

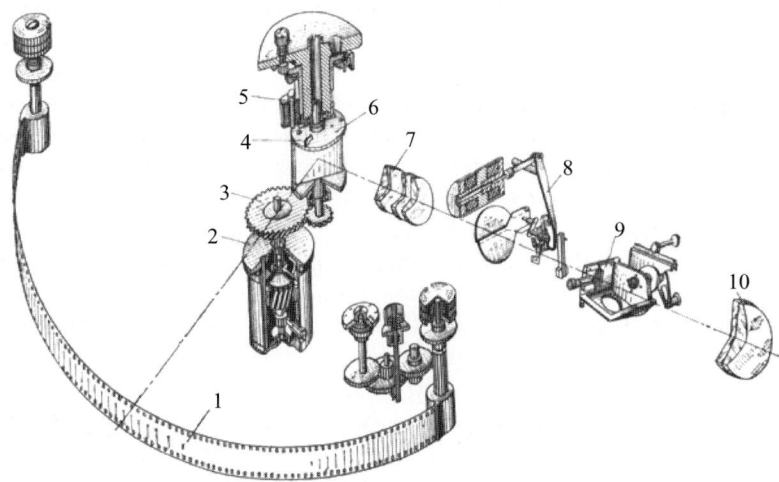

Fig. 3.8. Basic optical diagram of photorecorder SFR: 1 – film; 2 – electric motor; 3 – reducing gear; 4 – iron plate; 5 – electromagnetic sensor; 6 – mirror; 7 – objective; 8 – two-leafed shutter; 9 – adjustable slit; 10 – inlet objective

The device has three inlet objectives (10) with different focal distances for both short-range and long-range operations. The photorecorder mirror is set in motion by electric motor (2) through reducing gear (3). Two-leafed shutter (8) with an electromagnetic drive is located behind slit (9). The shutter automatically opens for given periods of time immediately before the start of the experiment. Camera SFR has a limited working angle (90°), hence, the phenomenon must be projected on the film by a rotating mirror. Electromagnetic probe (5) is located near the mirror. It serves to deliver a pulse for the initiation of component action that is consistent with the position of mirror (6) and to measure the mirror rotation speed using a tube tachometer and electron-beam tachoscope. When the mirror rotates into a certain position, iron plate (4) closes the magnetic circuit of the sensor, thereby inducing E.M.F. in its coil. Sensor (5) can be turned about the mirror axis and set in any position.

Thus, a pulse from the sensor can be produced in any given position of the mirror in its rotation, which allows us to record different phases of fairly long processes.

The photo recorder SFR is a semiautomatic device, in which the investigator manually sets the required speed of mirror rotation and presses the button for shooting. Other processes, such as the opening and closing of the shutter, the synchronization of accent lighting, and the closing of various circuits, are performed automatically. The slits can be attached immediately to the items, and then the full image of all the slits is scanned over the film. An experimental scheme of measuring wave and mass velocities using SFR, flash gaps, and several slits is shown in Fig. 4.7, Chap. 4. The device simultaneously measures velocities of the free surface and shock wave in several samples at a time.

In these recorders, the velocity of the beam motion across the film is up to 4–6 km/s. The flash duration is typically 10 to 50 nanoseconds depending on the gap thickness. The luminescence ceases in a few nanoseconds when the shock wave enters the transparent medium (this is typically organic glass blocks) which looses its transparency. To achieve high luminance, the gap is filled with argon. When the shock wave intensity is a few tens or more of GPa, good results are provided using air gaps.

Figure 3.9 presents the scheme and typical streak-camera image of an experiment on unloading of a shock-loaded sample into inert gases using a spherical-geometry shock-wave generator (see Chap. 2) and photo recorder SFR-2M with a slit unit positioned after the receiving objective. The experiment estimates the time of shock wave motion in gas after leaving the sample till the moment the shock wave reflects from a transparent window.

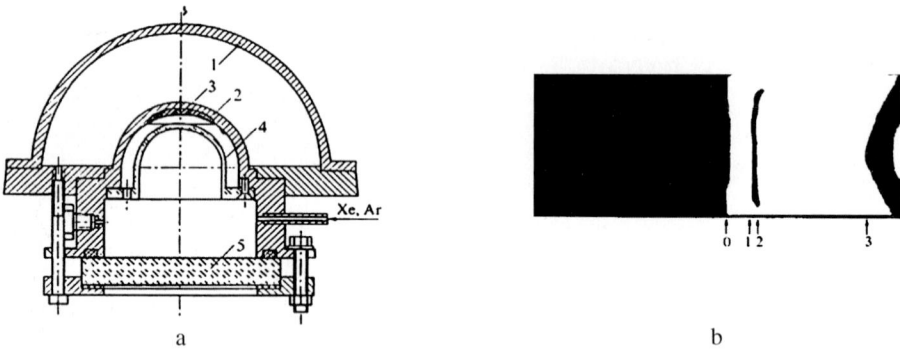

a   b

Fig. 3.9. Scheme (a) and typical streak-camera image (b) of the experiment on metal sample unloading into inert gas using spherical-geometry shock-wave generator and photorecorder. a: 1 – impactor; 2 – base plate; 3 – metal sample; 4 – organic glass cutoff; 5 – transparent window; b: 0 – shock wave release from the sample into the gas; 1 – shock wave reflection in the gas from transparent cutoff 4; 2 – shock wave release from cutoff 4; 3 – wave reflection from transparent window 5

The reference time marks for the shock wave velocity are estimated to be at the luminescence start (when the shock wave has arrived at the gas interface with the sample) and the cessation of luminescence or change in the luminescence intensity when the shock wave has reflected from the cutoffs. In the streak camera image, the indexes 0 and 1 denote the beginning and end of the shock wave luminescence in the gas on the gauge length.

A method of photographic scanning is also widely used. In this method, a light beam reflecting from a rotating mirror continuously moves across the film, so that the luminous item draws a continuous line on the film. The slope of the line is used to estimate the item velocity. Figure 3.10 presents a streak camera image of shock wave motion in the outlet tube of an explosive gas-dynamic compressor.

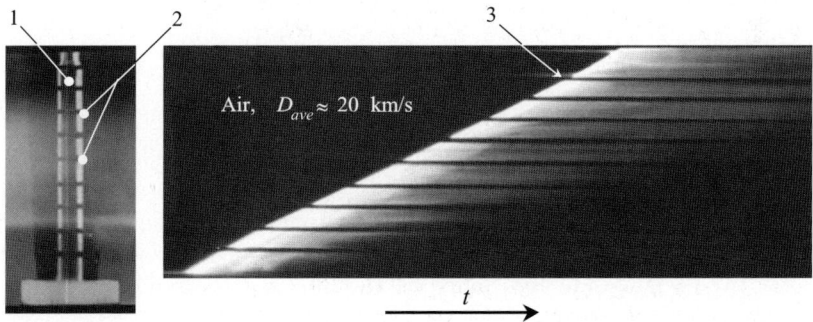

Fig. 3.10. Shock wave motion in the outlet tube of the explosive gas-dynamic compressor: 1 – outlet tube; 2 – marks; 3 – beginning of near-wall acceleration

Photographing with Optical Pins

The sensors presented in Fig. 3.11 for recording $x - t$ diagrams of the motion and shape of the shock front are segments of light guides with polished end surfaces [18]. The input ends of the segments are fixed in holes of a holder, whose surface is a control surface, and the output ends are fastened in holes of a plane panel in one or more rows perpendicular to the scanning direction. The indicator assembly is positioned above the control surface. It is fabricated, as a rule, from transparent material (for instance, organic glass) and composed of layers separated with gas-filled gaps. The number of the layers can be different (from zero to ten). Polystyrene monoliths with polymethylmethacrylate shell are typically used for the light guides. The outer diameter of the filament is from 0.15 mm to 1.5 mm, the optical transmission is ≈ 0.7 m^{-1}. The panel is positioned perpendicularly to the optical axis of the recorder. When the wave approaches the gaps above the surface of the input end, two short light flashes occur which are transmitted to the panel and then recorded on the film.

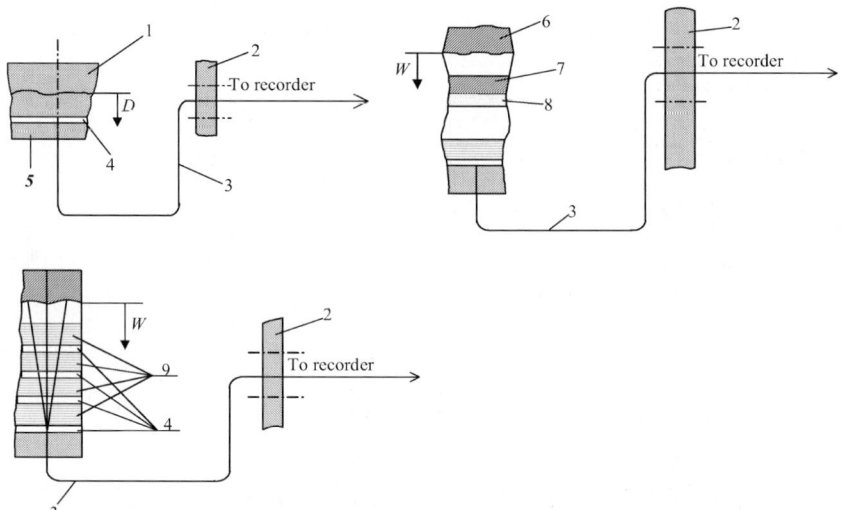

Fig. 3.11. End type sensors: 1 – medium; 2 – panel; 3 – light guide; 4 – gap; 5 – holder; 6 – plate; 7 – base plate; 8 – NaCl; 9 – organic glass

With increasing flash times "blackening" extends across the scan (the spot transforms into a strip). The maximum film exposure time depends on the diameter of the light guide end image on the film and the scanning rate:

$$t_{\max} = \frac{\beta d_{\lg}}{V_s} = \frac{d_{i.\lg}}{V_s},$$

where β is the optical recording system magnification coefficient; $d_{\lg}$, $d_{i.\lg}$ are the light guide diameter and image on the film, respectively; and V_s is the scanning rate.

The principle of operation of the end sensors is based [19] on the phenomenon of gas luminescence in gaps and the luminescence cutoff when the shock wave passes through them, in other words the same as the method of flash gaps. The shock wave-gap interaction generates a series of flashes. That is, each light guide ensures recording a series of points of the $x-t$ diagram of the shock wave motion in the indicator. The lower limit of the sensor operability is 50–80 kbar. Schematics of the sensors for recording the $x-t$ diagrams of the plate free surface motion are presented in Fig. 3.12.

Open and closed sensor types are used. To make the measurement quantity per shot larger, one seeks to use sensors of a diameter as small as possible. However, this shortens the flash duration and reduces the sensitivity of the method.

When plate velocities are slow, the exposure proves to be insufficient for the flash recording. To enhance luminance, the gap can be filled with inert gas (argon). By doing this, the minimum plate velocity that can be recorded is reduced to 1.5–1.6 km/s (for a 0.3-mm light guide diameter). To enhance

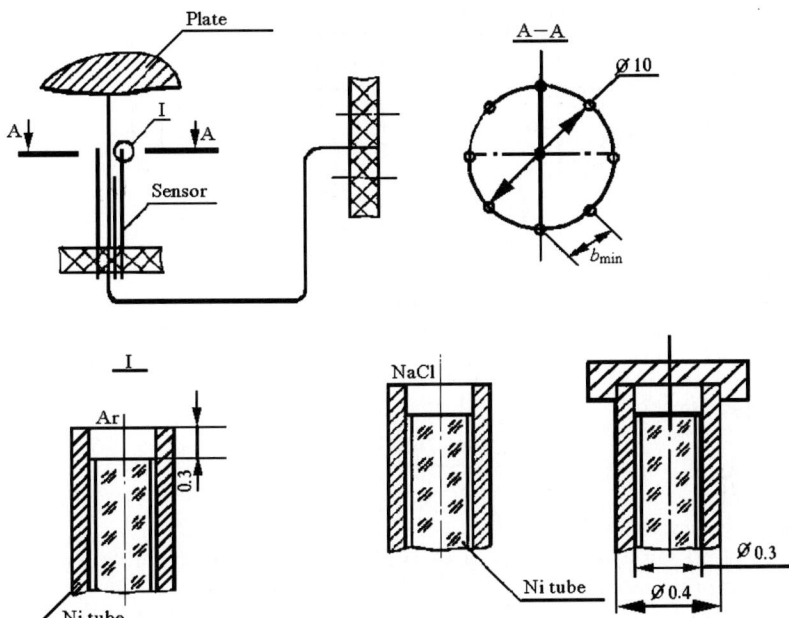

Fig. 3.12. Sensors for recording $x - t$ diagrams of the plate motion

the measurement accuracy, one seeks to increase the number of levels and the number of sensors at each level. When the sensors interact with flying plates, craters form in them, from which jets are ejected, and the plate surface around the crater swells. The minimum distance between the sensors, b_{min}, is a function of the perturbed region diameter, sensor and crater diameters, plate and sensor material densities, plate material hardness, and plate velocity.

Figure 3.13 depicts diagrams of loop sensors [18] for measurements in the lower velocity and pressure region. The sensors are light guide segments which contain a loop of a small bending radius in one of its parts. The loops are placed onto the control surfaces, while the input and output light guide ends are put into the light source and panel holes, respectively. The light source (of explosive type) is triggered synchronously with the phenomenon under study, the light from it is passed through the light guide and recorded on the film as a blackening band. When the shock wave arrives at the sensor, the light transmission changes or full luminescence cutoff occurs, which results in a change in band blackening density or its disappearance.

Errors of Time Interval Measurement
with Photorecorder SFR-2M

The error in measurement of the time interval depends [20] on the error in measurements of the SFR scanning rate, V_s, distance between film blackenings,

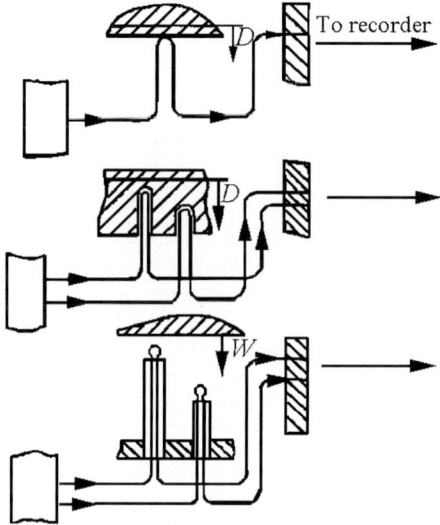

Fig. 3.13. Loop sensors

l, distance in the print, L, and aperture spacings in the print and film, M, m. Using film, the calculation of time is estimated as:

$$t = \frac{l}{V_s} = \frac{L}{V_{s.print}} = \frac{L}{kV_s} = \frac{L}{\frac{M}{m}V_s} \,,$$

whereas the calculation using print is given as:

$$t = \frac{L}{V_{s.print}} = \frac{L}{kV_s} = \frac{L}{\frac{M}{m}V_s} \,.$$

When calculating the total error, systematic errors are also included: eye resolution (0.2 mm) and SFR time resolution (0.01 μs at $V_s = 3.75$ mm/μs). Independent errors are listed in Table 3.1.

The limiting total methodological error in a single measurement is

$$3\sigma_t = \sqrt{\sum (3\sigma_i)^2 + \sum \delta_i} \,,$$

where σ_i, δ_i are experimentally estimated random and systematic errors. The relative error for the film is

$$\frac{3\sigma_t}{t} = 3\sqrt{\left(\frac{\sigma_l}{l}\right)^2 + \left(\frac{\sigma_{V_s}}{V_s}\right)^2} \,,$$

and that for the print is

$$\frac{3\sigma_t}{t} = 3\sqrt{\left(\frac{\sigma_L}{L}\right)^2 + \left(\frac{\sigma_M}{M}\right)^2 + \left(\frac{\sigma_m}{m}\right)^2 + \left(\frac{\sigma_{V_s}}{V_s}\right)^2} \,.$$

Table 3.1. Parameters and sources of error in the time interval measurement using streak camera SFR-2M

Parameter	Error source	Relative error $\dfrac{3\sigma_t}{t}$, %
Calculation by film		
	Replacement of Pascal's limacon by cylinder	0.5
Scanning rate V_s	Tachometer imprecision	0.1
	Film shrinkage on treatment	0.4
Distance l in film	Slit transfer from preliminary picture	$\dfrac{0.1}{l} \cdot 100$
	Inaccurate count in microscope BMI-1	$\dfrac{0.05}{l} \cdot 100$
Image quality (effect of different exposure)	Calculation by the blackening edges	$\dfrac{0.06}{l} \cdot 100$
	Calculation by the blackening centers	$\dfrac{0.03}{l} \cdot 100$
Calculation by print		
Scanning rate V_s	Replacement of Pascal's limacon by cylinder	0.5
Distance L in print	Slit transfer from preliminary picture	$\dfrac{1}{L} \cdot 100$
	Inaccurate measurement with ruler	$\dfrac{0.5}{L} \cdot 100$
Parameter	Error source	Relative error $\dfrac{3\sigma_t}{t}$, %
Image quality	Calculation by the blackening edges	$\dfrac{0.11k}{L} \cdot 100$
	Calculation by the blackening centers	$\dfrac{0.03k}{L} \cdot 100$
	Inaccurate aperture spacing	0.3
Magnification coefficient k	Inaccurate measurement with ruler (for 200-mm distance)	0.25
	Photographic enlarger optics distortion	1.1
	Nonuniform photographic paper drawing-out	0.9

Thus, for long time intervals the relative error is $3\sigma_t \sim t$, while for short ones it is essentially independent on time, $3\sigma_t \sim 1/V_s$. Evaluation of $3\sigma_t$ indicates that:

- for intervals $t > 10$ μs the relative error is constant and is equal to $3\sigma_t/t = 1.7\%$ irrespective of scanning rate (within 2.25–3.75 mm/μs) for photographic print.
- for short time intervals $3\sigma_t$ decreases with increasing scanning rate, while the relative error rapidly grows with decreasing t. For example, for $t = 1$ μs $3\sigma_t/t = 4.8\%$ ($V_s = 3$ mm/μs);

- the lowest error of a single SFR-2M measurement is $3\sigma_t = 0.03$ μs (for $t = 0$).

3.1.4 Electro-optical Technique

A shortcoming of recording with oscilloscopes is that the number of pulses that can be recorded by one device, as a rule, is no higher than 3 or 4, and it is often impossible to interpret the oscillograms when several sensors are actuated essentially simultaneously. Another method for recording time intervals is an electro-optical method [21]. This method enables recording the actuation of several tens of sensors by a single device, even when they are actuated simultaneously. This is made possible by independent operation of the measuring channels.

The electro-optical method involves streak camera recording of light pulses from a transducer. In the case under discussion the transducer is an electric spark apparatus transforming electric pulses from sensors to light pulses with special spark gaps. The spark gaps are aligned (sometimes between two parallel lines) and projected onto the photo-recorder frame, so that the line with the spark gap image in the frame is perpendicular to the direction of the photo-recorder sweep. Accuracy of the time interval measurement with the electro-optical method depends on a number of factors, including the photo-recorder resolution, sizes of the luminous spark gap image in the photo-recorder frame, spread in time of the spark gap breakdown, etc.

The method was further improved, based on the proposal to replace the spark discharges with a line of miniature Kerr capacitors. An embodiment of the technical proposal was a 50-channel light pulse generator [22]. An electric-to-light pulse converter in the device under discussion consists of a 52-channel nitrobenzene Kerr cell with slit type capacitors. The light pulses are produced via modulation of light beams passing through the Kerr cell channels by electrical pulses. The working slits of all the Kerr cell capacitors are located in the vertical plane at 1.5-mm spacing and projected onto the photo-recorder frame, so that the line with the working slit image is perpendicular to the sweep direction. For simplicity, the optical scheme (Fig. 3.14) depicts four capacitors only.

Polarizer 2 and analyzer 4 are common to all the Kerr cell channels and made of triacetate base polaroid film. Light source (1) is an IFT-800 flashtube (luminous part size 7 mm in diameter and 80 mm in length). The maximum luminance of the flashtube is $5 \cdot 10^9$ cd/m^2. Collecting lens (5) directs light beams which have passed through the cell channels into photo-recorder objective (6). Given the geometric characteristics of the parallel-plane electrodes of capacitors (3) (0.12 mm gap between the parallel planes, 18 mm electrode length), the electrical potential required for complete opening of the Kerr cell is about 400 V. The sensors are connected to the device with individual radio-frequency cables RK-50-2-11. In operation with contact sensors, 200-V dc is applied across cables extending from the device. When the contact sensors

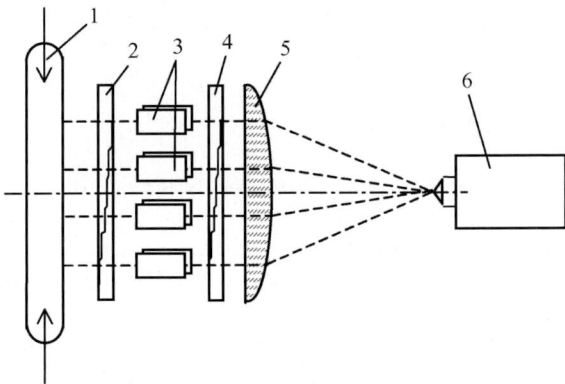

Fig. 3.14. Optical scheme of 50-channel light pulse generator [9]: 1 – light source; 2 – polarizer; 3 – capacitors; 4 – analyzer; 5 – collecting lens; 6 – photo-recorder

close, light pulses are produced in the cables that go to the relevant Kerr cell channels (opened for the time of the pulse action) at amplitudes that are nearly doubled. When the device is operated with piezoelectric sensors, the cables are mated on the side of the device, and the power supply is turned off.

In the device, the two extreme Kerr cell channels are used for time scaling of the photo-recorder frame simultaneously with the recording of the trans-duced electrical pulses from the sensors. Time scale tags are produced through modulation of light beams from an IFP-800 flashtube by stabilized high-frequency voltage [21] from an internal generator at frequency 0.1, 1, 2, 5 or 10 MHz, depending on the photo-recorder sweep rate taken.

The 50-channel light pulse generator, together with the SFR photo-recorder, can be successfully used to measure time intervals in the range from 10^{-6} to $5 \cdot 10^{-3}$ s. Luminance and luminosity duration of the IFP-800 flash-tube are varied through changes in the capacitor bank capacity and ballast resistance connected into the circuit of the capacitor bank discharge to the flashtube. The error in time interval measurement depends on the resolution of the photo-recorder and time tag quartz generator timing and is no higher than 0.1% of the full frame time at any SFR sweep rate. Thus, when the fastest sweep rate is 3750 m/s (the full frame time is 100 μs), the measurement error is no more than 0.1 μs.

The accuracy of the streak-camera method can be increased by an or-der of magnitude by using the two-channel timer (DV-2) [23], which when operated in combination with the photorecorder SFR, ensures mapping not only the process under study, but also the time scale and time tag relating to some specific phase of the process under study on the SFR working frame. Figure 3.15 presents a typical streak-camera image produced by the 50-channel light pulse generator.

Fig. 3.15. A typical light pulse streak-camera image with optical calibration by time tags at 10 MHz frequency

3.1.5 Method of Closed Contacts

The method of closed electrical contacts (CC) [7] has been developed for measuring shock wave velocities in conductors and, as the name implies, involves the recording of shock arrival time through the closing of electrical contacts. The electrical contacts are spaced at specified intervals along the direction of shock propagation. The CC method [24] involves the use of a series of electrically conductive plates possessing a hard rough contact surface. An electronic circuit diagram of the CC method appears in Fig. 3.16. In preparation for the measurement of the velocity of a planar shock wave, a CC is formed as two plates are pressed together with an effort of $\sim$10 N (screw clamp). Prior to the shock wave's arrival, they contact at a relatively small number of contact points because of roughness of the plates. The surface area of the contact points is, as a rule, a few hundredths of the apparent (geometrical) plate contact area S. Hence, even when the plates are pressed together with a relatively strong force there is still a gap between the plates, the width d is about equal to surface microroughness height. In the region of the contact points, a conductivity bridge appears as the oxide film degrades due to either plastic deformation of microshelves or to electric breakdown of the oxide film and formation of a thin metal bridge ($\sim$1-μm-diameter). Resistance, R_{bp}, and inductance, L_{bp}, depend on the resistance and inductance of parallel conductive bridges; the capacitance C_{bp} is higher, when S is greater and when d is smaller.

As the shock wave reaches one of the CC contacting surfaces, it begins to move into the gap at velocity W, which leads to an increase in C_{bp} (hence, a decrease in $V \sim 1/C_{bp}$) and later, after the gap is closed and the oxide film is

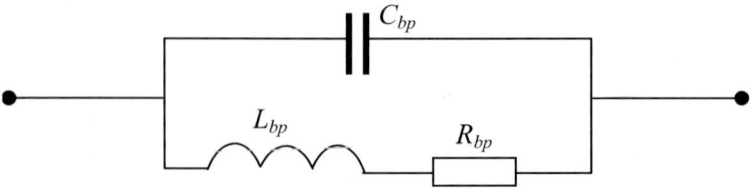

Fig. 3.16. Equivalent electrical circuit diagram for the method of closed contacts

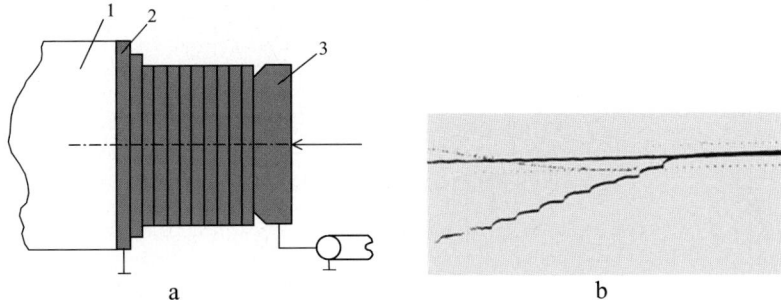

Fig. 3.17. Scheme of shock wave velocimeter (**a**) and typical oscillogram (**b**): 1 – HE charge; 2 – first plate; 3 – lead pressing the sensor on HE. The frequency of the time marks on the oscillogram is 200 MHz

destroyed by the shock wave, there is a decrease in R_{bp}. Thus, the CC acts as a capacitive transducer between the shock front and the given interface of the specimen. When the shock front and CC coincide, the CC electric resistance abruptly drops. Direct current is passed through a series of CC with the total voltage across them being measured, as the shock wave passes through the series an abrupt voltage drop is recorded. When the shock wave arrives at the next CC, another voltage drop takes place.

Figure 3.17 shows an example experiment recording the shock velocity in a package of aluminum foils using the method of closed contact, as well as the oscillogram of the experiment.

The measurement accuracy of D depends on the measurement of the time intervals of the shock wave passage through each aluminum foil.

3.2 Methods for Continuous Recording of Material Velocity Profiles

A comprehensive study of wave process kinetics, including relaxation phenomena that may occur as a consequence of the loading, has become possible thanks to the development of experimental diagnostics enabling us to acquire continuous information as physical changes take place in the nature of the shock wave [1–3]. In this section we consider the primary methods for continuously recording material velocity profiles.

3.2.1 Capacitive Sensor

The capacitive (capacitor) sensor is designed for continuous recording of metal specimen surface velocity $W(t)$ [25, 26]. The electric circuit and an example of setting up the experiment to study elastic-plastic waves in steel using a

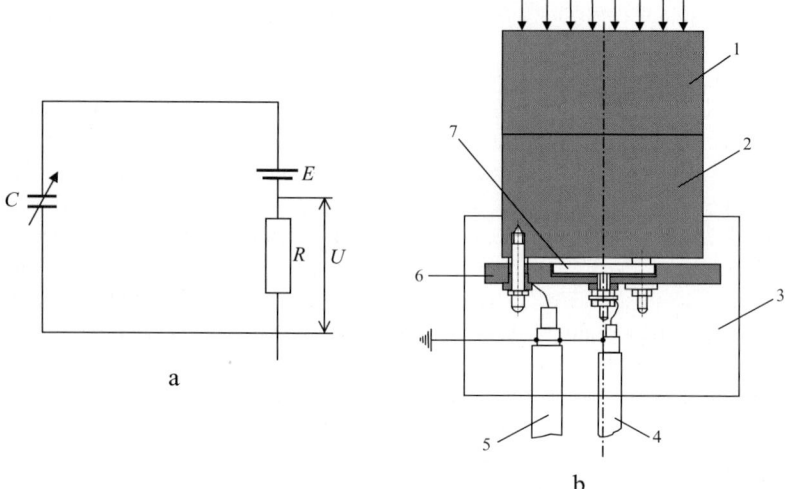

Fig. 3.18. Basic diagram of the capacitive sensor (**a**) and setting up the experiments recording the sample free surface velocity (**b**): 1 – HE charge; 2 – sample, a movable sensor electrode; 3 – base plate; 4, 5 – power and measuring cables; 6 – safety ring; 7 – immovable sensor electrode

capacitive sensor are shown in Fig. 3.18. An immovable electrode (7) is positioned at distance x_0 from the free surface of the sample (2). This electrode, in combination with the sample, creates measuring capacitor C. A voltage E is applied on the capacitor from an E.M.F. source through a load resistance R, which is small enough that the time constant RC is much shorter than the characteristic recording time. To maintain a homogeneous electric field in the measuring electrode region, safety ring (6) is used.

When the sample surface has been set in motion, the measuring capacitance begins to change, and a recharge current appears in the sensor circuit, whose force is proportional to the rate of change in capacitance, which in turn is related to the sample surface velocity $W(t)$.

If the rate of change in the capacitance, $(1/C)(dC/dt)$, is much less than $1/RC$, i.e. $R(dC/dt) \ll 1$, then $U = ER(dC/dt)$ with an accuracy that is second order to the independent variables (U = electrical potential difference, or voltage). Since $C = \alpha/(x_0 - \int_0^t W(t)dt)$, where $\alpha = \varepsilon S/4\pi$, we determine from the two last relations the free surface velocity as a function of time:

$$W(t) = Z(t)\left[x_0^{-1} + \int_0^t Z(t)dt\right]^{-2} , \text{ where } Z(t) = \frac{U(t)}{\alpha RE}.$$

If $\int_0^t W(t)dt \ll x_0$, then the expression for velocity will become

$$W(t) = \frac{U(t)x_0^2}{\alpha RE}.$$

Thus, having measured $U(t)$ (voltage across resistance R (Fig. 3.18a)), we find the desired function $W(t)$.

The method is contact-free, so its resolution is limited, in principle by the time difference of the recorded load pulse arrival at the sample surface in the region controlled by the sensor. The diameter of the measuring electrode and the distance between it and the sample surface range from 5 to 25 mm and from 1 to 6 mm, respectively, depending on the required resolution and total recording time. The actual experimental resolution of the 5-mm-diameter sensor was between 10 and 20 ns [16]. Due to the low signal level, the susceptibility of the method to noise is sufficiently high, which limits its applicability. Increasing U, R, and the ratio of immovable electrode diameter d to inter-electrode gap x (d/x) results in a higher signal from the sensor. The voltage U is limited by the electrical strength of the air gap. Insulating the electrode from the sample free surface with a thin dielectric layer allows the usage of greater voltages (up to a few kilovolts). Increases in the load resistance R are limited because of the need to match cable and wave resistances in order to avoid frequency distortion of the signal. The immovable electrode diameter can not be increased considerably, since this would result in more severe effects relating to the nonplanar wave fronts. When the voltage across the source is 3 kV, the characteristic capacitive sensor signal level is from a few to several tens of millivolts. The error of the $W(t)$ measurement with the capacitive sensor method is estimated to be 3–5% [16].

The capacitive sensor method is widely used to study metal elastic-plastic properties and spallation strength [26]. Examples of the capacitance sensors signal showing the characteristic of elastic-plastic waves appears in Fig. 3.19.

3.2.2 Magnetoelectric Method

The magnetoelectric method for recording mass velocity profiles $u(t)$ in dielectric materials is based on the generation of E.M.F. in a conductor moving in magnetic field [1, 27] (Fig. 3.20). The magnetoelectric sensor is a Π-shaped

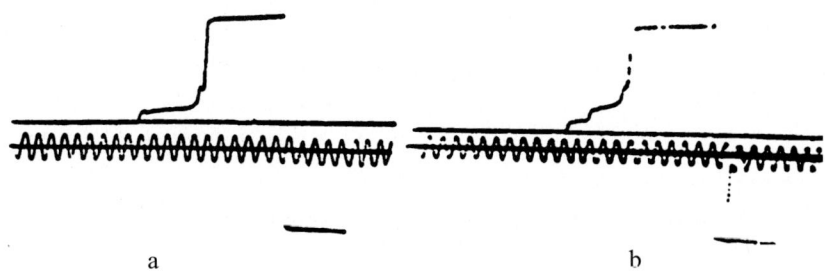

a   b

Fig. 3.19. Characteristic oscillograms of recording the signal from the capacitive sensor (time marks in 2 μs): **a** – steel 3 sample 90 mm long; **b** – stepped steel 3 sample ($h = 5$ mm)

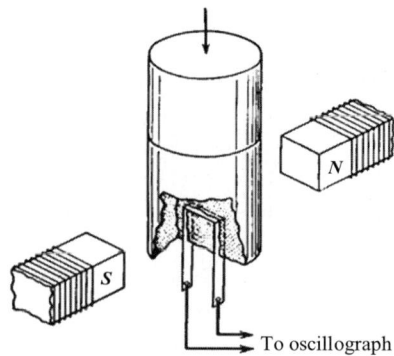

Fig. 3.20. Diagram of mass velocity profile recording with the magnetoelectric method

strip of thin conducting foil. The experimental assembly is placed in a homogeneous magnetic field produced by stationary electromagnets or an expendable solenoid. The sensor is positioned so that its crossbar, which is a sensitive element, is perpendicular to the field lines and parallel to the shock or detonation wave front. When the shock or detonation wave passes through the material, the sensor moves with the material or explosion products surrounding it; E.M.F. arising in the sensor due to the magnetic field lines is recorded by an oscilloscope. The sensor is assumed to be inertialess, in other words its velocity is equal to the materials mass velocity.

When the conductor moves in the magnetic field, induction electromotive force E is induced in it, which is related to conductor velocity u, length l, and magnetic field induction B as $E(t) = -Blu$.

In early experiments [1], the magneto-electric sensors were fabricated from copper foil 0.3 mm thick. High lag time of the sensors led to oscillogram front cut and an apparent decrease in mass velocities. Experimental verification [28, 29] revealed that ∼0.1-mm thick aluminum sensors to be optimal. Further thickness reduction is unreasonable, since explosion products frequently tear the sensors, and with very thin sensors the self-conductance effect of the EP becomes noticeable. Characteristic oscillograms of the mass velocity profiles in materials appear in Fig. 3.21.

Various sensor types were proposed and experimentally verified during the development of the magnetoelectric method: Π-shaped, stepped sensors allowing simultaneous recording of wave and mass velocities, stirrup sensors in the form of a thin-foil disk and wire leads connected to it. [30] discusses multichannel measuring systems, in which the charge to be studied is composed of several HE disks with plane magnetoelectric sensors between them. In another type a plane sensor set is integrated into a single package positioned in a demountable charge at an angle to the impinging flow. The package elements are at different depths in the flow, and the depth is readily adjustable by

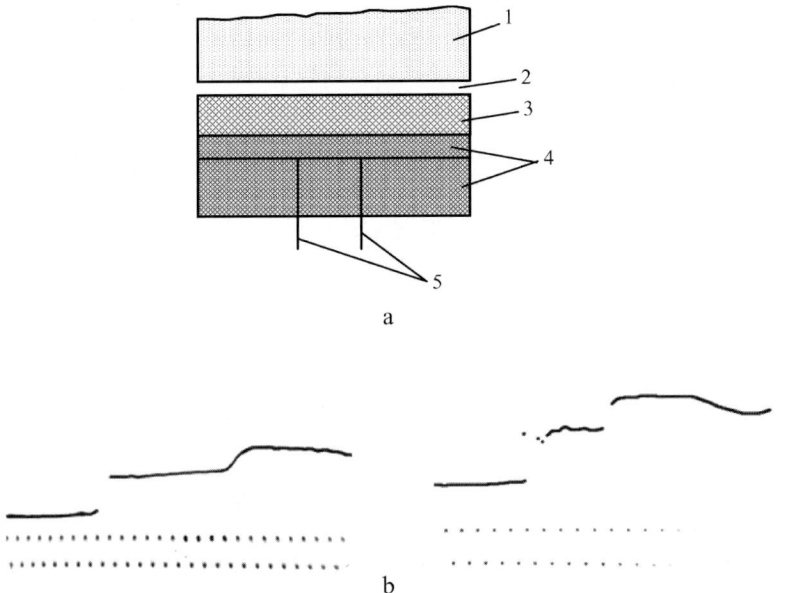

Fig. 3.21. Scheme of setting up the experiment (**a**) and oscillograms of two-wave profile in KCl (**b**): 1 – HE charge; 2 – air gap; 3 – base plate; 4 – material under study; 5 – sensor

changing the distance between the elements and the angle the package is placed at.

The multi-channel systems are fairly informative; however their use is not always warranted. The sensors embedded in the charge, coated with a protective layer may cause undesirable flow perturbations. In unsteady flows, for example, during the initiating shock wave evolution, the perturbations can noticeably distort the process flow and even stop a chemical reaction which has commenced. The sensor integration into a single package allows avoidance of these problems, however the sloping position of the sensors, as direct experiments suggest [31], upsets the 1D flow nature and results in an underestimation of actual velocities by ~10%.

The time resolution of the magnetoelectric method depends on a number of factors: sensor material and thickness, distance between leads, impinging wave or detonation front curvature, speed of operation of recording devices. In precision experiments [32] with 25-μm-thick sensors, with careful assembly of the devices, it is possible to achieve $\tau_r \sim 10$ ns time resolution.

The electromagnetic technique is widely used to study condensed HE detonation, estimate parameters of two-wave shock front profiles (an evidence of first-order phase changes (KCl, KBr, BN)), and investigate unloading waves [33].

3.2.3 Electromagnetic Method

The electromagnetic method was developed for measuring material particle velocity in dielectric or conductors under the stimulus of shock waves [34, 35]. The method measures the E.M.F. $E(t)$ induced in a measuring coil during (or when) a thin conducting foil, which is parallel to the coil plane, moves in the direction of the coil's axis. The whole system resides in a magnetic field. If the foil is quite large, magnetic image theory can be used to derive equations relating $E(t)$ to the foil movement.

A schematic view detailing the electromagnetic method with an axisymmetric electromagnetic sensor for the measurement of a conducting sample's free surface velocity is presented in Fig. 3.22. Metal disk 3, whose diameter is less than the charge diameter, is put at the center of the upper surface of HE charge 1. A layer of mineral oil 2 is poured onto the free surface end of the charge to avoid fast EP expansion, which would induce electrical noise. Plexiglas disk 4 is placed on plexiglas stands above the metal disk. The lower surface of the disk serves as a reference plane. Distance l between the upper surface of disk 3 and the reference plane is carefully measured for subsequent comparison with the integral of the experimentally measured free surface velocity.

Permanent magnet (6) is placed in a hollow on the upper side of disk (4). A one-turn coil (7) made of insulated copper wire is put in an annular groove. In the figure, z_0 is the distance from the initial position of the surface of disk (2) [plane $z_f(0)$] to the mean plane of the coil, r_0 is the radius of the inner

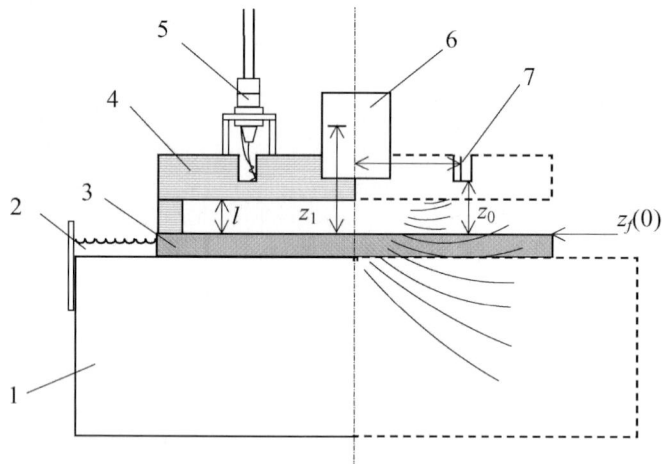

Fig. 3.22. Arrangement of the axisymmetric electromagnetic sensor: 1 – HE charge; 2 – mineral oil; 3 – metal disk; 4 – plexiglas disk; 5 – coaxial connector; 6 – permanent magnet; 7 – single-turn coil

edge of the coil plus a quarter of the wire diameter, z_1 is the distance from the surface of disk (2) to the center of the magnet.

Only the permanent magnetic field shown in Fig. 3.22 originally exists in the system. The conducting sample's movement toward the magnet distorts the field and results in a decrease of the magnetic flux penetrating the measuring coil. Due to this, the E.M.F. $E(t)$ induced in the measuring coil is related to conducting surface velocity $W(t)$. The desired velocity profile is found by numerical solution of the full electrodynamics problem, the initial data for which, besides the measured electromagnetic induction signal, are z_0, z_1, r_0, and spatial distribution of magnetic induction flux B_r under the specific experimental conditions.

The E.M.F. $E(t)$ appears in the coil, whose equivalent circuit is presented in Fig. 3.23. E.M.F. $E(t)$ arising in the coil is transformed by the coil into voltage signal $V(t)$, which enters a cable typically 15 m long. The signal and all reflections appearing in the cable yield signal $V_s(t)$ at the exit of the cable loaded by mated resistance R_s and input oscilloscope impedance. The signal oscillogram is photographed (Fig. 3.24). The graduation grid (in the lower part of the oscillogram) is photographed in advance, and the film is displaced before filming the oscillogram of the signal $V_s(t)$ to avoid image superposition, which can lead to a loss of information or misinterpretation. Only one graduation line corresponding to a voltage equal to -0.1 of maximum beam deflection in the electron-beam tube display is superimposed on the oscillogram. The same line referred to below as line "-0.1" is present on the previously filmed grid, which allows us to relate the oscillogram to the grid (make superposition) during the digital data analysis. Measurements are taken on the grid and the working oscillogram. The measured data is processed on the computer to plot the time function of the voltage characterizing the conducted experiment. The following times are marked on the oscillogram: 1 – detonation wave arrival at the HE-metal interface; 2 – shock wave arrival at the metal free surface and

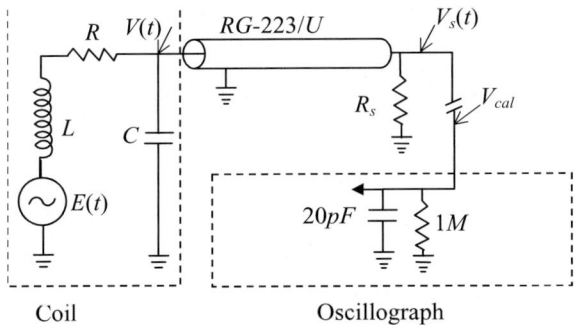

Fig. 3.23. Equivalent circuit of the axisymmetric electromagnetic sensor measuring coil and circuit of its connection to the oscilloscope (the input circuit of the oscilloscope amplifier is also shown)

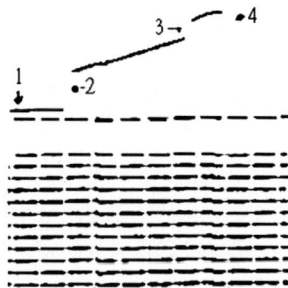

Fig. 3.24. Oscillogram from the experiment on studying the free surface velocity of an aluminum plate. The distance between the zero oscillogram line and the reference line "−0.1" is 0.7 V. The distance between the time tags is 0.5 μs

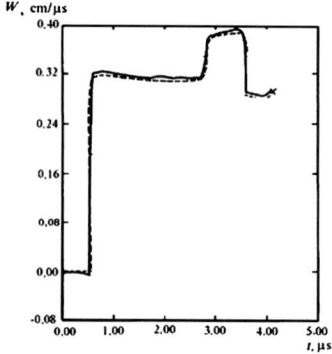

Fig. 3.25. Plots of function $W(t)$ constructed by results of analysis of the oscillogram in Fig. 3.24. The *dashed line* is the plot constructed by $V(t)$; *solid line* is the plot constructed by calculated function $E(t)$

relevant acceleration of the surface; 3 – second wave arrival at the free surface (on the rarefaction wave return to the HE-metal interface); 4 – time of metal plate free surface impact on the reference plane and relevant deceleration of the surface.

Figure 3.25 plots the free surface velocity $W(t)$ (constructed from the oscillographic record) of a 6.3 mm thick aluminum plate acted upon by a detonation wave propagating through a shattering charge of HE (TG 36/64, 102 mm in length). For this particular experiment, the distance l (from Fig. 3.22) was 10 mm. In Fig. 3.25 a $W(t)$ curve was been constructed from the $E(t)$ data, and another by its non-corrected plot $V_S(t)$. The times 2, 3, and 4, from the oscillogram found in Fig. 3.24 are marked in the plots.

The choice of magnet material comes from the requirement of minimizing secondary effects. In the event that the magnet's material is conducting, the eddy currents induced in the conducting plate, and in turn, the eddy currents induced in the magnet, leads to reduction in the signal. However, if the magnet is made from a nonconducting material, as in [35], the signal increases up to the limit imposed by the magnetic moment reduction as a result of the demagnetizing eddy current field in the plate.

Electromagnetic sensors are simple to fabricate and allow the estimation of particle velocities in dielectrics [34] and conductors. The resulting free surface velocities of the conductive or dielectric samples which are prepared in a special manner can be obtained with a reasonable degree of accuracy. The permanent magnet requires no special power supply, which is responsible for its high immunity to electromagnetic interferences [35]. measured the oscillatory motion of steel flask walls under action of internal electrical explosive loading using the electromagnetic sensor.

The estimated error of the method is 2%, the limiting resolution is one nanosecond per 1 cm of inductive sensor diameter (without accounting for the shock front deviation from the sensor plane). The data obtained is harder to process than those obtained with other electromagnetic methods; however given modern computers it presents no major difficulties.

3.2.4 Inductive Method

The inductive method [36] enables continuous recording of the velocities of condensed media subjected to high shock pressures.

Consider the principle of operation of the inductive method. Let a turn of radius R_1 with negligible wire cross section be connected to a stabilized direct current source and be located in condensed dielectric medium 1 (Fig. 3.26) at a height of h_0 above a conducting half-space 2 with electrical conductivity $\sigma \to \infty$. The assembly is then placed in a stationary magnetic field established in the entire space. The media magnetic permeabilities are $\mu_1 = \mu_2 = \mu_0$, where μ_0 is vacuum magnetic permeability. If the magnetic field changes for some reason in dielectric medium 1, an induction electromotive force (E.M.F.) E_1

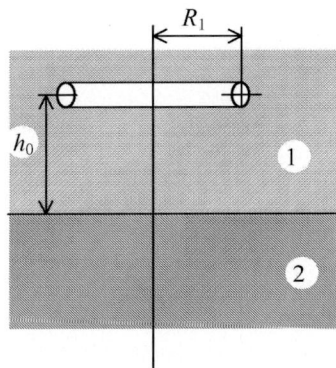

Fig. 3.26. Diagram of inductive probe: 1 – condensed dielectric medium; 2 – conducting half-space

will appear in the turn. A plane wave, whose front is parallel to the media interface, will propagate from the bottom upwards through the system. As the wave front moves through the conductor, no E.M.F. appears in the turn. However, when at some time $\tau = 0$ the wave front arrives at the interface between the dielectric and conductor, the former will be set in motion and concentric eddy currents will appear on the conducting medium surface. This will cause a change in the field which will induce an E.M.F. E_1 in the turn which depends on dielectric-conductor interface velocity $u(t)$. The induction E.M.F. E_1 in the turn with direct current I_0 is related to the velocity $u(t)$ as $E_1(t) = I_0 \alpha_1(t) u(t)$, where coefficient α_1 is a function not only of current distance $h(t)$ from the turn to the conductor surface, but also of initial distance h_0.

For the induction E.M.F. to be greater, not one, but several turns, i.e. an inductance coil, are used. The induction E.M.F. in the coil will be $E = I_0 \alpha u$, where $\alpha = \alpha_1 N^2$, N is the number of turns, and α_1 is a coefficient for equivalent turn.

The experimental schemes appear in Figs. 3.27a,b [36]. Plane shock waves were generated in the material with cylindrical HE charges 120 mm or 200 mm in diameter. The charges produced a wave front that was essentially planar within the central 100 mm diameter region, having variations in shock front arrival time of no more than 0.1 µs. The process under study lasted for 1–2 µs. The explosion took place ~0.1 s after feed of current $I_0 \approx 400$ A to the sensor circuit. The sensor was fabricated from 8 turns of 1-mm-diameter insulated copper wire.

In the experimental setup illustrated in Fig. 3.27a, a rectangular plane shock wave was passed from aluminum base plate (1) into a lead or bismuth sample (2) and then into the plexiglas sample (3). Thicknesses of samples (2) and (3) were 5 mm and 6 mm, respectively, and their diameters were 100 mm. Sensor (4) detected the signal. The second experimental setup

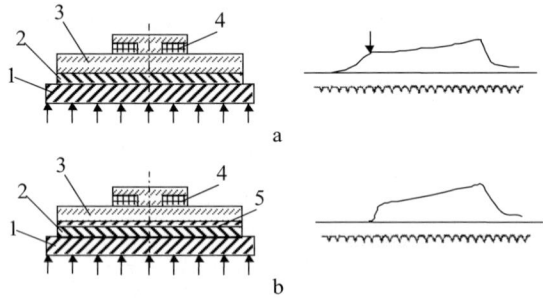

Fig. 3.27. Schemes of setting up the experiments on measuring metal-dielectric interface velocity (**a**) and dielectric medium velocity (**b**): 1 – aluminum base plate; 2 – sample; 3 – plexiglas disk; 4 – inductive probe; 5 – copper plate 0.1–0.3 mm thick

(Fig. 3.27b) differs from the first, in that copper plate (5) with a diameter of 100 mm and a thickness of 0.3 mm is placed at the lead (bismuth) – dielectric interface. The presence of the copper plate does not appreciably alter the interface velocity *vis-a-vis* the first experimental setup. The experiments used two explosive device types with known shock wave parameters behind the shock front in base plate 1 ($u_{bp} = 1.46$ and 2.33 km/s, respectively), ensuring an initial pressure $P \approx 40$–80 GPa in samples 2.

Figures 3.27a,b also present oscillograms of the first and second experimental series with bismuth conducted on explosive device 1. The time tag frequency is $f = 10$ MHz.

Figure 3.28 presents experimental curves $u(t)$ derived using the above relations from oscillograms $V(t)$ of the second experimental series with Bi and Pb while taking into consideration a small empirical correction for the copper electrical conductance finite thickness. The derived $u(t)$ from the oscillogram $V(t)$ presents no problems on computer. As mentioned above, a shock wave of constant pressure (velocity) was generated following the wave front. The obtained experimental curves $u(t)$ in fact have an essentially rectangular profile. The velocities obtained agree with those found using known metal Hugoniots and the parameters of explosive devices I and II within ±2%.

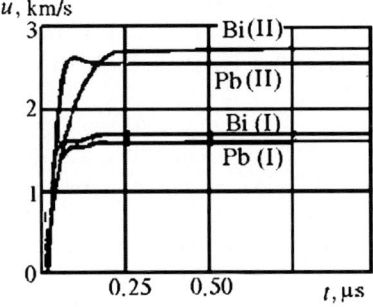

Fig. 3.28. Experimental curves $u(t)$ derived by re-calculation of the $V(t)$ oscillograms from Fig. 3.27

Experimental verification of the inductive method suggests [36] that in shock-wave processes the behavior of copper is close to that of an ideal conductor. This fact allows the continuous recording of the metal-dielectric interface velocity with a fairly good approximation when electrical conductance of the metal studied is close to that of copper. To record the interfacial velocity between the dielectric and the conductor of a low electrical conductance (bismuth, graphite, etc.) or dielectric medium velocity, it is necessary to put either a copper or an aluminum foil (from 0.1 mm to 0.3 mm in thickness depending on foil material and shock compression pressure), which is thin gas-dynamically, but relatively thick electromagnetically, on the interface or in the dielectric. In the special case, when the dielectric medium is air, the method allows the continuous recording of the free surface velocity of the studied material. When it is necessary to avoid the minor systematic velocity underestimation due to copper (aluminum) electrical conductance finiteness, which is typically no higher than 3%, a correction is introduced to the measured data which is determined either computationally or by special verification experiments. The experiments conducted to date using the inductive method (under conditions known in advance from other measurements) suggest that the experimental velocities obtained using the correction for finiteness of the copper (aluminum) electrical conductance differ from the expected results by as little as ±3–4%.

Because the inductive method allows the continuous measurement of condensed medium's velocity, it can be used to study complex shock-wave processes, for example: elastic-plastic waves, phase transformations, artificially generated loading, expansion waves, etc.

An example in Fig. 3.29 using this method presents an oscillogram of an experiment recording the separation of elastic and plastic waves in quartzite of 2.65 g/cm^3 density by composition TG 50/50 contacting the HE charge (200 mm in diameter and length), when the elastic wave has traveled 55 mm. In this experiment, the induction E.M.F. is caused by the motion of a 0.2-mm-thick aluminum foil introduced into the quartzite at $S = 55$ mm. As seen from Fig. 3.29, the elastic wave is 0.8 μs ahead of the plastic wave. The quartzite mass velocity in the elastic precursor increases from 0.3 km/s at its front to 0.45 km/s before the plastic wave front; behind the front, $u = 1.35$ km/s.

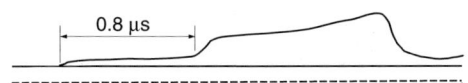

Fig. 3.29. Oscillogram of mass velocity in quartzite [36]

3.3 Methods for Continuous Pressure Profile Recording

Information about the state of stress of shock-compressed solids is of a great significance in metrology, high-pressure physics, construction of an equation of state, description of wave processes. It can be acquired by a number of methods that allow for the continuous measurement of mechanic stresses using pressure sensors operated on different principles.

3.3.1 Piezoelectric Pressure Transducers

The piezoelectric pressure transducers transform mechanical forces into an electric signal. The function of this sensor relies on the piezoelectric effect, which consists of a polarization appearance in a non-centrosymmetric dielectric under mechanical loading. The appearing polarization increment J is proportional to mechanical stress (pressure) P. In one-dimensional loading [11]: $J(x, t) = kP(x, t)$, where k is the material piezoelectric modulus, x is a coordinate, t is time.

In the general case, $k = k(P)$. However, in a given application, sensors are typically chosen for which $k \approx$ constant in the pressure range under consideration. Single-crystal piezoelectric quartz (SiO_2) and ferroelectrics, such as, lithium niobate ($LiNbO_3$) single crystal and polyvinylidene fluoride (PVDF) polymer with monomer formula (CH_2–CF_2) are most widespread materials used as sensors [37]. There have been attempts to use barium titanate and lead zirconate-titanate piezoceramics [38], bismuth germanate and lithium germanate single crystals [39] for these purposes, but these materials are currently not used for different reasons.

Historically, the quartz transducer was the first device with nanosecond time resolution that was used for direct measurements of shock-wave pressure profiles in solids [37]. Then transducers fabricated from lithium niobate and PVDF were developed. The time history of the piezoelectric pressure transducer method is illustrated in Fig. 3.30 [37]. As implied in the figure, it takes 10 to 15 years for each transducer type to be developed and placed into practice.

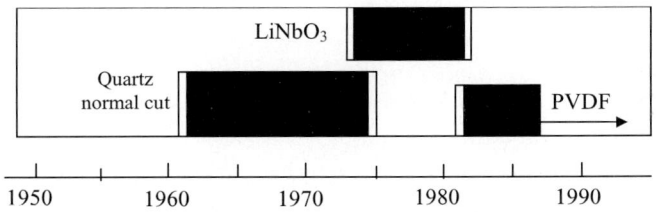

Fig. 3.30. Diagram depicting the historical timeline for the development of the piezoelectric transducer method

The sensor of the quartz transducer is a synthetic quartz disk, whose plane surfaces are perpendicular to the reference axis x. Metal electrodes 5–10 μm thick are attached to these surfaces. Silver is typically used for the electrode material. The transducer with a guard ring [37] is used most frequently. The electrode configuration on the sensitive element of the transducer is shown in Fig. 3.31.

The guard ring serves to remove the electric field heterogeneity in the crystal region confined by the measuring electrode in the shock wave acting on the transducer. Transducers, whose sensitive element has a continuous electrode, are used less frequently. Note that the sensitive element geometry in the lithium niobate transducers is the same. Here y and z crystal cuts are used in the main.

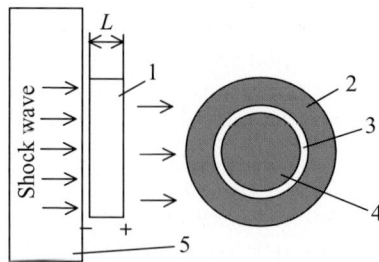

Fig. 3.31. The configuration of the quartz transducer with guard ring: 1 – quartz; 2 – guard ring; 3 – insulating pad; 4 – center electrode; 5 – sample

In experiments, the transducer is attached to the surface of the (Fig. 3.31). The shock wave passes from the sample through the interface into the transducer and induces polarization in its sensitive element material. The polarization change is accompanied with an electric current in the transducer's external circuit, with the current bearing information about the magnitude of the load.

The theory describing the quartz transducer was developed by Graham, Neilson, and Benedick [40]. According to [40], the expression for current $I(t)$ in the shorted external circuit of the transducer is $I(t) = kSD/L[P(0,t) - P(L,t)]$, where k is piezoelectric modulus, S is the measuring electrode area, D is the elastic wave velocity in quartz, L is the sensor thickness, and $P(0,t)$ and $P(L,t)$ are pressure amplitudes at the location of the front and back sensor electrodes.

The relation was derived based on the following assumptions:

– compression is one-dimensional;
– sensor deformation is insignificant (elasticity region);
– electric field is one-dimensional;
– permittivity is constant;
– there is no electric conductivity;
– piezoelectric polarization is directly proportional to mechanical stress.

It follows from the above relation, that the pressure profile at the interface between the sample and the transducer can be recorded during the elastic wave passage through quartz $[P(L,t) = 0]$. This typically lasts for 0.5–1.0 µs. The sample pressure amplitude is calculated from the measured pressure accounting for the dynamic impedances of the sample and quartz. In planar geometry, where quartz transducers are in contact with the sample, pressures are re-calculated by relation:

$$P_{samp} = \frac{Z_{qu} + Z_{samp}}{2Z_{qu}} P_{qu} \, ,$$

where P_{samp} and P_{qu} are pressures in the sample and quartz, respectively, and Z_{samp} and Z_{qu} are dynamic impedances of the sample and quartz.

The quartz transducer time resolution is several nanoseconds [37, 40, 41]. The upper-range of measured pressures is 4 GPa, which is somewhat lower than the Hugoniot elastic limit (6 GPa) [37]. The lithium niobate transducers are not inferior to the quartz transducer in terms of the speed of operation, but have a lower upper-range in measured pressures (1.4 GPa for z-cut and 1.8 GPa for y-cut), yet a significantly higher piezoelectric modulus: by a factor of 4.6 for z-cut and by a factor of 6.5 for y-cut. In the pressure range below 4 GPa the x-cut quartz piezoelectric modulus is a function of pressure. The dependence on pressure is weak (according to [41], $K(P) = (2.00 \pm 0.097P) \cdot 10^{-3} C/(m^2 \cdot GPa)$). The pressure amplitude is typically calculated using the average piezoelectric modulus of $K = (2.2 \pm 0.2) \cdot 10^{-3}$ $C/(m^2 \cdot GPa)$.

In the early 1980s, F. Bauer reported the development of a pressure transducer based on the polymer ferroelectric PVDF [42,43]. Presently, this transducer type is widely used for dynamic pressure measurements, as reflected by the number of publications on this subject. It is beyond the scope of this monograph to do a literature review of this transducer type, instead we mention a series of papers [44–50] by F. Bauer and associates from Sandia National Laboratories on the transducer development and standardization, which resulted in a transducer reproducing characteristics within 3%, with the measured pressure range up to 35 GPa and a time resolution ≤10 ns.

The standard transducer design is presented in Fig. 3.32.

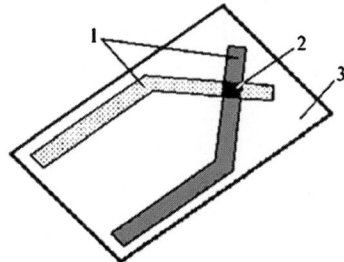

Fig. 3.32. Design of the standardized transducer: 1 – electrodes; 2 – active zone; 3 – PVDF film

The transducer sensor is located at the intersection of two metal electrodes applied on opposite sides of the piezoelectric film ∼25 μm thick. The sensitive zone is typically ∼2 × 2 mm. The electrodes are applied by vacuum deposition: first a platinum layer ∼0.05 μm thick, then a gold or aluminum layer ∼0.25 μm thick. The characteristic length of the current leads (electrodes) is ∼20 mm. The residual polarization of the piezoelectric film in the active zone is 9.2 ± 0.2 μC/cm². The piezoelectric film is covered with an insulating layer of Teflon 12.7 μm thick. Thus, the transducer package is a polytetrafluorethylene-PVDF-polytetrafluorethylene ∼50 μm in total thickness. Currently, pressure measurement is frequently made with the PVDF transducer configured in such a way that the electrical circuit is close to being shorted. For this purpose a resistive load of nominal value 0.01–0.5 Ω is connected between the transducer leads [37]. Under such conditions, the recorded current is proportional to the time derivative of pressure [37, 42–50]. The following scheme is adopted to obtain the final result, i.e. the function $P(t)$.

The measured drop in voltage $V(t)$ across the resistive load is used to estimate the current density $I(t)$ from the transducer. Integration of $I(t)$ results in charge density $Q(t)$ generated by the transducer. The gauge function $P(Q)$ is then used to estimate pressure: $P(t) = P(Q)Q(t)$.

By way of example Fig. 3.33 presents an oscillogram of the current from the transducer and its integration result. In this case the transducer recorded both a shock wave and its release wave. The maximum pressure amplitude is 7.8 GPa [42].

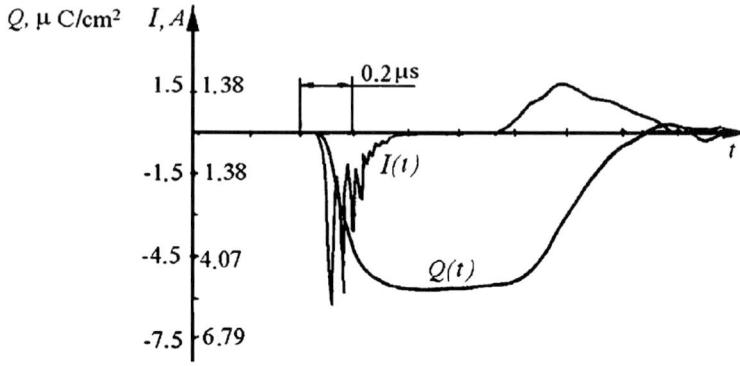

Fig. 3.33. Oscillogram of transducer current $[I(t)]$ and result of its integration $[Q(t)]$

The mode of an open measuring circuit is used less frequently. To implement the mode, an integrating capacitor is connected in the circuit, and $Q(t)$ is obtained straightforwardly. For both types of circuits (open or closed) the method for obtaining $P(t)$ are the same.

From the above discussion, it is apparent that the transducer must be calibrated for use in experiments. In order to calibrate the transducer, the dependence of transducer-generated charge density Q on pressure P is experimentally determined in the required pressure range. This is done using a loading device with well-known characteristics. [37, 44–49], for example, a light-gas gun is used for this purpose. The impactor and the target are fabricated, as a rule, from the same material. The materials are z-cut quartz, z-cut sapphire, and a number of other materials with well-known Hugoniots. The resulting experimental curve $Q(P)$ or $P(Q)$, which is more convenient in the processing of experimental data, is mathematically approximated by

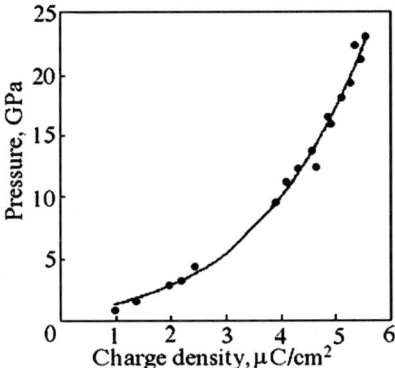

Fig. 3.34. Gauge function of PVDF transducer

some function, which is then used as a gauge function. The power function $Q = \text{const } P^a$ $(a < 1)$ [30, 50], third- and fifth-degree polynomials [46,51] are taken for the approximating functions. In some cases, for the experimental points to be described better, the pressure range is broken into segments and each segment is described by a polynomial.

By way of example, Fig. 3.34 presents the gauge function of a PVDF transducer [52].

There can be two measurement setting types for the PVDF transducer. In one type the transducer is placed on the material to be studied, in the other type the transducer is placed at the interface of two media: the material to be studied and media made acoustically consistent with the transducer (acoustic-diversion transducers).

When positioned in the material under study, the transducer will be either a soft or hard pad depending on the PVDF-material acoustic hardness ratio. It is well known that a pressure equal to the impinging wave front is eventually established in such systems and, hence, the transducer will measure pressure immediately in the material under study. The upper value of measured pressures that has been achieved for the experimental setup under discussion is 61 GPa [52].

For the acoustic-diversion transducer the measured pressure will be the diverted material pressure. Hence, here it is possible to measure higher pressures in materials of hardness greater than that of PVDF. For example, at 61 GPa pressure in PVDF the pressure in iron will be more than 100 GPa. In this case, knowledge of the Hugoniot of the material to be studied is required to evaluate the pressure amplitude.

In conclusion of this section, note that piezoelectric pressure transducers are advanced devices widely used for measuring dynamic pressures. Their advantages include nanosecond time resolution, the fact that no power sources are needed, and their applicability to a wide range of pressures.

3.3.2 Piezoresistive Pressure Transducers

Piezoresistive tools for measuring pressure (mechanical stress) are based on the use of sensors that are constructed from materials whose conductivity changes under action of shock loading. Materials most well known for this attribute are manganin, ytterbium, sulfur, and carbon. Manganin pressure sensors have found wide applicability among the techniques for measuring both static and dynamic pressures. Presently, the principal method for continuous recording of mechanical stresses (or pressure) involves the use of the manganin sensor [2, 16, 53–55].

Usefulness of manganin sensors is based on: (1) a high sensitivity in the electrical resistance of the manganin alloy (84% Cu, 12% Mn, 4% Ni) to pressure, (2) a low sensitivity to temperature changes, and (3) an essentially linear dependence of the change of relative manganin resistance $\Delta R/R_0$, with respect to mechanical stress σ (pressure P). In general, $\Delta R/R_0 = K\sigma$, or $\sigma = (1/K)(\Delta R/R_0)$, where R_0 is the initial sensor resistance, and K is the experimentally determined piezoelectric sensitivity factor. Figure 3.35 presents the most frequently used configurations of the sensitive elements of manganin sensors. Choice of sensor design depends primarily on the conditions, under which pressure is to be recorded. The most natural application of maganin sensors is in the measurement of the pressures that occur in planar impact. In this case, the sensitive element is formed as a straight-line segment of flattened wire or foil (Figs. 3.35a,b,d). To increase resistance and decrease occupied area, it is sometimes bent in a T-shaped or a zigzag manner, with very-low-resistance copper foil leads being attached by spot-welding. The sensor is typically manufactured as a 30–50-μm-diameter planar zigzag wire (Fig. 3.35c) or 10–30-μm-thick foil strip occupying 0.1–1-cm^2 area.

When recording the pressure profiles that occur with sliding detonation, or the principal normal stresses in planes parallel and perpendicular to the wave front [56,57], only the linear type sensors (Fig. 3.35a,b,d) are used. These are positioned such that the sensor will be parallel to the shock front. When recording the pressure in axisymmetric shock waves, an annular sensitive element (Fig. 3.35e) is used, the plane of which is parallel to the shock front, with the leads running perpendicular to it.

The thickness of the sensor's sensitive element is typically a few hundredths of millimeter. At the locations where the leads are fastened to the sensitive element, the thickness of the leads adds to the overall thickness of the sensor, resulting in total sensor thickness of 0.04–0.1 mm.

Use of the manganin sensor dictates the use of a compound sample with the sensor inserted between sample components (typically plates). When needed, the sensor may be separated (isolated) from the sample by insulating pads made of fluoroplastic, lavsan, mica, or epoxy resin (Figs. 3.35a,c). Direct current is passed through the sensor and voltage is recorded across it with the aid of an oscilloscope. Voltage drops with increasing mechanical stress (pressure) acting on the sensor. To improve the signal-to-electrical

Fig. 3.35. Typical designs of manganin sensors and their installation in experimental devices: 1 – sensor; 2 – leads; 3 – sample; 4 – insulation; 5 – cylindrical sample; 6 – annular sample

noise ratio, and to avoid overheating of the sensor, pulsed sources of current (5–10 A, ∼100 μs in duration) are used [53]. The sensor is connected to a resistance bridge (Fig. 3.36) or some other type of differential recording circuit (Fig. 3.37) in order to avoid constant signal component dependence on initial sensor resistance R_0, and thereby enhance measurement accuracy. A two-point connection circuit is used for relatively high-resistance (5–50 Ω) sensors.

In some cases, it is reasonable to use sensors with an initial resistance of a few tenths or hundredths of an Ohm. These sensors, in particular, offer the advantage that their readings are less sensitive to the shunting effect from environmental electrical conductivity. Such low-resistance sensors are connected into a potentiometric measuring circuit using a four-point circuit configuration (Fig. 3.38).

Because of the destructive nature of shock waves it is impossible to calibrate each sensor used. For this reason, pressure measurement is based on the unique relationship between the change in relative electrical resistance $\Delta R/R_0$ and shock compression pressure P (stress), and this relationship is

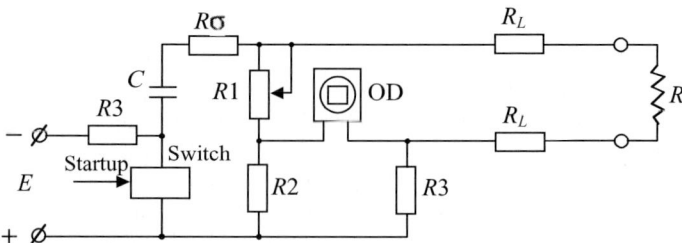

Fig. 3.36. Bridge circuit for pressure profile recording: R – sensor; R_L– connecting wire resistance; OD – output device (oscilloscope)

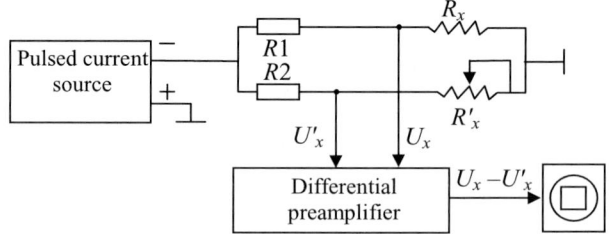

Fig. 3.37. Differential recording scheme: R_x – sensor; R'_x – sensor-equivalent resistance; U_x – drop in voltage across the sensor; U'_x – drop in voltage across sensor-equivalent resistance

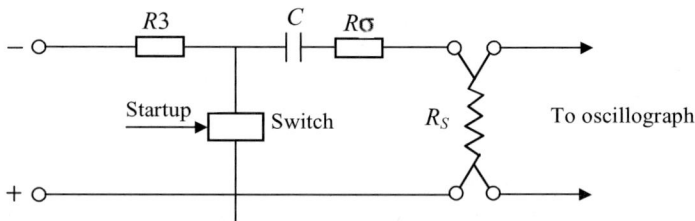

Fig. 3.38. Potentiometric circuit for pressure profile recording

assumed to be valid for all sensors made of this particular type of manganin. A verification curve can be constructed from the results of experiments with sensors positioned in reference materials of well-known compressibility. In the experimental assemblies, shock waves are actuated, whose kinematic parameters are estimated independently. For instance, the electromagnetic method or laser interferometric method can be used to measure free surface velocity W, or mass velocity u, and then pressure can be estimated using the laws of conservation. The following relationships are valid when the same material is used for both impactor and sample in planar geometry:

$$u = 1/2W; \qquad P - P_0 = \rho_0 Du \,,$$

where P_0, and ρ_0 are initial pressure (typically equal to 1 kg/cm^2 and taken equal to zero) and density, D is wave velocity in the material at pressure P, and W is impact velocity. Certain, physical phenomena which are known to occur at a given pressure, for example the phase transition of NaCl, can also be used for calibration [53]. Calibration measurements for manganin and constantan are presented most thoroughly in [16]. This paper also specifies the residual increments of the sensor's electric resistances, $\Delta R_{res}/R_0$, on unloading.

For pressures up to 50 GPa, the curve $\Delta R/R_0\,(P)$ is approximately linear and is described by the following the second-degree polynomial:

$$P = 356.2\frac{\Delta R}{R_0} + 42.7\left(\frac{\Delta R}{R_0}\right)^2,$$

which represents a fit to the experimental data with an accuracy of 5%. More precise measurements lead to a third-degree polynomial:

$$P = 340\frac{\Delta R}{R_0} + 70\left(\frac{\Delta R}{R_0}\right)^3; \quad P = 370\frac{\Delta R}{R_0} - 13.2\left(\frac{\Delta R}{R_0}\right)^2 + 44.2\left(\frac{\Delta R}{R_0}\right)^4$$

and even a fourth-degree polynomial [36, 43]:

$$P = 0.572 + 29.59\frac{\Delta R}{R_0} + 95.2\left(\frac{\Delta R}{R_0}\right)^2 - 312.74\left(\frac{\Delta R}{R_0}\right)^3 + 331.77\left(\frac{\Delta R}{R_0}\right)^4$$

Conducting the calibration of the manganin sensors under static versus dynamic (shock-wave) conditions gives somewhat different values for the piezo-electric sensitivity factor K. For static conditions, $K = \Delta R/R_0\,(P) = 0.0241$ –0.0238 GPa^{-1} in the 0.1–2.5 GPa pressure range, whereas under dynamic conditions, $K = 0.028$ GPa^{-1} in the 0.5–9.0 GPa pressure range. Thorough measurements of manganin conductivity reveal a difference in the piezoelectric sensitivity factor K under low versus high pressure conditions. This difference is attributed to the existence of an elastic-plastic transition in manganin at pressures of about 0.9 GPa.

A disadvantage of manganin sensors is the dependence of the piezoelectric sensitivity factor K on the sensor's environment and its loading history. The results of [58] reveal that $K = (2.3–2.7)\cdot 10^{-3}$ GPa^{-1} for a sensor embedded in epoxy resin, $K = (2.3–2.7)\cdot 10^{-3}$ GPa^{-1} in glass, and $K = (2.7–5.7)\cdot 10^{-3}$ GPa^{-1} in aluminum oxide. Such changes in the piezoelectric sensitivity factor are accounted for by considering differences in the level of strain in the different media.

Special measurements demonstrate [59] that for pressures of 7–10 GPa or higher, the change in manganin resistance is essentially reversible and independent of whether the dynamic compression is shock, stepped, or isentropic. The unloading path to zero pressure entails a small degree of hysteresis in the manganin sensor readings. The irreversible component of the manganin electrical resistance increment is attributed to cold-work hardening, and is no more than 2.5% of the initial resistance. Annealing manganin leads to an increase in the amplitudes and sensor reading hysteresis of the same value.

Electrical resistance as a function of pressure for manganin alloy (Cu-3% Ni-12% MN-0.25% Al-0.2%) is plotted in Fig. 3.39. Subtraction of the irreversible component of the resistance change from experimental curve 1 yields curve 2, which is close to the curve produced from measurements taken under hydrostatic compression (Fig. 3.39b).

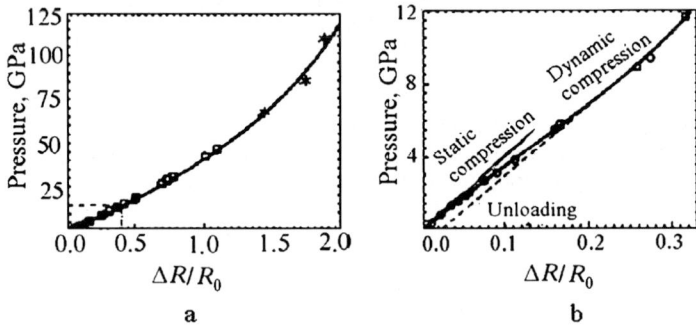

Fig. 3.39. Shock wave pressure vs. relative change in manganin sensor resistance

When insulating material pads are introduced between sensor and sample, sluggishness in the sensor readings can occur. This sluggishness will produce distortions in the signal. The magnitude of the distortion will depend on the time required for the establishment of insulation pressure (during multiple wave reflections in the pads) that is in equilibrium with the compressive stress in the subject material perpendicular to the sensor plane. The lag effect is considerable when recording low-intensity waves. Numerical simulation of the load pulse evolution in an elastic-plastic body with liquid-like pads [16] suggests that the inertial distortions introduce a systematic error (Fig. 3.40) with a characteristic time of 0.1 μs. With increasing pressure, profile distortions decrease due to an increasing sound speed in the insulating pads.

When wave profiles are recorded at two or more locations within a sample, distortions build up as the wave passes from one sensor to another. Consequently, the highest accuracy in measurement is achieved when one or more sensors are positioned at a single sample location, and insulating pads with minimal thickness are used. The problem of the buildup of distortion when sensors are positioned in several locations within a sample can be avoided by placing the sensors at several points on a stepped sample, as shown in Fig. 3.41a. One should not, however try to avoid the problem of distortion buildup by placing the sensors on an oblique section as shown in Fig. 3.41b [16] (use of an oblique section will not work because manganin sensors record normal mechanical stresses [55,56]).

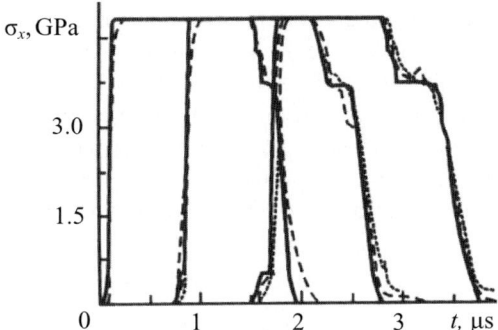

Fig. 3.40. Results of numerical simulation [16] for shock wave evolution in elastic-plastic material (aluminum AD1), when the pressure profiles are recorded by sensors located with insulation inside the sample: — — sample without pads; - - - - sample with insulation material (fluoroplastic) pad; · · · – sample with two insulating pads

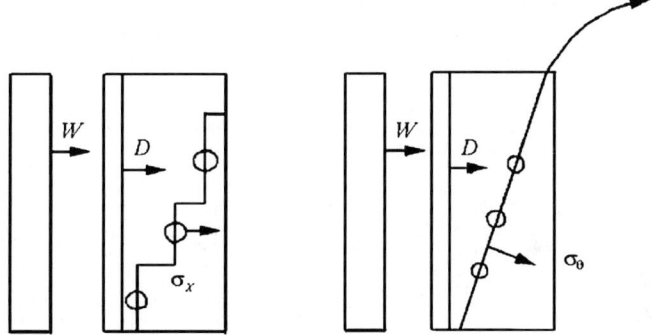

Fig. 3.41. Layout of several manganin sensors positioned in the sample section. O – location of the sensor sensitive element

Apparently, the electrical resistance of the sensor changes not only under pressure, but also due to the deformation of the sensitive element. Under non-uniform loading conditions, changes in the length of the sensor's sensitive element leads to an additional increase in electrical resistance. To discriminate between these factors, measurements are taken using materials with an appreciably different response to changes in pressure versus changes in length. By recording two independent oscillograms, one related to pressure and the other related to length of the sensitive element, the actual pressure profile $P(t)$ can be readily calculated. For this purpose sensors made of constantan, whose electrical resistance piezofactor is much lower than that of manganin [55], are used.

Manganin pressure sensors have found wide use in shock-wave experiments. They are used to take measurements at low (down to 77 K) and elevated (up to 600 K) temperatures, in studies of the plastic properties of materials and of polymorphic phase transformations, in measurements of compression pulse evolution in reacting explosives, and to estimate dynamic loading parameters under explosive processing conditions [16, 55].

It should be noted that manganin is not the only piezoresistive material used as a pressure sensor. In the range of low pressures (up to 1.5–2.0 GPa) ytterbium sensors are used, whose piezosensitivity is considerably higher than that of manganin. The applicable pressure range of ytterbium sensors is limited by a polymorphic phase transformation that occurs at pressures of about 4 GPa [60].

3.3.3 Dielectric Pressure Sensors

The dielectric pressure sensor [61, 62] is a plane capacitor with the sensitive element being composed of thin metal coatings that have a dielectric film placed between them (Fig. 3.42). The dielectric sensor undergoes changes in its electrical capacitance C_0 with compression. It is assumed that the change in sensor capacitance is due to a change in film thickness only, with changes in dielectric constant and shock polarization processes being neglected. The sensor is positioned within the sample and connected to a high-resistance measuring circuit, so that the capacitor discharge time constant is much longer than the duration of the recorded process (Fig. 3.43).

An initial polarizing voltage U_0 (of sufficient magnitude that any contrubution due to shock polarization has no appreciable influence on the measure data) is delivered to the sensor. Under this high-resistance loading, the coating charge remains essentially unchanged, and changes in the potential difference across the coatings are proportional to the changes in the capacitance of the measuring capacitor.

Dielectric sensors are more sensitive than the manganin sensors and offer some advantages over the latter in lower pressure regimes. Pressure measurement with a dielectric sensor consists of recording the change in the sensor capacitance under shock compression. A sensor of capacitance C_0 (Fig. 3.43)

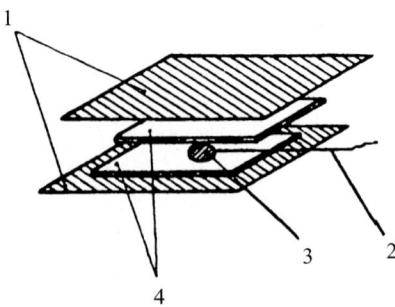

Fig. 3.42. Dielectric sensor arrangement: 1 – sample; 2 – lead; 3 – dielectric; 4 – electrode

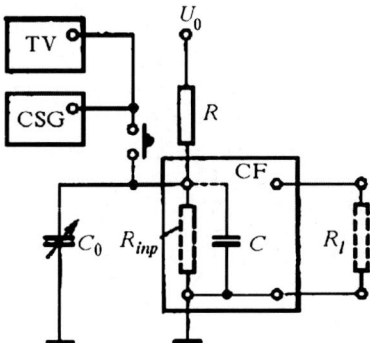

Fig. 3.43. Circuit for recording signal from the dielectric sensor with cathode follower (CF). CSG – calibration signal generator; TV – tube voltmeter

is charged from a constant voltage source U_0 through resistance R. Resistance R is chosen sufficiently high that the effect of recharging the C_0 capacitor is insignificant throughout the recording time. That is, $RC \gg t_r$, where C is total circuit capacitance (sensor capacitance + connecting cable capacitance + sensor lead measuring circuit input capacitance) and t_r is pressure recording time; $R \gg R_{inp}$.

During compression, the capacitor charge $Q = C_0 U_0 \approx CU$ (subscript "0" pertains to the unloaded state) remains essentially unchanged (providing no dielectric conductance), and the change in capacitance is due to dielectric compression which leads to a change in voltage ΔU across the sensor electrodes. The signal is recorded with a high-resistance input device, whose resistance R_{inp} also satisfies the condition $R_{inp} C_{inp} \gg t_r$.

Assuming that conditions allow, the dielectric sensor can be connected with a short cable segment through either cathode or emitter. This allows the signal to be transmitted through a cable that is consistent with the output circuitry to significant distances. If the cable segment is short but inconsistent with the recording circuit input resistance, cable transients can be neglected if the time of wave reverberation is much shorter than the characteristic time of the phenomenon to be recorded.

The change in the potential difference across the sensor electrodes is determined as:

$$\Delta U = -\int_0^P \frac{C_0 U_0}{C^2} \frac{dC}{dP} dP \ .$$

The amplitude of the sensor electrical signal is proportional to the polarizing voltage and inversely proportional to the added capacitance:

$$\Delta U = -\left(U_0 \frac{\Delta C_g}{C_0} \right) \left(1 + \frac{\Delta C_g}{C_0} \right)^{-1} \ .$$

To relate the changes in the relative sensor capacitance $\Delta C/C$ to pressure $P(t)$, calibration curves for each dielectric film material are constructed. The method for constructing the calibration curve is similar to that used for manganin sensors. The signal from the dielectric sensor increases linearly with increasing initial capacitance and stress irrespective of dielectric thickness.

A calibration curve of $\Delta U(P(t))$ form is constructed for the dielectric sensors by comparing the potential difference ΔU across their electrodes during compression with the parameters of load $P(t)$ either recorded simultaneously with another method or estimated computationally.

Figure 3.44 presents calibration curves for some dielectric sensor types. In each particular case, the polarizing voltage depends on the shock polarization.

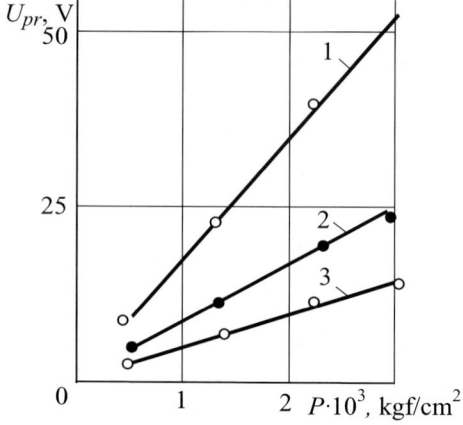

Fig. 3.44. Signal vs. pressure for the dielectric sensor with ether-cellulose film (1), organic glass (2), and mica (3). $U_0 = 700$ V

Reliable operation of the dielectric sensor is ensured under the condition that suppression of dielectric shock polarization is maintained. Shock polarization of polar dielectrics is caused by the rotation of dielectric dipoles under the action of the shock wave. Preliminary sensor dipole orientation is ensured by applying a polarizing voltage U_0 across the coatings of the capacitor to such an extent, that the dielectric is brought to "saturation" and the following shock wave results in no additional polarization. This is confirmed by the fact that a change in the polarizing voltage sign does not change the signal (modulus) at the same shock wave level.

Thus, changes in capacitance only depend only on changes in geometry (compression and expansion) and material dielectric permeability with changing pressure (in the absence of conductance). The results should be interpreted with certain care, since the high initial electrical field strength and its growth under compression can lead to a complete or partial dielectric breakdown and thus an inadequately recorded electric signal.

The time resolution and applicability of dielectric sensors is the same as that of the manganin sensors, since their design (using insulation for the measuring element) is essentially the same. The dielectric sensor has been used to record pressures in metals, plastics, composite and porous materials, glass in planes parallel and perpendicular to the shock front under axisymmetric loading for pressures up to 10 GPa [63].

An advantage of the dielectric sensor is the simplicity of its manufacture and operation, material accessibility, a high measured signal level (high sensitivity), and noise immunity. The major disadvantage of the dielectric sensor is that the behavior of the dielectric in loading and unloading is not completely understood. The error of pressure measurement with the dielectric sensors is no greater than 10%, which should be considered quite satisfactory, since none of the alternative methods developed to date offer a higher degree of accuracy. Improvement in our understanding of the of rheology of both sensor materials and reference materials under shock loading could enable a reduction in the above uncertainty figure by up to 2 to 5%.

3.3.4 Polarization Pressure Sensors

As a shock wave passes through a material, the material exhibits a change in the thermodynamic field variables. If the material possesses strength and viscosity, the material is compressed in a non-hydrostatic manner producing a stress tensor with varying values of dynamic yield strength. Anisotropy in the stress tensor can lead to anisotropies in physical characteristics of the material. For example, the shock wave can produce a spatial electronic charge redistribution or a shock polarization. In [64], the researchers showed how a shock wave moving through a polar dielectric within a capacitor produces E.M.F. in an electronic circuit containing the capacitor [4]. If there are no other E.M.F. sources and the arising E.M.F. is independent of coating material, the shock polarization effect can be attributed to dielectric polarization behind the shock front [64]. The shock polarization effect has been observed in a number of materials: ionic crystals, alloyed dielectrics (silicon, germanium), polycrystalline ferroelectrics and piezoelectrics, polar dielectrics (of the polymethylmethacrylate type), etc. [4].

The shock polarization effect has been used for the development of shock wave pressure sensors [4, 65, 66]. A threshold type sensor is based on the principle of a sign change in the initial polarization current at some pressure greater than some threshold value (Fig. 3.45). The sensor is a plane dielectric-filled capacitor. Threshold pressures for some alkali halide crystals are as follows: 5.1 ± 0.7 GPa for RbI; 6 ± 1.9 GPa for KI; 8.5 ± 0.5 GPa for KCl; 9.4 ± 1.6 GPa for KBr; 10.8 ± 0.8 GPa for NaCl; 12.5 ± 2.7 GPa for CsI; and 26.2 ± 7 GPa for NaF.

The polarization effect can be used to produce a marker for shock front arrival time. Use of this type of sensor (instead of a contact sensor) is discussed in [66]. The advantage of this sensor is its stability, and higher accuracy.

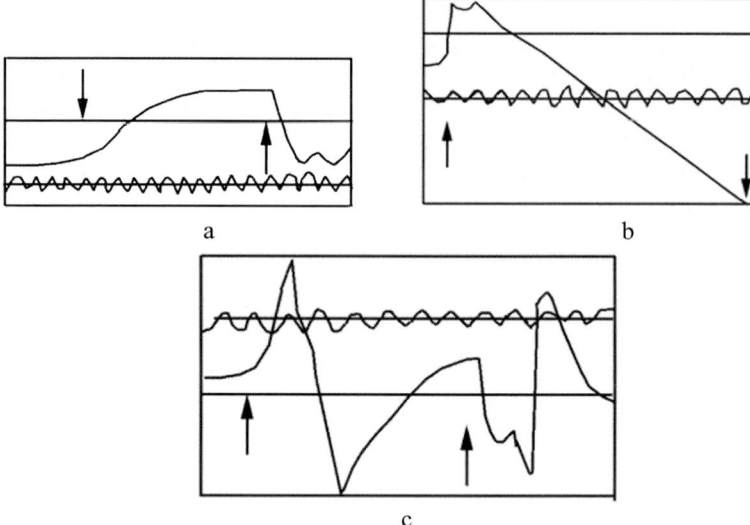

Fig. 3.45. Oscillograms of electric signals from threshold type electric sensors: **a** – polyamide resin, **b** – potassium bromide, **c** – dibutylphthalate (the arrows point to the times of shock wave coming into and going out of the sample)

3.4 Shadow Photorecording of Dynamic Processes

Imaging methods may applicable as diagnostics in a wide range of problems, yet their efficacy depends on such factors as the specific nature of the experiment, requirements for space and time resolution, and spectral sensitivity. Even in the simplest case, visualization of a process requires an *a priori* estimation of the range of certain parameters characterizing the process (velocity, characteristic sizes, self-luminescence and background luminescence). This estimate allows for the diagnostic setup scheme to be specified, which in turn, determines the requirements for the individual components of the dignostic scheme (for example, light source, size of the luminescence body, duration, intensity). In general, complex imaging methods supply information about the radiation field variables associated with the process under study, and this information enables a visulization of the process under the given experimental conditions.

3.4.1 Scheme of Shadow Photography in a Diverging Beam

In applications of this method, the silhouette image of an object illuminated by a homocentric light source is cast upon a base plate of photographic material. If the illuminated body is optically heterogeneous, beams of light will be deflected as they pass through the heterogeneity. If the refractive index of the body is n, the image of the body will be displaced on the base plate by a value proportional to the second derivative of n (the angular deviation is proportional to gradient n). A schematic of the method, which is applicable to investigations involving bodies of "medium" optical heterogeneity, is shown in (Fig. 3.46). This method is useful for providing quantitative measurements aimed at the calculation of a heterogeneity parameter distribution, and for providing qualitative visual assessments of gas-dynamic flows accompanying the object/effect under studied.

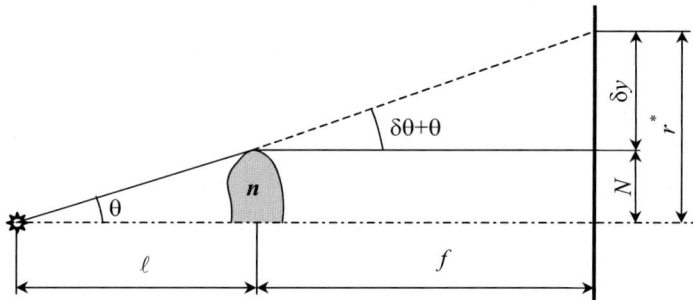

Fig. 3.46. Dvorzhak scheme of beam passage through optical heterogeneity

The geometry of the source–object–base plate configuration can be adjusted to suit the problem at hand. Irrespective of its physical state (plasma, solid, dielectric), the boundaries of the object under study are detected on the base plate through illumination re-distribution. If the gradient of the refractive index n, (i.e. the optical heterogeneity) is infinite, the illumination reaching the base plate changes only at the boundaries of image of the object, (all beams of light passing through the object are deflected by an angle $\delta\theta$). Assuming that $\delta\theta \to 0$ and $\tan(\theta + \delta\theta)\,\ell/r \approx 1$, the relative change in the base plate illumination is proportional to the ratio of the beam displacement δy to the shadow image size r^* as:

$$\frac{\delta_y}{r^*} = \frac{\tan(\theta + \delta\theta)}{r}\left(\frac{1}{f} + \frac{1}{\ell}\right)^{-1}.$$

Since $f + \ell = $ const, it follows that the maximum relative change in base plate illumination is achieved when the object is equidistant from the source and the base plate ($\ell = f$). The contrast in the image can be improved by increasing the distance between the source and the base plate ($\ell + f$).

Introduction of any additional optical element in the scheme reduces the quality of the image by producing severe distortions on the periphery of the image (clarity requires homocentricity of the light source). Physically, additional sources of light result in a change in the caustics: the point source is replaced by a source of finite (extended) size. Consider the simple case, where a parallel-sided plate δ_0 thick is placed between the beams of the point source and the base plate, the vertex of the caustic for beam refraction is displaced from the point source closer to the plate by distance $(n_{pl} - n_0)/(n_{pl})\delta_0$. In this case, it is the narrow beam in the vicinity of the normal beam ($\alpha = 0$) that remains homocentric, and with increasing angle the propagation difference between the parallel incident and refracted beams increases.

The geometric parameters that contribute to distortions when constructing the image of a body of finite size using the shadow method depend on the distance between the source and the body (Figs. 3.47 and 3.48).

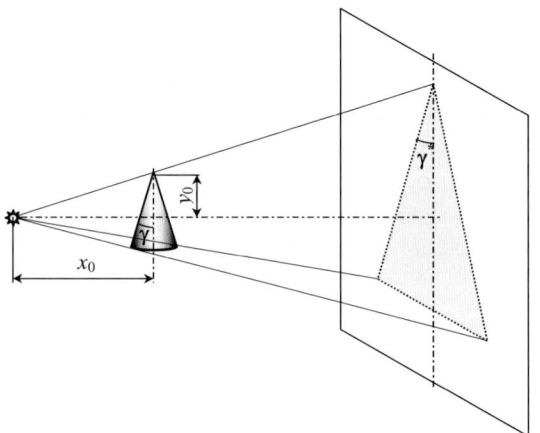

Fig. 3.47. Dvorzhak scheme for photographing a cone

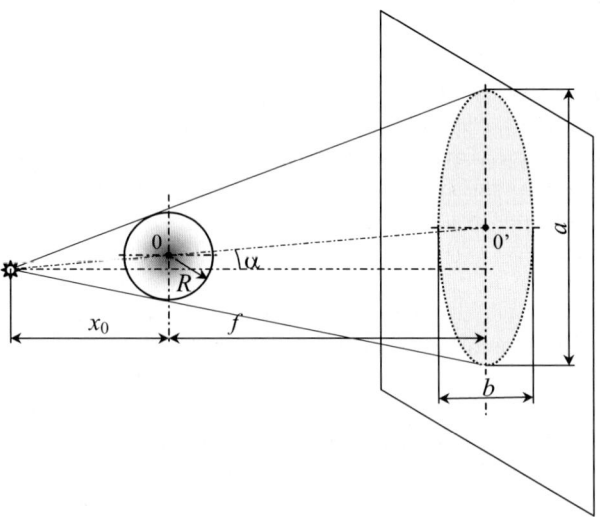

Fig. 3.48. Dvorzhak scheme for photographing a sphere

Item: a cone of γ half-angle.

Image: a triangle with half-angle at vertex: $\tan \gamma^* = \dfrac{\tan \gamma}{\sqrt{1 - \left(\frac{y_0}{x_0}\tan\gamma\right)^2}}$

Note that the cone axis is parallel to the base plate and intersects the beam normal to the base plate.

Item: a sphere of radius R.

Image: an ellipse with semiaxes a, b: $a = Rk\dfrac{\sqrt{1 - \frac{R^2}{x_0^2} + \tan^2\alpha}}{1 - \frac{R^2}{x_0^2}}$;

$b = Rk\sqrt{\dfrac{1 + \tan^2\alpha}{1 - \frac{R^2}{x_0^2} + \tan^2\alpha}}$, where $k = x_0 + f/x_0$ is optical magnification of the scheme.

A feature of the method is that because of Fresnel diffraction, irrespective of the physical parameters of the body, its shadowgraph boundary contains alternating zones of maximum and minimum illumination. On the base plate, the boundary of the shadowgraph is accompanied with a diffraction halo (first diffraction maximum illumination). The geometric shadow boundary is determined by the halo position: $\Delta_1 = 1.2(\lambda f/2k)^{1/2}$, where λ is the wavelength, f is the distance from the item to the base plate, and k is the optical magnification. The quality of the diffraction pattern depends on the "point" feature of the light source and the exposure time t_e. The requirements for the parameters are given using the method of estimation of gas density behind the wave front by measuring its shadow image [67].

3.4.2 Scheme of Shadow Photography in Studies of Solid Body Loading

The shadow photography scheme is useful for studying the processes that occur with the arrival of shock waves at a free surface [68]. For example, the appearance of a fine "dust" cloud in front of a flying shock-driven metal plate is an indication of free surface instability. The occurrence of this phenomenon is directly related to the degree of free surface roughness and can be experimentally recorded using streak-camera or pulse X-ray methods. In applying these methods, no direct images of escaping particles are produced. Instead, their presence is determined from the luminescence associated with the impact of these particles on an optical receiver occurring before the intense luminescence produced by the sample's impact (streak-camera method) or from the photographic density produced on X-ray film with known calibration (X-ray method).

The recording scheme appears in Fig. 3.49. An attribute of this method is that a wave reflected in the plate results in the generation of compression waves in air, which are recorded at the time of the source operation.

The processes captured in Fig. 3.49 were recorded using a short-duration point light source (1-mm luminescence body diameter, luminescence duration

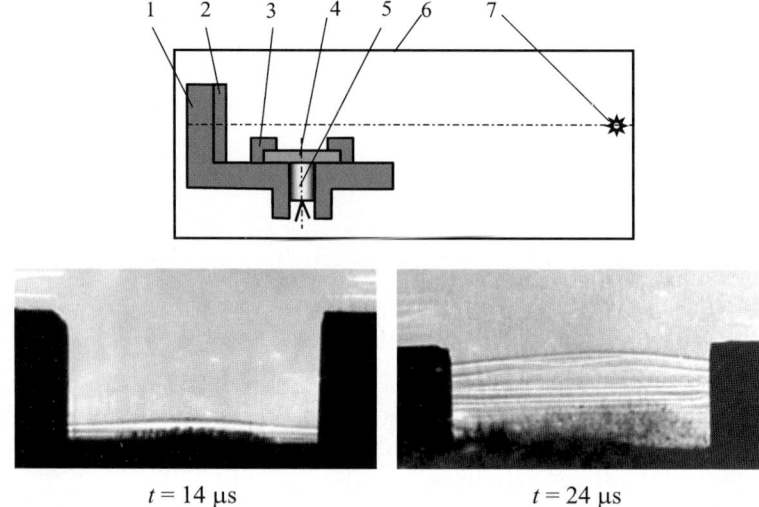

$t = 14\ \mu s$ $\qquad\qquad\qquad$ $t = 24\ \mu s$

Fig. 3.49. Scheme of the recording and photography of the shock wave release onto the free surface: 1 – base; 2 – photographic film T-42; 3 – pressure disk; 4 – plate to be studied; 5 – electric detonator; 6 – lightproof chamber; 7 – point light source

at a level of 0.35–0.5 μs). The diagnostic control set included: timing unit, generator G5-75, photomultiplier FEU-97, and frequency meter ChZ-34.

The surface roughness of the loaded steel plate was ∇80. The boundaries of the "dust" cloud (ejecta) can be seen in the photographs taken at different times t following the firing pulse. The obtained results are of interest for the development of numerical models aimed at predicting the formation of ejecta.

3.4.3 Scheme of Shadow Photography in Reflected Light Using a Camera

Subsection 3.4.1 notes that the introduction of even a parallel-sided plate to the scheme (for example, as a back window or shielding glass) affects the light source homocentricity (Fig. 3.50). In this case the incident beam caustic is the point source itself, while the refracted beam caustic is the curve with its peak at point S^*:

The distance SS^*, by which the beam refraction caustic is displaced toward the plate, is about one third the plate thickness.

On the whole, because of the effects of diffraction and aberrations in the light path optical element, the ability of the optical system to produce separate images of two points (or lines) that are close to one another is limited (resolution). In addition to distortion brought about by the rectifying glass, a variety of other factors can serve to reduce image quality: photographic film nonflatness; image shifts causing image blurring or even smearing; limitations

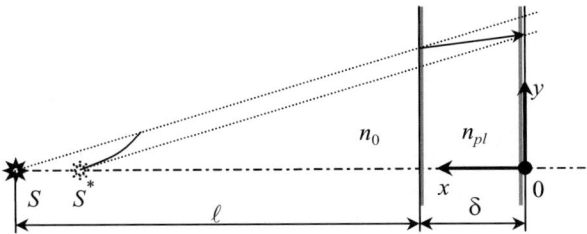

Fig. 3.50. Caustic displacement in a real scheme

in the imaging properties of photographic objectives, photochemical processing technology, etc.

The resolution of an optical system composed of i elements depends on that of each circuit element and is less than the worst of them, i.e. for an objective-photographic negative system, the resolution depends on that of the objective. For photographic cameras AFA-41 with objective MPO-2 used by VNIIEF in ballistic experiments, the resolution on the edge of type 42 photographic film ($R_f = 110$ mm^{-1}) is $R_\Sigma = 9.95$ mm^{-1}, which in terms of resolution determines the minimum image width, $\ell = 0.5 R_\Sigma^{-1} = 0.05$ mm, of an item component. Image resolution also depends on the photographic material, chemical-photographical processing technology and can be estimated experimentally by photographing in the test object space.

The image quality of a photograph also depends on the photographic image contrast or smear caused by displacement of the photographed item during its exposure. The difference in the optical densities between the image of the item and its surrounding background has to be, at a minimum, higher than the brightness-difference threshold of vision. Maximum optical density of a high-quality photographic negative should be 0.8–1.0; the minimum density should be higher than that of the fog density by 0.3. For type 42 film, the fog density is 0.18–0.2 and the minimum permissible photographic negative density is ≈0.5. Permissible smear Δ should not be larger than the sizes of the imaged item on the photographic negative: $\Delta \leq \lambda$, i.e. it should not degrade the total system resolution. For a given model speed this is achieved by an appropriate exposure time. Providing the closed aeroballistic complex is limited in width, the light source and the camera are positioned on one side of the model flight axis. In this case the shadowgraph on the reflecting base plate (Fig. 3.51) is simultaneous with the main image of the model. This scheme ensures the production of shadowgraphs of the body under conditions of noticeable ionization as well, where the model is invisible on the main image because of backgrounds. The base plate illumination from the point light source should not therewith be higher than the background.

Figure 3.52 presents a characteristic photograph from the experiment conducted at VNIIEF aeroballistic gallery with studied model velocity $W \approx$ 1.8 km/s.

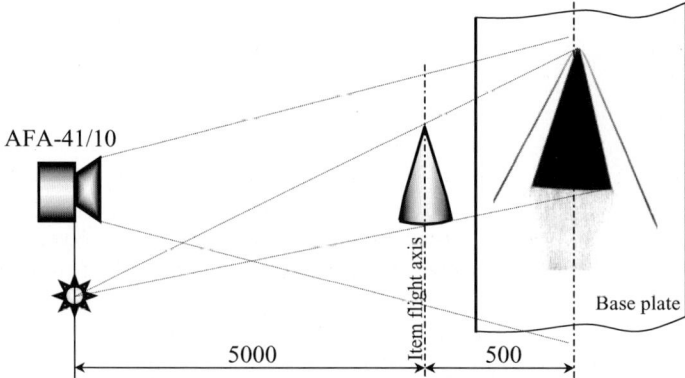

Fig. 3.51. Scheme of recording the item and its shadowgraph

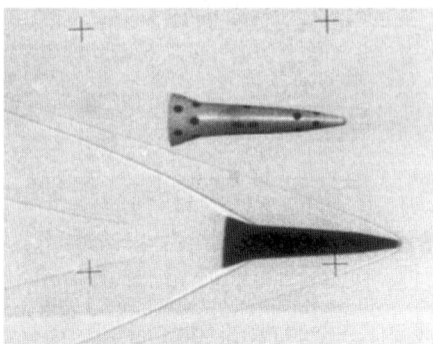

Fig. 3.52. Photograph of the axisymmetric body and its shadowgraph

3.4.4 Scheme of Shadow Photography in Passing Light with an Electrooptical Transducer

When recording processes that have an intense self-luminescence with the shadow method, it is necessary to limit the exposure time. During the exposure, the illumination of the base plate by the point light source used in construction of the image must not be greater than that of the background illumination produced by the body to be recorded before the base plate. There are two fundamental methods for limiting the exposure time: either by fast shot change or by using high-speed shutters. The electrooptical transducer (EOT), whose time resolution can be theoretically up to $\sim 10^{-12}$ s, is appropriate for using as the high-speed shutter. The simplest form of EOT is a semiconductor device operated on the principle of photoelectron emission, in which there is a receiving base, i.e. a cathode with a light-sensitive semitransparent layer sprayed onto its surface, and an output base, i.e. an anode with a luminescent coating. The image of the body under study is focused on the photocathode. In accordance with the distribution of image luminance, the incident light causes the emmision of electrons, which are accelerated in an electric field (several tens of kilovolts in the potential difference) and strike the luminescent base plate. Upon impact, the electron kinetic energy is converted into light. The luminescent base plates can be used to observe the electron image visually, which allows pre-adjustment of the device before the experiment. The photographic recording can be accomplished using a number of different methods, the simplest of which is to use roll film drawn by the light-sensitive layer to the base plate. A shortcoming of the method is the destruction of the EOT surface by photoemulsion-released material. This recording method is least sensitive.

Interference in the electrooptical focusing and the base plate self-resolution (20–40 line pairs per mm) dependent on the base plate luminophore grain size, grain distribution in layer, layer thickness, properties of binders, etc. are added to the total geometric resolution of negative material and objective (since the EOT as a photographic shutter is placed in the camera chain). At the time of the recording by the electrooptical photographing system its resolution (i.e. the dynamic one) decreases additionally by a factor of 1.5–2.0 and can be estimated in photographing a test object. For most systems with a 26-mm scene, 1.0–1.5 μs exposure, 1–10 gain factor, the resolution is no better than 16 line pairs per mm.

The luminescent base plate luminance is estimated as [69]:

$$L = k_1 k_2 \bar{I} U \alpha m_{bp} \ ,$$

where $k_1 = 0.641 \cdot 10^3$ lm/W; k_2 is the luminescent base plate efficiency (from 5 to 10%); $\bar{I}$ is the photocathode illumination intensity; α is the photocathode sensitivity ($\approx$50 μA/lm); m_{bp} is the ratio of the illuminated cathode area to the base plate image area; U is the potential difference between the EOT cathode and anode. That is, the EOT performs as a luminous flux amplifier ($L/\bar{I} > 1$). The electrooptical photographic system is composed, as a rule, of the following operational units: control board, objective, photographic film.

Figure 3.53 presents a complex diagram of optical-physical measurements using EOT base cameras. The shadowgraph is constructed on a semitransparent diffusing-surface base plate, the EOT cameras and object to be photographed are positioned on opposite sides of the base plate. Figure 3.54 gives the shadowgraph of detonating cord DSh-V EP expansion at $t = 500$ μs after the firing time produced in an EOT camera experiment with exposure time $\tau \approx 1$ μs.

3.4.5 Point Accumulative Radiation Source

In explosive and fragmentation experiments it is reasonable to use an expendable illumination source when performing the optical recording of the process under study. In this case the requirements for the performance of the whole circuit can be reduced to those for the source parameters. Depending on available devices (objective type, availability or unavailability of a fixed diaphragm,

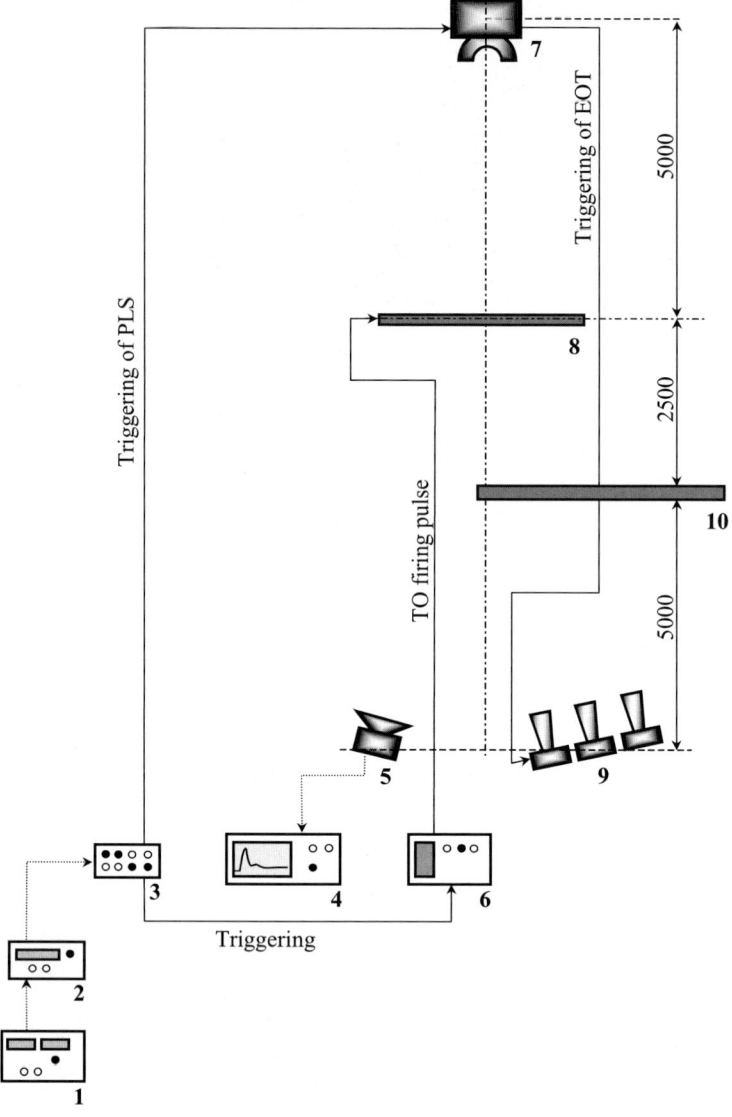

Fig. 3.53. Apparatus layout and control circuit in the shadow photorecording experiments: 1 – pulse generator G5-75; 2 – programmable timer "Volna"; 3 – signal switchboard; 4 – oscilloscope S9-8; 5 – photoreceiver; 6 – firing gear; 7 – pulsed light source (PLS); 8 – test object (TO); 9 – photographic cameras; 10 – shadow projection base plate

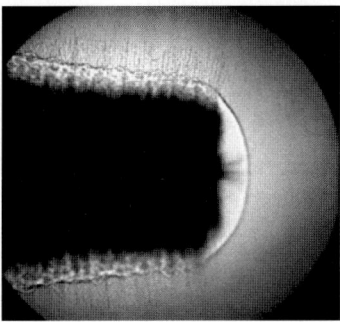

Fig. 3.54. The shadow photograph taken with EOT camera

etc.), infrastructure (fixed distance from the object to be photographed to the photogrammetric camera positions, scheme of the experiment), appropriate given intensity illumination durations and illumination zones can be ensured through selection and positioning of the expendable source.

The point accumulative radiation source (Fig. 3.55) is composed of a blasting cap (6), plastic explosive (5) designed to match the diameter of aluminum plate (4) to be launched, epoxy chamber (3) composed of a cylindrical segment for the plate acceleration and a hemispherical segment, outlet (2), translator (1). The translator is a quartz rod 2–3 mm in diameter, 40–50 mm in length. The outlet end of the translator is connected to the hole of a metal barrier of 1 mm diameter. The translator is inserted into a lid with threaded connection to the outlet channel. The translator is fixed onto the lid with epoxy compound, with the fixing widths making up less than 2% of the translator length. The distance ranged from 2 to 10 mm; accordingly, the extendable gauge length of the translator ranged from 8 to 0 mm. Before the experiment the camera and outlet channel were filled with xenon.

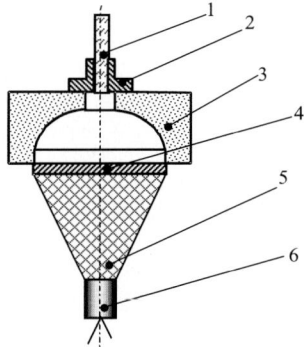
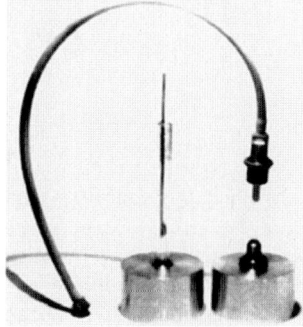

Fig. 3.55. Scheme of the point explosive light source

The dynamic experiments were configured with the translator located a distance of 5 mm from the camera pole to ensure a stable shock wave speed in xenon of 15 ± 1 km/s ($T = 80$–90 kK), and to meet the requirement for the pulse half-width to be less than 500 ns. The illumination characteristics (calibration) of the device were estimated using the method of comparison with similar parameters of pulsed light source IIS-8 developed by VNIIEF.

Dynamic tests (Fig. 3.56a) were used to assess the quality of pictures of flow fields around a flying bullet using a small-size explosive emitter (Fig. 3.56b). The explosive emitter was positioned at a distance twice longer than the design gas-discharge source was. The latch hook shooting with bullet (2) through cardboard cut-plate was positioned at a distance of 10 meters from base plate (1) and oriented so that the bullet would fly across the base plate at a distance of 10 cm from its surface. The light source was triggered by firing system (4), which was actuated by a signal issued by contact sensor (5) pierced by the bullet. Photographic film 0.5×0.5 m in size was positioned on the base plate. In the photograph (Fig. 3.57) made from the original negative one can clearly see the head and tail shock waves, rarefaction waves, turbulent trace, stagnation zone, and microfragments of the pierced cut-plate. The high resolution of the image is ensured by the luminous body pointwiseness (2 mm), short duration (200 ns), and high luminous intensity (up to 8.5 Mcd). These parameters enabled practical use of the emitter in VNIIEF production experiments aimed at studies of the nature of flow around a supersonic model passing through a dusty medium. Figure 3.58 presents the photograph made from the original negative of a flight of a cylindrical projectile 170 mm in length, 23 mm in diameter, with a conical front (30° angle) using the accumulative point light source.

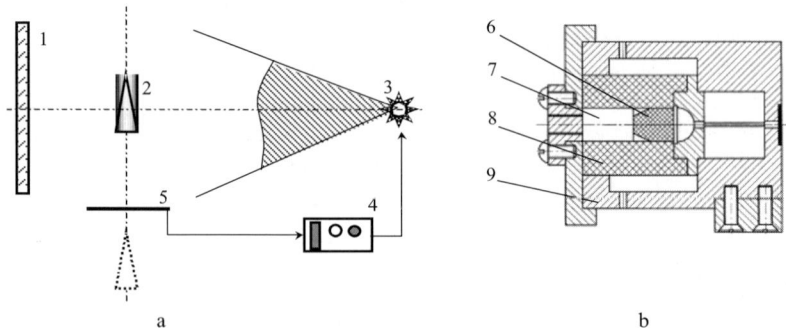

a   b

Fig. 3.56. Scheme of experiment (**a**) and explosive source design (**b**): 1 – base plate; 2 – bullet; 3 – explosive emitter; 4 – firing system; 5 – contact sensor; 6 – blasting cartridge; 7 – detonator; 8 – charge; 9 – casing

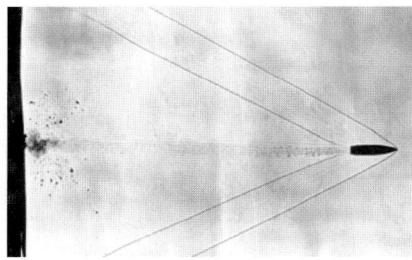

Fig. 3.57. Shadow photograph of the bullet that has pierced the barrier

Fig. 3.58. Axisymmetric body passage through dusty atmosphere

3.4.6 Explosive Thermochemical Light Source

For the solution of a number of experimental problems, a silhouette recording of the process on the background of a uniformly illuminated large-size base plate (of S more than 1 m^2) is required. The recording challenge reduces to one of providing appropriate exposures at the required speed of photographing with a given camera, and rapid tapering-off (in 1 or two frames) to uniform zone lighting.

Conventional attempts at resolving the technical challenges have included the use of argon glow lamps and a large space filled with inert gas, in which an emitted shock wave is produced by the detonation of a large-mass HE charge. Since shock wave velocities in this method (produced by 15–17 kg of condensed HE) are 8–10 mm/µs, the total required luminescence time is no greater than 500 µs for a three-meter vessel.

The problem can also be resolved by using several tens of grams of the same HE mixed with a material chosen to serve as an emitter. Performance of the explosive light sources (ELS), in which fine aluminum powder is combined with the HE [70], is described below.

The role of the HE reduces to the initiation of chemical reactions in Al powder and its relatively rapid de-excitation in comparison to lighting charges. As compared to the lighting charges the luminous intensity of these sources is higher due to larger emitting area that results from scatter of the burning particles. In its turn, the shell in the form of powder decreases air shock wave pressure near the charge. For the HE explosive cartridge, 30 g of an octogen base composition is used.

The above described scheme has been used in experiments requiring high-speed photography of objects moving against a large base plate background employing different modes of operation of a start-stop high-speed framing camera (0.5–4.0 ms filming time, 31.5–250.0 thousands of frames per second). Figures 3.59 through 3.61 present the records from these sources.

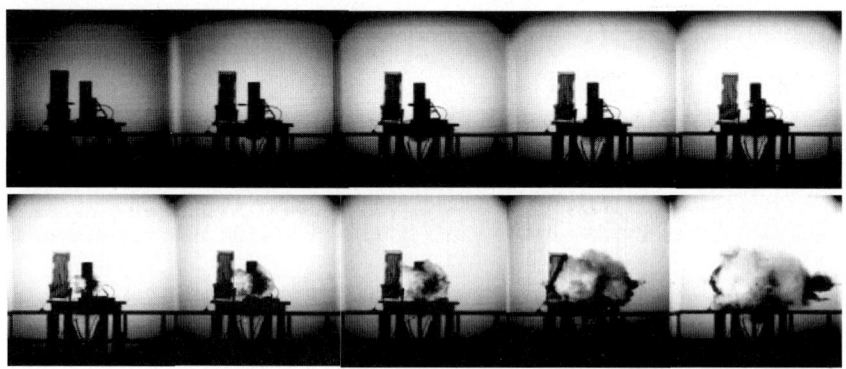

Fig. 3.59. Recorded piercing of an active barrier by a bullet (impactor caliber diameter = 12.5 mm, 2 explosive light sources: 30-g octogen, 15-g Al weight, the distance to the base plate is 1 m)

3.4.7 Gas-discharge Pulsed Light Source

Items moving at a high velocity (up to Mach 7) in conditions of the aeroballistic gallery are typically photographed using a pulsed light source (PLS) based on the gas-discharge xenon lamp IFK-500. This PLS has a number of shortcomings (long light pulse duration, low discharge temperature) that result in image quality that is not particularly high. An alternative PLS has been proposed (Fig. 3.62), whose principle of operation is based on an aperiodic discharge in air that is limited to one direction.

Figure 3.63 presents a photograph from an experiment (with model velocity $W = 1680$ m/s), where a source with discharge-confining lamp and a source with lamp IFK-500 were actuated sequentially within a single stereopost. Figure 3.64 presents photographs from experiments using the discharge-confining lamps at different model velocities.

The luminance body confinement allows us to obtain the main and shadow images simultaneously. When the model velocities are high ($W > 2$ km/s) in the conditions in which intense burning begins, the shadow picture provides visualization of the item tested.

Fig. 3.60. Recorded firing of the active barrier (barrier diameter $\emptyset = 80$ mm, HE charge $m \approx 1.2$ kg, 5 explosive light sources: 72-g hexogen, 12-g Al weight, the distance to the base plate is 1 m)

Fig. 3.61. Recording in the inverted experiment (barrier diameter $\emptyset = 150$ mm, $\alpha = 90°$, velocity $W = 800$ m/s, 2 explosive light sources: 30-g octogen, 10-g Al weight, the distance to the base plate is 1 m)

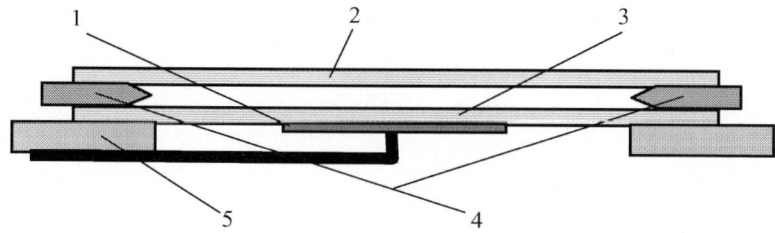

Fig. 3.62. PLS based on aperiodic discharge in air: 1 – trigger electrode; 2 – emitting glass; 3 – supporting glass; 4 – discharger electrodes; 5 – insulator

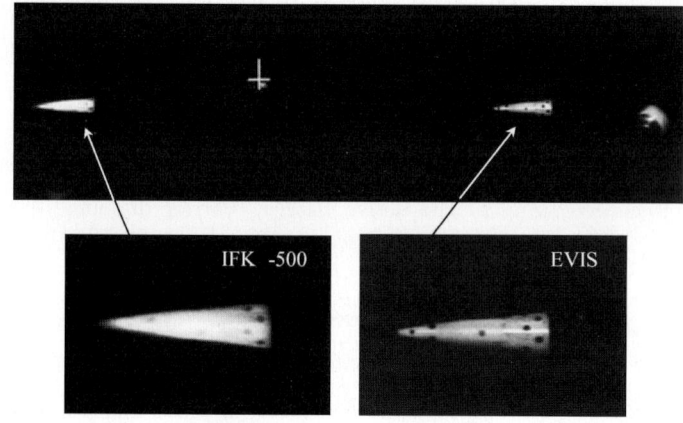

Fig. 3.63. Photographs taken during the experiment with different light sources

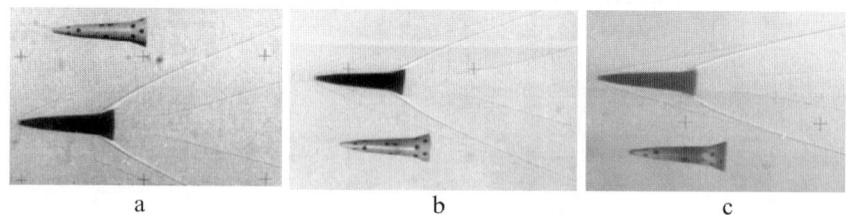

Fig. 3.64. Examples of using gas-dynamic emitters in ballistic tests: **a** – $W = 1094$ m/s; **b** – $W = 1132$ m/s; (**c**) $-W = 1835$ m/s

3.5 Pulsed X-ray Method

Pulsed X-ray radiation has found a wide utility in studies of various shock-wave processes. The idea of the pulsed X-ray method is that during the motion of an item in some medium, or during the process under study proceeding inside it, for example, an explosive process in high explosive or a shock process in material loading the item is exposed to a short, 10^{-8}–10^{-6}-s, X-ray pulse. Radiation that has passed through the item is recorded on X-ray film, thereby producing an X-ray image of the object at a specific point in time. Short radiation pulse duration ensures non-blurred pictures and acceptable measurement accuracy even when velocities of the boundaries to be recorded (shock front, cavity boundaries, etc.) are several hundreds or thousands of meters per second.

An advantage of the pulsed X-ray method is the possibility of viewing fast processes. In the X-raying no distortions are introduced to the material flows to be studied or characteristics to be measured, as there are no measuring elements, i.e. sensors, and different types of receivers interacting with the medium. Another advantage of the method is the possibility that it affords for recording the material state following the fronts of different-intensity shock waves, i.e. the states following the passage of peak pressures in the wave fronts, when other medium-interacting recording elements have been disintegrated. The same is true for the study of material state on repeated loading along with possible intermediate unloading states. Finally, the pulsed X-ray method offers a unique possibility unavailable from other techniques for examining the internal structure of the item from its X-ray image at any recorded time of the fast process.

3.5.1 Setting up X-ray Experiments

In X-ray methods, the principal components are: the pulsed X-ray radiation source, the item to be X-rayed, and the X-ray image recording system (Fig. 3.65). To implement the pulsed X-ray method, VNIIEF has developed and then improved special X-ray facilities. These include two classes of X-ray radiation generators: X-ray apparatuses with fine-focus X-ray tubes of ERIDAN type [71, 72] developed in the late 1940s and iron-free betatrons of BIM-234 type [73] developed in the mid 1960s.

Highly sensitive films with luminescent and metal intensifying screens are widely used for the recording of X-radiation. When the image can not be fitted onto one X-ray film, X-ray film packs with several (4, 9, 36) films are used.

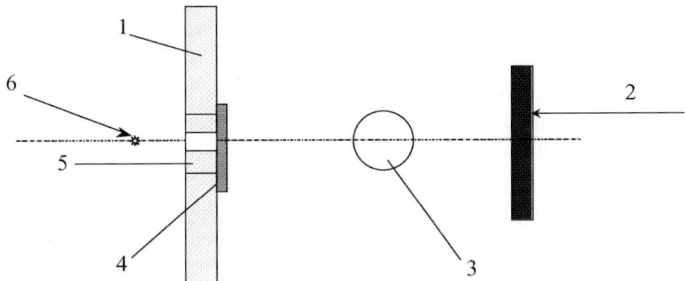

Fig. 3.65. A typical scheme of setting up the X-raying experiment: 1 – casemate armor; 2 – recording system; 3 – experimental assembly; 4 – Al base plate; 5 – collimator; 6 – γ-radiation source

X-ray imaging is performed with a diverging X-ray beam, therefore the size of the item (experimental assembly) image in the X-ray film base recording system is always greater than the size of the item itself. Hence the X-ray film plane must be perpendicular to the X-ray beam axis, otherwise the image magnification factor will be different for different film segments, and this will create additional difficulties in the resultant image processing.

VNIIEF has developed electron-optical recorders (EOR) [74] and recording systems based on electric-field-controlled sensitivity X-ray film [75] for special X-ray studies.

As a rule, X-ray radiation intensity is insufficient to produce a high-contrast image if the source of the radiation is far removed from the body being imaged. Consequently, the distance between the X-ray radiation source and the recording system is reduced to a few meters. In order to avoid damage to the X-ray source, it is surrounded by a protective structure, and the recording system is protected through the use of X-ray film packs with special protective devices for the EOT recorders and the electric-field-controlled sensitivity X-ray film positioned near the firing point.

In the experiment, only one image of the item (process) under study is recorded at a time. To study process evolution, multiple experiments must be conducted with images taken at other times. Comparison between the images so produced provides information about the dynamic field of the processes under study.

To acquire more information from one experiment, as early as the 1950s it was suggested that the X-raying should be made with several X-ray beams from different directions. In response to this suggestion, VNIIEF developed an eight-beam X-ray apparatus [76]. Since all eight beams operate independently, this allows time intervals to be varied over a wide range during the shot.

Another method to produce several X-ray images at different times has been proposed by developers of iron-free betatrons. The idea is that in the betatron, the electron beam can be released several times onto the target. It is at the time of the release of electrons onto the target that the betatron generates X-ray radiation. The basic difference between this method and the above-mentioned method is that here the item to be studied is X-rayed from one direction, rather than from eight different. As a result, X-ray images are superimposed on each other, which assists in their interpretation.

This situation persisted till the late 1980s, when VNIIEF developed a multi-frame X-ray image recording system GRIM [77] and some time later electrooptical X-ray image recorders (EOR) [78].

The principle of operation of the GRIM facility relies on the fact that the sensitivity of some X-ray film types becomes hundreds of times higher when exposed to a pulsed electric field.

The betatron generates either one long (~3 μs) or three short X-ray pulses, while the X-ray images are recorded on a few (from 3 to 10) one-type X-ray film packs positioned one after the other in the path of X-radiation. In each pack there is an X-ray film placed between two electrodes from which an electric field is applied. In all packs, the films are exposed by each X-ray pulse, but the image is recorded only onto the film which has the electric field pulse applied simultaneous with the corresponding X-ray pulse.

The electrooptical converter recording system uses γ-converters based on single crystals CsJ(Tl) or CsJ(Na) which ensures a higher recording sensitivity. EORs recording two and four fast processes have been developed [78]. The optical system records the process onto photographic film. PZS matrix base recorders have been developed which allow image numbering in units of luminance.

3.5.2 X-ray Image Processing

An important step in the pulsed X-ray method is processing of the experimental data. First these were visual methods, but beginning in the 1980s, digital processing methods came to be widely used. The idea of the latter is to digitize the experimental images and to use sophisticated computer algorithms for data processing.

The experimental results that are captured on the X-ray film are photometered and digitized, thereby creating a matrix (for example in units of film darkening) that is a digital representation of the recorded image. The photometry data array is input into a computer and processed by special computer programs.

A large complex computer program separately treats characteristic zones, object contours, shock and detonation wave fronts, and many others by certain indications using equidensity or gradient methods. Programs for image correction, such as the reconstruction/restoration of missing internal structures have been developed.

Processing methods using numerical simulations of the formation of an X-ray image have been developed in recent years [79]. Its idea here is that first the image formation system parameters are adjusted by a known image fragment and then one of hypotheses on the state of the item to be studied at the X-raying time is verified by numerical computation. If the image from the numerical simulation is the same as that recorded in the experiment, then the hypothesis of the item state at the X-raying time is accepted, otherwise the hypothesis is rejected and a new one is verified.

3.5.3 Directions of the Radiographic Studies

The pulsed X-ray method is widely used in studies of a great variety of physical phenomena and fast processes. X-ray radiation is of a high penetrating power and short duration; therefore it is most efficient to use it to study states in internal layers of material or elements behind a barrier during their motion or deformation, for example, in shock loading.

A major direction of X-ray studies is in HE detonation research. Shock initiation of HE, formation of detonation conditions and detonation wave propagation are studied.

An original method for EP studies was proposed and implemented by VNIIEF in the late 1940s. The experiments were conducted with charges that came to be known as "zebra" charges because lead foil layers were positioned inside the HE in such a way that a "striped" dark/light pattern is recorded in exposure to X-rays. Using this method, The EP field variables are estimated from the foil displacement relative to the detonation wave front. The process for setting up and analyzing the results from the detonation experiments are discussed in detail in Chap. 8.

The X-ray image presented in Fig. 3.66 was taken in an experiment aimed at studying the dispersion of metal shell fragments following the detonation of a cylindrical charge.

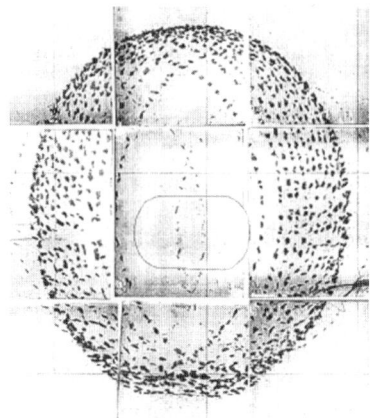

Fig. 3.66. Dispersion of a cylindrical charge metal shell fragments

The X-ray method has been used to study the behavior of soil under the action of HE charge explosion on or under the ground surface. Figure 3.67 presents an X-ray image produced by apparatus ERIDAN-3 at the time of HE charge explosion on the surface of the ground. The region of shock-compressed ground material, ground cavity, and the positions of copper wire in both unperturbed and shock-compressed soil are clearly seen.

Two-frame X-ray apparatus 2ERIDAN-3 records two phases of a fast process. Figure 3.68 presents the X-ray images of a shaped charge jet taken in one experiment on that apparatus.

As mentioned above, the principle of operation of apparatus GRIM is based on the fact that the sensitivity of some X-ray film types becomes hundreds times higher under the action of a pulsed electric field. This principle has enabled the acquisition of multiple-frame (up to 10) images of fast processes. Figure 3.69 presents three X-ray images obtained in one experiment at X-raying complex BIM234-GRIM in which the compression of a metal sphere under the action of HE explosion was studied.

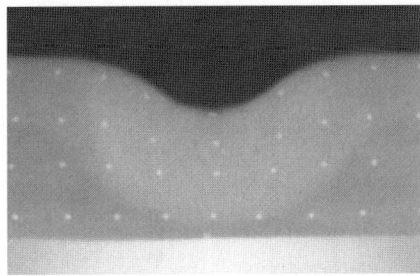

Fig. 3.67. Radiograph of HE charge explosion on the ground surface

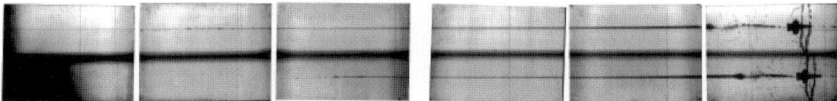

Fig. 3.68. Two phases of shaped jet. The jet length is ∼2.4 m; the time interval between the X-ray shots is 34 μs

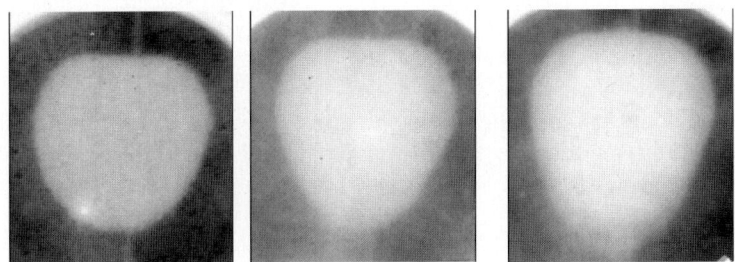

Fig. 3.69. Three phases of metal sphere compression under action of explosion

Investigation of internal structure and material density distribution requires meeting some additional conditions in setting up the experiment. The one-to-one relation between radiation energy striking the X-ray film and the X-ray film darkening density, on the one hand, and X-ray radiation flow attenuation with optical density of the X-rayed material, on the other hand, allows us to transfer from the darkening density on the X-ray image to the material density in the item under study. In [80,81] a demonstration is made for the use of X-ray images in a study of EP dispersion in a Π-shaped HE charge and EP density distribution.

The pulsed X-ray method has been successfully used to study quasi-isentropic compressibility of gaseous hydrogen and its isotope deuterium in the pressure range above 100 GPa. To achieve such pressures, the gas is pre-compressed to ∼25 MPa pressures in cylindrical-geometry devices. The schematic view of one of the devices appears in Fig. 3.70. Deuterium (6) is compressed by steel cylindrical shells (3) and (4) converging to the center under the force of explosion products from the cylindrical HE charge (5) fired by

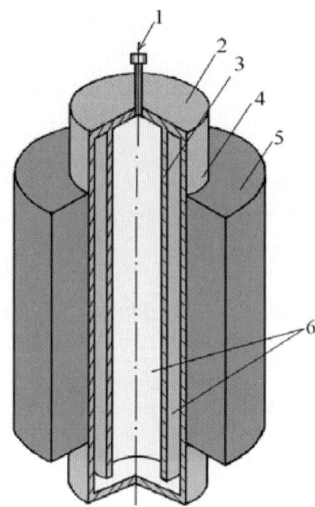

Fig. 3.70. Experimental device for studying gaseous deuterium compressibility: 1 – gas feeding tube; 2 – flange; 3 – internal shell; 4 – external shell; 5 – HE; 6 –gaseous deuterium

a special focusing system. The gas in the internal cavity is compressed through the gaseous deuterium layer, and the cavity is protected against direct HE action on material by shell (3), which prevents metal particles releasing into the gas cavity.

Two independent hard gamma radiation sources [73] with beams crossing at an angle of 135° (Fig. 3.71) are used to determine the time history of the steel shell motion and to determine the minimum gas compression radius.

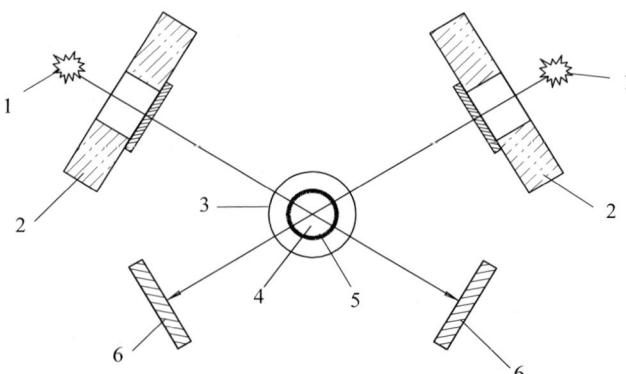

Fig. 3.71. Scheme of the experiment for measurement of gaseous deuterium compressibility: 1 – sources; 2 – shield; 3 – HE charge; 4 – gas; 5 – shell; 6 – recorders

In an experimental series, the $R(t)$ diagram of the motion of shell (5) is determined, from which the cavity size at the time of maximum gas compression ("shutdown" time) is estimated. Under the assumption of conservation of mass of compressed deuterium its average density for the cylindrical device is evaluated from the expression $\rho = \rho_0 (R_0/R_t)^2$, where R_0 is the shell size in the initial state, R_t is the shell size at the X-raying time, and ρ_0 is the initial material density. The thermodynamic field variables, i.e. compressed deuterium pressure P and temperature T, are estimated from gas-dynamic computations. X-ray images of the shells for the experimental device with 16-kg HE mass in the initial state and at the time of maximum compression are presented in Fig. 3.72. The initial state of the gas is as follows: $P_0 = 255$ atm, $T_0 = 241$ K, $\rho_0 = 0.0407$ g/cm^3. As seen from the radiographs, the technique allows satisfactory recording of the positions of the steel shells as they compress the gaseous deuterium.

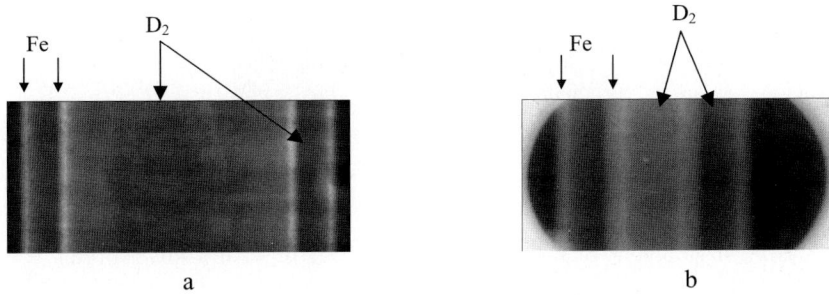

Fig. 3.72. Radiographs of the shells in the experimental device with 16-kg HE mass: in the initial state (**a**) and at the time of maximum deuterium compression (**b**)

One of the new directions in pulsed X-ray studies is in hydrodynamic instability research. Experimental and computational studies have been conducted on the instability of the free surface of a body (a sample of alloy AMg6) with the arrival of a shock wave. In applying this method, the initial stage of the instability development, i.e. incipient ejecta, has been studied. Figure 3.73 shows the experimental assembly sketch. On the surface of the AMg6 plate there are regular longitudinal grooves of different sizes.

Radiographs taken in three experiments under different loading conditions (Fig. 3.74) provide examples of perturbation growth. The studied plate images were read out in a special manner and subjected to digital processing in order to determine plate shapes and measure the perturbation amplitudes. Results of the experiments are presented in [82].

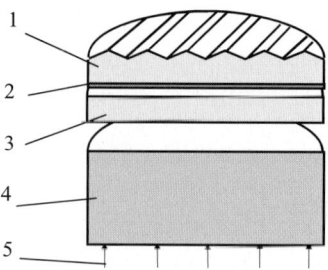

Fig. 3.73. The sketch of the experimental assembly: 1 – plate to be studied; 2 – buffer; 3 – impactor; 4 – HE; 5 – plane detonation wave

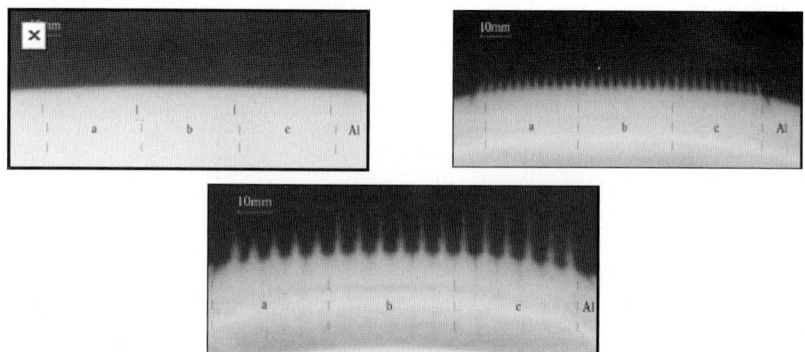

Fig. 3.74. Development of perturbations in the plate with a given profile of the grooves

References

1. Altshuler, L.V., "Use of Shock Waves in High-Pressure Physics," *Uspekhi Fizicheskikh Nauk*, Vol. 85, No. 2, 1965, pp. 197–258, [English trans., *Soviet Physics Uspekhi*, Vol. 8, No. 1, 1965, pp. 52–91].
2. Graham, R.A., and Asay, J.R., "Measurements of Wave Profiles in Shock-Loaded Solids," High Temperatures – High Pressures, Vol. 10, No. 4, 1978, pp. 355–390
3. Keeler, R.N., and Royce, E.B., "Shock Waves in Condensed Media," *in* Proc., International School of Physics "Enrico Fermi", Course XLVIII, Physics of High Energy Density, Caldirola, P., and Knoepfel, H., eds., Academic Press, New York, 1971, pp. 51–150, [Russian trans., Mir Publ., Moscow, 1974, pp. 60–170].
4. Mineev, V.N., and Ivanov, A.G., "Electromotive Force Produced by Shock Compression of a Substance," *Uspeki Fizicheskikh Nauk*, Vol. 119, No. 1, 1976, pp. 75–109, [English trans., *Soviet Physics Uspekhi*, Vol. 19, No. 5, 1976, pp. 400–419].
5. Glushak, B.L., Zharkov, A.P., Zhernokletov, M.V., Ternovoi, V.Ya., Filimonov, A.S., and Fortov, V.E., "Experimental Investigation of the Thermodynamics of Dense Plasmas Formed from Metals at High Energy Concentrations," *Zhurnal Eksperimentalnoi i Teoreticheskoi Fiziki*, Vol. 96, No. 4, 1989,

pp. 1301–1318, [English trans., *Soviet Physics JETP*, Vol. 69, No. 4, 1989, pp. 739–749].

6. Ashaev, V.K., Doronin, G.S., and Levin, A.D., "Detonation Front Structure in Condensed High Explosives," *Fizika Goreniya i Vzryva*, Vol. 24, No. 1, 1988, pp. 95–99, [English trans., *Combustion, Explosion, and Shock Waves*, Vol. 24, No. 1, 1988, pp. 88–92].

7. Gatilov, L.A., Ibragimov, R.A., and Kudashov, A.V., "Structure of a Detonation Wave in Cast TNT," *Fizika Goreniya i Vzryva*, Vol. 25, No. 2, 1989, pp. 82–84, [English trans., *Combustion, Explosion and Shock Waves*, Vol. 25, No. 2, 1989, pp. 206–208].

8. Trunin, R.F., (ed.), Properties of Condensed Materials at High Pressures, RFNC-VNIIEF, Sarov, Russia, 1992.

9. Bancroft, D., Peterson, E.L., and Minshall, S., "Polymorphism of Iron at High Pressure," *Journal of Applied Physics*, Vol. 27, No. 3, 1956, pp. 291–298.

10. Kurakin, N.I., Danilenko, V.V., Kozurek, N.P., et al., "Electrocontact Method for Recording $x - t$ Diagrams," *Khimicheskaya Fizika*, 1993, No. 5.

11. Poplavko, Yu.M., Physics of Dielectrics, Vyshcha Shkola Publ., Kiev, 1980.

12. Borisenok, V.A., Morozov, V.A., Novitsky, E.Z., et al., "Dynamic Compressibility of Single Crystal ADTGS and its Electrical Response to Shock Action," *Kristallografiya*, Vol. 37, No. 4, 1992, pp. 971–978.

13. Lee, L.M., Williams, W.D., Graham, R.A., and Bauer, F., "Studies of the Bauer Piezoelectric Polymer Gauge (PVF$_2$) Under Impact Loading," Shock Waves in Condensed Matter – 1985, Gupta, Y.M., ed., Plenum Press, NY, 1986, pp. 497–502.

14. Borisenok, V.A., Morosov, V.A., and Novitsky, E.Z., "PVDF as a Working Medium of Shock Wave Gauges," *proc., 10th Int. Conf. High Energy Rate Fabrication*, Lubljana, Yugoslavia, 1989, pp. 428–430.

15. Dubovik, A.S., Photographic Recording of Fast Processes, Nauka Publ., Moscow, 1964, p. 341.

16. Kanel, G.I., Razorenov, S.V., Utkin, A.V., and Fortov, V.E., Shock-Wave Phenomena in Condensed Matter, Yanus-K Publ., Moscow, 1996, [see also, Kanel, G.I., Razorenov, S.V., and Fortov, V.E., Shock-Wave Phenomena and the Properties of Condensed Matter, Springer-Verlag, New York, 2004].

17. Salamandra, G.D., Photographic Methods for the Study of Fast Processes, Nauka Publ., Moscow, 1974.

18. Danilenko, V.V., Kozeruk, N.P., and Telichko, I.V., "Fiber-Optic Sensors for Gas Dynamics Studies," *in* Fast HE Initiation. Peculiar Detonation Conditions, Tarzhanov, V.I., ed., RFNC-VNIITF, Snezhinsk, Russia, 1998, pp. 145–153.

19. Kozeruk, N.P., "Optical Signal Generation in Fiber-Optic Measurement Systems in Shock Phenomena Diagnostics," *in* Fast HE Initiation. Peculiar Detonation Conditions, Tarzhanov, V.I., ed., RFNC-VNIITF, Snezhinsk, Russia, 1998, pp. 154–166.

20. Danilenko, V.V., and Kozeruk, N.P., "On Errors in Time Interval Measurements with Streak Camera SFR-2M," *Zhurnal Nauchnoyi i Prikladnoy Fotografii i Kinematografii*, Vol. 34, No. 5, 1989, pp. 335–340.

21. Bolotov, A.A., and Chernyshev, V.K., "A Method for Producing Calibrated High Frequency Light Pulses for Time Scaling on Working Frame of Ultrahigh-Speed Photorecorders," *in* Filming Technology, Its Use in the Industry and Research. Collected Papers, No. 2, Moscow, 1966.

22. Bolotov, A.A., Lovyagin, B.M., Manulov, N.A., and Sakkeus, I.K., "50-Channel Light-Pulse Generator," *Pribory i Tekhnika Eksperimenta*, 1975, No. 3, pp. 198–200, [English trans., *Instruments and Experimental Techniques*, Vol. 18, No. 3, Part 2, 1975, pp. 909–911].

23. Bolotov, A.A., Lovyagin, B.M., and Ilyin, N.V., "Timer DV-2 to SFR type Photorecorder," *Zhurnal Nauchnoy i Prikladnoy Fotografii i Kinematografii*, 1977, No. 6, pp. 415–419.

24. Kholm, R., Electrical Contacts, Izdatelstvo Inostrannoy Literatury Publ., Moscow, 1961.

25. Ivanov, A.G., and Novikov, S.A., "Capacitive Data Transmitter Method for Recording the Instantaneous Velocity of Moving Surfaces," *Pribory i Tekhnika Eksperimenta*, 1963, No. 1, pp. 135–138, [English trans., *Instruments and Experimental Techniques*, 1963, No. 1, pp. 128–131].

26. Ivanov, A.G., Novikov, S.A., and Sinitsyn, V.A., "Investigation of Elastic-Plastic Waves in Explosively Loaded Iron and Steel," *Tverdogo Tela*, Vol. 5, No. 1, 1963, pp. 269–278, [English trans., *Soviet Physics – Solid State*, Vol. 5, No. 1, 1963, pp. 196–202].

27. Zaitsev, V.M., Pokhil, P.F., and Shvedov, K.K., "Electomegnetic Method for Measurement of Explosion Product Velocity," *Doklady Akademii Nauk SSSR*, Vol. 132, No. 6, 1960, pp. 1339–1340.

28. Dremin, A.N., Savrov, S.D., Trofimov, V.S., and Shvedov, K.K., Detonation Waves in Condensed Matter, Nauka Publ., Moscow, 1970, p. 169.

29. Zubarev, V.N., "The Motion of Explosion Products Behind the Front of a Detonation Wave," *Zhurnal Prikladnoi Mekhaniki i Tekhnicheskoi Fiziki*, 1965, No. 2, pp. 54–61, [English trans., *Journal of Applied Mechanics and Technical Physics*, 1965, No. 2, pp. 45–50].

30. Urtiew, P.A., Erickson, L.M., Hayes, B., and Parker, N.L., "Pressure and Particle Velocity Measurements in Solids Subjected to Dynamic Loading," *Fizika Goreniya i Vzryva*, Vol. 22, No. 5, 1986, pp. 113–126, [English trans., *Combustion, Explosion, and Shock Waves*, Vol. 22, No. 5, 1986, pp. 597–614].

31. Ko, J.F., "Improper Utilization of Electomagnetic Velocimeters in High Explosives," *Khimicheskaya Fizika*, 1995, No. 12, pp. 68–77.

32. Kheis, B., "A System for Nanosecond-Resolution Measurements of Material Particles in Shock and Detonation Waves," *Pribory dlya Nauchnykh Issledovaniya*, 1981, No. 4, pp. 92–102.

33. Altshuler, L.V., Pavlovskii, M.N., and Drakin, V.P., "Peculiarities of Phase Transitions in Compression and Rarefaction Shock Waves," *Zhurnal Eksperimentalnoi i Teoreticheskoi Fiziki*, Vol. 52, No. 2, 1967, pp. 400–408, [English trans., *Soviet Physics JETP*, Vol. 25, No. 2, 1967, pp. 260–265].

34. Fritz, J.N., and Morgan, J.A., "An Electromagnetic Technique for Measureing Material Velocity," *Review of Scientific Instruments*, Vol. 44, No. 2, 1973, pp. 215–221.

35. Novikov, S.A., Kashintsov, V.I., Fedotkin, A.S., Sinitsyn, V.A., Bodrenko, S.I., and Koltunov, O.I., "Measurement of the Velocities of Current-Conducting Shells with a Sensor of the Electromagnetic Type," *Fizika Goreniya i Vzryva*, Vol. 22, No. 1, 1986, pp. 71–74, [English trans., *Combustion, Explosion, and Shock Waves*, Vol. 22, No. 1, 1986, pp. 67–70].

36. Zhugin, Yu.N., and Krupnikov, K.K., "Induction Method of Continuous Recording of the Velocity of a Condensed Medium in Shock-Wave Processes," *Zhurnal*

Prikladnoi Mekhaniki i Tekhnicheskoi Fiziki, 1983, No. 1, pp. 102–108, [English trans., *Journal of Applied Mechanics and Technical Physics*, Vol. 24, No. 1, 1983, pp. 88–93].

37. Graham, R.A., Solids Under High-Pressure Shock Compression, Springer-Verlag, New York, 1993.
38. Astanin, V.V., Mineev, V.N., Obukhov, A.S., and Romanchenko, V.I., Electric Measurements of Shock Wave Parameters with Manometric Sensors: Preprint. Institute of Strength Problems, Kiev, 1985.
39. Novitskii, E.Z., Korotchenko, M.V., Volnyanskii, M.D., and Borisenok, V.A., "Investigation of the Dynamic Piezoelectric Moduli of Single Crystals of $Bi_{12}GeO_{20}$, Li_2GeO_3, and $LiNbO_3$," *Fizika Goreniya i Vzryva*, Vol. 16, No. 1, 1980, pp. 99–105, [English trans., *Combustion, Explosion, and Shock Waves*, Vol. 16, No. 1, 1980, pp. 93–98].
40. Graham, R.A., Neilson, F.W., and Benedick, W.B., "Piezoelectric Current from Shock-Loaded Quartz — A Submicrosecond Stress Gauge," *Journal of Applied Mechanics*, Vol. 36, No. 5, 1965, pp. 1775–1783.
41. Davison, L., and Graham, R.A., *Physics Reports*, Vol. 55, No. 4, 1979, pp. 255–379.
42. Bauer, F., "Ferroelectric Properties and Shock Response of a Poled PVF_2 Polymer and of VF_2/C_2F_3H Copolymers," Shock Waves in Condensed Matter – 1985, Gupta, Y.M., ed., Plenum Press, NY, 1986, pp. 483–496.
43. Bauer, F., "PVF_2 Polymers: Ferroelectric Polarization and Piezoelectric Properties Under Dynamic Pressure and Shock Wave Action," *Ferroelectrics*, Vol. 49, Nos. 1–4, 1983, pp. 231–240.
44. Lee, L.M., Williams, W. D., Graham, R.A., and Bauer, F., "Studies of the Bauer Piezoelectric Polymer Gauge (PVF_2) Under Impact Loading," Shock Waves in Condensed Matter – 1985, Gupta, Y.M., ed., Plenum Press, NY, 1986, pp. 497–502.
45. Graham, R.A., Bauer, F., and Anderson, M.V., "Properties of the Piezoelectric Polymer PVDF film under High Pressure Shock Compression," *in* Book of Abstracts, ISAF-90, 1990, p. 883.
46. Graham, R.A., Anderson, M.U., Bauer, F., and Setchell, R.E., "Piezoelectric Polarization of the Ferroelectric Polymer PVDF from 10 MPa to 10 GPa: Studies of Loading-Path Dependence," Shock Compression of Condensed Matter – 1991, Schmidt, S.C., Dick, R.D., Forbes, J.W., and Tasker, D.G., eds., Elsevier, Amsterdam, 1992, pp. 883–886.
47. Bauer, F., Graham, R.A., Anderson, M.U., Lefebvre, H., Lee, L.M., and Reed, R.P., "Response of the Piezoelectric Polymer PVDF to Shock Compression Greater than 10 GPa," Shock Compression of Condensed Matter – 1991, Schmidt, S.C., Dick, R.D., Forbes, J.W., and Tasker, D.G., eds., Elsevier, Amsterdam, 1992, pp. 887–890.
48. Reed, R.P., Graham, R.A., Moore, L.M., Lee, L.M., Fogelson, D.J., and Bauer, F., "The Sandia Standard for PVDF Shock Sensors," Shock Compression of Condensed Matter – 1989, Schmidt, S.C., Johnson, J.N., and Davison, L.W., eds., Elsevier, Amsterdam, 1990, pp. 825–828.
49. Bauer, F., "Properties of Ferroelectric Polymers Under High Pressure and Shock Loading," *Nuclear Instruments and Methods in Physics Research B*, Vol. 105, Nos. 1–4, 1995, pp. 212–216.
50. Bauer, F., "PVDF Gauge Piezoelectric Response Under Two-Stage light Gas Gun Impact Loading," Shock Compression of Condensed Matter – 2001,

Furnish, M.D., Thadhani, N.N., and Horie, Y., eds., AIP Press, Melville, NY, 2002, pp. 1149–1152.

51. Hodges, R.V., McCoy, L.E., and Toolson, J.R., "Polyvinylidene Floride (PVDF) Gauges for Measurement of Output Pressure of Small Ordnance Devices," *Propellants, Explosives, Pyrotechnics*, Vol. 25, No. 1, 2000, pp. 13–18.

52. Chartagnac, P., Decaso, P., Jimenez, B., Bouchu, M., Cavailler, C., Delaval, J., "Dynamic Behaviour of PVF_2 Gauges in the 0–600 kbar Range," Shock Compression of Condensed Matter – 1991, Schmidt, S.C., Dick, R.D., Forbes, J.W., and Tasker, D.G., eds., Elsevier, Amsterdam, 1992, pp. 893–896.

53. Fuller, P.J.A., and Price, J.H., "Electrical Conductivity of Manganin and Iron at High Pressures," *Nature*, Vol. 193, No. 4812, 1962, pp. 262–263.

54. Khristoforov, B.D., Goller, E.E., Sidorin, A.Ya., and Livshits, L.D., "Manganin Probe for Measuring Shock Pressures in Solids," *Fizika Goreniya i Vzryva*, 1971, No. 4, pp. 613–615, [English trans., *Combustion, Explosion, and Shock Waves*, Vol. 7, No. 4, 1971, pp. 525–527.

55. Kanel, G.I., "Using Manganin Sensors for Measurement of Condensed Matter Shock Compression Pressure, VINITI, N 477–74 Dep. 1974.

56. Dremin, A.N., and Kanel, G.I., "Compression and Rarefaction Waves in Shock-Compressed Metals," *Zhurnal Prikladnoi Mekhaniki i Tekhnicheskoi Fiziki*, 1976, No. 2, pp. 146–153, [English trans., *Journal of Applied Mechanics and Technical Physics*, Vol. 17, No. 2, 1976, pp. 263–267.

57. Batkov, Yu.V., Novikov, S.A., Sinitsyna, L.M., and Chernov, A.V., "Study of Shear Stresses in Metals at a Shock Front," *Problemy Prochnosti*, 1981, No. 5, pp. 56–59 [English trans., *Strength of Materials*, Vol. 13, No. 5, 1981, pp. 601–605].

58. Lyle, J.W., Schriever, R.L., and McMillan, A.R., "Dynamic Piezoresistive Coefficient of Manganin to 392 kbar," *Journal of Apllied Mechanics*, Vol. 40, No. 11, pp. 4663–4664.

59. Kanel, G.I., Vakhitova, G.G., and Dremin, A.N., "Metrological Characteristics of Manganin Pressure Pickups Under Conditions of Shock Compression and Unloading," *Fizika Goreniya i Vzryva*, Vol. 14, No. 2, 1978, pp. 130–135, [English trans., *Combustion, Explosion, and Shock Waves*, Vol. 14, No. 2, 1978, pp. 244–248.

60. Grady, D.E., and Ginsberg, M.J., "Piezoresistive Effects in Ytterbium Stress Transducers," *Journal of Applied Physics*, Vol. 48, No. 6, 1977, pp. 2179–2181.

61. Fot, N.A., Alekseevskii, V.P., and Yarosh, V.V., "Dielectric Pulsed-Pressure Pickup," *Pribory i Tekhnika Eksperimenta*, 1973, No. 2, pp. 199–201, [English trans., *Instruments and Experimental Techniques*, Vol. 16, No. 2, Part 2, 1973, pp. 567–569.

62. Stepanov, G.V., Elastic-Plastic Material Deformation Under Action of Pulsed Loads, Naukova Dumka Publ., Kiev, 1979.

63. Batkov, Yu.V., Novikov, S.A., Permyakov, V.V., and Chernov, A.V., "Peculiarities in Measurement of Pressure Pulses with a Dielectric Sensor," *Zhurnal Prikladnoi Mekhaniki i Tekhnicheskoi Fiziki*, 1981, No. 2, pp. 103–105, [English trans., *Journal of Applied Mechanics and Technical Physics*, Vol. 22, No. 2, 1981, pp. 227–228].

64. Tyunyaev, Yu.N., Mineev, V.N., and Lisitsyn, Yu.V., "Threshold Type Polarization Sensor for Pulsed Pressure Measurement," proc., 1st *All-Union Symposium on Pulsed Pressures*, VNIIFTRI, Moscow, 1974, pp. 53–56.

65. Ivanov, A.G., Lisitsyn, Yu.V., and Novitskii, E.Z., "Polarization of Dielectrics Under Shock Load," *Zhurnal Eksperimentalnoi i Teoreticheskoi Fiziki*, Vol. 54, No. 1, 1968, pp. 285–291, [English trans., *Soviet Physics JETP*, Vol. 27, No. 1, 1968, pp. 153–155].

66. Lebedev, N.N., Model, I.Sh., and Kuznetsov, F.O., "Recording of the Velocity of High-Intensity Shock Waves with Piezoelectric Transducers," *Pribory i Tekhnika Eksperimenta*, 1968, No. 3, pp. 183–185, [English trans., *Instruments and Experimental Techniques*, 1968, No. 3, pp. 696–698].

67. Semenov, A.N., "Simple Optical Methods for Supersonic Flow Study," *in* Aerophysical Studies of Supersonic Flows, Nauka Publ., Leningrad, 1967.

68. Gerasimov, S.I., and Kholin, S.A., "Optical recording of Precesses Associated with Shock Wave Release to the Plate Free Surface," *Voprosy Atomnoi Nauki i Tekhniki. Seriya: Teoreticheskaya i Prikladnaya Fizika*, 2000, No. 2–3, pp. 21–23.

69. Folkart, K., "Spark Light Sources and High-Frequency Spart Cinematography," *in* Physics of Fast Processes, Vol. 1, Zlatin, N.A., ed., Mir Publ., Moscow, 1971.

70. Gerasimov, S.I., Faikov, Yu.I., and Kholin, S.A., Accumulative Light Sources, RFNC-VNIIEF, Sarov, Russia, 2002.

71. Toner, G., "Pulsed X-Ray Engineering," in Physics of Fast Processes, Vol. 1, Zlatin, N.A., ed., Mir Publ., Moscow, 1971, pp. 336–381.

72. Ziuzin, V.P., Manakova, M.A., and Tsukerman, V.A., "Sealed Sharp-Focusing Pulse X-Ray Tubes," *Pribory i Tekhnika Eksperimenta*, 1958, No. 1, pp. 84–87, [English trans., *Instruments and Experimental Techniques*, 1958, No. 1, pp. 92–95],

73. Pavlovskii, A.I., Kuleshov, G.D., Sklizkov, G.V., Zysin, Yu.A., and Gerasimov, A.I., "High-Current Ironless Betatrons," *Doklady Akademii Nauk SSSR*, Vol. 160, No. 1, 1965, pp. 68–70, [English trans., *Soviet Physics – Doklady*, Vol. 10, no. 1, 1965, pp. 30–32].

74. Butslov, M.M., Stepanov, B.M., and Fanchenko, S.D., Electrooptical Converters and Their use in Scientific Research, Nauka Publ., Moscow, 1978.

75. Kovtun, A.D., and Makarov, Yu.M., Pulsed X-Ray Method, USSR Inventors Certificate N 519667. MKI G 03 B 42/02, Bulletin of Inventions, 1976, N 24.

76. Tsukerman, V.A., and Manakova, M.A., "Sources of Short X-Ray Pulses for Investigating Fast Processes," *Zhurnal Tekhnicheskoi Fiziki*, Vol. 27, No. 2, 1957, pp. 391–403, [English trans., *Soviet Physics – Technical Physics*, Vol. 2, No. 2, 1957, pp. 353–363].

77. Kovtun, A.D., Belyaev, G.K., Makarov, Yu.M., Motornov, A.P., Nikonov, N.A., and Pavlunin, A.N., "Multi-Frame Recording of High-Speed Precesses Using Single X-Ray Source," proc., 22nd Int. Congress on High-Speed Photography and Photonics, Santa Fe, NM, 1996, pp. 900–902.

78. Burtsev, V.V., Yelfimov, S.E., Makarov, Yu.M., Ryzhkov, A.V., "Four-Channel Module Electrooptical X-Ray Image Recorder ChINARA," proc., 16th Scientific Technical Conf. High-Speed Photographing, Photonics, and Metrology of Fast Processes, Moscow, 1993, p. 32.

79. Tolstikova, L.A., and Kovtun, A.D., "Material Density Estimation by X-Ray Image Using the Apparatus of X-Raying Numerical Simulation," *in* Advanced Methods for Designing and Refinement of Ordnance Devices, RFNC-VNIIEF, Sarov, Russia, 2000, pp. 281–286.

80. Batkov, Yu.V., Kovtun, A.D., Novikov, S.A., Skokov, V.I., and Tolstikova, L.A., "Mechanism of Formation of a Fast Gas Jet," *Fizika Goreniya i Vzryva,* Vol.

37, No. 5, 2001, pp. 98–103, [English trans., *Combustion, Explosion, and Shock Waves*,Vol. 37, No. 5, 2001, pp. 580–584].

81. Komrachkov, V.A., and Panov, K.N., "Research into the Effect of Loading Pressure on Material Density Distribution Following Initiating Shock Wave Front in Octogen Base HE," *proc., 3rd Int. Conf. Khariton Scientific Lectures*, RFNC-VNIIEF, Sarov, Russia, 2001, pp. 70–75.

82. Lebedev, A.I., Igonin, V.V., Nizovtsev, P.N., et al., "Study of Soild Free Surface Instability Under Shock Effect," Trudy, Vol. 1, RFNC-VNIIEF, Sarov, Russia, 2001, pp. 590–597.

4

Determination of Hugoniots
and Expansion Isentropes

M.V. Zhernokletov

4.1 Determination of Hugoniots

Equations (1.21) or (1.26), expressing the laws of conservation of mass, momentum, and energy at a shock discontinuity reveal that for known initial conditions, the problem of finding all five unknown shock front field variables (P, V, D, u, E) reduces to experimentally recording any two of them. These two are subsequently sufficient for construction of the Hugoniot. Straightforward and accurate measurements of compressed condensed matter pressure and density are difficult to perform. The methods most commonly applied, therefore, involve construction of the Hugoniot in terms of the kinematic parameters D and u.

In dynamic experiments, the simplest method for estimating the steady shock velocity D involves the use of electrocontact or optical methods. Sensors positioned in the path of shock propagation record the time of appearance of electrical or optical signals and a known distance between the sensors is used to calculate the average shock velocity taken as prompt.

The second kinematic parameter (material mass velocity, u, following the shock front), is quite difficult to measure. The mass velocity can be recorded directly, for example, with the electromagnetic method discussed in Chap. 3. However, this method has limitations in terms of pressure amplitudes and is inapplicable for metals or when any material becomes a good conductor.

Hence, indirect methods using general laws appearing in the Riemann problem are involved to estimate the mass velocity.

4.1.1 Spallation Method

The idea of early experiments [1] that measured condensed matter Hugoniots is as follows. A thin plate made of material under study is attached to the end of a fairly long HE charge. When the detonation wave has reached the boundary, the Riemann problem takes place resulting in a shock wave propagating at velocity D through the material under study. Either a reflected shock wave

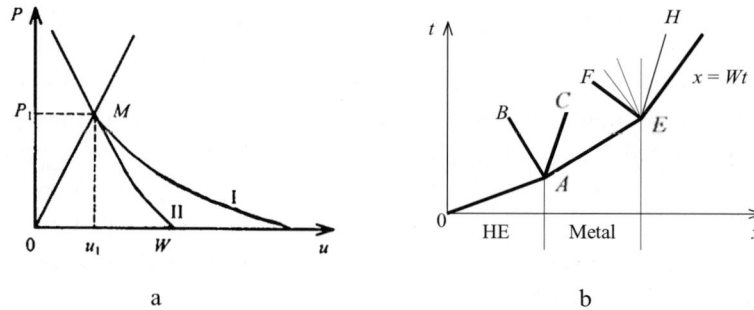

a   b

Fig. 4.1. Spallation method: (**a**) $P - u$ diagram. I – explosion product deceleration curve; II – compressed metal expansion isentrope; $0M$ – wave beam; $P = \rho_0 Du$; (**b**) $x - t$ diagram; $0A$ – detonation wave in HE; AE – shock wave in barrier; AB – reflected shock wave or the head of expansion wave in EP; AC – interface; EFH – expansion wave

or an expansion wave, depending on material stiffness, propagates in the HE; the HE-barrier interface moves at velocity u_1 equal to the compressed material mass velocity. The $P - u$ and $x - t$ diagrams taking place in this experiment setup are plotted in Fig. 4.1

The next Riemann problem occurs, when the shock wave reaches the free surface of the barrier. The expansion wave will run through the compressed material. The free surface velocity will be estimated by the relation

$$W = u_1 + \Delta u \ ,$$

where Δu is the material velocity gain in the expansion wave from state P_1 to state $P_0 \approx 0$.

In principle, to estimate the material velocity increment in the expansion wave, it is necessary to use the functional form of the equation of state. However, the rule of doubled velocity, $W = 2u_1$, has been proven valid within a good accuracy (1–2%) for many metals at shock pressures up to 50 GPa. In other words, the material velocity gain during rarefaction is equal to mass velocity in shock wave. This approximation is comprehensively studied in [2,3], which calculate the isentropes thoroughly using the Mie-Grüneisen equation of state and evaluate pairs D and Δu from D and W by sequential iteration.

First, in approximation $\Delta u = (1/2)W$, the experimental curve $D = f(W)$ was transformed to the appropriate curve $P - V$ through direct application of the Rankine-Hugoniot equations. Then the isentrope was calculated by the Mie-Grüneisen equation and Δu was evaluated. The estimated u_1 and calculated Δu were used to determine correction curve $W = f(W/u_1)$ that was used for subsequent approximations. One or two iterations suffice to ensure convergence to Δu.

The calculations conducted for a number of solids and shock pressures in the range up to 50 GPa show that ratio $\Delta u/u_1$ ranges from 1.0 to 1.03.

Zeldovich and Raizer [4] found the doubled-velocity rule from the general equations for the shock wave and rarefaction wave when considering arrival of a not very strong shock wave at the free surface. In a small-amplitude shock wave the entropy change and compression are slight, and volume increments can be represented as $V_1 - V_0 = (\partial V/\partial P)_S P_1$, where S is entropy prior to compression. Then mass velocity in the shock front is $u_1 = \sqrt{P_1 (V_0 - V_1)} \approx -\left(\frac{\partial V}{\partial P}\right)_S^2 P_1 = \frac{P_1}{\rho_0 c_0}$. In the same approximation the compressibility change can be neglected, that is derivative $(\partial V/\partial P)_S$ can be taken as constant. In view of Eq. (1.34), we find that

$$\Delta u = \int_0^{P_1} \left(-\frac{\partial V}{\partial P}\right)^{\frac{1}{2}} dP \approx \left(-\frac{\partial V}{\partial P}\right)_S^{\frac{1}{2}} P_1 = \frac{P_1}{\rho_0 c_0} = u_1 .$$

The experimental verification in [5] shows that the doubled velocity rule, $W = 2u_1$, is valid for lead up to $u_1 = 0.5$ km/s (wave front pressure 15 GPa), for iron and copper up to $u_1 \approx 1.5$ km/s (pressure ≈ 80 GPa), for aluminum and molybdenum up to $u_1 \approx 2.5$ km/s (pressure 60 and 210 GPa, respectively).

Thus, in this method the experimental measurements reduce to estimating the wave velocity D and material free surface velocity W. A set of the D and W pairs is obtained by varying the HE charge length or power.

Plate motion under the action of EP is usually attended by plate fragmentation. For this reason this measurement method is often called the "spallation" method.

With increasing pressure a deviation from the doubling rule occurs, which reduces the accuracy of the estimated mass velocity because of the need to introduce computational corrections. The spallation method is approximate, and as a consequence has not found wide acceptance (as reflected through the published literature) by Soviet investigators. Judging from the American literature, however, [2,3] the method is basic in experiments aimed at obtaining an estimation of metal Hugoniots using explosive shock-wave generators.

4.1.2 Deceleration Method

A direct method for the estimation of velocity u is that developed in the late 1940s in the Soviet Union by the authors of [1]. The deceleration method involves measurement of the velocity W_{imp} of an impactor prior to its impact on the target under study, and of the shock wave velocity D in the target.

Consider the process of an impactor impacting a target using the $x - t$ and $P - u$ diagrams presented in Fig. 4.2. For the impactor and the target fabricated from the same material, the impactor deceleration against the barrier produces two waves of equal intensity (AB and AE in the $x - t$ diagram) propagating on either side of the impact surface. By virtue of the impact symmetry, $u = W_{imp}/2$. The interface itself (line AC) has the same velocity. The other parameters of the shock wave in the target are found by substitution of u and wave velocity D into conservation law Eqs. (1.21).

By varying impactor velocity W_{imp} and taking appropriate measurements of wave velocities in the target, one obtains a set of points that determine the position of the single compression Hugoniot of the reference material under study.

When the impactor and target are made of different materials, the equality of the mass velocity in the target and the impactor does not hold. This difficulty is overcome by fabrication of the impactor from a material with a known Hugoniot. In this case the mass velocity estimation is based on the laws obeyed by the Riemann problem at the impactor-target interface.

Figure 4.2b depicts the impactor deceleration Hugoniot as curve AI going out of initial state $P = 0$, $u = W_{imp}$.

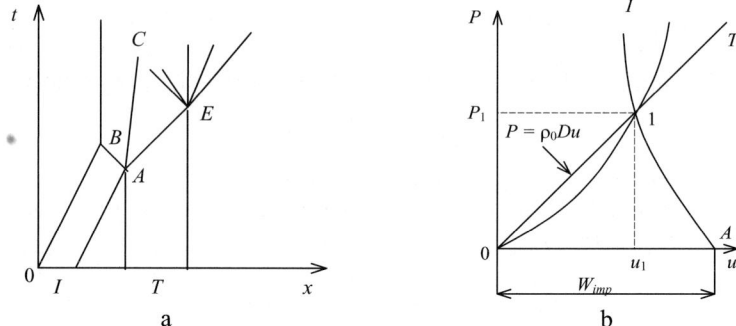

Fig. 4.2. Deceleration method: (a) $x - t$ diagram; (b) $P - u$ diagram

Initial atmospheric pressure P_0 is neglected. The measured wave velocity in the target determines the slope of wave beam $0T$, which satisfies the equation $P = \rho_0 Du$ (ρ_0 is initial target material density). Wave beam-impactor deceleration curve intersection point 1 determines the desired pressure P_1 and mass velocity u_1 in the material under study. These values, D and u_1, and equations of conservation are then used to find the other compression parameters. By analogy with the previous case, by changing the impactor velocity and measuring the wave velocity in the material under study, we obtain a number of points on its Hugoniot.

In principle, the deceleration method is applicable for waves of intensities as high as is wished. However, with increasing W_{imp}, the application of the method is complicated by technical difficulties in maintaining relatively modest heating of the sample and highly symmetric acceleration of the impactor.

The deceleration method has been used to determine iron and aluminum Hugoniots when steel impactors were accelerated to velocities up to $W_{imp} = 18$ km/s. Aluminum and iron are basic materials used as reference materials in the method discussed in the following section.

4.1.3 Reflection Method

The reflection method developed by the authors of [6] has found wide application in studies of shock compressibility of various materials. In the foreign literature it is referred to as the impedance-matched method. In this method, the shock wave is translated into the material under study through a base plate fabricated from a material of a known equation of state or, at least, of a known Hugoniot.

If the shock wave parameters in the base plate are known, it is sufficient to measure the shock-wave velocity in the material under study. The wave parameters in the base plate are estimated either in a preliminary experimental series or in the experiment, in which D in the sample under study is recorded.

Figure 4.3a illustrates the shock wave transition from the base plate to the sample under study in $P - u$ coordinates, Fig. 4.3b depicts (in $x - t$ coordinates) the case, where the reflected wave is a shock wave, and Fig. 4.3c shows depicts the case, where the reflected wave is a rarefaction wave.

The Riemann problem appearing during the shock wave reflection from the boundary of the base plate with the sample to be studied leads to the generation of two waves. The shock wave always propagates through the sample; the reflected wave in the base plate can be either a shock wave, if the dynamic impedance of the material under study is higher than that of the base plate, or an expansion wave for the inverse relationship between impedances (see Fig. 4.3).

The intersection of the wave beam of the measured wave velocity D_{bp} in the base plate with its Hugoniot $0\,A$ is used to estimate the parameters of initial state 1 of the base plate. Upon the shock wave reflection from the sample a new state arises, which is located either on recompression Hugoniot 1–2 or on expansion adiabat 1–2′ . The experimental recording of wave velocity D_0 in the sample gives wave beam slope $P = \rho_0 D_0 u$. The intersection point of the beam with curve $P_{Hbp} - 1 - P_{Sbp}$ determines pressure and shock velocity in the sample, which are equal to the pressure and velocity at the base plate-sample interface. Strictly speaking, the computation of the expansion wave P_{Sbp} or recompression wave P_{Hbp} in plane $P - u$ requires knowledge of the equation of state of the base plate material in the parameter range of our concern. However, when pressure differences in the reflected shock waves or rarefaction waves relative to the wave passing through the base plate are not too large, the curve $P_{Hbp} - 1 - P_{Sbp}$ is approximated with reasonable accuracy by the specular reflection of Hugoniot $0A$ from point 1 relative to axis P [7]. Although no strict thermodynamic justification has been found, the "specular" hypothesis simplifies estimation of shock-wave parameters in the sample under study. Therefore in practice one seeks to select base plate materials such that the dynamic impedances are close to that of the materials under investigation.

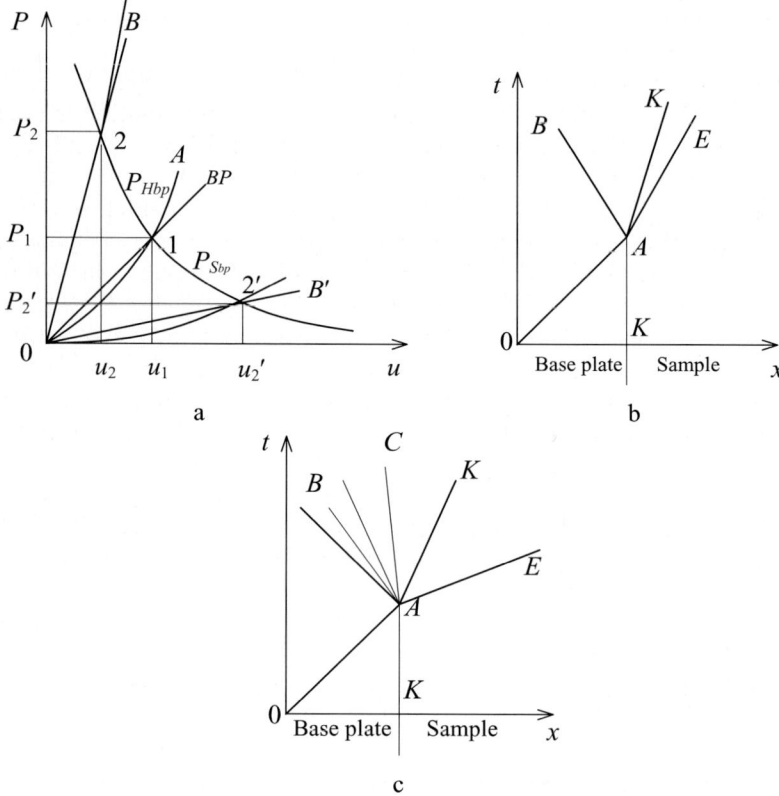

Fig. 4.3. Diagrams of the reflection method: (**a**) $P - u$ diagram: $0BP$ is the wave beam $P_{bp} = \rho_{bp}D_{bp}u$; $0A$ is the base plate Hugoniot, P_{Sbp} is the isentrope of base plate expansion from state 1; P_{Hbp} is the adiabat of the base plate deceleration from state 1; $0B$, $0B'$ are the wave beams of the samples under study; (**b**), (**c**) are the $x-t$ diagrams: $0A$ is the shock wave in the base plate, AE is that in the sample, AB is the reflected wave in the base plate, KAK is the interface; BAC is the expansion wave in the base plate

It should be noted that the interpretation of the results of spherical experiments with converging shock waves should include their specific features [8], strictly speaking, due to the unsteady nature of the shock wave in the sample: the relation $u = (1/2)W$ strictly holds on the impact surface, i.e. within base plate radius r_{bp} longer than radius r_m, at which the wave velocity in the samples is measured. However, as numerous hydrodynamic computations for iron shell impacts on an iron target indicate, convergence equally causes an increase in iron shell velocity $W(r)$ "in free flight" and mass velocity following the shock front in iron base plates and iron samples. Therefore it is assumed that mass velocity is $u\,(r_m) = (1/2)W(r_m)$ within radius r_m as well. Moreover, computations show that the average wave velocity D recorded in iron samples essentially coincides with local velocity D (r_m) within the nominal radius at the sample midpoint r_m.

Another situation arises, when the shock wave transfers from the rigid base plate to samples of different metals. Because of different conditions of shock wave convergence in the iron sample and samples of other materials located on the internal surface of the base plate, u and D in them vary as functions differing from the rate of u and D increase in the iron samples. Without any corrections, it would be rigorous to compare velocities corresponding to states in infinitely thin samples and the base plate. Among several possible comparison options, below we take the one, where the states in the samples under study are found from the iron state corresponding to that within the base plate inner radius, i.e. at the base plate-sample interface. We illustrate this procedure by using the example of copper and tantalum. The states on the inner boundary of the base plate are found computationally from known parameters in the iron base plate corresponding to the state at the measurement radius. For the hemispherical shock wave generator MZ-3 (see [13] of Chap. 2), in which the measurement radius is $r_m = 29$ mm, mass velocity decreases by 4.3% during the transition to the states at the base plate inner radius $r_m = 31$ mm. In the computations the equations of state of iron, copper, and tantalum were given in the Mie-Grüneisen form. The average wave velocities in the copper and tantalum samples have to be computationally corrected by 2.5% and 2.1%, respectively, toward lower velocities.

For the device MZ-7 (see [13] of Chap. 2), iron mass velocity 6.5 km/s pertains to 17.6-mm radius. On the transition to the state at the 19.6-mm base plate inner boundary it decreases by 7.9%. The corrections to average wave velocities are 5.1% for the copper samples and 4.3% for the tantalum samples, as previously, toward the lower values.

4.1.4 Double Compression Adiabat

To acquire information about material thermodynamic properties in the range of densities greater than those attained in single shock compression, methods have been developed for recording states arising during recompression. Double compression data are obtained either by the method of barriers, or in experiments involving either head-on or oblique shock wave impact of the sample. The experimental scheme for determination of the double compression adiabat using the method of barriers is in essence a variation of the reflection method.

Refer back to Fig. 4.3. Arrival of the shock wave propagating in the sample under study with parameters P_1 and u_1 at the interface with a barrier harder in terms of dynamics is initially attended by the generation of two shock waves running in opposite directions on either side of the interface. As a result, the material under study is compressed doubly, namely, by the incident wave to pressure P_1 and by the reflected wave to pressure P_2

$$P_2 = P_1 + \rho_1 D_{12} \Delta u_{12} . \tag{4.1}$$

In so doing material density increases to

$$\rho_2 = \rho_1 \frac{D_{12}}{D_{12} - \Delta u_{12}} . \tag{4.2}$$

In the above expressions D_{12} is the reflected wave velocity relative to material before the reflected wave front and Δu_{12} is the mass velocity jump at the reflected wave front. To practically implement the method, it is necessary to know single compression Hugoniots both of the material under study and of the rigid barrier. For this case, it is sufficient to estimate shock wave velocity D_1 in the material under study and shock wave velocity D_2 in the barrier. D_1 is used to find P_1, u_1, ρ_1, and D_2 to find P_2 and u_2. Hence,

$$D_{12} = \frac{P_2 - P_1}{\rho_1 \Delta u_{12}} \quad \text{and} \quad \Delta u_{12} = u_1 - u_2 .$$

Substitute these values into Eqs. (4.1) and (4.2) to find the doubly compressed material's pressure and density. Using barriers of different dynamic stiffness and recording the resulting P, u, the run of the recompression adiabat can be determined for different pressure levels. Equations (4.1) and (4.2) are also valid for the determination of recompression states that occur in head-on or oblique shock impact. Incident wave parameters must be known in either case. Experiments with the head-on impact of two shock waves equal in their amplitude are simplest in their interpretation. In this case material density ρ_2 behind the shock front is estimated with the pulsed X-ray radiography method [9] or with other methods, in particular, the electromagnetic method [10] may be used for the experimentally measured value.

As in the impact region the material is at rest, the mass velocity increment, $\Delta u_{12} = \Delta u_1$, is also known. These values are used to find the reflection pressure and doubly compressed material density.

In the case of regular conditions of oblique wave impact, estimation of double compression characteristics requires knowledge of the wave angle of incidence α and angle of reflection β, in addition to the incident wave parameters. The angle of incidence α is given by the experimental conditions. Reference [9] estimates the angle of reflection with the pulsed X-ray radiography method, which detects prompt shock front positions inside the X-rayed sample. When these values are given as simple trigonometric relations presented in [9], it is possible to determine the reflected wave velocity D_{12} and mass velocity increment:

$$D_{12} = D_1 \frac{\sin \beta}{\sin \alpha} + u_1 \cos \alpha (\alpha + \beta); \quad \Delta u_{12} = u_1 \frac{\cos \alpha}{\cos \beta} ,$$

which are used to find P_2 and ρ_2 using Eqs. (4.1) and (4.2).

4.1.5 Requirements Imposed on the Experiments

To avoid systematic errors in experiments on Hugoniot determination, certain requirements should be met. First, pressure in the impacting plate, when it impacts on the target plate, should be essentially uniform. This is needed to ensure that the shock wave in the target plate and samples would not be affected by small compression or rarefaction waves.

The second condition is that the accelerated impacting plate should be sufficiently thick, so that the initial shock wave in the plate and samples would not be overtaken by the rarefaction wave generated by the rear surface of the impacting plate facing the explosive system. The condition for maximum ratio R of the target thickness (including the samples) to the impactor thickness can be expressed as

$$R = \frac{\left[D^{-1} + (\rho_0/\rho c)\right]_{imp}}{\left[D^{-1} - (\rho_0/\rho c)\right]_{T}},$$

where c is sound speed following the shock front, ρ_0/ρ is the ratio between densities on both sides of the shock front, and D is shock wave velocity. This ratio is easy to derive from rectilinear characteristics of a simple centered wave for 1D flow in Lagrangian coordinates (see Sect. 4.4). Finally, the accelerated plate should not experience spallation.

The above requirements on the impacting plate thickness and the accuracy conditions that depend on the sample sizes limit the maximum effective pressure inherent in a given explosive device. In principle, explosive devices that are very large (as compared to the impacting plate thickness) ensure plate velocities close to the rates of explosion product expansion to air, which are 7500–8000 m/s for hexogen base compositions. The relevant shock pressures in iron might be up to 300–320 GPa. In practice, accuracy requirements restrict pressures produced by plane explosive devices to ~200 GPa.

The next shock-experiment condition is that the base plate-mounted samples have a large enough ratio between width (b) and thickness (h) to avoid the effect of rarefaction waves from edges on the shock front. From the flow analysis we obtain the following expression for minimum possible sample width-to-thickness ratio:

$$\frac{b}{h} = 2\left[\left(\frac{c\rho}{\rho_0 D}\right) - 1\right]^{\frac{1}{2}}.$$

If plates are launched using systems like light-gas guns, the impacting plate velocity can be measured very accurately by recording the time of its free flight at relatively long segments of its path using electrocontact sensors or pulsed X-rays. Note that when studying relatively strong shock waves imparting velocities of several km/s to the surface, electrocontact sensors have to be covered with shielding caps separated from contacts by 0.2–0.3-mm gaps. The contact shielding is needed to prevent premature closing of the contacts that may be caused by the air-shock wave traveling in front of the impacting plate.

When using HE charges to drive the plates, an accurate measurement of their velocity involves certain difficulties, as the velocity of the flying plates is not constant at the time of impact, hence, the velocity for a certain position has to be estimated within a very short time period. When the impactor velocity is variable or increasing, in the deceleration method the wave velocity in the target and the impactor velocity should be measured for the same path point coinciding with the midpoint of the gage length [1]. Strictly commensurable values of W_{imp} and D_T can only be obtained by superposition of the target surface and gage length midpoint s and wave velocity measurement on infinitesimal gage lengths, which, of course, is impossible. For real gage lengths of several millimeters (Fig. 4.4), in wave measurements the impactor motion from the target surface to the gage length midpoint is replaced with shock wave motion in the target. For the experiment to be pure, it is necessary to introduce small corrections to measured wave velocities, since relative increments of impactor velocity during the impactor's motion in air and those of mass velocity in the shock wave during its propagation through the target do not coincide.

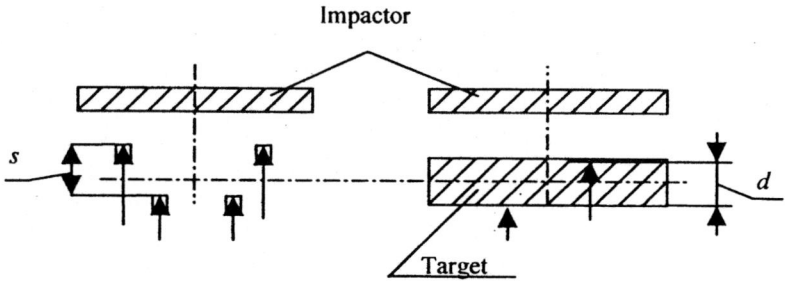

Fig. 4.4. Schematic of wave and mass velocity measurement with the deceleration method for: (**a**) impactor; (**b**) shock wave

A proper choice of the gage length is also necessary in the spallation and reflection methods. In the spallation method, a solid plate has to have passed the gage length earlier than the second shock wave arrival at the free surface. The mass velocity following the shock front and the velocity of the shock wave itself have to be measured at an identical distance from the charge. As seen from Fig. 4.5, velocities W are measured on thinner barriers, whose free surface is located at the same distance from the HE boundary as the gage length midpoint in the wave measurements.

Shock velocities in base plates and samples to be studied are typically measured using electrocontact or streak-camera methods. For a majority of Russian (mainly RFNC-VNIIEF) experiments on shock compression of condensed matter reviewed in [11], shock velocities are measured with the electrocontact method. The sensors are positioned at two levels that allows three

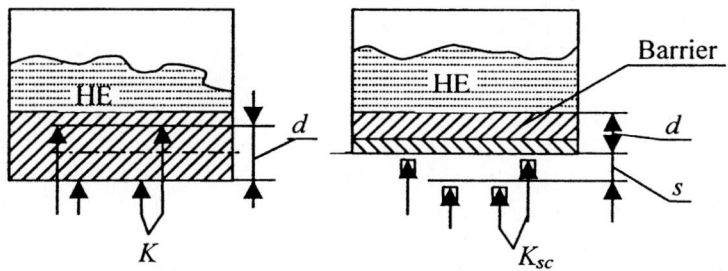

Fig. 4.5. Schematic of spallation method measurements for: (**a**) wave velocity; (**b**) spallation rate in barrier. K – electrocontact sensors, K_{sc} – screened sensors

different samples 12–14 mm in diameter to be measured on a single explosive launcher. The duration of the shock wave travel through the sample is
estimated as the difference between the time of closing of the electrocontact
sensor attached to the sample free surface and the averaged time of actuation
for the sensors attached to the base plate about the sample. To ensure reliability of the measured data, 5 to 8 experiments using the same explosive device
are required. Averaging the experimental data allows avoidance of accidental
gross errors and reduction in the average error.

In practice, a major source of error in the measurement of shock velocity is
skewness and curvature of the shock front surface. Skewness and curvature in
the shock front are directly related to irregularities in impactor surface shape
and orientation at the time of the impact on the target. Moreover, skewness
and curvature are inherent not only in HE base launchers, but also in light-gas
guns. For these reasons, shock wave arrival times according to sensors lying
in a plane may be recorded with a considerable spread; sometimes variations
in the arrival time may be greater than the time that would be required for
the wave to travel the distance between two electrical contacts. Consequently,
to interpret the arrival times properly, it is necessary to know the shock front
shape and orientation.

The shock front shape is determined using various electric contact layouts
in targets, among which of the highest practical interest are those employed at
Livermore National Laboratory in the light-gas gun experiments. Two schemes
called the "snowflake" and "wheel" appear in Fig. 4.6. The method of analysis for data obtained using them for determining shock front distortions is
discussed in detail in [12]. The schemes contain 13 contacts each, 7 of them
are at the lower level and 6 at the upper. At the internal, R_a, and external,
R_b, radii there are 6 contacts at each, however, in the "wheel" layout both
lower and upper contacts are present within either radius. In the "snowflake"
scheme, the exceptional role is that of the lower central contact as a reference
contact to find the wave front curvature, and if it fails, the front distortion
determination becomes problematic.

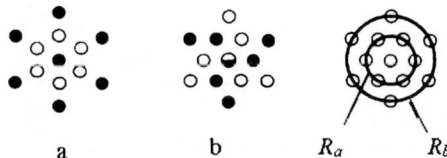

Fig. 4.6. Layouts of electric contacts in the targets: (**a**) "snowflake", (**b**) "wheel", ●, ○ – lower and upper contacts, – either lower or upper contact

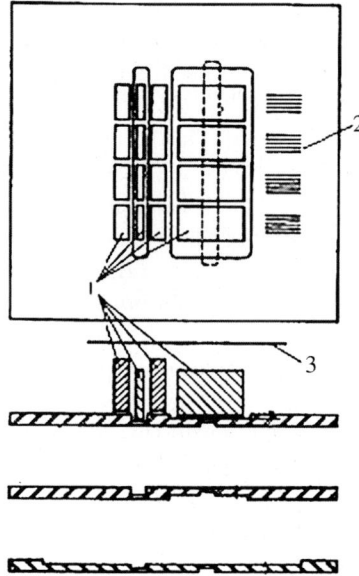

Fig. 4.7. Recording scheme and target designs used to measure wave and mass velocities with the streak-camera method: 1 – luminous acryl blocks; 2 – peep-holes; 3 – apertured plate

The time interval measurement with photorecording devices – mechanical photorecorders, photodiodes, photocells, and electrooptical transducers – is based on gas luminosity in the gaps during the shock wave passage through them, on a change in luminous intensity when the shock wave arrives at the free surface or when the flying body is decelerated at its impact on a transparent barrier. A measuring scheme [13] using photorecording devices appears in Fig. 4.7. The top of the figure shows positions of the luminous acrylic blocks, which provide four independent series of the experimental points for D as a function of W_{imp}. The left part of the system is used for the shock velocity measurement, the right part for the moving plate velocity measurement. Each block row is viewed through four or five peepholes.

The bottom of the figure depicts cross sections of the target constructions used in different pressure ranges. The first two constructions are designed for the low pressure range; the upper construction has a groove for measurement of W_{imp}. At a higher pressure the grooves for the W_{imp} measurement contract not very rapidly and can be made as deep as the target thickness permits.

In the high-pressure experiments the target thickness is restricted by the fact that the moving plates are very thin (0.9 mm).

Figure 4.8 presents a magnified streak-camera image for an experiment conducted at a relatively low pressure on the target of the uppermost construction (see Fig. 4.7). In this streak-camera image, time increases downward; hence, the first marks on the left side mark the arrival of the shock wave at the bottom of the narrow groove. The relevant base marks specify the time of its arrival at the upper face of the plate. As the luminous blocks on the right side are at a lower level, the base marks for the W_{imp} measurement appear earlier. The mark shift corresponds to the difference between the times of shock wave and impacting plate passage through a small gap cut out at the plate bottom.

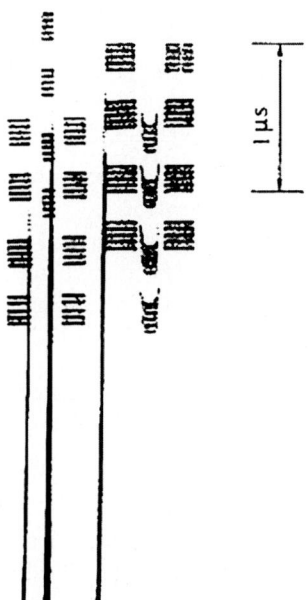

Fig. 4.8. Streak-camera image from the experiment according to the scheme shown in Fig. 4.7

No matter how simple such a system may be, the proper depth of the grooves should be selected as carefully as possible to ensure optimal accuracy. The groove width and side position always remained identical in these experiments. In low-pressure shock waves, the marks for the W_{imp} measurement are compressed due to the material side displacement under action of the shock wave, which is considerably ahead of the impacting plate.

Thus, the moving plate velocity can be measured only on the path length of about 1 mm. This restriction plays no great role, as in coordinates $D-u$ the Hugoniot has relatively little sensitivity to mass velocity measurement errors at low pressures.

One may estimate the error in measurement of pressure, density, and energy on the Hugoniot. For simplicity, make the estimations for the deceleration method in impact of the impactor on the target fabricated from the same material. In these experiments, the measured values are initial density, ρ_0, shock wave velocity, D, and impactor velocity, W_{imp}, the mass velocity is determined as $u = W_{imp}/2$. Assume that the experimental errors are random and, thus, the total experimental error is equal to the square root of the total error due to the measurements of individual parameters. Then the laws of conservation yield:

$$\frac{\delta P}{P} = \left[\left(\frac{\delta \rho_0}{\rho_0} \right)^2 + \left(\frac{\delta D}{D} \right)^2 + \left(\frac{\delta W_{imp}}{W_{imp}} \right)^2 \right]^{1/2} ;$$

$$\frac{\delta V}{V} = \left\{ \left(\frac{\delta \rho_0}{\rho_0} \right)^2 + (\sigma - 1)^2 \left[\left(\frac{\delta D}{D} \right)^2 + \left(\frac{\delta W_{imp}}{W_{imp}} \right)^2 \right] \right\}^{1/2} ;$$

$$\frac{\delta E}{E} = 2 \frac{\delta W_{imp}}{W_{imp}} ,$$

where $\sigma = \rho/\rho_0$ is the compression ratio.

In all the equations, the relative error in u and W_{imp} is considered to be the same, i.e., $\Delta W_{imp}/W_{imp} = \delta u/u$. If the initial impactor and target densities are different, then $u \neq 1/2 \, W_{imp}$ and it is necessary to make a slight correction to estimate u:

$$u = \frac{W_{imp}}{2} \left[1 - \frac{a_0}{2(a_0 + a_1)} \left(\frac{\rho_0 - \rho_{0imp}}{\rho_{0imp}} \right) \right] ,$$

where a_0, a_1 are coefficients in linear D–u relation for the target.

More complex computations are required to estimate errors in the experiments performed with the reflection method. We recommend the paper by Mitchell and Nellis [14] that presents a comprehensive study of errors, and includes, among other things, wave front skewness and curvature, to those readers, who are interested.

4.2 Recording of Material Expansion Isentropes from the State Following Shock Compression

Analysis of many modern high energy density physics problems (anti-meteorite protection of spacecraft, effect of powerful electromagnetic radiation on substances, development of gas nuclear engines and heterogeneous nuclear reactors, assessment of mechanical effects in emergency modes of nuclear reactors, etc.) testifies that it is difficult to separate one specific state range for their solution. In some cases information is needed about material properties over a wide range of the phase diagram, which begins with a highly compressed condensed state and stretches up to an ideal gas, including the two-phase liquid-vapor region.

The following methods are widely used to acquire information about thermodynamic material properties in heated expanded states:

- material shock loading with subsequently recording of the expansion isentrope with the method of barriers;
- method of isobaric expansion or method of exploding wires;
- method of vaporized particle deceleration against a witness plate, whose velocity is measured with a laser interferometer.

Studying the expansion isentropes of shock-compressed materials essentially complements test information on shock compressibility and allows a fuller judgement about material properties at high pressures and temperatures. In some cases the expansion isentrope study is the only source of information about states of severely heated material which has expanded and enables the final expansion states to be related to the states following the shock front.

Experimental studies of the isentropic expansion of shock-loaded material began back in the 1960s. The developed methods can be divided into two types: discrete and continuous.

The first method is that of barriers, where the expansion isentrope is determined by measured shock wave intensities in barriers of less dynamic hardness than that of material under study. The second type involves sensors inside the material under study, which allows us to continuously record parameters of shock-loaded material in the expansion wave. Selection of one or another method depends on the working pressure range and diagnostic capabilities.

4.2.1 Method of Barriers

The method is schematically presented in terms of $P - u$, $x - t$ and $P - x$ diagrams in Fig. 4.9. The shock wave generated by deceleration of the EP-driven impactor drives it to a compression state "a" and irreversible heating takes place on a barrier made of material under study M. The shock wave arrives at interface I with dynamically softer barrier B which causes the generation of a centered Riemann wave R shown by a fan of C-characteristics in the $x - t$ diagram. In this wave, the shock-compressed material adiabatically expands from state "a" to state "i". The pressure profile corresponding to time t^* in the $x - t$ diagram appears in Fig. 4.9c. This expansion of the material under study generates in barrier B a shock wave propagating at velocity D_i. Recorded D_i and the known barrier Hugoniot h_i can be used to estimate pressure P_i and barrier mass velocity u_i, which coincides with the relevant characteristics of

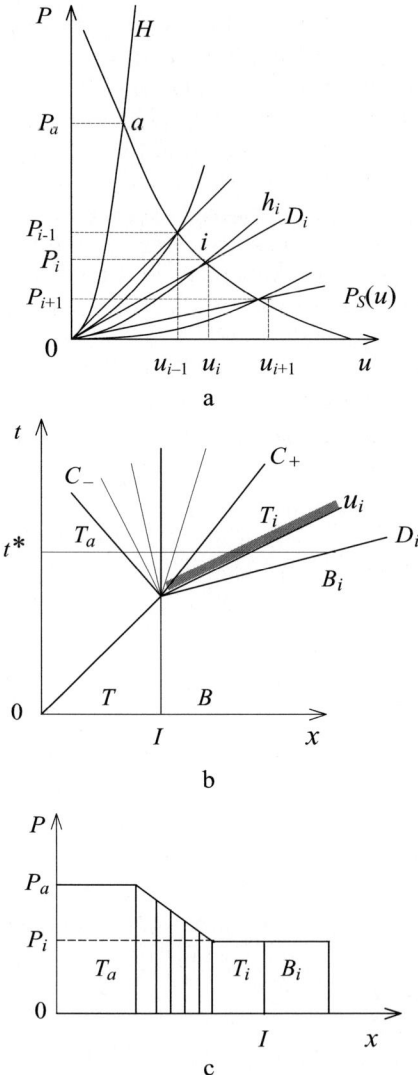

Fig. 4.9. Method of isentropic expansion: (**a**) $P - u$ diagram of the experiment; (**b**) $x - t$ diagram; (**c**) $x - P$ diagram for time t^*

the expanding material under study by virtue of flow continuity at interface I.

By using barriers with different dynamic stiffnesses and recording shock wave velocities D_i, one can follow the path of the expansion isentropes $P = P_s(u)$ in a discrete manner from the state on the Hugoniot to various pressures. Launchers of various velocities allow a variation of the entropy in the shock wave passing through the sample and, thereby, recording various isentropes covering the selected region in the phase diagram.

The transition from hydrodynamic variables $P-u$ to thermodynamic variables $P - V - E$ can be performed using (1.38) and calculation of Riemann integrals expressing the laws of conservation for a given self-similar flow type.

References [5,15] report on the expansion isentropes of pre-compressed continuous samples of lead, iron, and copper using the method of barriers. For the barriers "soft" condensed materials are used: light metals, teflon, plexiglass, polyethylene, and polystyrene foam of different initial density. Shock wave velocities in the barriers are estimated experimentally, and the transition to pressures and mass velocities is performed using their known $D - u$ relations.

References [16–19] report on the adiabatic expansion of copper, lead, bismuth, and uranium samples, which are fabricated from fine powder to intensify the dissipation effects in the shock front. As in the previous case, soft condensed media serve as barriers in the high dynamic pressure range. The region of considerable rarefaction and reduced dynamic pressures, which is of specific interest in plasma studies, is produced by using inert gases – argon and xenon – at 0.1–100 bar pressure and air at atmospheric pressure as the barriers. Maximum initial gas pressure is thereby selected from the condition of obtaining states close to those achieved with the softest condensed barriers.

By way of example, Fig. 4.10 presents the data for copper isentropic expansion over a wide range of initial states of shock compression for pressures up to 1.4 TPa. According to (1.38) the experimental $P - u$ curves are converted to the thermodynamic variables $P - V$. We find that it was possible to evaluate the expansion data over a wide range, i.e. four orders of magnitude

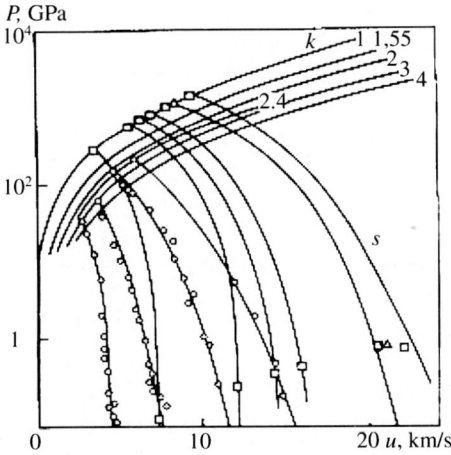

Fig. 4.10. Copper Hugoniots and expansion isentropes: k – Hugoniots for samples of different porosity; S – expansion isentropes of shock-compressed samples; points – experiment; lines – computation by equation of state [16]

in pressure and two in density, from a strongly compressed metallic liquid to a quasi-ideal Boltzmann plasma and metal vapor. The obtained data has been used to develop semi-empirical equations of state [16] which cover a wide-range of state-space.

4.2.2 Method of Isobaric Expansion

Ordinary static methods for material thermophysical property measurements are limited in temperature and pressure through chemical reactivity of the sample at high temperatures and high-temperature vessel strength. The upper limits are usually about 2000 K and 0.2 GPa.

The method of isobaric expansion developed by the authors of [20] used an improved facility for thermophysical measurements which allowed measurements of pressures up to 0.4 GPa and of temperatures up to 8000 K. Heated metal density can be about twice as low as that under standard conditions. The method is based on heating the sample in the form of a rod $\sim$25 mm in length and 1 mm in diameter by current contained in a high-pressure cell filled with gas and simultaneously measuring supplied energy, volume expansion, and surface temperature. The pressure in the cell is measured to within $\pm$0.5 MPa with a pressure gage having an upper limit of 0.7 GPa. The sample volume is approximately $\sim$0.1% of the gas volume in the cell, therefore even two-fold and three-fold expansion of the sample essentially occurs isobarically until it remains sufficiently slow, such that sound waves are unable to reverberate in the gas volume. Inert gas not only ensures an isobaric medium, but also prevents chemical reactions on the surface and thus considerably extends the temperature range; melted metals can be conveniently studied as a consequence of the increase in their boiling point.

The sample is heated by fast discharge ($t < 100$ μs) of a capacitor bank through the sample. In doing so the discharge should be quite fast, so that the liquid metal "column" remains cylindrical during the measurement, while the current pulse should be quite slow ($t > 10$ μs), so that the material state remains close to equilibrium and no nonuniform heating due to the skin effects occur. The capacitor bank used in the facility allows energy storage up to 60 kJ at 20 kV. Current flowing through the sample is $\sim$30 kA. This current is sufficient to heat, for example, tantalum to its liquid state ($T \sim 3270$ K) for about 17 μs.

The optical diagnostic system is schematically depicted in Fig. 4.11. The system is composed of three parts and allows the simultaneous photographing of the whole sample, recording the sample diameter change and surface luminance at three different wavelengths. The whole sample is photographed 10 μs after initiation of the shunting spark discharger.

Volume expansion is measured using two shadowgraphy systems with laser radiation and narrow-band filters to remove the intense self-radiation of the sample. One system employs a modulated-quality-factor ruby laser supplying

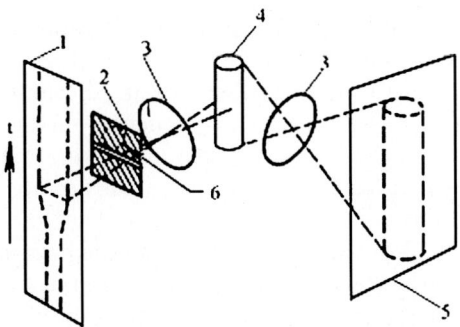

Fig. 4.11. Scheme of the optical system: 1 – recorded trace; 2 – slit; 3 – projection optics; 4 – sample; 5 – shadowgraph; 6 – pyrometer field of vision

a 20 ns light pulse, which illuminates the sample from the rear and produces the photograph of the whole rod at any given time.

Another observation channel is used for dynamic recording of wire expansion. In this channel, the illumination is made using a 1 W argon laser operated continuously at a 514.5 nm wavelength. Before the laser beam illuminates the cell, it is broadened with a cylindrical lens. The sample's shadow is focused onto a slit which is mapped onto the electrooptical transducer photocathode. A 2 nm band-pass interference filter which cuts off the sample thermal radiation and ruby laser radiation is positioned in front of the EOT. A high-speed photorecorder is employed to obtain a streak-camera image of the diameter change in the selected sample cross section. If the sample expansion is purely radial, then the volume ratio $V(t)/V(0)$ equals the squared ratio of the relevant diameters. Taking the above images at specific times allows us to ensure that the sample volume ceases to increase when the current is turned off and the sample diameter remains uniform along the sample length.

The sample image produced on the slit diaphragm is used to record the sample surface temperature with a three-channel pyrometer. Light from the region indicated in Fig. 4.11 in the slit plane is transferred through fiber waveguides to filters, whose bandpass centers lie at 450, 600 and 700 nm, respectively, while the bandwidths are the same, 50 nm in all three filters. The availability of three channels enables temperature calculation with different methods. Thus, a luminance temperature can be found for each spectral region and ratios of intensities from two channels can be used to calculate two color temperature values independently.

Enthalpy is measured using four-point probe measurements of current and voltage drop. Then the enthalpy time history can be related to the temperature time history to estimate specific heat for temperatures up to 7000–8000 K. Voltage and current as functions of time in combination with the volume time history can be used to find specific electrical resistance at the above temperatures.

As the authors themselves note, measured temperature data are most doubtful for the above discussed method. The difficulties in the measurements can be due to anomalous sample surface heating because of the skin effect, unknown liquid metal emissivities, and opaqueness of the diffusive boundary layers of heated inert gas and cooled dense metal vapor. Hydrodynamic and magnetohydrodynamic instabilities are also undesirable, because these can garble the measured data.

Despite these difficulties, the method of isobaric expansion is very important for acquisition of thermodynamic information regarding metal properties in the vicinity of their critical points and comparison of the obtained results to shock-compressed metal isentropic expansion data. Experiments with exploding wires have been conducted for Ta, Mo, U, Pb, W, V, Ir, Al, and Cu. Processing of these data allowed determination of enthalpy density H versus temperature and slopes of the curves for high-temperature refractory metal melting.

Figure 4.12 presents experimental data [21] on electric explosion of uranium wires under pressure and uranium properties at standard pressure as well as their description by the model of hard and soft spheres.

4.2.3 Recording Shock-Induced Vaporization Using Laser Interferometry

The method for studying vaporization processes during expansion of preliminarily shock-compressed metals using laser interferometry is discussed in detail in [23, 24]. The experimental device used for these purposes is presented

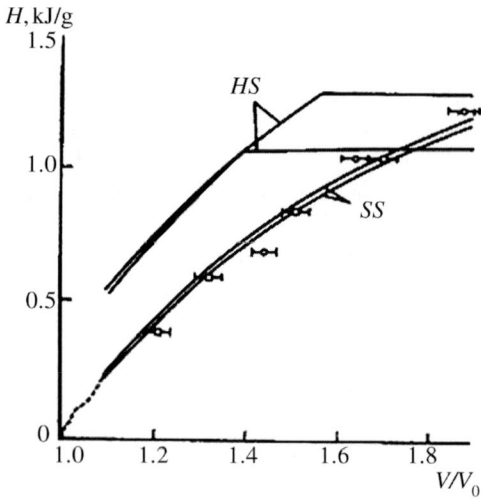

Fig. 4.12. Experiments on electric explosion of uranium wires under pressure [21] and their description by the model of hard (HS) and soft (SS) spheres [22]. The *dashed lines* represent static data

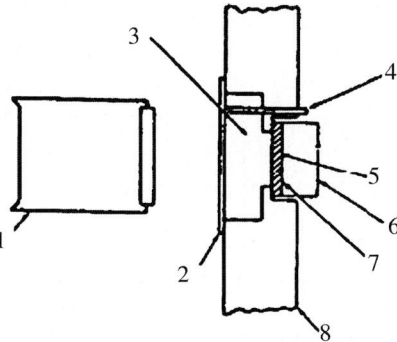

Fig. 4.13. The device for studying vaporization processes in expansion wave: 1 – projectile with tantalum impactor; 2 – foil made of material to be studied; 3 – vacuum gap; 4 – aluminum ring; 5 – aluminum buffer; 6 – LiF window; 7 – laser beams; 8 – electrical shorting pins

in Fig. 4.13. A two-staged light-gas gun accelerates a tantalum impactor, which generates a planar shock wave in it and a thin target upon impact with that target. The impactor velocity and target material are selected so that the target material is partially vaporized when the shock wave arrives at the free surface. The vaporization products expand into an evacuated (50 mtorr) gap and are decelerated on the interferometric sensor composed of a thin aluminum buffer superimposed on a window made of single crystal LiF.

The aluminum buffer/transparent window interface velocity time history is recorded by the laser interferometer. Although parameters of the equation of state are not straightforwardly estimated in such original experiments, comparison of 1D computations to computations with hydrocodes and various versions of multiphase equations of state to experimental records enables an assessment of equation-of-state reliability and the determination of a consistent set of key EOS parameters.

Figure 4.14. presents experimental mass velocity profiles obtained in experiments with lead and cadmium [23].

According to the computations, whose records are presented in the upper two plots, when the lead and cadmium isentropes enter the region of liquid/vapor coexistence, the vaporized material makes ~9% of the mass. Either case is characterized by an initial mass velocity peak, which corresponds to the impact of vapor-liquid expansion products on the aluminum buffer. This is followed by a decrease and a second peak of the mass velocity rise associated with the impact of the tantalum plate on the buffer. The numerical computations suggest that a spall layer can form in the impactor, resulting in a third velocity peak. It is seen from the upper plots that the computed velocity profiles are in a good agreement with the experimental profiles.

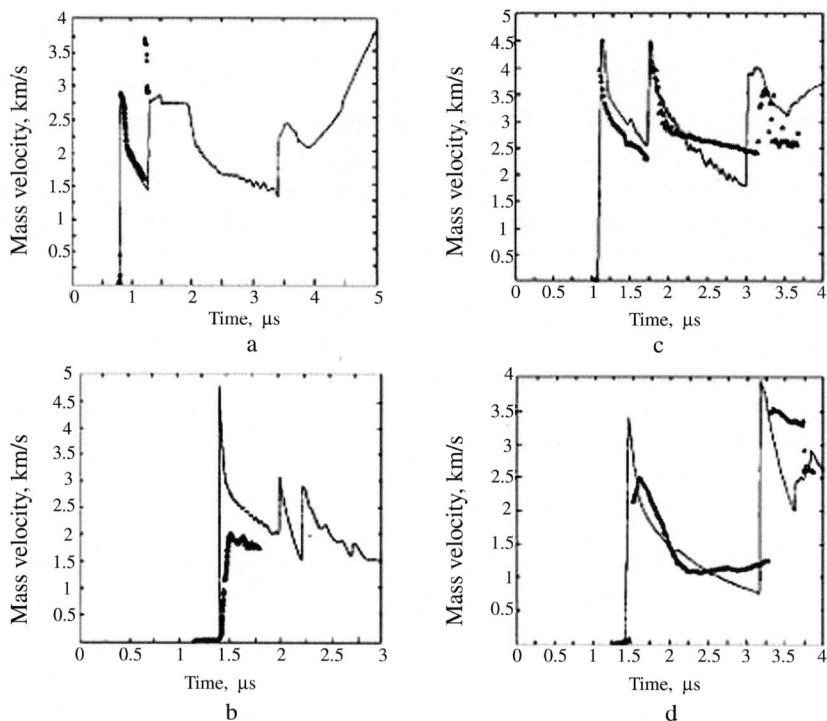

Fig. 4.14. Comparison between the experimental and computed vaporization product deceleration rates: (**a, b**) Pb; (**c, d**) Cd. The *upper* two plots correspond to 9% vapor concentration and the *lower* to 35% concentration. (——— computation; ▲ – experiment)

Another situation is observed for higher vapor concentrations produced during sample expansion from the high pressure states. As the two lower plots suggest, there is a significant discrepancy between the experimental and computed profiles; this is particularly noticeable for lead. The results are reproduced well for the same experimental setups. As the developers of the method suggest in [23, 24], mixed liquid/vapor models have to be developed to eliminate the discrepancy between the experimental data and numerical computations for high vaporization concentrations, and there are grounds to believe that the method will be a good means for the validation of equation of state models and gas-dynamic codes.

4.2.4 Recording Centered Rarefaction Wave Structure

Obtaining reliable information about material isentropic expansion with the method of barriers requires that a great number of explosive experiments be conducted. In the relatively low pressure range this information can be acquired in a single experiment using internal methods for recording mass velocity or pressure (electromagnetic and manganin sensors).

Having recorded the mass velocity profile with the electromagnetic method or the pressure profile with the manganin sensor, it is possible to correlate the thermodynamic parameters (P, V, E) for expanding materials.

In a simple rarefaction wave, mass velocity, sound speed, pressure, and density are uniquely related. For the centered wave bordering on a constant flow region these relations are simplest. From the condition of constancy of one of the Riemann invariants and a simple form of the other family of characteristics it immediately follows that knowledge of any one of u, P, ρ, c on the line intersecting the fan of the characteristics is sufficient to determine all other values and, hence, to obtain the relationship of most interest: $P(\rho)$.

Consider now the method for the transition to thermodynamic parameters from the expansion isentrope using a record of the mass velocity profile in a rarefaction wave obtained with an electromagnetic sensor. Figure 4.15 presents the oscilloscope trace [25] of mass velocity obtained with an electromagnetic sensor placed in a teflon sample. The centered rarefaction wave has formed during the arrival of 12.2-GPa amplitude shock wave at the sample free surface. The shock and rarefaction wave motion diagram in coordinates of $x - t$ is identical to that presented in Fig. 4.20.

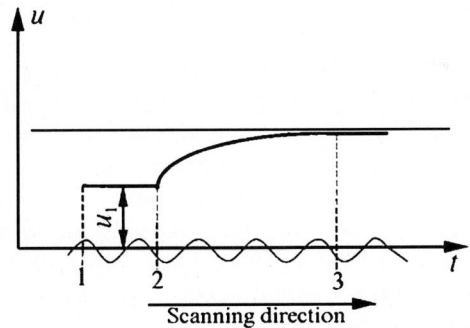

Fig. 4.15. Oscilloscope trace of mass velocity following the shock front (1–2) and in the rarefaction wave (2–3)

The sensor trace is found by integrating the velocity profile:

$$x = x_0 + u_1(t_2 - t_1) + \int_{t_2}^{t_3} u\,dt \ .$$

Under the assumption that the rarefaction wave is self-similar, the coordinates of the intersection of the characteristics with the sensor trace allow determination of $c = (x - x_1)/(t - t_1) + u(x/t)$ and, thereby, sound speed as a function of mass velocity. The density change is found from the condition of constancy of the first Riemann invariant, $u + \int cd\rho/\rho = $ const., when $\rho\left(u\right) = \rho_1 \exp \int_{u_1}^{u} du/c(u)$ is obtained through integration. Following these operations, $P(\rho)$ is found in parametric form from the expression $P\left(u\right) = \int_{u}^{u_1} \rho(u)c(u)du$. The use of internal methods for estimating thermodynamic parameters in the rarefaction wave is primarily limited to low pressures and temperatures as long as shock-compressed material conductivity can be neglected. For conducting media the electromagnetic method is inapplicable. Rarefaction wave profile recording with manganin sensors yields good results for metals up to 35 GPa pressures, at higher pressures the sensor insulation is ruptured, and thicker insulation considerably deteriorates the sensor time resolution. From these standpoints, the method of barriers offers a great advantage.

4.3 Sound Speed in Shock-Compressed Material

Sound speed, $c = V\left(-\partial P/\partial V\right)^{1/2}$, following the shock front determines the slope of the Poisson adiabat running through point P, V on the Hugoniot, i.e. the initial compressed material behavior upon unloading and under a weak shock recompression wave. Sound speed is used to verify the EOS of material.

At the shock front $u + c > D$ (see Chap. 1), i.e. perturbations arising behind the shock front can overtake the shock front and attenuate it. For this reason knowledge of sound speed is needed for properly setting up shock-wave experiments. Measuring elastic and volumetric sound speeds following the shock front as functions of the wave amplitude allows determination of opaque solid melting under shock loading (Sect. 5.3). Finally, the sound speed in a solid at high pressures is of interest for a number of geophysics problems.

The first methods for recording sound speeds following powerful shock waves were developed from 1948 to 1958. They were called "side" and "overtake" unloading [26]. Either method records the rarefaction wave "head" movement in the shock-compressed material.

4.3.1 Method of Side Unloading

A schematic of the experiments developed in Russia by the authors of [26] is presented in Fig. 4.16. The rarefaction wave source is at the interface between the cylindrical sample side surface and the base plate surface. Consider two shock front positions at times t and $t + \Delta t$. Let point 0 separate the loaded front part from the unloaded. For a time Δt the shock front will move a distance $D\Delta t$, while the metal particles originally at point 0 will move following the wave front by $u\Delta t$. For the same time Δt the unloading waves will cover a sphere of radius $c\Delta t$ with its center at 0_1. Because the motion of the unloading wave appears in the shock wave flow about the inside angle is similar, the boundary point path is a straight line which forms a constant unloading angle

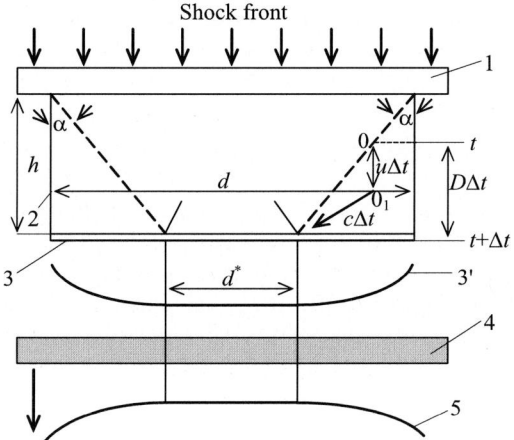

Fig. 4.16. Scheme of the side unloading method: 1 – base plate; 2 – sample; 3 – thin light material layer flying off the surface of the material under study; 3' – the same layer in flight; 4 – barrier made of transparent plexiglass; 5 – scheme of streak-camera recording of the time difference in the layer impact on the plexiglas barrier; h, d – sample height and diameter; d^*– diameter of the non-unloaded part of the sample

α with the shock wave propagation direction. At the time the shock wave arrives at the end of the cylinder, the expansion waves cover a peripheral zone $(d - d^*)/2 = h\mathrm{tg}\alpha$ wide. From the geometric explanations in Fig. 4.16 it is easy to obtain the expression for sound speed:

$$c = D\sqrt{\mathrm{tg}^2\alpha + \left(\frac{D - u}{D}\right)^2}\,.$$

Thus to measure sound speed, with known shock wave parameters D and u, it is necessary to experimentally estimate loaded region size d^* and use it to find the tangent of unloading angle α. Experimentally, the boundary of the loaded specimen is found by the change in velocity of the specimen free surface with the arrival time of the specimen's free surface at a plexiglass plate placed some distance from the specimen. This actually reduces to streak-camera recording of the time difference between the impact of a thin layer pre-applied on the sample with a transparent plexiglass barrier. The authors of [26] detected a wave set propagating at velocities greater than the isentropic sound velocity for compressed volumes with the same method. These were interpreted by the authors of [26] as waves characterizing the elastic phase of the compressed material upon expansion. A plastic wave runs at a lower velocity through the unloaded sample than the elastic wave. The plastic wave velocity is of major interest, since it is at this velocity that perturbations attenuating the shock front propagate . Existence of elastic rarefaction waves dramatically impedes accurate determination of the plastic wave boundary by the streak camera images in the experiments, whose setup is illustrated in Fig. 4.16. For this reason, the method of side unloading has not found wide application.

4.3.2 Method of Overtake Unloading

The idea of the "overtake" method [26] is the recording of the attenuation from the shock wave generated in the target by the impact of a thin plate moving at velocity W_{imp}. The wave processes occurring in this experiment are depicted in plane path (x) – time (t) coordinates in Fig. 4.17. For simplicity consider the case, where an impactor and a target are fabricated from the same materials.

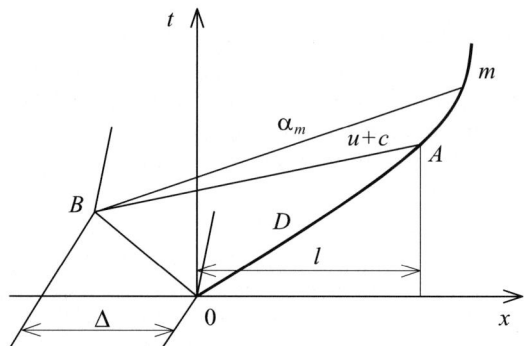

Fig. 4.17. $x - t$ diagram of thin plate impact on a target. $0Am$ – shock wave path in the target; B – centered rarefaction wave pole; A – beginning of shock wave attenuation

Shock waves $0A$ and $0B$ propagate through both bodies from impact point 0, and because the impact is symmetric, $u = W_{imp}/2$. When the shock wave has arrived at the rear side of the plate, at point B, a centered rarefaction wave forms, whose leading characteristic propagates through the material at velocity $u + c > D$ and overtakes the shock front at point A. From that time on, the shock wave amplitude is attenuated and the front path begins to bend. From the diagram it is easy to relate the sound speed in the material under consideration to the shock wave parameters, impactor thickness Δ, and distance l, from which the first characteristic transits to the front path:

$$c = (D - u)\frac{l + \Delta}{l - \Delta} .$$

This relation shows that with known shock wave kinematic parameters the sound speed evaluation reduces to an estimation of distance l, at which the wave begins to be attenuated. This distance is estimated either with the electrocontact method, as in [26], or with the optical method using transparent indicator media discussed in detail in the following section.

Because of the faster elastic rarefaction wave presence, estimation of distance l, from which the shock wave attenuation begins, involves noticeable experimental error. Hence, as demonstrated in [26], it is reasonable to note the time t_m of the shock wave arrival at a given point m, which is located at the curved segment of the shock wave path. The coordinates of the point m are used to find the characteristic slope $\alpha_m = u_m + c_m$. If mass velocity, u_m, is measured at that point, then c_m is determined from the slope of the characteristic α_m. In this case, the sound speed will be related to the shock wave parameters for the loaded region and the coordinates of the point m as:

$$c_m = \frac{x_m + (2 - \sigma) - \Delta/\sigma}{Dt_m - \Delta} D - u_m, \quad \text{where} \quad \sigma = \frac{D}{D - u}.$$

Thus, in the "overtake" unloading method, the sound speed measurement reduces to the experimental determination of the shock front path in coordinates $x - t$ at the point where noticeable attenuation and mass velocity measurement at one or more points of the path.

The interpretation of the experiments for the impact of plates made of different materials involves no basic difficulties, although the $x - t$ diagram becomes somewhat more complex because of the formation of a contact discontinuity, during which the α characteristics are refracted. Here not only the impactor Hugoniot, but also the correlation between mass velocity and sound speed in the isentropic rarefaction wave must be known.

4.3.3 Indicator Method for Sound Speed Measurement

Considerable progress in sound speed measurement with the "overtake" method for unloading was achieved using shock front emission in transparent media as a pressure sensor and time interval recorder. This possibility in inherently transparent materials was pointed out as early as in the paper by Kormer and associates [27]: "In circumstances where material following shock front is characterized with a high absorptivity, luminosity decay can serve as a measure of sound perturbation propagation velocity with the overtake method for unloading." However, the idea was implemented experimentally [28] by US investigators from Los Alamos Laboratory 15 years later.

The major merit of the method is that it is highly sensitive, as the extremely strong dependence of intensity from thermal radiation caused by material compression in a shock wave is used here. When taking the measurement in the entire spectral range, the temperature dependence of total black body radiation energy (M) is described with Stefan-Boltzmann equation, $M = \sigma T^4$, where σ is a constant. From experiments it is known that the shock compression pressure dependence of temperature is nearly linear for many materials. Hence, the thermal radiation intensity is proportional to the fourth power of pressure, i.e. a minor change in the shock front pressure causes a considerable change in the photoreceiver-recorded signal corresponding to the radiation energy.

In the method for sound speed measurement proposed by the authors of [28], rarefaction waves are generated with the usual method, i.e. by an impact of a rapidly flying thin plate on the target. The target thickness l, at which the rarefaction wave generated from the impactor's rear side overtakes the shock front, and is estimated with optical indicators. If the indicator is placed on a target of thickness x less than l, then before the time the first characteristic rarefaction wave arrives at the shock front the indicator will report constant radiation intensity, which will then begin to decay. Either a shock wave or a rarefaction wave can be reflected back from the interface, depending on the impedance ratio of the indicator and target. When these waves interact with the overtaking rarefaction wave, the velocity of the latter can either increase or decrease. However, the study in [28,29] indicates that the dependence of the time interval between the appearance of constant-intensity radiation and the onset of radiation decay is a linearly decreasing function of target thickness both for shock wave and rarefaction wave reflection from the indicator. Hence, having taken measurements for various target thicknesses and extrapolating $\Delta t = f(x)$ to zero, we arrive at a thickness whereby the rarefaction wave overtakes the shock wave just at the target surface. If the target is fabricated in the form of a stepwise wedge and covered with a layer of transparent indicator, then $\Delta t = f(x)$ and, hence, the desired target thickness l can be estimated in a single experiment.

Figure 4.18 illustrates the experimental scheme and the $x - t$ diagram in Lagrangian coordinates [28] for estimating the thickness l using the optical indicator. In the scheme, the wedge-shaped target has three steps (there may be more of them). If we restrict our consideration to the characteristic leading wave, then it can be easily found that the local (Lagrangian) rarefaction wave velocity is determined as $c^L = D(l + \Delta)(l - \Delta)$. The sound speed in the target subjected to shock compression will be determined by the expression $c = c^L(\rho_0/\rho) = (D - u) \cdot (l + \Delta)(l - \Delta)$ obtained in the previous section.

In the $x - t$ diagram the solid lines represent the shock waves and the rarefaction waves generated from the impact of an aluminum plate 2 mm thick onto a target made of the same material. The characteristic C^Ls going to the left are caused by reflections of the shock wave from the target-indicator interface and correspond to the case where the indicator acoustic stiffness is less than the target material stiffness. The dotted lines denoted as D, C_2^L, and C_3^L correspond to the shock waves and characteristic rarefaction waves in the indicator. The dot-and-dash line connects the points the shock wave overtakes the rarefaction waves in the indicator. As seen from Fig. 4.18, if there had been no indicator in the system, the rarefaction wave would have overtaken the shock wave at a distance of 7 mm from the original impact surface. From the above $x - t$ diagram it follows that its construction requires data for the indicator Hugoniots and pressure or mass velocity functions of sound speeds.

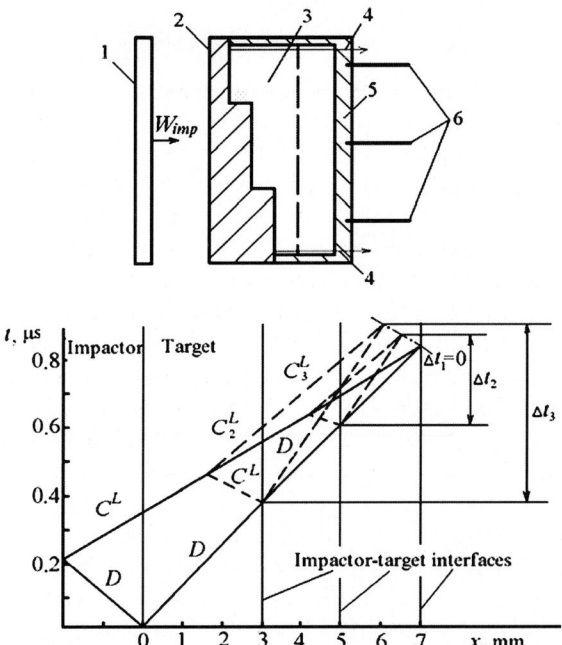

Fig. 4.18. Scheme (**a**) and $x - t$ diagram (**b**) of the experiment on sound speed measurement with optical indicator. 1 – impactor; 2 – target; 3 – liquid indicator; 4 – electric contacts; 5 – light guide holder; 6 – light guides

If the indicator itself is used in the experiment, then the procedure of estimating sound speed in it is considerably simplified. Figure 4.19 presents a typical streak-camera image of the shock front emission in liquid argon, which can be considered as an indicator, and its darkening densitogram obtained in [30]. The shock wave was transferred to liquid argon through a metal base plate upon impact by a thin plate.

The densitogram shows that in time Δt after the shock wave is transferred to the condensed gas the recorded emission intensity is decreased, which is due to the rarefaction wave arrival, the second rise in the luminous intensity corresponds to the shock wave release to atmospheric air. The time interval Δt is determined directly from the film photometric measurements. The procedure to further estimate the sound speed in the liquid gas reduces to a construction of the experimental $x - t$ diagram. To do this, it is necessary to know the thicknesses of the impactor and the target, shock compression and sound speed parameters for each, and finally, the shock wave parameters for the material under study. The sound speed in a gas is related to the slope of the Hugoniot and the Grueneisen factor as shown in [13]:

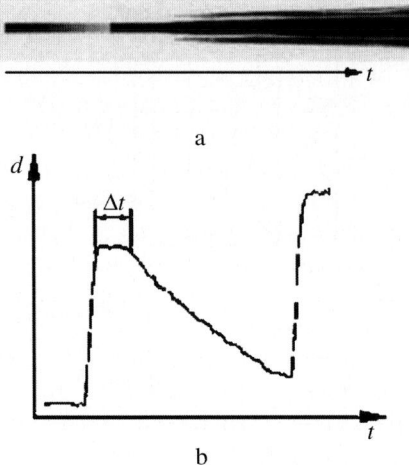

Fig. 4.19. Experimental streak-camera image (**a**) and film darkening densitogram (**b**)

$$c^2 = -V^2 \left(\frac{dP_H}{dV}\right)\left[1 - \left(\frac{\Gamma}{V}\right)\frac{(V - V_0)}{2}\right] + V^2\left[\left(\frac{\Gamma}{V}\right)(P_H - P_0)\right]\bigg/2$$
$$+V^2(P - P_H)\left[\left(\frac{\Gamma}{V}\right) + d\ln\left(\frac{\Gamma}{V}\right)/V\right].$$

The first two terms in this expression describe sound speed on the Hugoniot and the last one the sound speed as a function of pressure on the isentrope as it departs from the Hugoniot.

If the sound speed and material Hugoniot are estimated in independent experiments, this expression can be used to determine the Grüneisen factor. Indeed, if $P = P_H$, we have:

$$c^2 = -V^2 \left(\frac{dP_H}{dV}\right)\left[1 - \left(\frac{\Gamma}{V}\right)\frac{(V - V_0)}{2}\right] + V^2\left[\left(\frac{\Gamma}{V}\right)(P_H - P_0)\right]\bigg/2$$

and

$$\Gamma = 2\left[c^2 + V\left(\frac{dP}{dV}\right)\right]\bigg/V\left[\left(\frac{dP}{dV}\right)(V_0 - V)/2 + \frac{P}{2}\right].$$

Recall that when the analytical function $P_x(\delta) = \rho_0 c_0^2/n(\delta^n - 1)$ is used for the cold compression curve in the Mie–Grüneisen equation of state, we obtain Eq. (1.32) for the isentropic sound speed c in the shock-compressed state.

Many materials can be used as indicators. The principal requirements for an indicator material are transparency, radiation emission as a function of pressure in the shock wave with a very narrow front width, and the possibility of an impedance match between the material under study and the indicator material. These requirements are well met by liquid methane halogen derivatives. As of now, a large amount of experimental information has been accumulated on their characteristics and behavior under shock compression conditions which allows their use for studying shock and detonation waves in condensed matter. It should be also noted that liquid indicators have two advantages over solid indicators: first, they readily provide a perfect interface with the target, and second, no complications relating to the effect of elastic-plastic flow in the indicator arise. Some characteristics of these materials are presented below.

Bromoform. This is a transparent liquid of $\rho_0 = 2.89$ g/cm^3 density with a sound speed $c = 0.928$ km/s at $t = 20°$C. Its Hugoniot is experimentally estimated in many papers. The available data are computationally processed and used most extensively in [31]. According to [31], the shock wave velocity as a function of mass velocity is expressed as D [km/s] $= 1.116 + 1.548\ u - 0.0183\ u^2$.

Processing the available sound speed data over $u = 1.5$–6.5 km/s mass velocity range resulted in the following [31]:
c [km/s] $= 1.95 + 1.289\ u - 0.027\ u^2$.
The shock front temperature dependence can be represented as [30]:
T [kK] $= 0.293 + 0.424\ u + 0.4469\ u^2$.
Carbon tetrachloride. Its initial density is $\rho_0 = 1.585$ g/cm^3. According to [29], its sound speed is $c = 0.917 + 0.302P - 2.19 \cdot 10^{-3}P^2$ for the pressure range from 13 to 20 GPa.

According to [32], D as a function of u is $D = 0.94 + 2u - 0.1\ u^2/0.94$.

Among other methane halogen derivatives, of interest are chloroform, dichloromethane, dibromomethane, and di-iodomethane [29].

4.3.4 Sound Speed Measurement with Electromagnetic and Manganin Sensors

The authors of [33,34] estimated sound speeds in shock-compressed dielectrics using electromagnetic and manganin sensors. The $x - t$ diagram illustrating the typical setup for an experiment using internal Lagrangian type sensors appears in Fig. 4.20.

Maximum velocity of the rarefaction wave propagation through shock-compressed material under study is calculated as:

$$c_1 = u_1 + \frac{[L - u(\Delta t_1 + \Delta t_2)]}{\Delta t_2} = \frac{L}{\sigma \Delta t_2}\ ,$$

where L is sample thickness from the sensor to the free surface; D_1 and u_1 are shock wave characteristics; Δt_1 is the time of the shock wave travel across the sample from the sensor to the electric contact positioned on the sample free

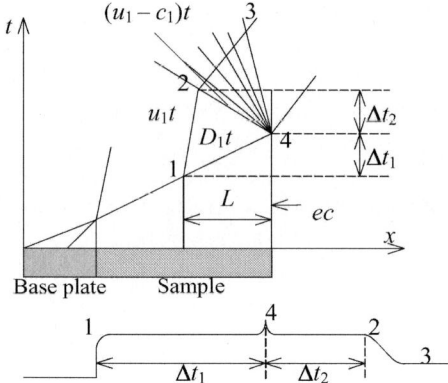

Fig. 4.20. $x - t$ diagram of wave processes obtained using internal sensors:1, 4 – shock wave; 2, 3, 4 – rarefaction wave; ec – electric contact

surface; Δt_2 is the difference between the time of the electric contact closing and that of the rarefaction waves arrival at the internal sensor; σ is the sample compression in the shock wave.

4.4 Hugoniot Classification by Experimental Data Processing Method

As already noted, two variables appearing in the conservation Equations (Eq. 1.21) are sufficient to estimate the Hugoniot state parameters when initial measurement conditions are known. The dependency describing the experimental points on the Hugoniot can be represented as a relationship between any pair of the variables, for example, $P - u$, $P - V$, $P - \rho$, $P - \eta$. However, using kinematic parameters D and u as variables measured in the experiments is especially convenient for the primary experimental data description.

As early as the 1950s, in the first shock compressibility studies, it was found that Hugoniots for many metals and other materials could be represented with fairly good accuracy by the linear $D-u$ relationship, $D = a_0 + a_1 u$, though no strict theoretical justification for the linearity has thus far been shown. It is, however, validated to a certain extent by calculations with model potentials performed for Na, K, and CsI in [35].

As the high-pressure range was investigated, material classes and groups were studied, experimental data statistics were augmented, Hugoniots were obtained whose configurations in $D - u$ variables differed significantly from linear ones. Appropriate explanations are invoked to account for the deviation from linearity. Thus, for example, in a large group of metals, including some of the alkali metals, alkaline-earth and transition metals, a peculiar kind of Hugoniot with kinks separating highly compressible states from much less compressible states were detected. These may be attributed to an outer electron transition to free inner orbits [36, 37]. Many materials are characterized with $D - u$ dependencies of a stepwise shape, attributable to the phase transitions occurring at different pressure levels.

Analysis of the literature suggests that shock Hugoniot systematization by the following types [38] is most suitable to the analytical representation.

1. Linear Hugoniots approximating the experimental data by the previously presented relation: $D = a_0 + a_1 u$, $a_0 > 0, a_1 > 0$.
2. Parabolic Hugoniots describing the experimental data by equation $D = a_0 + a_1 u + a_2 u^2$, $a_0 > 0, a_1 > 0, a_2 < 0$.
3. Parabolic Hugoniots with $a_2 > 0$.
4. Hugoniots with kinks. The lower branches of these Hugoniots are almost always linear, while the upper can be either linear or parabolic.
5. Hugoniots with a stepwise shape. Like in the previous case, the lower branches are usually linear, while the upper can be either linear or parabolic.

The authors of [38] performed statistical analysis for both domestic and foreign experimental data using the above Hugoniot classification for 56 metallic elements. Reasoning from the analysis, most reliable $D - u$ dependencies are found and the measure of accuracy of their position in the plane of kinematic variables is determined. The authors of [39] obtained Hugoniot relations specified more exactly in the form of dependencies $D(u)$ characteristic of the first, second and third Hugoniot types for 25 metals with smooth Hugoniots, taking into consideration additional experimental data. A portion of their statistical analysis results is presented in Tables 4.1–4.5 specifying regression factors a_0, a_1, a_2 and ranges of approximations either in pressure P_{max} or in mass velocity u_{max}. For the elements marked with asterisks in Table 4.1, initial linear intervals are separated. However, if all available information is taken into consideration, the Hugoniots of the metals refer to the second or third type. The coefficients in trinomials for the elements with separated linear segments appear in Tables 4.2 and 4.3. Figures 4.21–4.24 present the $D - u$ diagrams [40] for metals pertaining to groups 2, 3, 4, and 5.

4.5 Semi-Empirical Equations of State

Information about continuous and porous material Hugoniots, sound speeds in shock-compressed materials, and release isentropes from shock-compressed states obtained from dynamic experiments is used to construct semi-empirical equations of state (EOS). The development of semi-empirical models is normally accomplished through implementation of the following steps. Based on intuitive considerations, the generic form of the proposed analytical EOS is selected. This generic form often includes a number of free constants that must be adjusted so that the experimental data is optimally matched. There

Table 4.1. Hugoniots, group 1

Element	P_{max}, GPa	ρ_0, g/cm^3	a_0, km/s	a_1
Li	71	0.53	4.77	1.065
Na*	34	0.97	2.63	1.208
K	90	0.86	1.99	1.175
Cu*	444	8.93	3.91	1.500
Ag*	157	10.49	3.15	1.651
Au	603	19.30	3.01	1.576
Be*	91	1.85	7.99	1.128
Mg	177	1.74	4.51	1.250
Zn*	198	7.14	3.14	1.489
Cd*	152	8.64	2.47	1.642
Al	208	2.71	5.33	1.357
In*	155	7.28	2.43	1.551
Pb*	990	11.35	1.98	1.568
V*	128	6.10	5.07	1.186
Nb*	186	8.59	4.44	1.192
Ta*	224	16.65	3.40	1.230
Mo	1041	10.21	5.10	1.262
W	542	19.22	4.00	1.255
Re	625	21.02	4.17	1.349
Co*	167	8.82	4.67	1.342
Ni*	438	8.87	4.54	1.507
Rh*	216	12.43	4.74	1.426
Pd*	221	11.99	3.96	1.652
Ir	661	22.48	3.93	1.533
Pt*	687	21.42	3.59	1.571

Table 4.2. Hugoniots, group 2

Element	P_{max}, GPa	a_0, km/s	a_1	$a_2 \cdot 10^2$, s/km
Cu	927	3.90	1.534	−0.96
Ag	460	3.14	1.744	−4.0
Be	162	7.96	1.212	−2.74
Zn	839	3.11	1.534	−1.01
Cd	863	2.46	1.734	−4.43
In	362	2.43	1.588	−1.79
Pb	990	1.98	1.568	−3.26
Ni	1019	4.53	1.545	−0.95
Pd	221	3.96	1.652	−4.53
Pt	687	3.59	1.591	−2.81

Table 4.3. Hugoniots, group 3

Element	P_{max}, GPa	a_0, km/s	a_1	$a_2 \cdot 10^2$, s/km
Na	99	2.63	1.193	+0.87
V	343	5.08	1.144	+2.40
Nb	409	4.45	1.117	+3.85
Ta	1136	3.42	1.193	+2.12
Co	434	4.63	1.288	+4.58
Rh	498	4.75	1.356	+4.19

Table 4.4. Hugoniots, group 4

Element	ρ_0, g/cm^3	a_{0l}, a_{0u}	a_{1l}, a_{2u}	$a_{2u} \cdot 10^2$	D_k	u_k	u_{max}
La	6.15	2.06	1.012	–	3.12	1.05	5.3
		1.39	1.702	−4.56			
Pr	6.81	2.11	0.779	–	3.14	1.32	5.2
		0.76	1.858	−4.3			
Nd	7.00	2.17	0.853	–	3.15	1.19	5.1
		1.42	1.450	–			
Sm	7,50	2,22	0.806	–	3.15	116	5.0
		1.84	1.062	6.62			
Gd	7.93	2.20	0.947	–	3.27	1.16	7.1
		1.80	1.370	–			
Er	9.05	2.29	0.947	–	3.87	1.67	4.8
		1.58	1.370	–			
Yb	6.93	1.43	0.865	–	2.12	1.03	5.2
		0.87	1.419	–			
Dy	8.52	2.25	0.926	–	3.62	1.48	4.9
		1.84	1.157	3.09			
Ho	8.73	2.29	0.934	–	3.44	1.22	4,8
		2.15	0.986	5.26			
Lu	9.74	2.20	1.00	–	4.08	1.88	–
		0.98	1.65	–			
Ca	1.52	3.44	0.968	–	7.01	3.69	7.00
		2.40	1.248	–			

Note: Subscripts "l" and "u" denote the coefficients at the lower and upper Hugoniot branches, respectively, D_k and u_k are kink coordinates.

Table 4.5. Hugoniots, group 5

Element	ρ_0, g/cm^3	a_{0l}, a_{0u}	a_{1l}, a_{2u}	$-a_{2u} \cdot 10^2$	D_k	u_{kl}, u_{ku}	$u_{\max}$
Fe	7.85	4.63	1.334	3.37	5.06	0.32	8.0
		3.66	1.79			0.86	
Ti	4.50	5.22	0.767	–	5.72	0.65	9.4
		4.72	1.130			0.88	
Zr	6.51	3.83	0.914	–	4.63	0.88	5.2
		3.24	1.301			1.07	
Hf	13.16	2.95	1.069	–	3.89	0.88	4.4
		2.42	1.325			1.11	
Eu	5.19	1.72	0.848	–	2.43	0.84	4.0
		1.02	1.303			1.08	

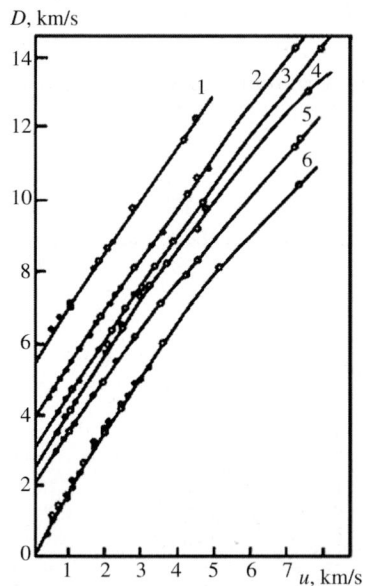

Fig. 4.21. $D - u$ diagrams for group 2 metals: 1 – Ni ($D + 1$); 2 – Cu; 3 – Zn; 4 – Cd; 5 – Pb; 6 – Ce ($D - 1$)

are many variations of the semi-empirical EOS, some of them are discussed in reviews [22, 40, 41]. Below we consider several semi-empirical EOS that have been used at VNIIEF and implemented into hydrodynamic computer codes.

For many years VNIIEF used the Mie–Grüneisen form of EOS (see Chap. 1) with the analytical dependence for the cold compression curve, $P_x(\delta) = \rho_0 c_0^2 / n(\delta^n - 1)$, for gas-dynamics computations of various problems. The analytical expression for the Hugoniot with this dependency $P_x(\delta)$ is Eq. (1.27) and the equation for the isentrope passing through point P_1, P_{1T},

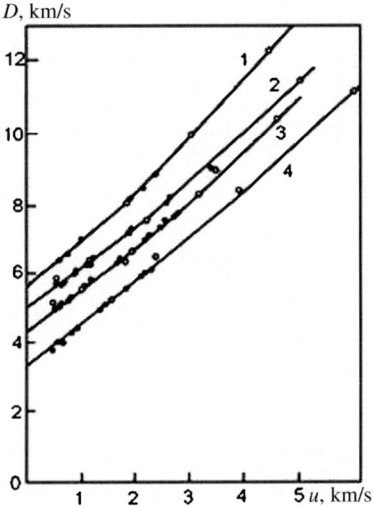

Fig. 4.22. $D - u$ diagrams for group 3 metals: $1 - $ Co $(D + 1)$; $2 - $ V; $3 - $ Nb; $4 - $ Ta

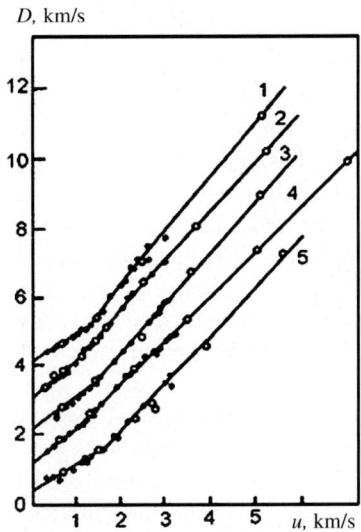

Fig. 4.23. $D - u$ diagrams for group 4 metals: $1 - $ Pr $(D + 2)$; $2 - $ La $(D + 1)$; $3 - $ Nd; $4 - $ Gd $(D - 1)$; $5 - $ Y $(D - 3)$

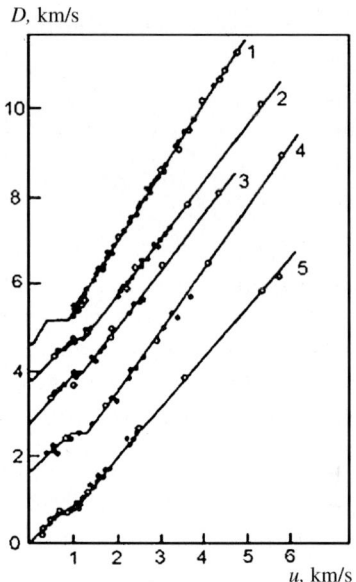

Fig. 4.24. $D - u$ diagrams for group 5 metals: 1 – Fe; 2 – Zr; 3 – Hf; 4 – Eu; 5 – Ti ($D - 5$)

σ_1 on the Hugoniot is Eq. (1.33). At VNIIEF this EOS has been called the "limiting-density EOS" since, if $\eta_1 \rightarrow h$, $P_1 \rightarrow \infty$, i.e. h is condensed matter which limits the compression in a shock wave. However, if $\sigma \rightarrow h$, P_x will be a constant and the main contribution to total pressure and energy will be made by their thermal components. It should be noted that the limiting-density EOS was developed by E.I. Zababakhin [42] and has been widely used until now in gas-dynamics applications, so his name in its own right might be assigned to the equation. The simplicity of the EOS and minimum quantity of experimental data needed for selection of its three adjustment parameters, c_0, n, and h, make it convenient for evaluation of the material state under various shock effects.

When estimating EOS parameters, material initial density ρ_0 and porosity (k) are considered given. The limiting compression h is estimated either by an experimentally studied compression range or found from computational models (for example, the Thomas-Fermi model with corrections) at pressures of several tens of thousands of GPa. The two remaining parameters, c_0 and n, can be found from minimizing the function

$$F = \sum_{i=1}^{N} \left[P_{ie} - P_{i.calc} \left(\sigma_i \right) \right]^2 ,$$

where P_{ie}, σ_i are experimental pressures and compressions, $P_{i.calc}$ is the calculated Hugoniot, N is the number of points.

Minimizing the function F leads to the following set of equations:

$$\sum_{i=1}^{N} [P_{ie} - P_{i.calc}(\sigma_i)] \frac{\partial P_{i.calc}(\sigma_i)}{\partial c_0} = 0;$$

$$\sum_{i=1}^{N} [P_{ie} - P_{i.calc}(\sigma_i)] \frac{\partial P_{i.calc}(\sigma_i)}{\partial n} = 0.$$

The set of equations are nonlinear in c_0 and n and can be solved using an iterative method. It should be noted that the form of the limiting-density EOS yields different parameters n, h, and c_0, depending on the pressure or compression range taken for the description. In Table 4.6, the limiting-density EOS parameters for a number of materials are given with their applicable pressure range and the computed deviation from experimental Hugoniot values. As seen from the table, for the metals presented the values of c_0 somewhat differ from those of a_0 which arise from the $D - u$ dependencies.

Table 4.6. EOS parameters with limiting densities for some metals and polymers

Material	P_0, g/cm^3	c_0, km/s	n	h	Maximum range in P, GPa	Mean-square deviation $\frac{P_{calc}-P_e}{P_e}100\%$
Metals						
Aluminum	2.71	5.30	3.998	4.0	220	0.4
Tungsten	19.35	4.09	3.42	3.5	500	0.4
Iron	7.85	3.74	5.642	4.5	$0.35 \le P \le 320$	0.6
Copper	8.93	3.90	4.832	4.5	1000	0.5
Molybdenum	10.2	5.17	3.582	4.5	1000	0.5
Nickel	8.87	4.80	4.507	5.2	1000	2.8
Tin	7.28	2.58	4.824	4.5	300	3.4
Lead	11.34	2.41	3.95	5.2	1000	8.7
Tantalum	16.38	3.45	3.249	3	1000	1.4
Zinc	7.14	3.20	4.757	5.2	800	2.7
Polymers						
Caprolon	1.14	3.40	3.52	3.2	60	1.8
Paraffin	0.90	2.81	5.16	4.5	50	1.9
Polycarbonate	1.19	2.38	4.6	3.3	25	9.7
Polyethylene	0.92	2.87	4.427	3.0	50	0.4
Polymethylmethacrylate	1.18	2.72	4.324	2.5	≤ 20	1
(Plexiglas)	1.18	3.54	3.08	3.5	$20 \le P \le 100$	0.7
Polystyrene	1.05	2.35	5.066	3.0	≤ 20	0.6
	1.05	0.84	5.276	2.38	$20 \le P \le 50$	0.15
Polytetrafluoroethylene (teflon)	2.19	2.045	4.972	2.35	180	0.4

For example, Fig. 4.25 presents in $P - u$ coordinates experimental data for the isentropic expansion of lead from two points on the Hugoniot as well as its description from the limiting-density EOS using the parameters from Table 4.6. The limiting-density EOS is limited for a number of reasons. First, the power dependence for the elastic pressure is satisfactory over a range of densities although not too high and, as a rule, it does not give the proper asymptotic law $P_x \sim \rho^{5/3}$ following from the statistical atom model for high pressures.

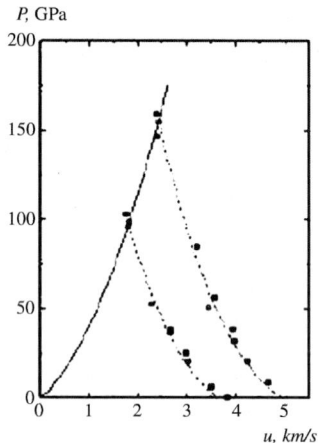

Fig. 4.25. Lead Hugoniot (———) and expansion isentropes originating from the states at $P = 98.5$ GPa (- - -) and $P = 154.6$ GPa (-·--·--·-), ■ – experiment

In the $\delta < 1$ range it is also limited, since if $\delta \to 0$, P_x tends to be negative, which disagrees with physical reality, where P_x should be zero when the atoms are far apart.

The EOS most flexible in its form and most physically justified is the EOS [43] called OSA from VNIIEF. The EOS pressure and energy are represented as a sum of potential, thermal nuclear and thermal electronic components. The Grüneisen factor for the lattice is described as a function of density and its analogs for electrons and nuclei heat capacities are taken as constants. Electronic heat capacity is taken as a linear function of temperature.

The following experimental information is used for estimating the EOS parameters and making them consistent: continuous material Hugoniot given by the most probable $D - u$ dependency; porous material Hugoniots; sound speed in shock-compressed materials; data for the isentropic expansion of materials from shock-compressed state.

Further refinements of the EOS OSA involves making the vibrational and electronic heat capacities functions of density and temperature. The updated EOS OSA version has been called ROSA [43]. For EOS ROSA an algorithm has been developed to calculate other thermodynamic parameters from the

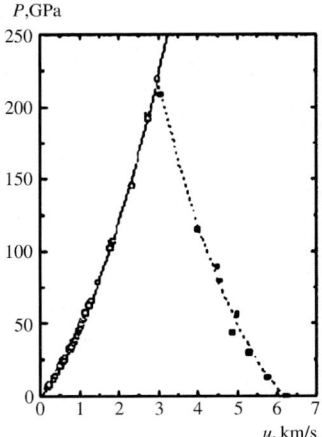

Fig. 4.26. Continuous copper Hugoniot ($\square$ – experiment, —— – computation) and expansion isentrope ($\blacksquare$ – experiment, - - - – computation)

given P, ρ or E, ρ using analytical dependencies. For example Fig. 4.26 presents experimental data on the Hugoniot and expansion isentrope of copper with their EOS ROSA description, more examples of the usage of EOS ROSA are given in [44]. Using the EOS ROSA proves to be fairly computationally efficient and differs slightly from the simpler EOS OSA model in its characteristic form.

References

1. Altshuler, L.V., Krupnikov, K.K., Ledenev, B.N., Zhuchikhin, V.I., and Brazhnik, M.I., "Dynamic Compressibility and Equation of State of Iron Under High Pressure," *Zhurnal Eksperimentalnoi i Teoreticheskoi Fiziki*, Vol. 34, No. 4, 1958, pp. 874–885, [English trans., *Soviet Physics JETP*, Vol. 7, No. 4, 1958, pp. 606–614].
2. Rice, M.N., McQueen, R.G., and Walsh, J.M., "Compression of Solids by Strong Shock Waves," *Solid State Physics – Advances in Research and Applications*, Vol. 6, 1958, pp. 1–63.
3. McQueen, R.G., and Marsh, S.P., "Equation of State for Nineteen Metallic Elements from Shock-Wave Measurements to Two Megabars," *Journal of Applied Physics*, Vol. 31, No. 7, 1960, pp. 1253–1269.
4. Zeldovich, Ya.B., and Raizer, Yu.P., *Physics of Shock Waves and High-Temperature Hydrodynamic Phenomena*, Nauka Publ., Moscow, 1966, [English trans. Academic Press, NY, Vol. 1 (1966), Vol. 2 (1967); Reprinted in a single volume by Dover Publ., Mineola, NY 2002].
5. Zhernokletov, M.V., Simakov, G.V., Sutulov, Yu.N., and Trunin, R.F., "Expansion Isentropes of Aluminum, Iron, Molybdenum, Lead, and Tantalum," *Teplofizika Vysokikh Temperatur*, Vol. 33, No. 1, 1995, pp. 40–43, [English trans., *High Temperature*, Vol. 33, No. 1, 1995, pp. 36–39].

6. Altshuler, L.V., Krupnikov, K.K., and Brazhnik, M.I., "Dynamic Compressibility of Metals Under Pressures From 400,000 to 4,000,000 Atmospheres," *Zhurnal Eksperimentalnoi i Teoreticheskoi Fiziki*, Vol. 34, No. 4, 1958, pp. 886–893, [English trans., *Soviet Physics JETP*, Vol. 7, No. 4, 1958, pp. 614–619].

7. Bugaeva, V.A., Evstigneev, A.A., and Trunin, R.F., "Analysis of Calculation Data on the Adiabats of Expansion for Copper, Iron, and Aluminum," *Teplofizika Vysokikh Temperatur*, Vol. 34, No. 5, 1996, pp. 684–690, [English trans., *High Temperature*, Vol. 34, No. 5, 1996, pp. 674–680].

8. Altshuler, L.V., Trunin, R.F., Krupnikov, K.K., and Panov, N.V., "Explosive Laboratory Devices for Studying Material Compression in Shock Waves," *Uspekhi Fizicheskikh Nauk*, Vol. 166, No. 5, 1996, pp. 575–581.

9. Altshuler, L.V., and Petrunin, A.P., "An X-Ray Investigation of the Compressibility of Light Materials Under the Action of Shock-Wave Impacts," *Zhurnal Tekhnicheskoi Fiziki*, Vol. 31, No. 6, 1961, pp. 717–725, [English trans., *Soviet Physics – Technical Physics*, Vol. 6, No. 6, 1961, pp. 516–522].

10. Altshuler, L.V., and Pavlovskii, M.N., "Magnetoelectric Method for Determining the Density Behind the Front of Colliding Shock Waves," *Zhurnal Prikladnoi Mekhaniki i Tekhnicheskoi Fiziki*, 1971, No. 2, pp. 110–114, [English trans., *Journal of Applied Mechanics and Technical Physics*, Vol. 12, No. 2, 1971, pp. 268–272].

11. Trunin, R.F., Gudarenko, L.F., Zhernokletov, M.V., and Simakov, G.V., Experimental Data on Shock Compression and Adiabatic Expansion of Condensed Materials, RFNC-VNIIEF, Sarov, Russia, 2001.

12. Erskine, D.J., "Improved Arrangement of Shock-Detecting Pins in Shock Equation of State Experiments," *Review of Scientific Instruments*, Vol. 66, No. 10, 1995, pp. 5032–5036.

13. McQueen, R.G., Marsh, S.P., Taylor, J.W., Fritz, J.N., and Carter, W.J., "The Equation of State of Solids from Shock Wave Studies" in *High Velocity Impact Phenomena*, Kinslow, R., ed., Academic Press, New York, 1970, pp. 293–417 (with appendices on pp. 515–568), [Russian trans., Nikolayevsky, V.N., ed., Mir Publ., Moscow, 1973, pp. 299–427].

14. Mitchell, A.C., and Nellis, W.J., "Shock Compression of Aluminum, Copper, and Tantalum," *Journal of Applied Physics*, Vol. 52, No. 5, 1981, pp. 3363–3374.

15. Zhernokletov, M.V., Zubarev, V.N., and Sutulov, Yu.N., "Porous-Specimen Adiabats and Solid-Copper Expansion Isentropes," *Zhurnal Prikladnoi Mekhaniki i Tekhnicheskoi Fiziki*, 1984, No. 1, pp. 119–123, [English trans., *Journal of Applied Mechanics and Technical Physics*, Vol. 25, No. 1, 1984, pp. 107–110].

16. Altshuler, L.V., Bushman, A.V., Zhernokletov, M.V., Zubarev, V.N., Leontev, A.A., and Fortov, V.E., "Unloading Isentropes and the Equation of State of Metals at High Energy Densities," *Zhurnal Eksperimentalnoi i Teoreticheskoi Fiziki*, Vol. 78, No. 2, 1980, pp. 741–760, [English trans., *Soviet Physics JETP*, Vol. 51, No. 2, 1980, pp. 373–383].

17. Glushak, B.L., Zharkov, A.P., Zhernokletov, M.V., Ternovoi, V.Ya., Filimonov, A.S., and Fortov, V.E., "Experimental Investigation of the Thermodynamics of Dense Plasmas Formed From Metals at High Energy Concentrations," *Zhurnal Eksperimentalnoi i Teoreticheskoi Fiziki*, Vol. 96, No. 4, 1989, pp. 1301–1318, [English trans., *Soviet Physics JETP*, Vol. 69, No. 4, 1989, pp. 739–749].

18. Zhernokletov, M.V., Medvedev, A.V., and Simakov, G.V., "Expansion Isentrope and Equation of State of Molybdenum at High Energy Densities, *Khimicheskaya Fizika*, Vol. 14, No. 2–3, 1995, pp. 49–55.

19. Zhernokletov, M.V., "Shock Compression and Isentropic Expansion of Natural Uranium," *Teplofizika Vysokikh Temperatur*, Vol. 36, No. 2, 1998, pp. 231–238, [English trans., *High Temperature*, Vol. 36, No. 2, 1998, pp. 214–221].

20. Gathers, G.R., Shaner, J.W., and Brier, R.L., "Improved Apparatus for Thermophysical Measurements on Liquid Metals up to 8000 K," *Review of Scientific Instruments*, Vol. 47, No. 4, 1976, pp. 471–479.

21. Gathers, G.R., Shaner, J.W., and Young, D.A., "Experimental, Very High-Temperature, Liquid-Uranium Equation of State," *Physical Review Letters*, Vol. 33, No. 2, 1974, pp. 70–72.

22. Bushman, A.V., and Fortov, V.E., "Model Equations of State," *Uspekhi Fizicheskikh Nauk*, Vol. 140, No. 2, 1983, pp. 177–232, [English trans., *Soviet Physics Uspekhi*, Vol. 26, No. 6, 1983, pp. 465–496].

23. Asay, J.R., and Trucano T.G., "Experimental Measurements of Shock-Induced Vaporization in Cadmium and Lead," Shock Compression of Condensed Matter – 1989, Schmidt, S.C., Johnson, J.N., and Davison, L.W., eds., Elsevier, Amsterdam, 1990, pp. 143–146.

24. Asay, J.R., "The Use of Shock-Structure Methods for Evaluating High-Pressure Material Properties," *International Journal of Impact Engineering*, Vol. 20, No. 1–5, 1997, pp. 27–61.

25. Zhernokletov, M.V., and Zubarev, V.N., "Determination of Expansion Isentropes After Shock Compression," *in* Combustion and Explosion, Nauka Publ., Moscow, 1972, pp. 565–568.

26. Altshuler, L.V., Kormer, S.B., Brazhnik, M.I., Vladimirov, L.A., Speranskaya, M.P., and Funtikov, A.I., "The Isentropic Compressibility of Aluminum, Copper, Lead, and Iron at High Pressures," *Zhurnal Eksperimentalnoi i Teoreticheskoi Fiziki*, Vol. 38, No. 4, 1960, pp. 1061–1073, [English trans., *Soviet Physics JETP*, Vol. 11, No. 4, 1960, pp. 766–775].

27. Kormer, S.B., Sinitsyn, M.V., Kirillov, G.A., and Urlin, V.D., "Experimental Determination of Temperature in Shock-Compressed NaCl and KCl and of their Melting Curves at Pressures up to 700 kbar," *Zhurnal Eksperimentalnoi i Teoreticheskoi Fiziki*, Vol. 48, No. 4, 1965, pp. 1033–1049, [English trans., *Soviet Physics JETP*, Vol. 21, No. 4, 1965, pp. 689–700].

28. McQueen, R.G., Hopson, J.W., and Fritz, J.N., "Optical Technique for Determining Rarefaction Wave Velocities at Very High Pressures," *Review of Scientific Instruments*, Vol. 53, No. 2, 1982, pp. 245–250.

29. Gogulya, M.F., and Dolgoborodov, A.Yu., "Indicator Method for Study of Shock and Detonation Waves," *Khimicheskaya Fizika*, Vol. 13, No. 12, 1994, pp. 118–127.

30. Grigorev, F.V., Kormer, S.B., Mikhailova, O.L., Mochalov, M.A., and Urlin, V.D., "Shock Compression and Brightness Temperature of a Shock Wave Front in Argon. Electron Screening of Radiation," *Zhurnal Eksperimentalnoi i Teoreticheskoi Fiziki*, Vol. 88, No. 4, 1985, pp. 1271–1280, [English trans., *Soviet Physics JETP*, Vol. 61, No. 4, 1985, pp. 751–757].

31. Nikolayev, D.N., Pyalling, A.A., Khishchenko, K.V., *et al.*, "Thermodynamic Properties of Bromoform at High Pressures," *Khimicheskaya Fizika*, Vol. 19, No. 10, 2000, pp. 98–108.

32. Dolgoborodov, A.Yu., and Voskoboinikov, I.M., "Velocities of Wave Perturbations Behind a Shock-Wave Front in Aluminum," *Fizika Goreniya i Vzryva*, Vol. 25, No. 1, 1989, pp. 88–93, [English trans., *Combustion, Explosion, and Shock Waves*, Vol. 25, No. 1, 1989, pp. 80–85].

33. Altshuler, L.V., Pavlovskii, M.N., and Drakin, V.P., "Peculiarities of Phase Transitions in Compression and Rarefaction Shock Waves," *Zhurnal Eksperimentalnoi i Teoreticheskoi Fiziki*, Vol. 52, No. 2, 1967, pp. 400–408, [English trans., *Soviet Physics JETP*, Vol. 25, No. 2, 1967, pp. 200–205].

34. Pavlovskii, M.N., "Measurements of the Velocity of Sound in Shock-Compressed Quartzite, Dolomite, Anhydrite, Sodium Chloride, Paraffin, Plexiglas, Polyethylene, and Flouroplast-4," *Zhurnal Prikladnoi Mekhaniki i Tekhnicheskoi Fiziki*, 1976, No. 5, pp. 136–139, [English trans., *Journal of Applied Mechanics and Technical Physics*, Vol. 17, No. 5, 1976, pp. 709–712].

35. Ruoff, A.L., "Linear Shock-Velocity-Particle-Velocity Relationship," *Journal of Applied Physics*, Vol. 38, No. 13, 1967, pp. 4976–4980.

36. Altshuler, L.V., Bakanova, A.A., and Dudoladov, I.P., "Effect of Electron Structure on the Compressibility of Metals at High Pressure," *Zhurnal Eksperimentalnoi i Teoreticheskoi Fiziki*, Vol. 53, No. 6, 1967, pp. 1967–1977, [English trans., *Soviet Physics JETP*, Vol. 26, No. 6, 1968, pp. 1115–1120].

37. Gust, W.H., and Royce, E.B., "New Electronic Interactions in Rare-Earth Metals at High Pressure," *Physical Review B*, Vol. 8, No. 8, 1973, pp. 3595–3609.

38. Altshuler, L.V., Bakanova, A.A., Dudoladov, I.P., Dynin, E.A., Trunin, R.F., and Chekin, B.S., "Shock Adiabatic Curves of Metals. New Data, Statistical Analysis, and General Laws," *Zhurnal Prikladnoi Mekhaniki i Tekhnicheskoi Fiziki*, 1981, No. 2, pp. 3–34, [English trans., *Journal of Applied Mechanics and Technical Physics*, Vol. 22, No. 2, 1981, pp. 145–169].

39. Altshuler, L.V., Brusnikin, S.E., and Kuzmenkov, E.A., "Isotherms and Gruneisen Functions for 25 Metals," *Zhurnal Prikladnoi Mekhaniki i Tekhnicheskoi Fiziki*, 1987, No. 1, pp. 134–146, [English trans., *Journal of Applied Mechanics and Technical Physics*, Vol. 28, No. 1, 1987, pp. 129–141].

40. Fortov, V.E., Altshuler, L.V., Trunin, R.F., and Funtikov, A.I., Shock Waves and Extreme States of Matter, Nauka Publ., Moscow, 2000 [*see also*, Fortov, V.E., Altshuler, L.V., Trunin, R.F., and Funtikov, A.I., *High-Pressure Shock Compression of Solids VII*, Springer-Verlag, New York, 2004].

41. Altshuler, L.V., "Use of Shock Waves in High-Pressure Physics," *Uspekhi Fizicheskikh Nauk*, Vol. 85, No. 2, 1965, pp. 197–258, [English trans., *Soviet Physics Uspekhi*, Vol. 8, No. 1, 1965, pp. 52–91].

42. Zababakhin, E.I., Some Problems of the Gasdynamics of Explosions, RFNC-VNIITF, Snezhinsk, Russia, 1997, [English trans., RFNC-VNIITF, Snezhinsk, Russia, 2001].

43. Glushak, B.L., Gudarenko, L.F., Styazhkin, Yu.M., et al., "Semi-Empirical Equation of State of Metals with Variable Electron Heat Capacity," *Voprosy Atomnoi Nauki i Tekhniki. Seriya: Matematicheskoe Modelirovanie Fizicheskikh Protsessov*, 1991, No. 1, pp. 32–37.

44. Glushak, B.L., Gudarenko, L.F., Styazhkin, Yu.M., et al., "Semi-Empirical Equation of State of Metals with Variable Nuclei and Electron Heat Capacity," *Voprosy Atomnoi Nauki i Tekhniki. Seriya: Matematicheskoe Modelirovanie Fizicheskikh Protsessov*, 1991, No. 2, pp. 57–61.

5

Studies of Phase Transformations

B.L. Glushak and M.A. Mochalov

Depending upon thermodynamic state, materials can exist in a number of different physical phases (e.g., solid, liquid, or gas). Under certain conditions of temperature and pressure, it is possible (for some solids) to transition from one crystalline structure to another. Such transformations in solid materials are referred to as polymorphic transformations. Polymorphic transformations, melting and evaporation of solids are first-order phase transitions. Shock waves can be used to produce a wide range of material states (P, V, T) and establish the conditions for proceeding from one phase to another.

Phase transformations in shock waves, as well as phase transformations which may take place during expansion from a shock-compressed state, are studied using a variety of experimental techniques and diagnostics. Techniques and diagnostics include: experimentally determined $D - u$ and $P - u$ curves, measurements of flow parameters following the shock front, measurements of the history of free surface velocity, measurements of optical and electrical parameters, pulse X-ray diffraction analysis, and studies of structural changes in samples recovered following the application of shock-wave loading [1]. Many substances have been studied to investigate the issue of phase stability and the conditions under which any phase changes may take place, for example, many elements of the periodic table, metallic alloys, alkali metal halides, carbides and nitrides, oxides, rocks, and minerals.

In the remainder of the chapter we shall address a variety of methods for studying each type of the first-order phase transition in more detail.

5.1 Polymorphic Transformations in Shock Waves

For many solid materials, the propagation of a compressive shock wave of sufficient strength will result in a polymorphic phase transformation. Such transformations involve the rearrangement of the crystal lattice in such a way that new atomic equilibrium positions are arrived at, resulting in shorter inter-atomic distances. All polymorphic transitions (other than those which occur during unloading) occur with increasing pressure and result in a reduction of specific volume, that is, the process of polymorphic transformation takes the material from a "low" density phase to one of higher density. Polymorphic transformations have been observed in many metals, semiconductors, ionic compounds, and essentially all minerals and rocks.

Polymorphic transitions in shock waves are typically categorized as martensitic, being driven by material shear strains. The rearrangement of the crystalline structure by this mechanism provides for a "superfast" (on the order of 10^{-7}–10^{-8} s) transformation from one lattice type to another, accomplished through a cooperative shift of sets of atoms over short distances. This essentially athermal transformation consumes no activation energy and proceeds at relatively low temperatures. The results of original papers on this topic are generalized, analyzed, and interpreted in [1–6].

The rearrangement in crystalline structure that occurs during the compression of a solid leads to a change in material compressibility and is responsible for some rather interesting features of material behavior, both in compression and in subsequent expansion.

These features are due to the existence of convex segments residing on either side of a phase-transformation-induced kink in the P-V shock Hugoniot (see Fig. 5.1).

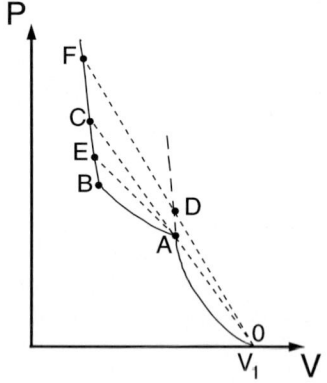

Fig. 5.1. Diagram of shock compression of a solid undergoing phase transition

Following Zeldovich [6], let us consider, qualitatively, the 1D pattern of flow that occurs when a solid body possessing a Hugoniot of the type depicted in Fig. 5.1 is subjected to a constant pressure P, applied to the surface of the body at the initial time. We shall consider the pressure to be sufficiently high that strength effects may be neglected, that is, the material is considered to behave hydrostatically. Segment $0A$ on the Hugoniot corresponds to the first (initial) phase Hugoniot. During compression, once state A has been achieved, transition from the first phase to the second denser phase begins, and this transformation is completed at state B. Thus, segment AB of the Hugoniot corresponds to a two-phase material state. Above state B the second-phase Hugoniot rises steeply.

Material compressibilities are different in the two phases, thus the slopes of the curves corresponding to the single-phase states at points A and B are different.

Let us consider the variety in shock wave structure that may exist when a medium capable of undergoing a polymorphic transformation is subjected to shock-wave loading of varying strength.

If the applied pressure P is lower than pressure P_A (the pressure at which phase transition commences), only one shock wave will propagate through the medium. Behind the shock front, the medium will be in the first (initial) phase.

If the applied pressure P is higher than P_C (the pressure corresponding to the intersection of the Hugoniot with the line $0AC$ that touches the Hugoniot at kink A), then once again, only one shock wave will propagate through the medium. Behind such a shock, the material exists in the state represented by point F and the material structure is wholly second phase. The transition from the first phase to the second occurs across the shock front. Direct shock compression to state F involves an intermediate state D, located at the intersection of the extrapolated first-phase Hugoniot $0D$ and the line segment $0F$. It is at point D that the phase transition begins, with the thickness of the transition zone, Δx being determined by the transition relaxation time. The resultant pressure profile for a shock of this magnitude is shown in Fig. 5.2. The point characterizing the state in the extended zone Δx of the wave front sweeps through the segment DF of Fig. 5.1.

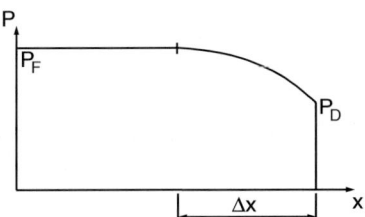

Fig. 5.2. Pressure profile in the shock wave with phase transition relaxation

Let us now consider the intermediate case, where the magnitude of the applied pressure is somewhere between P_A and P_C, for example, P_E. In this pressure range, the shock wave is split into two independent waves, each with its own velocity of propagation. These two waves propagate one after the other, and the medium is subjected to double compression. The first shock wave (of amplitude P_A), compresses the material to state V_A and the velocity of its propagation through unperturbed material is determined by the slope of straight line $0A$:

$$D_1 = V_0\sqrt{\frac{P_A - P_0}{V_0 - V_A}} \; .$$

The velocity of the first shock wave relative to the material behind it is

$$D_1 = (D - u) = V_A\sqrt{\frac{P_A - P_0}{V_0 - V_A}} \; . \tag{5.1}$$

The velocity of the second shock wave running through the compressed and moving material is determined by the slope of straight line AE (see Fig. 5.1):

$$D_2 = V_A\sqrt{\frac{P_E - P_A}{V_A - V_E}} \; . \tag{5.2}$$

From Eqs. (5.1) and (5.2), it follows that $D_1 > D_2$, that is, the first wave propagates at a higher velocity $((P_A - P_0)/(V_0 - V_A) > (P_E - P_A)/(V_A - V_E))$. Consequently, the two-wave structure is stable. With the passage of the second shock wave a phase transition occurs. At state E the material exists as either a two-phase material (if $P_E < P_B$) or as a wholly second phase material ($P_E \geq P_B$). Since the phase transformation takes place in finite time, the front of the second shock wave will be severely smeared, relative to the thin front of the first shock wave.

The pressure profile for the case involving the splitting of the shock into two independent waves is schematically plotted in Fig. 5.3. The distance between the two wave fronts increases with time since the wave velocities are different, but the pressure distribution in the second wave is time-independent and the profile in the second wave propagates unchanged in time.

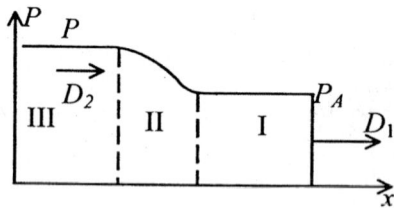

Fig. 5.3. Pressure profile in shock wave splitting. I – phase 1, II – combination of two phases, III – phase 2

Qualitatively, this two-wave pattern of the flow in a medium that has undergone phase transition due to compressive shock loading is similar to the patterns of flow seen in an elastic-plastic medium. The reason for the appearance of two waves in each of these cases is an anomaly in the material response resulting in a convex upward region of the Hugoniot. The pressure range $P_A < P < P_C$ represents a domain for the existence of a two-wave shock structure. The states in this domain are of great interest when studying phase transformations. This state-space is inaccessible by single shock-wave measurements in compression from the initial state V_0. Hence the conventional method for determining shock compressibility involving the measurement of average wave velocity produced by one-time loading of the sample cannot be used to study the states in this region. This method can, however, can be used to determine states P_A and P_C.

In shock wave experiments, the production of new phases is detectable if the transformation time is considerably shorter than the duration of the high pressure loading on the samples under study. Otherwise, the dynamic experiments will fix low-pressure phases or equilibrium metastable phase combinations. The above-discussed features, which are representative of the behavior of materials undergoing polymorphic phase transformation, are used as the basis for the most frequently used experimental methods.

The simplest of these involves the determination of the compression Hugoniot in $D - u$ coordinates using the reflection (impedance-matching) method developed to study shock compressibility [3]. In accordance with theory, the $D - u$ dependency should exhibit a discontinuity, such as is shown in Fig. 5.4. In the region where a two-wave configuration of the shock wave exists ($P_A < P < P_C$), the wave velocity of the first shock wave corresponding to state P_A is recorded. This velocity will be independent of the magnitude of the shock wave penetrating the material in the above pressure range.

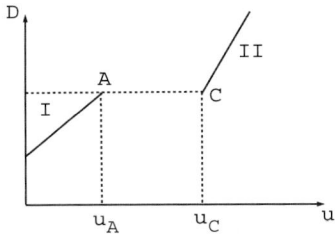

Fig. 5.4. $D - u$ dependency for media with polymorphic transition: I – phase 1, II – phase 2

With known experimental values of D_A and u_A, the position of point A in the $P - V$ plane can be found using the relationships $P_A = \rho_0 D_A u_A$; $V_A = (D_A - u_A)/\rho_0 D_A$. This method has been widely used to study rock behavior.

The most favorable conditions for obtaining quantitative measurement of the phase transition parameters are provided when a measurement of the two-wave configuration is made. This measurement may be made using sensors, either internal or external to the object under study. Since the waves are separated from each other in space, whatever diagnostic is chosen for the study, the recording provided by this diagnostic for any given Lagrangian point, should be continuous during a time span sufficient to acquire the entire wave structure.

Using external sensors, the two-wave shock profile may be recorded by means of measuring the velocity history of the free surface of a planar sample subjected to shock loading. The phenomenon of shock wave splitting in iron has been experimentally observed using several dozens of electro-contact sensors positioned at various distances from the free surface, i.e. an $x - t$ diagram of free surface motion such as that presented in Fig. 3.5 of Chap. 3, was found using this technique. This method was used to estimate the critical pressure, $P_A = 13$ GPa, required for the transition of iron from the α-phase to the ε-phase by determining the amplitude of the first shock wave measuring free surface velocity and exploiting the relationships between the various Hugoniot parameters. The complete two-wave structure of the shock wave in iron was recorded in [7] using a more advanced method involving the continuous recording of free surface velocity using a capacitive probe.

The experimental data obtained using continuous recording of free surface velocity (processed to give pressure) is plotted in Fig. 5.5, where region 1 represents the first shock wave, and region 3 represents the second shock wave. The region of shock front smearing, (region 2), extends over a time period $\tau \approx (2-3) \cdot 10^{-7}$ s, which characterizes the time required for the $\alpha \to \varepsilon$ phase transformation in iron.

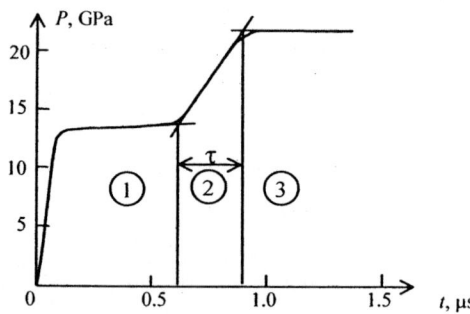

Fig. 5.5. Two-wave structure in iron

For measurement of free surface velocity in shock experiments, the Doppler velocimeter [4], has proven to be an excellent alternative to the capacitive probe.

Methods for the continuous recording of medium flow parameters using embedded sensors have considerably simplified the process of examining phase transformations occurring in shock waves, not only in compressive waves, but in rarefaction waves as well.

Manganin pressure sensors have been used to record two-wave pressure profiles in Sn, Fe, Ti, and Zr [1]. Electromagnetic velocity sensors have been used to determine mass velocity two-wave configurations in KCl, boron nitride, Si, and Ge [1]. Figure 5.6 plots the shock wave profiles in tin ($P_A = 8.9\,\text{GPa}$)

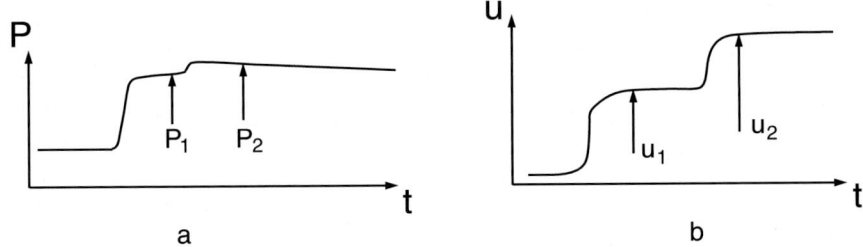

Fig. 5.6. Shock wave profiles: **a** – Sn (manganin sensor) and **b** – KCl (electromagnetic sensor). (P_2, u_2 are field variables of the second wave)

and KCl ($P_A = 2.0$ GPa). For KCl, the time required for phase transformation decreases with increasing amplitude of the secondary shock, P_2.

Continuous recording of the aforementioned wave profiles makes possible the construction of the Hugoniot for shock compression in the phase transition region in $P - V$ coordinates. Figure 5.7 presents such constructions for Fe and KCl.

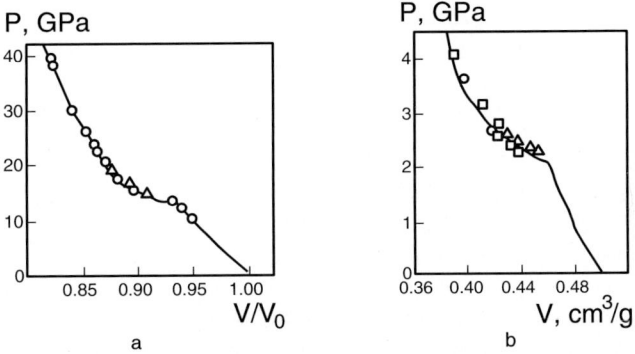

Fig. 5.7. $P - V$ diagram of the shock compression in the phase transition region according to [1]: **a** – Fe; **b** – KCl

Thus far we have limited our consideration to solid bodies in the gas-dynamic approximation, whereby we make the assumption that they are resistant to volume change only (no resistance to volume preserving shape changes). If the material undergoing the polymorphic transition is, however, elastic-plastic, then at a certain level of stress σ_x, a set of two compression waves will be generated at the shock front for reasons that are unrelated to phase transformation (Fig. 5.8). The first of these, called the elastic precursor, propagates at the longitudinal elastic perturbation velocity. If the material is elastic-plastic and the shock pressure exceeds that required for phase transformation σ_A, a three-wave profile may be observed (Figs. 5.8 and 5.9).

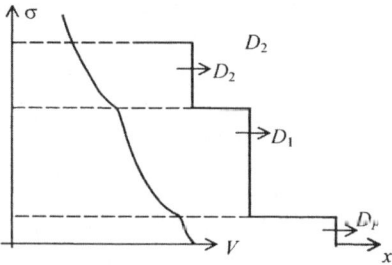

Fig. 5.8. Triple wave in elastic-plastic medium

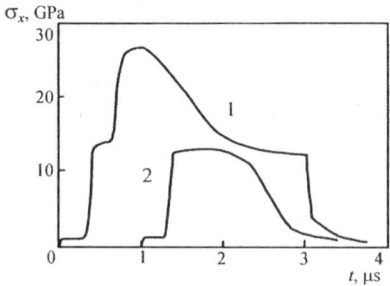

Fig. 5.9. Compression momentum profiles in armco iron: $1 - \sigma_x > \sigma_A$, $2 - \sigma_x < \sigma_A$

Such a three-wave shock profile is shown for armco iron in Fig. 5.9. The experimental data presented in Fig. 5.9 was recorded using a manganin gauge [4]. Figure 5.9 also shows the shock profile for a case wherein the shock wave amplitude σ_x is less than the phase transformation stress σ_A. In this case the wave profile reveals the elastic-plastic behavior of α-phase armco iron.

Many interesting results pertaining to phase transformations in a number of different metals have been obtained in the course of numerous experimental studies aimed at documenting structural changes in planar and ball-shaped samples subjected to shock wave loading and subsequently recovered intact after unloading [1]. The analysis of this data using complete equations of state for the phases, and taking into account phase transitions, has enabled the determination of the Hugoniot, unloading isentrope, and phase boundaries.

5.2 Expansion Shock Waves in Media that have Experienced Polymorphic Transition

When the form of the Poisson adiabat (or of the Hugoniot) in the $P - V$ plane includes a point of inflection as the consequence of phase transformation, this can lead to anomalous behavior involving rarefaction jumps (or rarefaction shocks) in the segments where $\partial^2 P/\partial V^2 < 0$ [6]. The first direct experimental proof of the existence of rarefaction jumps in this scenario was provided in [8] for iron and steel (Steel 3). According to this paper, atypical fracture phenomena were detected on the surfaces of cylindrical specimens subjected to HE charge firing. These atypical features included the observation of a very smooth spall surface of a peculiar core-like shape. These unusual spall surfaces were observed to be much smoother than those previously produced by the interaction of simple rarefaction waves. When the unloading is gradual and continuous (such as in the case when simple rarefaction waves interact), the zone over which the spall-producing tensile stress acts normally extends across a significant portion of the sample thickness, and the resultant spall surface is quite rough. The above noted spall features have been observed in steel St. 3, as well as in some of the stronger types of steel, such as 40 Kh and 30 KhGSA. No anomalous spall phenomena of this type have been observed in similarly designed experiments involving copper, brass, or aluminum. The production of smooth-surface spall is evidence of the fact that during the failure process, the tensile stresses occurred over a very narrow zone. Since such spalls have been detected only in iron and steels whose Hugoniots have an anomalous kink at the point of phase transition, their appearance can be related to the existence of rarefaction shock waves and the effects of their coupling.

Let us consider the mechanism that enables the formation of a rarefaction jump. Let the form of the Hugoniot in the $P - V$ plane be like that shown in Fig. 5.10a.

Let us assume that the Hugoniot represents a state of thermodynamic equilibrium (this implies that the phase transitions occur quite rapidly and that the thermal pressure and energy can be neglected). Observe that the Hugoniot depicted in Fig. 5.10a possesses two kinks: at point A – the beginning of the polymorphic transformation, and at point B – the termination of the polymorphic transformation.

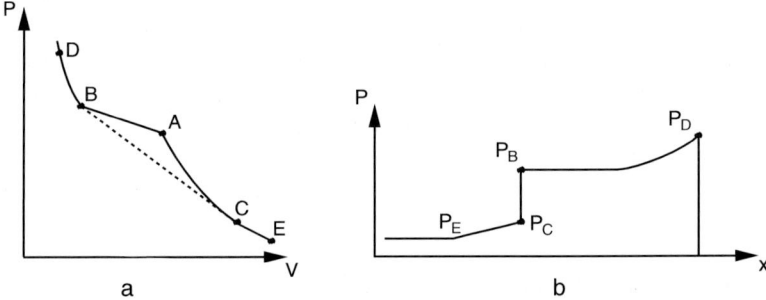

Fig. 5.10. Rarefaction shock wave formation: **a** – $P - V$ diagram; **b** – pressure distribution in the expansion wave

Let a region of expansion be produced at time $t = 0$ in material previously compressed to state D (Fig. 5.10a). Within this region of expansion, we expect pressure and specific volume to change gradually (and isentropically) as they approach their values at state E. For material possessing a normal Hugoniot (no kinks), the flow remains continuous as $t \to 0$, that is, there are no discontinuities in flow parameters at the initial time. Another situation arises, however, if the Hugoniot possesses anomalous properties, i.e. kinks (see Fig. 5.10a). We may use the Hugoniot singularities to consider the nature of the initial pressure profile and its subsequent evolution.

For a simple wave propagating to the right, C_+ characteristics, $dx/dt = u + c$, are straight lines in the $x - t$ plane. The material velocity is continuous at points A and B by virtue of the opposite Riemann invariant constancy requirement for simple waves ($J_- = u - \int dP/\rho c = $ const.) [6]. By virtue of the above assumptions (simple wave; isentropic process), sound speed is determined by the slope of the tangent to the Hugoniot at any given point. Hence, two sound speeds exist at point B: c_{B+0} if $P \geq P_B$ and c_{B-0} if $P \leq P_B$, with $c_{B+0} > c_{B-0}$. Consequently, two C_+ characteristics possessing different slopes in the $x - t$ plane, but carrying identical pressures emerge from point B. A consequence of this is that the region between these two characteristics is a region of constant pressure P_B (Fig. 5.10b).

The situation at kink point A is different. Here $c_{A-0} > c_{A+0}$, i.e. the characteristic possessing the greater pressure propagates at a lower velocity. Consequently, characteristics emerging from any two points located near point A, but on opposite sides of point A, will intersect. This will lead to the formation of a discontinuity, one whose amplitude grows with time. Discontinuities of this form will arise so long as the line BC connecting material states on opposite sides of the discontinuity lies below the Hugoniot DAE. The process terminates when the line drawn from point B touches the Hugoniot in segment AE. Thus maximum shock wave amplitude P_P depends on the tangency points B and C: $P_P = |P_C - P_B|$. The sound speed at tangency point C is equal to the speed of the discontinuity. The velocity of propagation of the discontinuity through the material before the discontinuity (point B), and after the discontinuity (point C) is determined by the following equations

$$(P_B - P_C)\, V_B^2 = (V_C - V_B)\,(D - u_B)^2 \;;$$
$$(P_B - P_C)\, V_C^2 = (V_C - V_B)\,(D - u_C)^2 \;.$$

At point C, $D - u_C$ is equal to local sound speed c (at that point straight line BC touches the first phase Hugoniot). At point B, $D - u$ satisfies $c_{B-0} < D - u_C < c_{B+0}$. Rarefaction shock waves have been recorded in tin, armco iron, a number of steels, KCl, and other materials. As an example, Fig. 5.11 shows the pressure profile produced by the propagation of two compressive shock waves, a rarefaction shock wave, and two simple rarefaction waves through iron (steel) subjected to explosive loading. As in the case of loading by impact (see Fig. 5.9), the wave profile recorded experimentally has singularities that

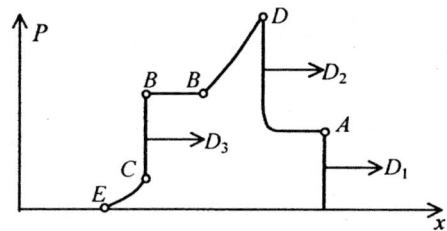

Fig. 5.11. Wave profile in iron explosive loading [9]: D_1 – compression shock wave of phase I; D_2 – phase transformation wave, D_3 – rarefaction shock wave

are consistent with our theoretical understanding of the problem. In the case of impact (Fig. 5.9), an elastic precursor has also been recorded. Here again the profile is composed of two (three, including the elastic precursor) compression shock waves, a rarefaction shock wave, and two simple expansion waves.

Experimental observations of pressure profiles possessing a two-wave configuration of compression (with the deviator neglected) and a rarefaction shock wave provide visual and compelling evidence for the viability of reversible polymorphic transition in the medium. Perhaps the most effective methods for recording a profile that demonstrates the presence of both shock and expansion waves, are those methods which involve the continuous recording of field variable (P or u) at some location embedded in the specimen.

Reverse transformation to the original phase will occur during the unloading of a shock-compressed solid body that has undergone polymorphic transformation. The process is hysteretic in nature. The material state at the onset of the reverse phase transformation differs from that at the onset of the polymorphic transformation brought about by the compression wave, not only in pressure, but also in specific volume: the reverse transformation pressure and specific volume are lower. A more detailed discussion of the hysteresis phenomenon can be found in [1]. Moreover, examples of the hysteretic nature of the reverse polymorphic transformation may be found in [1], where methods for continuously recording the wave profile at locations internal to the sample were applied in studies of Fe, Bi, Sn, clay shale, and ionic crystals KCl, KBr.

5.3 Melting of Solid Bodies in Shock Waves and Unloading Isentropes

Melting of the solid can occur during the passage of a compressive shock wave; it can also occur during isentropic expansion from the shock-compressed state. Whether melt occurs on compression or on release depends upon the strength of the shock, and upon the shape of the Hugoniot and melt curves.

Let us consider a set of experiments leading to first melt on release. In this set of experiments, the solid is subjected to a series of shock compressive loadings of varying magnitudes. The variation in shock compression is from "low" to "high". As the shock pressure is increased, a state will eventually be reached that is just below the melt line on shock up, but because the slope of the melt line is greater than the slope of the Hugoniot, the melt line will be crossed on release. For this material, if the shock loading is increased further, eventually a point will be reached wherein the solid melts on compression. These observances may be attributed to an increase in entropy on the melting curve with increasing pressure in the solid phase. This also explains the fact that the melting curve is higher than the isentrope passing through standard pressure. Given the nature of the shock compression – isentropic expansion process just described, the liquid phase mass fraction will increase with increasing pressure at the shock front.

First-order phase transitions are characterized by unusual thermodynamic properties resulting in Hugoniot kinks at the interface between regions composed of the original one-phase material, and regions composed of the two-phase material [2,5]. These kinked Hugoniots results in a loss of hydrodynamic stability, a consequence of which is the formation of a two-wave configuration (this can occur in the rarefaction shock wave as well as in the compressive shock wave). This two-wave configuration is manifested only over a certain range of pressures.

As was noted earlier, the scenario wherein melt occurs on release is not mandatory. It is also possible to melt on shock up. In this scenario, according to [2], hydrodynamic stability is maintained and the liquid phase is formed in the single shock front. Numerous measurements of shock-wave compressibility, involving pure elements as well as various chemical compounds, have failed to reveal Hugoniot kinks in the melting curves. This is in contrast to the case for polymorphic transformations.

In view of the above-discussed flow features associated with melting in shock waves, melt has to be recorded using experimental methods other than those used for the study of polymorphic transformations.

A visually convincing and accurate method for establishing the occurrence of melt in the shock-wave process involves the measurement of elastic longitudinal c_L and volumetric c_B sound speeds behind the shock front as functions of shock wave amplitude. Methods for measuring sound speed were discussed in the previous chapter and are not repeated here.

For elastic-plastic bodies (see Chap. 1), the ratio $c_L/c_B = \sqrt{3\,(1-\nu)/(1+\nu)}$ tends to unity as Poisson's ratio ν tends to 0.5. Hence the state, at which c_L and c_B are indistinguishable, corresponds to the liquid phase:

$$c_L - c_B = c_B \left[1 - \sqrt{\frac{3\,(1-\nu)}{1+\nu}} \right] \to 0 \quad \text{as} \quad \nu \to 0.5 \,.$$

Figure 5.12 shows experimentally measured c_L and c_B versus shock pressure for iron [10].

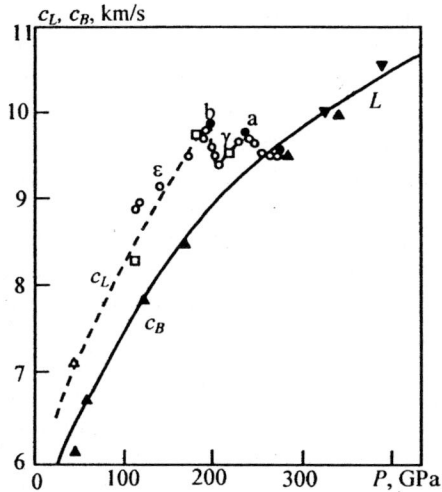

Fig. 5.12. Volumetric, c_B (—), and elastic, c_L(- - -), sound speeds vs. pressure in shock wave for iron: ∘, ▲, ▼, □ are experimental points

The non-monotonicities seen in $c_L(P)$ are interpreted as phase transitions: $\varepsilon \to \gamma$ at $P = 200$ GPa and $\gamma \to L$ at $P = 243$ GPa. The position of point a in Fig. 5.12 corresponds to the start of melting (entrance into the two-phase γ-L region). Judging from Fig. 5.12, it follows that the measurement of sound speeds can also be used to determine polymorphic transformation points in the megabar pressure range. Point b in the figure corresponds to the start of the $\varepsilon \to \gamma$ polymorphic transformation in iron.

Another experimental method for recording shock-compressed solid melting involves the optical measurement of temperatures by radiation luminance [4]. Using this method, melting curves are determined by the non-monotonicity of measured temperature versus pressure on the Hugoniot. This method does not always provide conclusive results. The results for iron, for example, taken by different authors proved to be inconclusive [4].

The most commonly used method for obtaining evidence of melting on the unloading isentrope involves the use of microstructural analyses involving various metallographic techniques, such as X-ray diffraction and electron-microscopy, applied to specimens recovered following the shock loading process.

A different method was used in [11] to detect melt. In this work, a composite sample made of the material under study and a material possessing a lower-melting point was constructed and loaded. The lower melting point material formed a solid solution or intermetallic compound when mixed with the metal under study. If melting occurred during the loading and subsequent expansion of the metal under study, the liquid phase would mix with the melt indicator and an intermetallic compound or solid solution would be formed upon solidification. Detection of intermetallics or solid solutions in the recovered samples by microstructural analyses unambiguously testified to the fact that melting had occurred in the material under study. For lead (with bismuth serving as the indicator) a lead-bismuth solid solution was observed for shock amplitudes of 20.5 GPa ($T_0 = -10°C$). It is most likely in this case that the melting of lead occurred in the rarefaction wave as opposed to the shock wave.

Let us consider a simple method for an approximate calculation of the melting line. In cases where there is no justified computation of the liquid and solid phase free energy available, various phenomenological and purely empirical relations are frequently used. In the Lindemann phenomenological model, for instance, melting takes place at a critical ratio of the average thermal atomic displacement to the equilibrium distance between nearest neighbors. If the critical ratio is the constant 0.5, then, by the Lindemann melt model,

$$\frac{d \ln T_{melt}}{d \ln \rho} = 2\Gamma - \frac{2}{3} \tag{5.3}$$

where T_{melt} is the melting temperature, and Γ is the Grueneisen oscillatory parameter (Grüneisen factor of the lattice).

The Grüneisen parameter Γ may be given as a function of specific volume:

$$\Gamma = \Gamma_\infty + \frac{\Gamma_0 - \Gamma_\infty}{\delta^M} , \tag{5.4}$$

where Γ_0, Γ_∞, and M are constants, and δ is the relative compression. One may then integrate Eq. (5.3) to obtain an equation for the melting curve in $\delta - T$ coordinates:

$$\frac{T_{melt}}{T_{0\,melt}} \left(\frac{\delta_{melt}}{\delta_{0\,melt}}\right)^{2\left(\Gamma_\infty - \frac{1}{3}\right)} \exp \left[\frac{2\left(\Gamma_0 - \Gamma_\infty\right)}{M\left(\delta_{0\,melt}\right)^M} - \frac{2\left(\Gamma_0 - \Gamma_\infty\right)}{M\left(\delta_{melt}\right)^M}\right] . \tag{5.5}$$

In Eq. (5.5), $T_{0\,melt}$ is the melting temperature at standard pressure, and δ_{0melt} is the relative density of liquid material at standard pressure.

From Eq. (5.5) and the equation of state of the solid phase (given, for example, in the thermal form), we can determine the melting curve in the $P-T$ plane. To estimate melting parameters at the shock front, it is sufficient to determine the position of the point of intersection of the melting curve and $P(T)$ along the Hugoniot in the $P - T$ plane.

Note that careful consideration of Eq. (5.5) should be made when selecting the dependency $\Gamma(V)$.

5.4 Evaporation of Shock-Compressed Solid Bodies in Expansion

Liquid-gas phase transition phenomena may be experimentally recorded in tests involving the adiabatic expansion of shock-compressed material. Shock-compressed condensed material expansion measurements, which are to date few in number, can be divided by the nature of the information acquired in them into two groups. The first of these involves the recording of final material parameters upon unloading [12, 13]. The second group involves the measurement of unloading isentropes throughout the entire intermediate parameter region from solid to plasma or gas [14]. In studies of the liquid-gas phase transition, the first test group should be preferred.

Using the preferred method, the material under study is compressed by shock waves of various intensities to pressures on the order of 10^5 MPa, and is subsequently released into air. Vaporization effects are detected through the measurement of the velocity of isentropic expansion [12, 13]. The experiments suggest that the onset of vaporization during unloading can be detected if the relaxation time of the metastable states appearing below the saturation curve is much shorter than the characteristic expansion time.

The velocity of the expansion of material into air W is experimentally determined and can be subsequently represented as a function of the shock wave mass velocity u. It is known that if the material remains in the condensed state upon unloading, the velocity on the free surface will differ little from a doubling of the mass velocity in the shock wave (see Chap. 1). Another situation arises, however, if evaporation takes place.

Consider Fig. 5.13, which contains a schematic $P - V$ diagram and several experimentally achieved release states. Note that some of the isentropes lead to final states that lie in a liquid-vapor two-phase region (the melting region is not shown in Fig. 5.13). During transition through the phase equilibrium boundary, the change in isentropic compressibility is [12]:

$$\left(\frac{\partial V_2}{\partial P}\right)_{S_2} - \left(\frac{\partial V_1}{\partial P}\right)_{S_1} = -\frac{T}{C_P}\left(\frac{dS_1}{dP}\right)^2 . \tag{5.6}$$

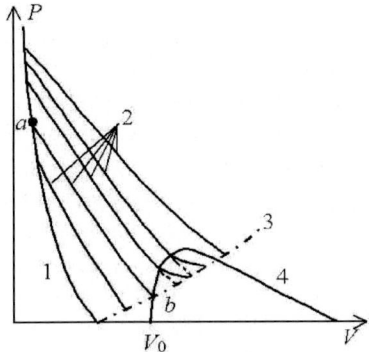

Fig. 5.13. $P - V$ diagram of the experiment: 1 – Hugoniot; 2 – unloading isentrope; 3 – curve of final states in unloading; 4 – two-phase region boundary

In (5.6), the total derivative refers to the phase equilibrium line, subscript "1" pertains to the liquid phase, and subscript "2" to the phase coexistence region.

Variations in the velocity of the expanding material is related to the derivatives of (5.6) as:

$$\left(\frac{\partial \bar{u}}{\partial P}\right)_S = \sqrt{-(\partial V/\partial P)_S} \,. \tag{5.7}$$

From Eqs. (5.6) and (5.7), it follows that the entrance of the isentrope into the two-phase region from the side of the liquid causes an additional velocity increase: $W = u_{fr} + u(P \approx 0)$, where u_{fr} is the mass velocity at the shock front. The influence of the shock compression parameters on material velocity during expansion to the various final states will be smoothly monotone in the case of unloading along the metastable isentropes (the dashed lines in Fig. 5.13), and have kinks corresponding to the onset of vaporization should that occur during equilibrium expansion (solid lines in Fig. 5.13). The kinks are due to the considerable difference between the material compressibility of the liquid and gaseous states.

A kink in the expansion isentrope leads to a kink in the experimental $W(u_{fr})$, which is evidence of metal vaporization. Moreover, the kink coordinates provide unambiguous experimental identification of isentrope ab (see Fig. 5.13), whose final state is located on the equilibrium line. The value of W_b, combined with knowledge of the Hugoniot of air, are used to estimate the pressure P_b, which coincides with the pressure of the expanding metal (see Sect. 1.5). The pressure at initial state a is found from the value of u_{fr} at the kink of curve $W(u_{fr})$. A detailed description of the experimental technique can be found in [12, 13].

In [12] and [13] strong shock waves were generated in specimens made of the material to be studied and measurements were recorded using carefully calibrated diagnostics. On shock wave arrival at the metal-air interface ($P \sim 10^5$ Pa), metal expansion speed was recorded using the streak-camera method. Velocity W is recorded in [12], and the air shock wave velocity is recorded in [13]. It is an easy matter to transfer to mass velocities ($W = u_{air}$) since the Hugoniot of air is well known.

The experimental data are illustrated in Figs. 5.14a, b, and c, where the function $\Delta = W - 2u_{fr}$ characterizes deviation from the law of doubling. The kinks in the $\Delta(u_{fr})$ plots testify to the accuracy of the above-discussed physical concepts regarding evaporation during the expansion to air of shock-compressed Pb, Sn and Cd. For comparison, Fig. 5.14d plots a similar function for aluminum, which exhibits a single curve with positive slope throughout the entire interval of states studied. Similar behavior is seen in $\Delta(u_{fr})$ for Mg ($P_{fr} \leq 87$ GPa) and Cu ($P_{fr} \leq 219$ GPa) [13].

The initial and final values for the parameters of the unloading isentrope ab during expansion to air are presented in Table 5.1.

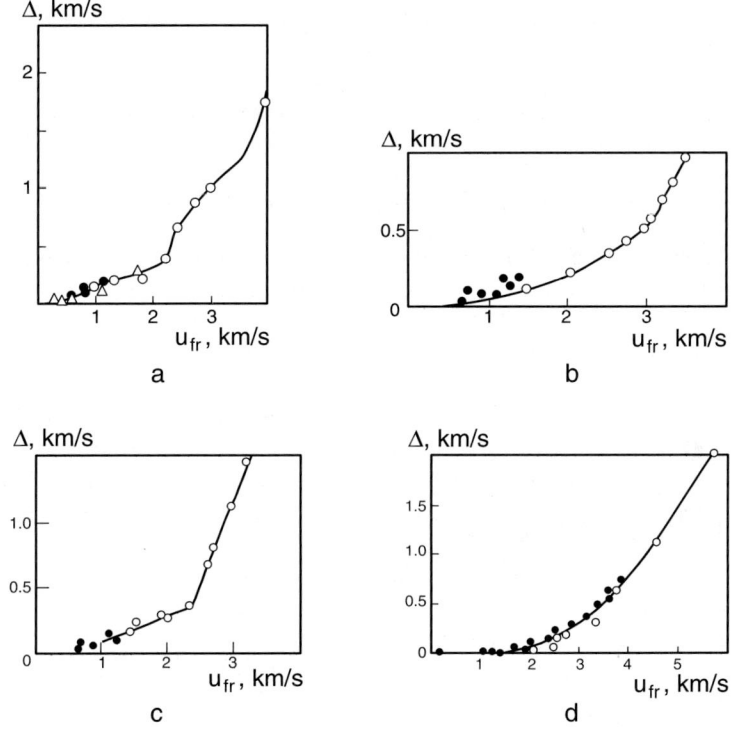

Fig. 5.14. $\Delta(u_{fr})$ for Pb (**a**); Sn (**b**); Cd (**c**); Al (**d**)

Table 5.1. Initial and final values of unloading isentrope parameters for several metals [13]

Metal	Parameters on the Shock Wave			Final States in Unloading		
	P_a, GPa	U_a, km/s	ρ_a, g/cm^3	P_b, MPa	W_b, km/s	ρ_b, g/cm^3
Cd	124.0	2.32	13.74	32.8	5.00	6.37
Sn	158.0	3.04	12.69	56.5	6.58	3.70
Pb	133.0	2.20	19.34	33.0	4.80	6.40

Note also that the existence of vaporization can be deduced from a recording of the shock-wave parameters in a base plate during the deceleration of a material that has expanded into an air gap [12], and through spectroscopic measurements of the fraction of evaporated material produced during expansion into a vacuum. If vaporization occurs, the shock amplitude in the base plate will be considerably smaller than would be the case involving superheated fluid deceleration.

5.5 Second-order Phase Transitions in Shock Waves

By applying high (megabar) pressures to a solid body, it is possible to induce not only first-order phase transformations as discussed above, but second-order phase transformations as well [15]. Hugoniot kinks have been detected using conventional methods for studying shock compressibility in rare earth and alkaline metals. The characteristic form of the Hugoniot in the $D - u$ plane is shown in Fig. 5.15 and in the $P - \eta$ plane in Fig. 5.16 ($\eta = V_0/V$).

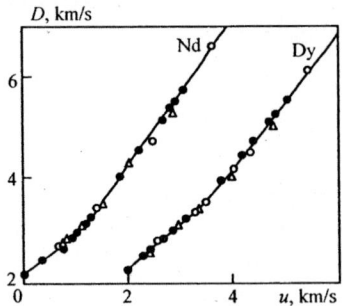

Fig. 5.15. $D - u$ curves for Nd and Dy ($u + 2$ km/s):●, ○, △ – experimental points

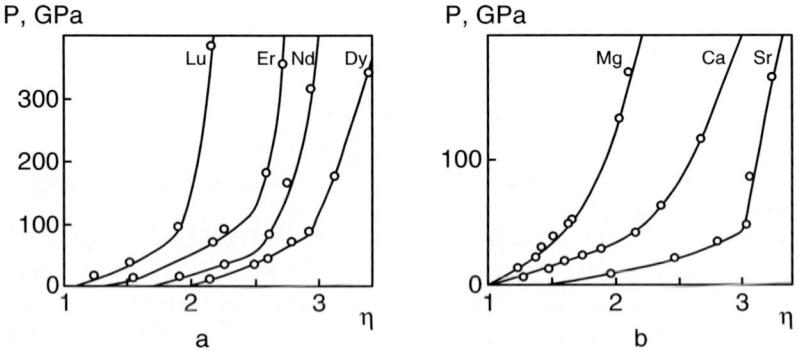

Fig. 5.16. Hugoniots of: **a** – rare earth metals Lu, Er ($\eta + 0.3$), Nd ($\eta + 0.6$), Dy ($\eta + 0.9$); **b** – alkaline earth elements Mg, Ca, Sr ($\eta + 0.5$)

For some materials, Hugoniot kinks have been observed at critical points in the $P - \eta$ plane. When this occurs, it is evidence of a decrease in the metal compressibility and a second-order phase transition that is related to a rearrangement of the electron energy spectra. Qualitatively, the physical nature of the kink can be explained as follows. The gently sloping Hugoniot segment is associated with outer electron shell compression and outer electron displacement into vacant inner electronic shells of the atom. The kink is associated with a termination of the form of electron transitions just described, and the formation of poorly compressible electronic configurations. Critical pressures, P_{cr}, at which Hugoniot kinks take place are given for a number of metals in [15]:

$$Nb - P_{cr} = 80\,GPa \qquad Zr - P_{cr} = 53\,GPa$$
$$La - P_{cr} = 22\,GPa \qquad Sr - P_{cr} = 33\,GPa$$
$$Nd - P_{cr} = 61\,GPa \qquad Gd - P_{cr} = 54.5\,GPa$$
$$Dy - P_{cr} = 54\,GPa \qquad Er - P_{cr} = 123\,GPa$$

5.6 Electrical Conductivity of Shock-compressed Materials

As was mentioned above, measurement of shock front temperature provides a useful diagnostic for the detection of phase transitions in materials subjected to shock compression. A number of studies involving measurements of the relevant optical absorption constants have revealed a fairly significant influence on the behavior of shock-compressed material that can be attributed to thermal electron excitation in the conduction band and the interaction of electrons with lattice phonons.

Data on conduction electron concentration and energy gap behavior are essential if one is to accurately account for the influence of free electrons on the behavior of materials at high temperature and pressure. The most straightforward, and perhaps only feasible means of obtaining this data, is through the direct measurement of electrical conductivities of shock-compressed material. Accurately accounting for this data is an important requirement, which must be placed upon predictive mathematical models.

Recording of electrical resistance changes during compression and unloading has proven to be a useful diagnostic for the occurrence of phase transition. The method has proven to be effective in metal-dielectric transitions as well as in dielectric-metal transitions. In some cases, it is possible to determine the electrical conductivity type, which is important for determining the thermodynamic state of the material and its composition.

The first publications reporting measurements of the electrical conductivity of shock-compressed materials appeared in the 1956–1968 time frame [16–22], and involved studies of molecular and ionic crystals, liquids, sulfur, and detonation products. The results of these studies showed that under the influence of shock wave compression, all of the materials transformed into reasonably good conductors. For example, at shock pressures on the order of 10^5 atm, water conductivity increases from $10^{-5}(\Omega \cdot cm)^{-1}$ to $0.2\ (\Omega \cdot cm)^{-1}$, with this increase being independent of the presence of impurities. At shock pressures on the order of 10^6 atm the conductivities of organic glass and paraffin increase some 15–20 orders of magnitude from their initial values of $10^{-15}(\Omega \cdot cm)^{-1}$ and 10^{-18} $(\Omega \cdot cm)^{-1}$, respectively. This increase may be interpreted as evidence of dielectric metallization. Note that the increase in electrical conductivity brought about by shock compression cannot be attributed solely to thermal ionization. It is associated with a change in the electronic band structure and a decrease in band gaps under compression, which leads to the appearance of free electrons and metallic conductivity in materials that were initially dielectric. Ionic crystal conductivity following the passage of a plane wave shock front is studied in [18–22]. Based on the temperature dependence of the electrical conductivity of NaCl, [19] reports an activation energy $\Delta W_0 = 1.2$ eV and infers the ionic nature of that conductivity. The conclusions drawn in [22] are somewhat different. In [22], simultaneous measurements of the electric conductivity and optical absorptivity of shock-compressed NaCl support the conclusion that the principal role in absorption and conductivity is played by free electrons.

Recent years have witnessed heightened interest in research on the conductivities of condensed inert gases. A relatively high measure of conductivity is reported in [23] for liquid argon at pressures ~20 GPa. An abrupt decrease in resistance (from 10^{13} Ω to 10^4 Ω) under static compression of xenon is reported in [24]. The authors of [25, 26] report a change in the conductivity of shock-compressed liquid nitrogen from $10^{-3}(\Omega \cdot cm)^{-1}$ at 20 GPa, to $10^2(\Omega \cdot cm)^{-1}$ at 60 GPa. The authors of [27–29] report on the electrical conductivities of liquid argon, krypton, and xenon behind a plane wave shock in the 10–90 GPa pressure range. A liquid oxygen conductivity of ~$8 \cdot 10^3(\Omega \cdot m)^{-1}$at pressures up to 43 GPa is measured in [30]. The conductivity of liquid hydrogen up to ~180 GPa pressures is studied in [31].

Experimental Technique

Experiments measuring the electrical conductivity of shock-compressed dielectrics often employ the scheme proposed in [16]. Its advantages and disadvantages are analyzed in [32]. A typical diagram of the experimental device, in which the measuring electrodes are positioned parallel to the shock front is presented in Fig. 5.17.

This design was used to measure the electrical conductivity of shock-compressed NaCl in [19]. The scheme employs the discharge of a large-capacity capacitor through sample-shunting resistance R_0. Passage of the shock wave leads to a drop in the resistance of the material under study and a concomitant drop in total resistance, which is accompanied by a change in voltage across the measuring electrodes, and that change is recorded by an oscilloscope. Crystal resistance R_x according to the measurement scheme just described was evaluated to be:

$$R_0 = V_x \Big/ (V_0 - V_x) \left(1/R_0 + 1/R_a\right)^{-1}, \tag{5.8}$$

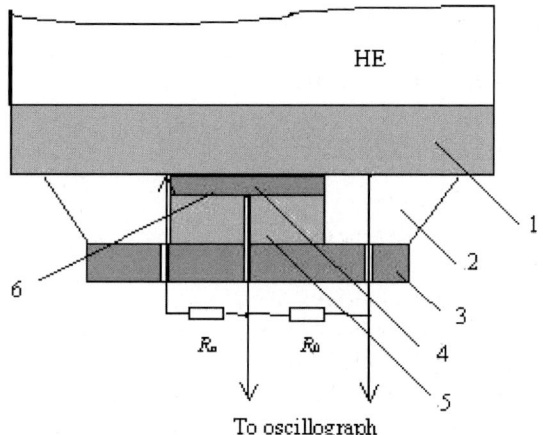

To oscillograph

Fig. 5.17. Diagram of ionic crystal electric conductivity measurement: 1 – base plate; 2 – compound; 3 – plexiglas; 4 – crystal; 5 – substrate; 6 – copper foil contact

where V_0 and V_x are voltages across the electrodes before and after the passage of the shock wave through the crystal. The resistance R_0 ranged from 0.8 to 6.0 Ω. A resistance R_a of 50 Ω was used to mark the time of the arrival of the shock wave at the surface of the crystal under study.

The specific conductance G was calculated as $G = L/R_x S$, where L is the shock-compressed crystal thickness, R_x is the measured resistance, and S is the contact area (this ranged from 0.2 cm^2 to 7.0 cm^2). When processing the experimental data, the authors of [19] used temperatures calculated from an equation of state.

More recently, a similar experimental design was employed in [20–22] to measure the electrical conductivity of ionic crystals of KCl, KBr, and NaCl (Fig. 5.18). Experimental measurement of the luminance temperature of the shock-compressed samples was also made [20–22].

Conductance

$$G = eN_e\mu \tag{5.9}$$

was of great importance in determining the mechanism of absorption in shock-compressed NaCl [22]. Here e is the electron charge, N_e is the concentration of free electrons, and μ is their mobility.

In the context of classical theory, the absorptivity is given by the Drude-Zener relation:

$$\alpha = \left(2N_e e^2/ncm\right)\left(q/\left(q^2 + v^2\right)\right), \tag{5.10}$$

where n is the refractive index, c is light speed, m is the effective mass of an electron, v is the incident frequency, and $q = e/2\pi m\mu$ is the damping parameter.

Independent experimental measurements enabled the authors of [22] to evaluate electronic mobility in shock-compressed ionic crystals as a function

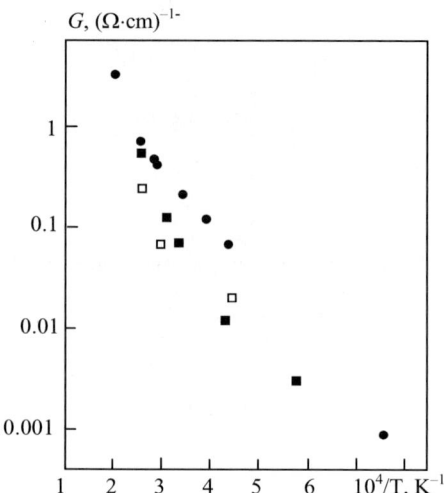

Fig. 5.18. Electric conductance of shock-compressed ionic crystals versus temperature. Experiment: • – NaCl; □ – KCl; ■ – KBr

of temperature using Eqs. (5.9) and (5.10). The experimentally determined mobility of KCl and KBr was $\mu \sim 0.4$ cm^2/s. For NaCl, this quantity was found to be $\mu \sim 2$ cm^2/s.

Analysis of the experimental data supplemented by computation, supports the proposition suggested in [22], that electrical conductivity and light absorption in shock-compressed ionic crystals are caused by the effects of free electrons thermally excited to the conduction band. According to [22], at the shock front the ionic crystal transforms from a dielectric to a semiconductor state with donor levels whose thermal dissociation leads to the appearance of free electrons in the conduction band. For NaCl, the experimental data obtained in [22] are described well by the theory if one assumes a donor concentration $N_d = 1.6 \cdot 10^{19}$ cm^{-3} and width of the energy gap between the donor levels and conduction band $\Delta W_0 = 2.4$ eV. For KBr and KCl crystals, good agreement between the experiments and computations is achieved with donor concentrations $N_d = 2.5 \cdot 10^{19}$cm^{-3} and $1 \cdot 10^{19}$ cm^{-3}, respectively, at $\Delta W_0 = 2.3$ eV.

A scheme with contacts positioned perpendicularly to the shock front is used for continuous recording of the electrical resistance of shock-compressed material. The scheme is employed in [33] to measure the electrical conductivity of boron, sodium chloride, and polytetrafluorethylene (Fig. 5.19).

In this scheme, the change in total resistance of the measuring cell is estimated as:

$$R(t_0) = V(t_0) R_{sh}/(V_0 - V(t_0)), \tag{5.11}$$

where V_0 is the initial voltage. Average resistance from the time of shock wave entry into the sample (at $t = 0$) to t_0 is

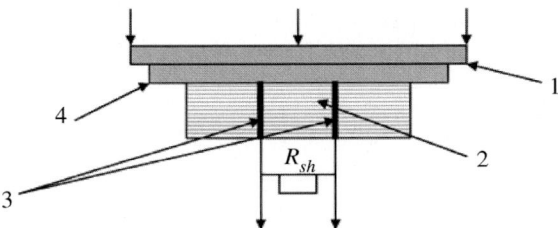

Fig. 5.19. Scheme of electric conductivity measurement with perpendicularly positioned electrodes: 1 – base plate; 2 – sample; 3 – electrodes; 4 – gasket (polytetrafluorethylene)

$$\rho\left(t_0\right) = R\left(t_0\right) a D t_0 / l \delta ,$$

where a is the width of the contacts, l is the distance between them, D is the shock wave velocity, and δ is the relative compression. If conductance is constant throughout the space behind the shock front, then voltage across the electrodes should vary with time as:

$$V\left(t\right) = V_0 / (1 + mt) ,$$

where $m = a D R_{sh} / \rho \delta$ is a constant, which depends on the sample geometry and specific resistance.

Results from the experiments discussed in [33] revealed a linear dependence of the logarithm of boron nitride conductance on temperature with a 3-eV activation energy in the 30–50 GPa pressure range.

In order to measure the electrical conductance of caprolon, [34] makes use of a similar scheme, but in this work the sample is situated between two coaxial cylindrical electrodes. Electrical resistance in this measuring cell is estimated as:

$$\rho = \frac{2\pi}{\ln \Phi_1 / \Phi_2} \frac{D}{\delta} \frac{R_{sh}}{Z_0} \frac{Z_x^2}{(dZ_x/dt)_t} . \tag{5.12}$$

In Eq. (5.12), Φ_1 and Φ_2 are inner and outer diameters of the sample, R_{sh} is the shunt resistance, D is the shock wave velocity, δ is the material compression, Z_0 is the initial oscillographic beam bias, Z_x is the signal amplitude at time t, and $(dZx/dt)_t$ is the beam path slope at time t. The recorded specific electric conductivity of caprolon ranged from $3 \cdot 10^{-3}$ $(\Omega \cdot \mathrm{cm})^{-1}$ at $P = 35$ GPa to $4.4 \cdot 10^{-2}$ $(\Omega \cdot \mathrm{cm})^{-1}$ at $P = 57$ GPa.

The major drawback of the above schemes arises when one attempts to measure inductances greater than 10^2 $(\Omega \cdot \mathrm{cm})^{-1}$. In this instance, the measuring circuit inductance impedes an abrupt change in current in the shock-compressed conductor. In order to measure high conductances, the authors of [35] developed a new technique, and applied this technique to efforts aimed at measuring the electrical conductivity of CsI, and later, the liquefied inert gases Ar, Kr, and Xe [27–29].

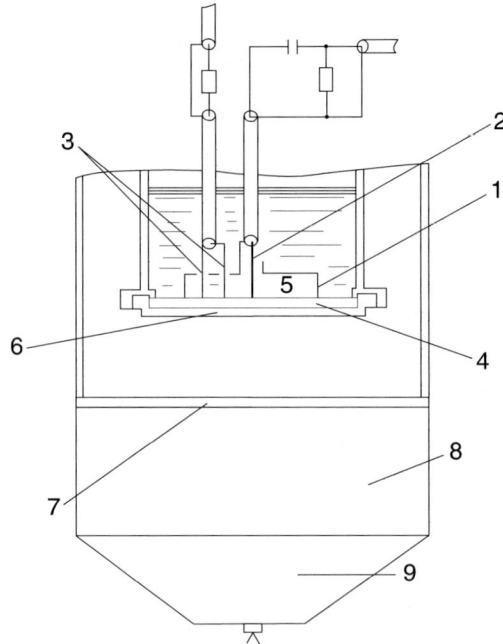

Fig. 5.20. Scheme of the experimental device

The scheme of the experimental device used to study the conductivity of liquefied inert gases appears in Fig. 5.20. Pulsed current $I(t)$ flowing through a sample of shock-compressed liquefied gas is recorded. The current flows between two copper electrodes: a coaxial electrode with an outer diameter of 40 mm and a thickness of 0.5 mm (1), and a central electrode possessing a diameter of 1 mm (2). Electrical potential difference $V(t)$ is measured using two copper probes of diameter 0.5 mm (3). The electrodes are positioned within the material under study at distances of $R_1 = 5$ mm and $R_2 = 15$ mm from the center. The discharge circuit closes, upon shock front arrival at the interface between dielectric plates (4) and the liquid sample (5). The dielectric impedes shunting of the sample conductance, which would otherwise be brought about by the metal base plate (6), through which the shock wave propagates into the experimental cell. For experiments involving pressures up to 40 GPa, polytetrafluorethylene and caprolon were used for the dielectric material, the ceramic Al_2O_3 was used for pressures greater than 40 GPa.

The samples were loaded through the deceleration of impactor (7), which was accelerated using the detonation of high explosive (8) through plane wave generator (9). Pressure was estimated from $P - u$ curves for liquid Ar, Kr, and Xe, applied to the region between the base plate and dielectric plate using the method of reflection. Cooling of structural elements down to the temperature of boiling for liquefied gas was taken into account.

Under the assumption that electromagnetic induction can be neglected in the conducting region, specific electrical conductivity was calculated as [35]:

$$G = [1/2 \, (D - u)] \, [I \, (t)/E_\Phi \, (t)] \, , \qquad (5.13)$$

where D and u are shock and mass velocities, respectively, and $E_\Phi(t)$ was estimated using the expression

$$E_\Phi \, (t) = \frac{V \, (t)}{\ln r_1/r_2} + \frac{\mu u I \, (t)}{2\pi} - \frac{\mu \, (Z - Dt)}{2\pi} \frac{dI \, (t)}{dt} \, , \qquad (5.14)$$

where $\mu = 4 \cdot 10^{-7}$ H/m is the material magnetic permeability.

In an independent series of experiments, luminance temperature was measured in the violet (430 nm), green (500 nm) and red (600 nm) spectral domains in gases at the shock front. The data enabled an estimation of the true temperature at the shock front in the gray body approximation. Measurements of the equilibrium temperature dependence with respect to electrical conductivity are given in Fig. 5.21 for argon, krypton and xenon.

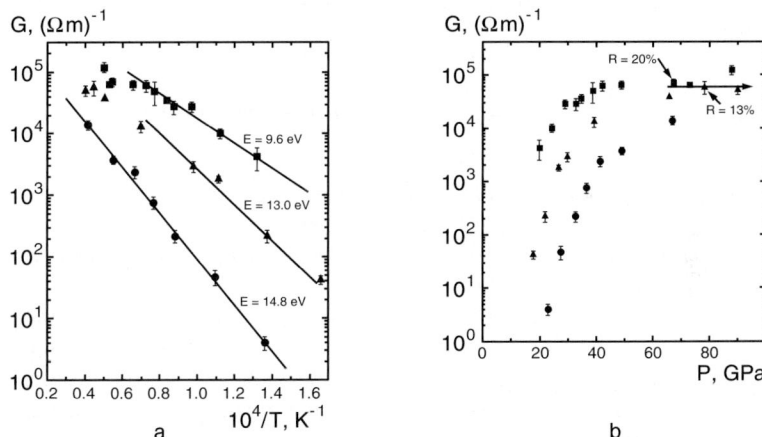

Fig. 5.21. Electric conductivity of • – Ar, ▲ – Kr, ■ – Xe as a function of temperature (a) and pressure (b) – is the approximation; R is the luminous reflectivity from the shock front

Segments of electric conductance that may be expressed in terms of linear functions of temperature can be singled out in the experimental plots of $G(1/T)$: for argon this is the case throughout the measurement region, and for krypton and xenon throughout the entire measurement region with the exception of three and four upper points, respectively. This observation leads to the inference that within these segments, electrical conductivity can be described by a function that is characteristic of semiconductors,

$$G = G_0 \exp \left(-E \, (T, V)/2kT\right) \, , \qquad (5.15)$$

where $E(T,V)$ is the energy gap between the valence band and conduction band, and G_0 depends weakly on temperature.

The experimental points in the linear segments were approximately fit using the least-squares method, with 90% confidence limits. Approximation dependencies are presented in Fig. 5.21a. They were used to estimate the energy gap width in each of the gases studied. The results: 14.8 ± 0.5 eV for argon, 13 ± 0.8 eV for krypton, and 9.6 ± 0.5 eV for xenon. The results obtained are in good agreement with the experimentally determined energy gaps: 13.4 eV for argon, 11 eV for krypton, and 9.22 eV for xenon, based on measurements of photoconductivity [36, 37].

The uncertainties in experimental measurements coupled with a strong dependence on temperature do not allow for an unambiguous determination of the role of compression at small band gaps. The experimental values may be characteristic only over the initial part of the $G(1/T)$ curve, followed by a decrease in the conduction activation energy.

Experimental electrical conductivity versus shock wave amplitude is plotted in Fig. 5.21b. Of particular interest are the results for liquid xenon. The data reveal an anomalous segment in the 42–78 GPa pressure range, where the electrical conductivity is constant at $\sim 6 \cdot 10^4 (\Omega \cdot m)^{-1}$. The limiting value of $\sim 6 \cdot 10^4 (\Omega \cdot m)^{-1}$ is also approached by liquid krypton at ~ 70 GPa pressure. For argon the pressure is presumably higher than ~ 100 GPa.

Similar behavior with respect to specific electrical conductivity was observed for liquid argon in the experiments of [25, 26]. In this work the saturation trend at ~ 100 GPa pressures is identified with a dissociation phase transition. In the liquid hydrogen experiments of [31], an association with hydrogen metallization under shock compression is made for the electric conductivity saturation at ~ 140 GPa.

Shock-wave compression provides unique possibilities for studying the transition of a dielectric to the metallic state, since density increase is attended by a significant temperature rise and an increase in the number of free electrons. This leads to a "smearing" of the insulator-metal transition, creating prerequisites for an observation of the continuous transition of dielectrics to the metal state. In this case the increasing number of free electrons should result in a higher luminous reflectivity from the shock front, which can be recorded experimentally. Light reflectivity $R \sim 13\%$ from the shock front in liquid krypton at ~ 76 GPa pressure is recorded in [38]. The reflection was measured using the shock front self-luminescence. A similar experiment involving liquid xenon recorded a $\sim 20\%$ luminous reflectivity from the shock front at 63.6 GPa pressure. Such high optical reflectivities correspond to pressures at which the saturation of electrical conductivity has been recorded in krypton and xenon (see Fig. 5.21b). This can be due to the formation of a well-reflecting metal surface at the shock front with an increasing number of free electrons, as is seen in ordinary semiconductors with increasing temperature.

References

1. Funtikov, A.I., and Pavlovskii, M.N., "Shock Compression of Solid Bodies and Polymorphous Transformations. Shock Waves in Solid Bodies," *in* Shock Waves and Extreme States of Matter, Nauka Publ., Moscow, 2000, pp. 138–159 [*see also*, Funtikov, A.I., and Pavlovsky, M.N., "Shock Waves and Polymorphic Phase Transformations in Solids," *in* Fortov, V.E., Altshuler, L.V., Trunin, R.F., and Funtikov, A.I., *High Pressure Shock Compression of Solids VII*, Springer-Verlag, New York, 2004, pp. 197–223].
2. Altshuler, L.V., "Phase Transitions in Shock Waves," *Prikladnoi Mekhaniki i Tekhnicheskoi Fiziki*, 1978, No. 4, pp. 93–103, [English trans., *Journal of Applied Mechanics and Technical Physics*, Vol. 19, No. 4, 1978, pp. 496–505].
3. Altshuler, L.V., "Use of Shock Waves in High-Pressure Physics," *Uspekhi Fizicheskikh Nauk*, Vol. 85, No. 2, 1965, pp. 197–258, [English trans., *Soviet Physics Uspekhi*, Vol. 8, No. 1, 1965, pp. 52–91].
4. Kanel, G.I., Razorenov, S.V., Utkin, A.V., and Fortov, V.E., *Shock Wave Phenomena in Condensed Matter*, Yanuk-K Publ., Moscow, 1996, [*see also*, Kanel, G.I., Razorenov, S.V., and Fortov, V.E., *Shock-Wave Phenomena and the Properties of Condensed Matter*, Springer-Verlag, New York, 2004].
5. Kuznetsov, N.M., "Shock Compression of Solid Bodies and Polymorphous Transformations. Some Issues of Polymorphous Transformations in Shock Waves," *in* Shock Waves and Extreme States of Matter, Nauka Publ., Moscow, 2000, pp. 174–198 [*see also*, Kuznetsov, N.M., "Some Questions of Phase Transition in Shock Waves," *in* Fortov, V.E., Altshuler, L.V., Trunin, R.F., and Funtikov, A.I., *High-Pressure Shock Compression of Solids VII*, Springer-Verlag, New York, 2004, pp. 247–273].
6. Zeldovich, Ya.B., and Raizer, Yu.P., *Physics of Shock Waves and High-Temperature Hydrodynamic Phenomena*, Nauka Publ., Moscow, 1966, [English trans. Academic Press, NY, Vol. 1 (1966), Vol. 2 (1967); Reprinted in a single volume by Dover Publ., Mineola, NY 2002].
7. Novikov, S.A., Divnov, I.I., and Ivanov, A.G., "Investigation of the Structure of Compressive Shock Waves in Iron and Steel," *Zhurnal Eksperimentalnoi i Teoreticheskoi Fiziki*, Vol. 47, No. 3, 1964, pp. 814–816, [English trans., *Soviet Physics JETP*, Vol. 20, No. 3, 1965, pp. 545–546].
8. Ivanov, A.G., and Novikov, S.A., "Rarefaction Shock Waves in Iron and Steel," *Zhurnal Eksperimentalnoi i Teoreticheskoi Fiziki*, Vol. 40, No. 6, 1961, pp. 1880–1882, [English trans., *Soviet Physics JETP*, Vol. 13, No. 6, 1961, pp. 1321–1323].
9. Ivanov, A.G., Novikov, S.A., and Tarasov, Yu.I., "Fragmentation Phenomena in Iron and Steel Caused by Explosive Shock Wave Interactions," *Fizika Tverdogo Tela*, Vol. 4, No. 1, 1962, pp. 249–260, [English trans., *Soviet Physics – Solid State*, Vol. 4, No. 1, 1962, pp. 177–185].
10. Funtikov, A.I., "Shock Compression of Solid Bodies. Iron Phase Diagram," *in* Shock Waves and Extreme States of Matter, Nauka Publ., Moscow, 2000, pp. 160–173 [*see also*, Funtikov, A.I., "Phase Diagram of Iron," *in* Fortov, V.E., Altshuler, L.V., Trunin, R.F., and Funtikov, A.I., *High-Pressure Shock Compression of Solids VII*, Springer-Verlag, New York, 2004, pp. 226–246].
11. Batkov, Yu.V., German, V.N., Osipov, R.S., Novikov, S.A., and Tsyganov, V.A., "Melting of Lead in Shock Compression," *Zhurnal Prikladnoi Mekhaniki i Tekhnicheskoi Fiziki*, 1988, No. 1, pp. 149–151, [English trans., *Journal of Applied Mechanics and Technical Physics*, Vol. 29, No. 1, 1988, pp. 139–141].

12. Altshuler, L.V., Bakanova, A.A., Bushman, A.V., Dudoladov, I.P., and Zubarev, V.N., "Evaporation of Shock-Compressed Lead in Release Waves," *Zhurnal Eksperimentalnoi i Teoreticheskoi Fiziki*, Vol. 73, 1977, pp. 1866–1872, [English trans., *Soviet Physics JETP*, Vol. 46, No. 5, 1977, pp. 980–983].

13. Bakanova, A.A., Dudoladov, I.P., Zhernokletov, M.V., Zubarev, V.N., and Simakov, G.V., "Vaporization of Shock-Compressed Metals on Expansion," *Zhurnal Prikladnoi Mekhaniki i Tekhnicheskoi Fiziki*, 1983, No. 2, pp. 76–81, [English trans., *Journal of Applied Mechanics and Technical Physics*, Vol. 24, No. 2, 1983, pp. 204–209].

14. Bushman, A.V., Glushak, B.L., Gryaznov, V.K., Zhernokletov, M.V., Krasyuk, I.K., Pashinin, P.P., Prokhorov, A.M., Ternovoi, V.Ya., Filimonov, A.S., and Fortov, V.E., "Shock Compression and Adiabatic Decompression of a Dense Bismuth Plasma at Extreme Thermal Energy Densities," *Pisma, Zhurnal Eksperimentalnoi i Teoreticheskoi Fiziki*, Vol. 44, No. 8, 1986, pp. 375–377, [English trans., *JETP Letters*, Vol. 44, No. 8, 1986, pp. 480–483].

15. Altshuler, L.V., and Bakanova, A.A., "Electronic Structure and Compressibility of Metals at High Pressures," *Uspekhi Fizicheskikh Nauk*, Vol. 96, No. 2, 1968, pp. 193–215, [English trans., *Soviet Physics Uepekhi*, Vol. 11, No. 5, 1969, pp. 678–689].

16. Brish, A.A., Tarasov, M.S., and Tsukerman, V.A., "Electrical Conductivity of Dielectrics in Strong Shock Waves," *Zhurnal Eksperimentalnoi i Teoreticheskoi Fiziki*, Vol. 38, 1960, pp. 22–25, [English trans., *Soviet Physics JETP*, Vol. 11, No. 1, 1960, pp. 15–17].

17. Brish, A.A., Tarasov, M.S., and Tsukerman, V.A., "Electric Conductivity of the Explosion Products of Condensed Explosives," *Zhurnal Eksperimentalnoi i Teoreticheskoi Fiziki*, Vol. 37, 1959, pp. 1544–1550, [English trans., *Soviet Physics JETP*, Vol. 10, No. 6, 1960, pp. 1095–1100].

18. Alder, B.J., and Christian, R.H., "Metallic Transition in Ionic and Molecular Crystals," *Physical Review*, Vol. 104, No. 2–15, 1956, pp. 550–551.

19. Altshuler, L.V., Kuleshova, L.V., and Pavlovskii, M.N., "The Dynamic Compressibility, Equation of State, and Electrical Conductivity of Sodium Chloride at High Pressures," *Zhurnal Eksperimentalnoi i Teoreticheskoi Fiziki*, Vol. 39, 1960, pp. 16–24, [English trans., *Soviet Physics JETP*, Vol. 12, No. 1, 1961, pp. 10–15].

20. Grigorev, F.V., Kirillov, G.A., Kormer, S.V., et al., "Study of Shock-Compressed Ionic Crystal Electric Conductivity and Absorption in 150–800 kbar Pressure Range," *proc., 2nd All-Russian Symposium on Combustion and Explosion*, Yerevan, 1965, pp. 259–262.

21. Grigorev, F.V., Kirillov, G.A., Kormer, S.V., et al., "Compressibility, Temperature, Electric Conductivity, and Absorptivity of Shock-Compressed Carbon Tetrachloride," *proc., 2nd All-Russian Symposium on Combustion and Explosion*, Yerevan, 1965, pp. 255–257.

22. Kormer, S.B., Sinitsyn, M.V., Kirillov, G.A., and Popova, L.T., "An Experimental Determination of the Light Absorption Coefficient in Shock-Compressed NaCl. The Absorption and Conductivity Mechanism," *Zhurnal Eksperimentalnoi i Teoreticheskoi Fiziki*, Vol. 49, 1965, pp. 135–147, [English trans., *Soviet Physics JETP*, Vol. 22, No. 1, 1966, pp. 97–105].

23. van Thiel, M., and Alder, B.J., "Shock Compression of Argon," *Journal of Chemical Physics*, Vol. 44, No. 3, 1966, pp. 1056–1065.

24. Nelson, D.A., Jr., and Ruoff, A.L., "Metallic Xenon at Static Pressures," *Physical Review Letters*, Vol. 42, No. 6, 1979, pp. 383–386.

25. Radousky, H.B., Nellis, W.J., Ross, M., Hamilton, D.C., and Mitchell, A.C., "Molecular Dissociation and Shock–Induced Cooling in Fluid Nitrogen at High Densities and Temperatures," *Physical Review Letters,* Vol. 57, No. 19, 1986, pp. 2419–2422.

26. Nellis, W.J., Radousky, H.B., Hamilton, D.C., Mitchell, A.C., Holmes, N.C., Christianson, K.B., and van Thiel, M., "Equation-of-State, Shock-Temperature, and Electrical-Conductivity Data of Dense Fluid Nitrogen in the Region of the Dissociative Phase Transition," *Journal of Chemical Physics*, Vol. 94, No. 3, 1991, pp. 2244–2257.

27. Gatilov, L.A., Glukhodedov, V.D., Grigorev, F.V., Kormer, S.B., Kuleshova, L.V., and Mochalov, M.A., "Electrical Conductivity of Shock Compressed Condensed Argon at Pressures from 20 to 70 GPa," *Zhurnal Prikladnoi Mekhaniki i Tekhnicheskoi Fiziki*, 1985, No. 1, pp. 99–102, [English trans., *Journal of Applied Mechanics and Technical Physics*, Vol. 26, No. 1, 1985, pp. 88–91].

28. Mochalov, M.A., Glukhodedov, V.D., Kirshanov, S.I., and Lebedeva, T.S., "Electric Conductivity of Liquid Argon, Krypton, and Xenon Under Shock Compression up to Pressure of 90 GPa," Shock Compression of Condensed Matter – 1999, Furnish, M.D., Chhabildas, L.C., and Hixson, R.S., eds., AIP Press, Melville, NY, 2000.

29. Urlin, V.D., Mochalov, M.A., and Mikhailova, O.L., "Liquid Xenon Study Under Shock and Quasi-Isentropic Compression," *High Pressure Research*, Vol. 8, No. 4, 1992, pp. 595–605.

30. Hamilton, D.C., Nellis, W.J., Mitchell, A.C., Ree, F.H., and van Thiel, M., "Electrical Conductivity and Equation of State of Shock-Compressed Liquid Oxygen," *Journal of Chemical Physics*, Vol. 88, No. 8, 1988, pp. 5042–5050.

31. Weir, S.T., Mitchell, A.C., and Nellis, W.J., "Metallization of Fluid Molecular Hydrogen at 140 GPa (1.4 Mbar)," *Physical Review Letters*, Vol. 76, No. 11, 1996, pp. 1860–1863.

32. Kiler, R., "Electric Conductivity of Condensed Matter at High Pressures," *in* High Energy Density Physics, Mir Publ., Moscow, 1974.

33. Kuleshova, L.V., "Electrical Conductivity of Boron Nitride, Potassium Chloride, and Polytetraflouroethylene Behind a Shock-Wave Front," *Fizika Tverdogo Tela*, Vol. 11, No. 5, 1969, pp. 1085–1091, [English trans., *Soviet Physics – Solid State*, Vol. 11, No. 5, 1969, pp. 886–890].

34. Kuleshova, L.V., and Pavlovskii, M.N., "Dynamic Compressibility, Electrical Conductivity, and Sound Velocity Behind a Shock Front in Kaprolon," *Zhurnal Prikladnoi Mekhaniki i Tekhnicheskoi Fiziki*, 1977, No. 5, pp. 122–126, [English trans., *Journal of Applied Mechanics and Technical Physics*, Vol. 18, No. 5, 1977, pp. 689–692].

35. Gatilov, L.A., and Kuleshova, L.V., "Measurement of High Electrical Conductivity in Shock-Compressed Dielectrics," *Zhurnal Prikladnoi Mekhaniki i Tekhnicheskoi Fiziki*, 1981, No. 1, pp. 136–140, [English trans., *Journal of Applied Mechanics and Technical Physics*, Vol. 22, No. 1, 1981, pp. 114–117].

36. Huang, S.S., and Freeman, G.R., "Effect of Density on the Total Ionization Yields in X-Irradiated Argon, Krypton, and Xenon," *Canadian Journal of Chemistry*, Vol. 55, No. 11, 1977, pp. 1838–1845.

37. Asaf, U., and Steinberger, I.T., "Photoconductivity and Electron Transport Parameters in Liquid and Solid Xenon," *Physical Review B*, Vol. 10, No. 10, 1974, pp. 4464–4468.

38. Glukhodedov, V.D., Kirshanov, S.I., Lebedeva, T.S., and Mochalov, M.A., "Properties of Shock-Compressed Liquid Krypton at Pressures of up to 90 GPa," *Zhurnal Eksperimentalnoi i Teoreticheskoi Fiziki*, Vol. 116, No. 2, 1999, pp. 551–562, [English trans., *Journal of Experimental and Theoretical Physics*, Vol. 89, No. 2, 1999, pp. 292–298].

6

Dynamic Strength of Materials

B.L. Glushak, O.A. Tyupanova, and Yu.V. Batkov

6.1 Experimental Methods for Studying Shear Strength

The methods used in practice for the acquisition and interpretation of experimental data pertaining to the resistance of materials to plastic-strain (shear-strength) are based on the concepts of the nature of motion of a material possessing strength and the features of high-rate deformation discussed in Chap. 1. These methods are briefly described below. Additional information on this topic is available to the reader in monographs [1–3], review [4], and the original papers cited in [1–4].

6.1.1 Comparison between the Shock Hugoniot and the Hydrostatic Compression Isotherm

Given two assumptions: (1) material strength is isotropic, and (2) the difference between isothermal hydrostatic compression and the mean stress $\sigma_{ii}/3$ is small, the dynamic yield strength is calculated to be the difference between the stress σ_1 on the elastic-plastic shock Hugoniot and the pressure P on the hydrostatic compression isotherm (given in terms of specific volume V (or strain ε)) as $Y_d = 3(\sigma_1 - P)/2$.

This approach to the estimation of Y_d is limited to cases wherein the stresses σ_1 are relatively low, and the temperature rise during shock compression is modest (resulting in a thermal component of pressure that is low in comparison to the total pressure).

For the aluminum alloy Al-2024, [5] recommends that for shock pressures in the 2.7 GPa $\leq \sigma_1 \leq$ 9.4 GPa range, one should use the following relations for $\sigma_1(\varepsilon)$ (the normal stress component in a plane wave shock) and $P(\varepsilon)$ (the hydrostatic compression) in calculations of the dynamic yield strength: $\sigma_1 = 0.18 + 72.06\varepsilon - 347.1\varepsilon^2$ (GPa); $P = 75.9\varepsilon + 201\varepsilon^2$ (GPa). This method for the evaluation of Y_d, among other things, requires a highly accurate determination of the curves $\sigma_1(\varepsilon)$ and $P(\varepsilon)$ since minor changes in ε can lead to significant changes in σ_1 and P.

6.1.2 Recording Shock Wave Decay

In this method, Y_d is evaluated from studies of the non-hydrodynamic decay of a shock wave moving across a sample constructed from an elastic-plastic material [6,7]. The scheme of the experimental setup is shown in Fig. 6.1. The sample is loaded by the deceleration of a free-flying plate impactor, preferably constructed from the same material as the target. The impact results in shock waves that travel in opposite directions through the plate and sample material. The shock wave traveling through the impactor will (upon arrival at the rear free surface) be reflected as a centered rarefaction wave including elastic and plastic release components.

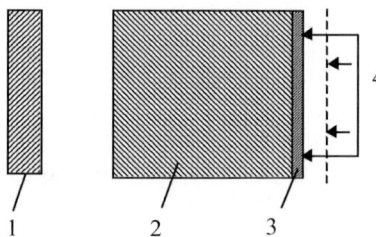

Fig. 6.1. Scheme of setting up experiments on non-hydrodynamic decay: 1 – impacting plate; 2 – material to be studied; 3 – rebounding indicator; 4 – electric pins for indicator velocity measurement

If the impactor and sample are fabricated from the same material, the wave moves across the system without reflection at their interface.

The elastic unloading wave overtakes the shock front in the sample before than the plastic rarefaction wave arrives and decreases its amplitude by interacting with it.

The magnitude of the decrease in the amplitude of the shock front depends on the magnitude of the elastic unloading wave $\Delta\sigma_{el}$. Thus, the recording of non-hydrodynamic shock wave decay carries information about $\Delta\sigma_{el}$, and this is related to Y_d by:

$$Y_d = \frac{1 - 2v}{2(1 - v)}\Delta\sigma_{el} .$$ (6.1)

In order to record shock wave parameters at a fixed distance x from the impact surface, a thin ($\Delta \approx 0.15$ mm) rebounding indicator is used. Since the indicator thickness is small, it is assumed that its velocity W_{ind} (typically measured using the electrocontact method) corresponds to the prompt value of σ_1 at the shock front for a given cross section. The transformation from flight velocity, W_{ind}, to mass velocity, u, at the shock front in the sample is made with the benefit of the results from tests specially calibrated to measure the flight velocity produced when a steady shock wave moves across the sample. From the values of u, the value of stress at the shock front in the sample, σ_1,

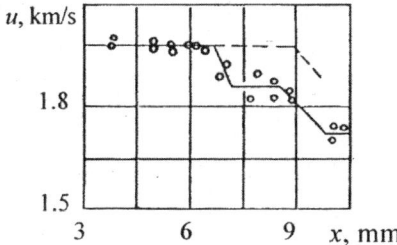

Fig. 6.2. Mass velocity at the shock front vs. distance traveled by the shock wave in copper (copper impactor thickness is 1.5 mm): - - - - calculation in the hydrodynamic approximation; ○ – experimental points

is calculated using the conservation of momentum and the $D - u$ relation of the material under study.

Figure 6.2 plots the calculated material velocity versus distance traveled by the shock wave, along with experimental points from [6]. The experimental data reveals a more rapid decay of the shock wave than that calculated using the hydrodynamic approximation.

Critical shear stresses may be evaluated by comparing the experimental shock wave dependencies to those predicted in a series of computations using different dynamic yield strengths. The value of Y_d, that produces the best agreement with the experimental data, is accepted as the desired value. Shock wave amplitude decay can also be estimated by recording the shock wave amplitude using a quartz gauge placed at different distances from the loaded surface [8].

6.1.3 Recording Stress Profile

In this method, the unloading wave stress profile $\sigma_1(t)$ is continuously recorded with piezoresistive (manganin, ytterbium) gauges embedded in the cross section of a sample pre-compressed by a steady shock wave.

Throughout the several reverberations that the applied shock wave loading may produce, the gauge is assumed to record true stress σ_1 in both compression and unloading, despite differences in shock impedance of the gauge material and the material under study. It should be noted, however, that there is no strict proof of the validity of this assumption (taking into account possible manganin hysteresis). In Fig. 6.3, both the elastic and plastic portions of the release wave can be clearly distinguished in the oscillographic records from the gauge.

If the interface between the elastic and plastic waves is clearly discernable, then the elastic unloading wave amplitude, $\Delta\sigma_{el}$, is straightforwardly estimated and the dynamic yield strength, Y_d, is calculated by relationship (6.1) for a given Poisson ratio, ν. In [9], the authors obtained $Y_d = 1.0$ GPa for

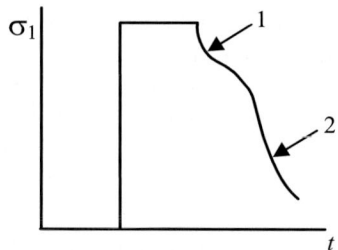

Fig. 6.3. Signal from the gauge: 1 – elastic release wave; 2 – plastic release

Al-2024 at $\sigma_1 = 9.45$ GPa, which is fairly close to the value of $Y_d = 1.1$ GPa calculated by the previously presented relationships for $\sigma_1(\varepsilon)$ and $P(\varepsilon)$.

Unlike the stress history record illustrated in Fig. 6.3, it is more often the case that no clear-cut interface between the elastic and plastic rarefaction waves is recorded. That is, the experimentally recorded stress profile has no characteristic kink relevant to the tail of the elastic rarefaction wave (despite the prediction of its presence using idealized models of rarefaction in metals). For this reason, Y_d of the shock-compressed material is often difficult to estimate directly from the profile.

In an attempt to increase the accuracy of the estimation of material shear strength, the authors of [10] implemented the following procedure. A set of profiles $\sigma_1(t)$ at several different internal cross sections located at varying distances from the loading surface of the sample across which the shock wave (and the following rarefaction wave) move is recorded. Deviation of the material state from an ideal hydrostatic state in the expansion wave is then sought. That deviation is given by:

$$\sigma_1 - P = \int \frac{a_{ph}^2 - c_v^2}{a_{ph}^2} d\sigma_1 , \tag{6.2}$$

where a_{ph} is the phase velocity at fixed stress level σ_1 determined from experimental data and c_v is the bulk sound speed calculated using the EOS. The maximum shear stress is then calculated using the relationship $\tau - 3(\sigma_1 - P)/4$, where $\sigma_1 - P$ is taken as the maximum of the values obtained in the integration.

6.1.4 Method for Measurement of Principal Stresses

The existing experimental methods for pulsed stress measurement using manganin or dielectric gauges has enabled the measurement, not only of stress σ_1 that is normal to the wave front, but also of transverse stresses σ_2 or $\sigma_3 = \sigma_2$ that lie in the shock front plane. Thus, the dynamic yield strength of the material compressed by a plane wave shock can be estimated directly from the experimental data (as the difference of two measured principal stresses,

$Y_d = \sigma_1 - \sigma_2 = \sigma_1 - \sigma_3$), without the need for additional independent information. It is assumed that the sensors measure true values of σ_1 and σ_2.

Presently, there is no rigorous proof of the validity of this assumption. Using current practices, studies suggests that the results of measurements of the principal stress σ_1 in the steady shock wave front using manganin gauges are highly accurate. In the slit where the gauge for measuring σ_2 is placed, material motion is complex in nature. Results of 2D computations and their comparison with experimental data suggest that the experimentally determined value of σ_2 is less than the true value of σ_2. In other words, the value of $Y_d = \sigma_1 - \sigma_2$ calculated directly from the measured data proves to be higher than the true value. The above difference, which is small for σ_1 in the gigapascal range, becomes significant at pressures of several tens of GPa. Results of experimental measurements taken in the gigapascal stress range that is presented below corroborate this assessment.

In the experiments of [11, 12], thin gauges made of manganin wire or foil enclosed in insulators are inserted into grooves in a sample made of the material to be studied. The gauges are placed either parallel or perpendicular to the moving steady shock wave front (Fig. 6.4).

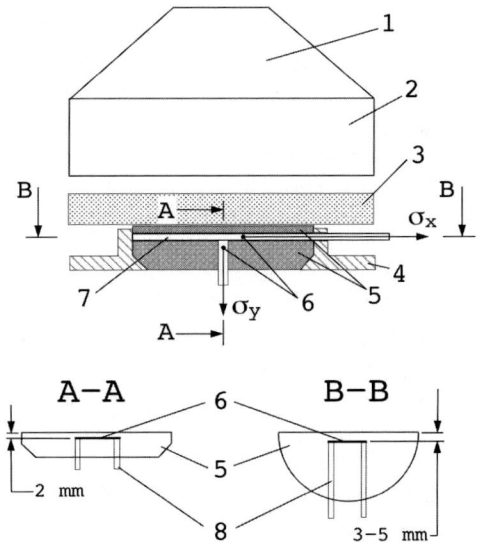

Fig. 6.4. Scheme of setting up the experiment on the principal stress measurement: 1 – plane-wave lens; 2 – HE charge; 3 – base plate (or screen); 4 – clamp; 5 – sample components; 6 – sensing element of the gauge; 7 – gauge insulation; 8 – leads

The geometric dimensions of the sample are such that there is no influence of side rarefaction waves or the rarefaction wave reflected from the free surface on material flow at the gauge position during the measurement. Measurement of σ_1 and σ_2 in a single experiment excludes the error resulting from variations in the values of σ_1 and σ_2 from one experiment to another on the same loading device with σ_1 and σ_2 measured separately. This is important since the difference $(\sigma_1 - \sigma_2)$, as a rule, is not large.

We shall consider in somewhat more detail the issue of measuring the lateral stress σ_2. Let us assume that the insulating material behaves like an ideal fluid. Under this assumption, it follows that:

(a) there are no shear stresses in the insulation layer;

(h) if the insulation layer is in mechanical equilibrium with the solid under study, then the hydrostatic pressure in the insulation is equal to stress σ_2 in the surrounding solid;

(c) the gauge calibration made for stress σ_1 is valid for stress σ_2.

The first two consequences are ensured by selecting polymers such as lavsan, polytetrafluorethylene, or polyethylene for the insulation material. If the calibration of σ_1 is made with polymer insulation, then it can be surmised to be acceptable for the measurement of σ_2. Records of gauge signals (see Fig. 6.5) suggest that stresses σ_1 and σ_2 taper off to a constant value.

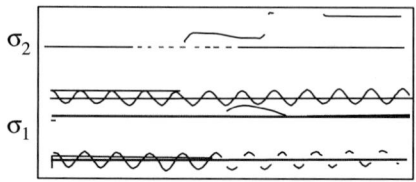

Fig. 6.5. Typical oscillogram of principal normal stresses σ_1 and σ_2. Time mark period is 1μs

Experimental evidence that the value of stress σ_2 measured using the method under discussion is correct in the gigapascal stress range is provided in the results of a number of studies. For example, the results obtained in the elastic deformation region of steels KhS38 [12], St.3, and 30KhGSA [13, 14], and of ceramics based on aluminum oxide, Al_2O_3 [15]. Further evidence is given in the results of the study of σ_θ for steel St.3 in the elastic and elastic-plastic deformation regions [13] (σ_θ is the stress along the normal to the cross sections oriented at angle θ to the shock front plane).

In the first case, Poisson's ratio calculated from $\sigma_2 = (v/(1-v))\sigma_1$ agrees with values in the literature obtained under static loading.

In the second case, the experimental curves of σ_θ agree with the well-known theoretical dependence

$$\sigma_\theta = \frac{\sigma_1 + \sigma_2}{2} \pm \frac{\sigma_1 - \sigma_2}{2} \cos 2\theta \ . \tag{6.3}$$

Computational dependencies are compared with the experimental data in Fig. 6.6.

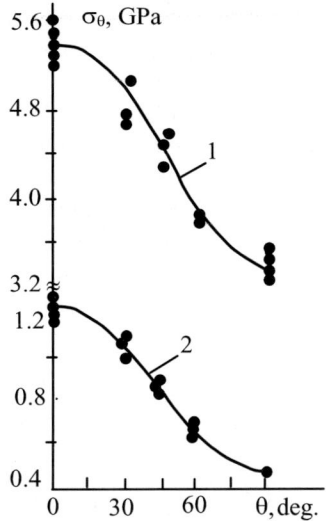

Fig. 6.6. Normal stress σ_θ in St.3 vs. angle θ: — – calculation by Eq. (6.3); 1 – elastic-plastic region; 2 – elastic region; • – experimental points

6.1.5 Self-consistent Method

The idea of the self-consistent method [16, 17] is illustrated in Fig. 6.7. In this method, functions $\sigma_1(V)$ or $\sigma_1(\varepsilon)$ are developed using experimentally determined particle velocities of shock-compressed materials in reloading and unloading from an "initial" shock-compressed state. States σ_u and σ_l, which represent the upper and lower limits of the yield surface, are obtained by extrapolation of the reloading and unloading curves $\sigma_1(\varepsilon)$ to a single isochore passing through the initial state of shock compression.

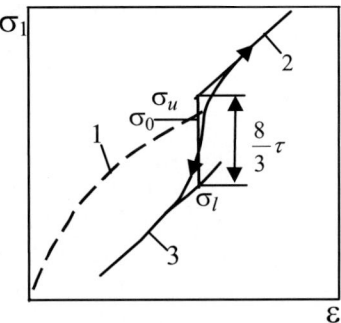

Fig. 6.7. Scheme of shear stress estimation with the self-consistent method: 1 – primary compression; 2 – secondary compression; 3 – unloading; σ_0 – shock-compressed state

We shall briefly describe the experimental process [17]. A compound impactor composed of two layers is impacted on a planar sample made of the material to be studied. Upon impact, shock waves move across the impactor front plate, and across the material under study. If the impactor back plate material is of a higher impedance than the impactor front plate material, then a shock wave will be reflected from the interface of the two plates and will move back across the front plate. Once the reflected shock wave reaches the impactor-sample interface, a second shock wave is generated in the sample.

If instead, the impactor back plate material is of lower impedance than the impactor front plate material, then an expansion wave will be reflected from the interface of the two plates and will follow the shock wave through the sample.

A laser interferometer (with the laser focused on the sample after passing through a radiation-transparent "window" in intimate contact with the sample surface), is used to record particle velocity versus time (Fig. 6.8). The results from [16] showing mass velocity as a function of time are presented in Fig. 6.9. Given that the stress in the window σ_{1win} can be determined as a function of the measured window mass velocity u_{win}, the stress and mass velocity in the sample (σ_{samp} and u_{samp}), may be estimated using the differential relationships [18]:

$$du_{samp} = \frac{1}{2}\left(du_{win} + \frac{d\sigma_{win}}{\rho_0 c}\right); \quad d\sigma_{samp} = \frac{1}{2}(d\sigma_{win} + \rho_0 c\, du_{win}) . \quad (6.4)$$

In Eq. (6.4), ρ_0 is the initial density of the sample, and c is the Lagrangian wave velocity corresponding to the increment $d\sigma_{samp}$. The material density of the sample can be determined from the well-known relation $d\varepsilon_{samp} = d\sigma_{samp}/\rho_0 c^2$, where $\varepsilon_{samp} = 1 - \rho/\rho_0$. In this manner the loading and unloading paths in the $\sigma_1(V)$ or $\sigma_1(\varepsilon)$ plane are found.

The dynamic yield strength is then calculated from the well-known relationship: $Y_d = 3(\sigma_u - \sigma_l)/4$ (see Fig. 6.7).

6.1.6 Taylor Method

G.I. Taylor proposed a simple method for the estimation of dynamic yield strength (see, e.g., [19]), in which the final length L_f of a rod following impact with a rigid barrier is related to the dynamic yield strength (Fig. 6.10).

Normal impact of a rod upon a rigid boundary produces high tractions on the surface of the rod contacting the rigid boundary. If the resultant stresses in the rod are higher than the yield strength of the rod material, a plastic front will propagate through the rod. An elastic wave, which runs ahead of the plastic front, decelerates the rod. For a given impact velocity, the higher the material yield strength, the faster the rod is decelerated. Under various assumptions, approximate analytical dependencies can be obtained that relate impact velocity and material properties to the rod geometry before impact and after the rod has come to rest.

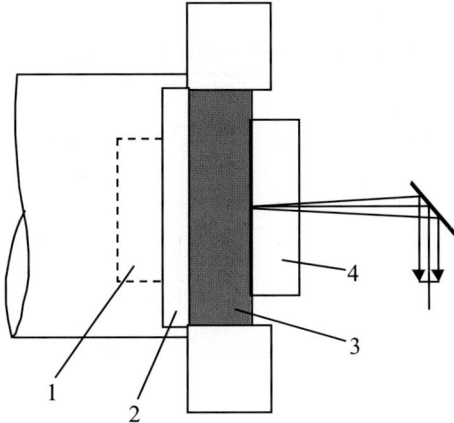

Fig. 6.8. Setup of the experiment using the self-consistent method [16]: 1 – impactor rear; 2 – impactor front; 3 – sample of material under study; 4 – transparent material barrier ("window")

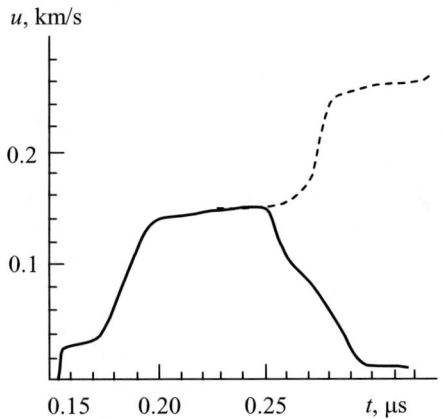

Fig. 6.9. Mass velocity vs. time [16] (aluminum sample): - - - - re-compression wave; —— – unloading wave

If the plastic front is assumed to localize near the rigid boundary (the value of h in Fig. 6.10 is small), then the rod deceleration rate is given by expression $\rho_0 L(dW_c/dt) = -Y_d$, where Y_d is the dynamic yield strength, and ρ_0 is the initial density of the rod material. The rate of decrease in the length of the rod is $dL/dt = -W_c$. By combining the two previous expressions, we obtain $dL/L = (\rho_0 W_c)/Y_d$, whose integration yields

$$\frac{L_f}{L_0} = \exp\left(-\frac{\rho_0 W_c^2}{2Y_d}\right) . \qquad (6.5)$$

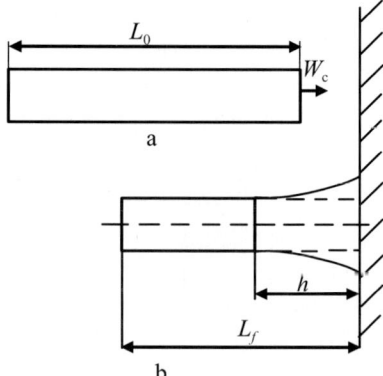

Fig. 6.10. Scheme of the experimental data recording in the impact experiment: **a** – rod geometry before the impact; **b** – rod geometry after the impact

Equation (6.5) shows that for $W_c = $ const., the material properties that are responsible for the ratio L_f/L_0 are density ρ_0 and yield strength Y_d. With increasing ρ_0 the ratio L_f/L_0 decreases, and conversely, with increasing Y_d the ratio L_f/L_0 increases. If plastic deformation occurs, an increase in impact velocity leads to a decrease in the ratio L_f/L_0. The above relationship holds for materials that behave as a rigid-plastic body. In general, the plastic wave front moves a considerable distance from impact surface. Computational analysis reveals that the distance, h, should be proportional to initial rod length L_0 and weakly dependent upon impact velocity. In this case $h/L_0 = $ constant, and the above-derived equation is transformed to [20]:

$$\frac{L_f}{L_0} = \left(1 - \frac{h}{L_0}\right) \exp\left(-\rho_0 \frac{W_c^2}{2Y_d}\right) + \frac{h}{L_0} \ . \tag{6.6}$$

Equation (6.6), along with measurements of W_c, h, and L_f can be used to estimate the dynamic yield strength.

The Taylor method produces plastic strain rates in the 10^3–10^5 s^{-1} range [20]. The analytical expressions given above should be viewed as being approximate in as much as they are related only to L_f, or to L_f and h, and do not reproduce the rod profiles in full measure after the impact. The complete post-shot rod geometry can be described adequately, only by numerical simulation methods that solve the 2D problem accounting for realistic strength properties of the rod material.

6.1.7 Measurement of Elastic Precursor Parameters

A number of experimental methods involving the continuous recording of interface velocity at a given Lagrangian point have been developed for determination of the elastic precursor. Here we shall discuss only those methods that are used most frequently in experimental practice.

In [21,22], the authors discuss a capacitive probe method that can be used to study the fine details of shock wave structure. This method is particularly well suited for studying the details of shock waves in an elastic-plastic body. Key elements of the experimental setup and the associated measurements appear in Fig. 6.11. The principles of operation of the capacitive probe and the methods used for processing the recorded signal are discussed in Chap. 3. A typical processed signal is shown in Fig. 6.11b.

Elastic-plastic wave parameters can also be measured with an electromagnetic velocity gauge. The gauge measures E.M.F. induced in a measuring coil with the conducting surface moving parallel to the coil plane (see [23,24] and Chap. 3).

The optical lever method [25,26] can be used for the same purposes. This method has not found wide utility and is not discussed here.

Since the beginning of the 1970s, a superior optical method, (laser interferometry – see [27] and Chap. 9), has been widely used in the study of shock-wave processes. The capacitive probe method and the laser interferometer are employed most frequently for measurements of mass velocity. Observed free surface velocities can be used to calculate the elastic precursor amplitude from the relationship $\sigma_{\mathrm{HEL}} = (\rho_0 W c_{el})/2$, where c_{el} is the elastic wave velocity, and W is the velocity of the free surface upon reflection of the elastic wave.

A second group of experimental methods continuously measures the stress history. Measurements of this type are taken with transducers possessing

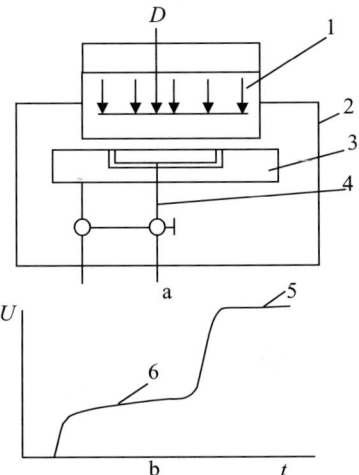

Fig. 6.11. Scheme of setting up capacitive probe measurements (**a**) and recorded signal (**b**) 1 – sample of material to be studied; 2 – metal base plate; 3 – measuring gauge; 4 – cables to power source and oscilloscope; 5 – plastic wave; 6 – elastic wave

elements that are sensitive to changes in pressure. This class of transducers includes piezoelectric and piezoresistive pressure gauges.

6.2 Dynamic Yield Strength of Materials in Shock Waves

6.2.1 Dynamic Precursor and Dynamic Strength of Metals

Studying the structure of compression waves and the various parameters associated with that structure enables a better understanding of the nature of material flow between the elastic and plastic wave fronts, and a better understanding of the nature of high-velocity elastic-plastic material deformation.

For this purpose, experimental studies often use methods enabling the continuous recording of free surface motion. An analysis of available results suggests a complex, incompletely characterized pattern of deformation in one-dimensional shock waves.

If $d^2\sigma_1/dV^2 < 0$ over some segment of the Hugoniot, a variety of discontinuities in both elastic and plastic response of the material are possible, the extent of which will depend on the specific rheological features of the material experiencing the high-velocity deformation. Possible wave profiles, similar to those presented in [28], are depicted in Fig. 6.12. Wave profile 1 illustrates the response of an ideal elastic-plastic material subjected to shock compression. Wave profile 2 illustrates the case where $(d\sigma_1/dV)_{\mathrm{HEL}} > \sqrt{(\sigma_1 - \sigma_{\mathrm{HEL}})/(V_{\mathrm{HEL}} - V)}$ at point σ_{HEL} of the Hugoniot. This phenomenon can be interpreted as an increase in shear stress occurring at the early stage of inelastic deformation.

Wave profile 3, representative of what one might see in copper, corresponds to the case, where the above-specified condition $((d\sigma_1/dV)_{\mathrm{HEL}} > \sqrt{(\sigma_1 - \sigma_{\mathrm{HEL}})/(V_{\mathrm{HEL}} - V)})$ is met in the elastic deformation segment. Plastic wave dispersion (profile 4) is considered to be evidence of viscoplastic flow. Experiments show that profiles of type 2 are characteristic of iron and steels. Profile type 3 is characteristic of metals possessing no pronounced yield strength.

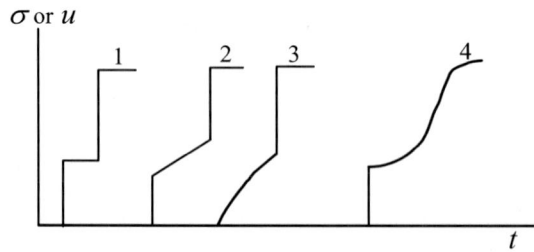

Fig. 6.12. Loading wave structure in ideal elastic-plastic medium (1); work-hardening medium (2); dispersion medium (3); viscous-plastic flow medium (4)

Numerous experiments testify to the fact that a two-wave configuration forms in most metals and their alloys in accordance with the basic concepts of elastic-plastic material behavior under shock loading. The elastic wave can have a clear-cut front (shock discontinuity) such as is seen in steels, or it may exhibit flow that is characteristic of a simple compression wave with dramatic sound velocity dispersion such as is seen in copper. The sound speed in such a wave is intermediate in magnitude, residing somewhere between the elastic and plastic sound speeds. At the elastic wave front, plastic strains are negligible and the elastic precursor amplitude σ_{HEL} characterizing the yield strength in uniaxial deformation under plane wave loading is related to the yield strength Y_d as: $\sigma_{HEL} = (1 - \nu)/(1 - 2\nu)Y_{dHEL}$. Thus, the dynamic yield strength can be calculated from the measured σ_{HEL} so long as ν is known. For standard conditions, the value of ν, is either taken from the literature, or calculated from a measurement of principal stresses in uniaxial deformation $(\sigma_2 = \sigma_3 = \nu/(1 - \nu)\sigma_1)$.

Values of σ_1 and σ_2 measured in the elastic region using the approximate method of principal stress measurement, and the subsequently calculated values of ν for two steels are presented in Table 6.1.

The calculated Poisson ratios are in a good agreement with the literature.

An experimentally observed feature of the elastic precursor is the decay of its amplitude with distance L traveled by the wave for unchanged loading conditions. This phenomenon has been observed for steels St.3 and St.20, aluminum alloy D16, tantalum, Armco-iron, and titanium (see [4]). The amplitude of the elastic wave versus distance traveled in steel St.3 [29] is plotted in Fig. 6.13.

At short distances from the surface of loading (produced either by the firing of HE in contact with the specimen or by plate impact), the elastic precursor

Table 6.1. Experimental values of principal stresses and calculated Poisson ratios in the elastic region [13]

Steel	σ_1	σ_2	ν
	GPa		
	1.27	0.49	0.28
	1.26	0.48	0.28
St.3 (annealed)	1.28	0.48	0.27
	1.18	0.48	0.29
	1.15	0.49	0.30
	1.97	0.86	0.30
	1.95	0.85	0.30
30KhGSA (quenched)	1.80	0.75	0.29
	1.85	0.78	0.30
	1.78	0.75	0.29
	1.87	0.82	0.31

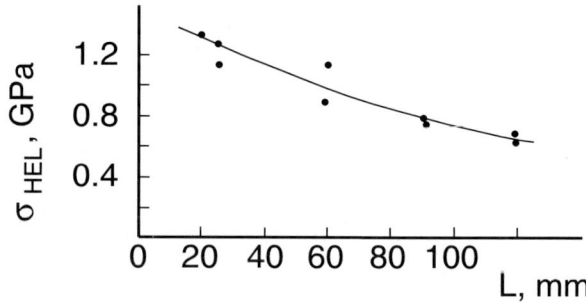

Fig. 6.13. Elastic wave decay in steel St.3: • – experimental points

amplitude decays at a rapid pace. The decay rate slows considerably as the distance traveled by the shock wave increases.

Under given loading conditions, there exists some particular distance, beyond which the elastic wave amplitude becomes essentially constant. For aluminum alloy V95 and steel St.20, stress amplitude σ_{HEL} is a constant at distance $l_0 > 4.0$ mm, when the impact interface mass velocity characterizing the plane wave intensity ranges from $u = 0.05$ km/s to $u = 0.25$ km/s for steel St.20 and from $u = 0.05$ km/s to $u = 0.4$ km/s for V95. The dependence of the amplitude σ_{HEL} on the distance traveled by the wave should be kept in mind when specifying the value of σ_{HEL}, that is, alongside the value of σ_{HEL}, the value of l_0 should be specified.

Table 6.2 presents σ_{HEL} at fixed l_0 for a number of metals [4]. The value of σ_{HEL} is estimated to within $\pm 10\%$ accuracy. The function $\sigma_{HEL}(l)$ should also be taken into account when setting up studies in which the sample to be studied is loaded by an elastic wave transmitted from an elastic-plastic body. The decay of the elastic precursor in tantalum, Al-1060, and Armco-iron is illustrated by the data of Table 6.3.

Since experimentally recorded wave profiles often differ from idealized representations, there can be some uncertainty in determining the value of the σ_{HEL} amplitude that should be used in calculations of dynamic strength. In some papers the magnitude of σ_{HEL} is taken to be the amplitude of the elastic wave front, which, as explained in the foregoing, is variable at short distances l. In [29,30] the elastic wave amplitude is characterized by its magnitude near the interface with the plastic region, which, according to [31], seems more appropriate for finding the dynamic strength. Noteworthy is the experimentally observed phenomenon [30], that when firing identical HE charges (i.e. under identical conditions of loading of steel samples), the stress σ_1 at the elastic wave front drops with distance traveled, whereas the stress σ_1 immediately in front of the plastic wave remains constant, i.e. independent of the distance traveled by the wave.

Estimated values of σ_{HEL} for a number of metals and their alloys are summarized in Table 6.4.

Table 6.2. Elastic wave amplitude in metals [4]

Metal	l_0, cm	σ_{HEL}, GPa
Al-2024	1.2	0.54
Al-6061	1.2	0.54
Chromium	0.5	1.6
Molybdenum	0.3	1.6
Nickel	1.2	1.00
Titanium	1.2	1.85
Tungsten	0.5	3.2
Iron: fine-grained hard	0.3	1.5
fine-grained soft	0.3	1.1
coarse-grained hard	0.3	1.4
coarse-grained soft	0.3	0.9
Annealed lead	1.2	0
Tantalum	1.0	1.87
Tantalum	0.373	2.1

Table 6.3. Elastic wave decay in metals [4]

Metal	σ_1	σ_{HEL}	l_0, cm
	GPa		
Tantalum	2.5	2.5	0.3
		1.3	1.0
		1.2	1.9
Al-1060	2.0	1.06	0.138
		0.77	0.303
		0.64	0.465
		0.53	0.968
Armco-iron	7.5	1.5	0.15
		1.2	0.25
		0.8	1.2
		0.65	2.4
		0.40	5.0

σ_{HEL} is calculated from the amplitude of the elastic wave at its interface with the plastic wave. For the metals studied, the dynamic yield strength, Y_d, as a rule, is higher than the static yield strength. That is, hardening takes place in weak shock waves or compression waves that are just elastic waves. Aluminum is fairly prone to hardening, both in the annealed state and in the quenched state. Significant hardening is also observed for most steels. The dynamic strength of steel varies over a wide range: from $\sigma_{HEL} = 0.59$ GPa for steel 12Kh18N10T to $\sigma_{HEL} = 1.88$ GPa for quenched steel 30KhGSA. The value of σ_{HEL} and the degree of hardening depend on production technology, thermal processing, and the presence of alloying additives. For the softest steels, the dynamic strength σ_{HEL} is significantly higher than the static yield strength $\sigma_{0.2}$, indicative of the occurrence of significant hardening.

Table 6.4. Dynamic strength of metals in state $\sigma_1 = \sigma_{HEL}$ [29, 30, 32]

Metal	$\sigma_{0.2}$	σ_{HEL}	$\sigma_{HEL}/\sigma_{0.2}$
	GPa		
Steel St.3	0.21	0.86	4.1
Steel 40Kh (quenched)	0.82	1.66	2.0
Steel 40 Kh (annealed)	0.42	1.24	2.9
Steel 30KhGSA (quenched)	1.45	1.88	1.3
Steel 30KhGSA (annealed)	0.47	1.32	2.9
Armco iron	0.15	0.73	4.9
Aluminum D1 (annealed)	0.155	0.237	1.53
Aluminum D16 (annealed)	0.13	0.263	2.02
Aluminum D16 (quenched)	0.27	0.422	1.56
Copper (annealed)	0.05	0.048	0.97
Brass LS59-1 (annealed)	0.125	0.20	1.60
Steel 30KhGSA			
$\perp$ HRC 21-23	0.239	0.887	3.11
$\perp$ HRC 35-36	0.820	0.980	1.20
‖ HRC 21-23	0.410	0.949	2.31
‖ HRC 35-36	1.43	1.07	0.75
Steel EP 712	0.74	0.81	1.09
	1.11	0.805	0.73
Steel 12Kh18N10T $\perp$	0.32	0.59	1.84
Steel 36NKhTYu ‖	0.97	1.12	1.15

Note: $\perp$ is the direction perpendicular to the rolling direction; ‖ is the direction parallel to the rolling direction.

This fact may be associated with above-barrier dislocation slip drag [32]. With increasing static strength, the difference between static and dynamic strength monotonically decreases. For high-strength steels, σ_{HEL} is close to $\sigma_{0.2}$ and can even be less than σ_{HEL}. This is the case, for example, with steels 36NKhTYu and 30KhGSA when loaded in the direction parallel to the rolling direction. No such hardening occurs for annealed copper.

The elastic wave amplitude as a function of the initial temperature of the sample is studied for steel 45 in [33]. Experimental results obtained by processing the measurements taken using the capacitive probe method are summarized in Table 6.5.

Judging from general physical considerations, one would expect the strength of metals experiencing no structural changes to drop with increasing initial temperature of the sample. However, it has been observed that probable structural changes occurring during shock compression of metals having elevated initial temperature can lead to anomalous behavior. For steel 45 at low initial temperatures, the elastic precursor amplitude and the stress σ_1 immediately in front of the plastic wave drop with increasing temperature. At the same time, in the range of initial temperatures $T = 240-300\,^{\circ}\mathrm{C}$, the dynamic yield measurements increase noticeably (which, by the way, agrees

Table 6.5. Elastic wave amplitude and dynamic yield strength versus initial temperature in steel 45 [33]

T, °C	Y'_d	Y''_d	σ_{HEL}
	GPa		
5	1.06	1.25	1.68
170	0.78	1.02	1.23
210	0.80	0.91	1.27
240	0.82	1.09	1.29
305	0.77	1.16	1.20
310	0.76	1.02	1.16
360	0.60	0.87	0.94

Note: Y'_d is the state at the elastic wave front; Y''_d is the state before the plastic wave front.

with the observances of temperature effects on the static yield strength of steel 45). In static experiments this phenomenon is attributed to structural changes in the steel near the blue brittleness temperature. Under dynamic loading, with an initial temperature $T > 300\,°C$, a significant decrease in the elastic wave amplitude is detected. A decrease in the elastic wave amplitude has been observed for annealed aluminum D16 and for annealed copper at initial sample temperatures of $T = 100\,°C$ and $200\,°C$ [30].

Experiments involving the explosive loading of steels and iron reveal a yield peak ("tooth") in the elastic wave front for relatively short sample lengths, L_{samp}, as shown in Fig. 6.14 [29]. The yield peak parameters are summarized in Table 6.6. Its amplitude is relatively small, but it is distinctly observed in the experiments. With increasing sample length, the elastic wave amplitude decreases and the yield peak disappears.

The experimental evidence points to complex rheological behavior for metals subjected to shock loading. Decay of the elastic precursor amplitude as the wave moves across the sample is evidence of relaxation processes. The con-

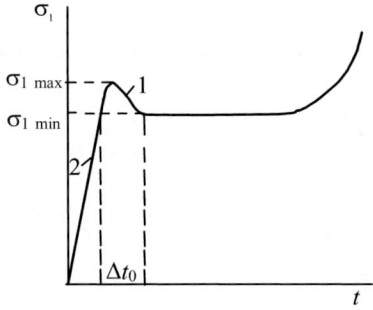

Fig. 6.14. Curve $\sigma_1(t)$ for shock wave in iron [30]: 1 – yield "tooth"; 2 – elastic wave front

Table 6.6. Parameters of dynamic yield "tooth" for steels [29]

Steel	$\sigma_{1\,max}$	$\sigma_{1\,min}$	Δt_0, μs	H_{samp}, cm
	GPa			
St.3	1.13	1.10	0.09	2.5
	1.26	1.21	0.13	2.5
	1.35	1.33	0.08	2.0
30KhGSA (annealed)	1.72	1.46	0.09	6.0
30KhGSA (quenched)	1.84	1.78	0.13	6.0
Armco iron	0.99	0.89	0.14	2.5
	1.06	0.98	0.21	6.0

stancy in the elastic wave amplitude beginning at some distance l_0 traveled by the wave suggests the completion of stress relaxation along this wave path.

Elastic precursor decay is not described by simple elastic-plastic deformation models. To improve the agreement between experimental data and computational prediction, attempts are being made to employ more complex rheological models that reflect actual material properties and deformation mechanisms to a greater extent. Dislocation models describe the nature of the elastic precursor decay qualitatively, and given an appropriate choice of parameters, quantitatively. However, as the bulk of papers addressing this issue note, with a minimum number of free constants and parameters in the equations describing plastic deformation, quantitative agreement with experimental data is achieved only under the assumption of a rapid dislocation multiplication rate. Initial dislocation density in such models is required to be on the order of 10^8–10^9 cm^{-2}, which is considerably higher than its true value in the materials that have been studied [34]. Therefore, we can surmise that dynamic yield strength depends not only on the evolution of the material dislocation structure, but also on other processes. In view of this observation, the relaxation dilation model is worthy of attention, further analysis, and improvement.

6.2.2 Shear Strength of Shock-compressed Metals

Experimental information regarding the shear strength of metals compressed by steady shock waves in the range $\sigma_1 > \sigma_{HEL}$ has been acquired using the following methods:

- recording of the rarefaction wave profile;
- measurement of principal stresses σ_1, σ_2;
- self-consistent method;
- recording of non-hydrodynamic shock wave decay.

Aluminum and Its Alloys

The results for aluminum and its alloys are summarized in Table 6.7. The experimental data from [11], complemented with the results of [13] for aluminum AD1 and AMg6, are described by linear analytical dependencies relating the measured principal stresses σ_1 and σ_2:

AD1 $\sigma_2 = (0.918 \pm 0.015)\sigma_1 - (0.495 \pm 0.177)$ (GPa);

AMg6 $\sigma_2 = (0.951 \pm 0.007)\sigma_1 - (0.127 \pm 0.100)$ (GPa).

The first of these expressions is valid in the 3.7 GPa $\leq \sigma_1 \leq$ 22 GPa range, the second in the 2.0 GPa $\leq \sigma_1 \leq$ 21.0 GPa range.

The above relations relate the primary experimental data. To determine actual values of Y_d, it is necessary (as specified in Sect. 6.1.4), to introduce corrections obtained by mathematical simulation. Note that [35] detects no noticeable difference in strength between aluminum AD1 and duralumin up to $\sigma_1 \approx 16$ GPa, and for these materials the general dependency of $Y_d(\sigma_1)$ is given (see Table 6.7).

The curve of dynamic yield strength, Y_d, along the Hugoniot is of a peculiar "bell-like" shape with a peak at $\sigma_1 \approx 0.4 - 0.5\sigma_{1melt}$ (σ_{1melt} is the principal stress at $T = T_{melt}$) [36]. In the ascending branch of the $Y_d(\sigma_1)$ curve, the dominant factor is isothermal hardening, in the descending branch the dominant factor is thermal softening.

Subjected to shock wave loading, aluminum and its alloys experience significant hardening: maximum Y_d is ~6 times as high as σ_{HEL}. The authors of [17] report significant hardening of aluminum alloy 6061-T6 in the stress range $\sigma_1 = 8 - 40$ GPa (shear strength increases by a factor of ~5).

According to [37], the conventional yield strength of alloy AMg6 in uniaxial compression under quasi-static conditions (deformation rate $\dot{\varepsilon} \approx 10^{-3}$ s^{-1}) is 205 MPa, which is less by about an order of magnitude than the maximum dynamic yield strength in shock waves.

Results from the static loading tests of [38] are quite close to the results for shock-wave experiments on the ascending branch of $Y_d(\sigma_1)$. With equal pressures P, the relevant yield strengths differ only slightly one from another, evidence of a significant role for pressure and relatively weaker effect of other hardening factors. Based upon measurements of principal stresses using a dielectric gauge, [39] reports no increase in the shear strength above that at the state $\sigma_1 = \sigma_{HEL_0}$ for the aluminum alloy V95 in the elastic-plastic region up to $\sigma_1 = 7$ GPa. The results of those measurements (according to [40]) might have been affected by inaccuracies in the pulsed stress measurement using the dielectric gauge when there was multiple shock-wave loading and wave reverberation in the gauge material.

Copper

Results for copper are summarized in Table 6.8.

Table 6.7. Results of shear strength studies for aluminum and its alloys [4]

Material (method)	σ_1	$\Delta\sigma_{el}$
		GPa
AMg6 (NHD)	30	6
	68	10
Al-2024 (NHD)	13	1.2
	9	0.8
Duralumin (SCM)	40	$Y_d - (1.5 \pm 0.75)$ GPa
Duralumin (SP)	\multicolumn{2}{c}{$Y_d = (0.06 \pm 0.018)\sigma_1 + (0.16 \pm 0.01)$}
	\multicolumn{2}{c}{up to $\sigma_1 = 16$}
AD1 (PS)	\multicolumn{2}{c}{$\sigma_2 = f(\sigma_1)$}
AMg6 (PS)	\multicolumn{2}{c}{$\sigma_2 = f(\sigma_1)$}
Al-2024 (NHD)	11	2.5
	34	6.5
	2.1	1.26
	2.1	1.26
	3.1	1.6
Al-2024 (SP)	6.0	2.6
	9.45	3.4
Al-2024 (annealed) (SP)	5.6	1.6
	3.05	0.85
Al-2024 (SH and IC)	\multicolumn{2}{c}{$\sigma_1 = 0.18 + 72.06\varepsilon_1 + 347.14\varepsilon^2$}
	\multicolumn{2}{c}{$P = 75.9\varepsilon_1 + 201.2\varepsilon^2 + 368.36\varepsilon^2$ $2.7 \leq \sigma_1 \leq 9.4$}
	σ_1	Y_d
		GPa
Al-2024 (SCM)	28	1.28
	40	1.64
	54	1.91
	68	1.84
Al-2024 (SCM)	72	1.47
	81	1.51
	88	1.68
Al-2024 (NHD)	31.3	1.4

Notes: NHD is non-hydrodynamic decay; SCM is self-consistent method; PS is principal stress measurement method; SP is stress profile recording method; SH and IC is method of shock compression and isothermal compression curve comparison.

In [13], the primary experimental data obtained using the method of principal stress measurement (in the 2.2 GPa – 22.0 GPa pressure range), can be described by the linear analytical relationship $\sigma_2 = (0.868 \pm 0.008)\,\sigma_1 - (0.129 \pm 0.108)$ (GPa).

Based on work involving the non-hydrodynamic decay method, recommended values of Y_d are given in [6,41]. It should be noted, however, that these

Table 6.8. Results of the copper shear strength study [4]

Method	σ_1	$\Delta\sigma_{el}$	Y_d
		GPa	
Non-hydrodynamic decay	34	9	2.5
	86	15	4.1
Non-hydrodynamic decay	122	10 ± 2	1.6 ± 0.4
Principal stress measurement-method	$\sigma_2 = f(\sigma_1)$		
Self-consistent method	91	–	1.24
(by unloading wave)	140	–	0.63

papers use Morland's computational scheme, which, as [42] demonstrates, is incorrect and can yield erroneous results. The discrepancy in the experimental data of Table 6.8 can be explained both by differences in the accuracy of Y_d obtained by different experimental methods, and by the inadequacy of the methods themselves. Nevertheless, the curve $Y_d(\sigma_1)$, as is the case for aluminum and its alloys, is bell-shaped. According to [37], the conventional yield strength of copper in compression ($\dot{\varepsilon} \sim 10^{-3}$ s^{-1}) is 0.29 GPa. It is therefore apparent that copper undergoes greater hardening under the effect of shock waves than does aluminum. Studies performed in [43] show that under quasi-static loading at a given temperature, yield strength increases linearly with pressure: $Y = Y_0 + \alpha P$, where Y_0 is yield strength at atmospheric pressure, and the temperature dependence can be represented as $Y_d = Y_0 \exp[bT/T_{melt}(P)]$ ($b < 0$). Hence, yield strength increases with pressure at $T = $ constant and decreases with temperature at $P = $ constant.

Iron and Steel St.3

Results of the measurements using the principal stress method for armco iron and steel St.3 in the stress range 1.3 GPa $\leq \sigma_1 \leq$ 25 GPa [11, 13, 14] may be described by the analytic functions:

$$\sigma_2 = (0.855 \pm 0.014)\sigma_1 - (0.238 \pm 0.192) \text{ (GPa) for armco iron;}$$
$$\sigma_2 = (0.718 \pm 0.012)\sigma_1 - (0.350 \pm 0.084) \text{ (GPa) for St.3.}$$

The experiments revealed no clear-cut anomalies in the $\sigma_2(\sigma_1)$ curve in the phase transition region. In steel St.3, the elastic unloading wave amplitudes estimated with the non-hydrodynamic decay method were: $\Delta\sigma_{el} = 8.0 \pm 2.0$ GPa at $\sigma_1 = 111$ GPa and $\Delta\sigma_{el} = 21 \pm 3.5$ GPa at $\sigma_1 = 184$ GPa [6]. Processing of the data using a model for an ideal elastic-plastic medium resulted in the following dynamic yield strengths: at $\sigma_1 = 64$ GPa, $Y_d = 1.55 \pm 0.25$ GPa; at $\sigma_1 = 122$ GPa, $Y_d = 1.9 \pm 0.4$ GPa; and at $\sigma_1 = 184$ GPa, $Y_d = 2.9 \pm 1.0$ GPa. These results reveal a trend toward hardening for soft iron (steel St.3) with increasing σ_1 far from the onset of melting on the Hugoniot ($\sigma_{1melt} \approx 250$ GPa) [44].

6.2.3 Shear Strength of Shock-compressed Nonmetals

Aluminum Oxide-based Ceramics

Understanding the behavior of high-strength ceramics has become very important in recent times due to the use of these materials for anti-projectile protection. Various Al_2O_3-based ceramics differ from one another in initial density (porosity k), impurity quantity, and method of production.

Results of studies of the Hugoniot elastic limit of these materials using the optical method are summarized in Table 6.9.

From the data presented in Table 6.9, it is apparent that Al_2O_3 based ceramics have σ_{HEL} significantly higher than σ_{HEL} of typical metals (see Table 6.2). As was seen for metals, elastic precursor decay is a phenomenon that occurs in general.

Curves of the difference in principal stresses ($\sigma_1 - \sigma_2$), are plotted against shock wave amplitude, σ_1, in Fig. 6.15 using the results of [15]. The data indicate that within the range of σ_1 considered, the behavior of the ceramics is characterized by a linear dependence of $\sigma_1 - \sigma_2$ on σ_1 up to $\sigma_1 = \sigma_{HEL}$, and then an essentially constant value of $\sigma_1 - \sigma_2$ thereafter. These observations are valid at least up to $\sigma_1 \approx 20$–25 GPa. Note that, according to [47], for a sample thickness $h = 3 - 4$ mm, the Hugoniot elastic limit of monocrystalline Al_2O_3 is 14.4–17.4 GPa, or about twice as high as that of polycrystalline Al_2O_3.

Table 6.9. Hugoniot elastic limit of several aluminum oxide (Al_2O_3) based ceramics [45, 46]

Ceramics	ρ_0, g/cm^3	Sample thickness, mm	σ_{HEL}, GPa	ν
AD 85	3.42 $k = 1.066$	3.2 6.3 9.6	6.5 6.1 ± 0.5 6.1	0.256
	3.72 $k = 1.055$	3.2 4.1	7.9 ± 0.5 9.5 ± 2	0.234
R-3142-1	–	8.0 9.0	8.1 ± 4 7.2	–
Al-995	3.81 $k = 1.04$	6.4	8.3	0.218
Hot-pressed	3.92 $k = 1.008$	3.2 6.0 9.6 12.8	16.0 ± 0.6 13.4 ± 0.5 11.2 ± 0.3 9.2 ± 0.3	0.243
Lucalox	3.97 $k = 1.002$	6.4 12.7	11.2 ± 0.2 11.1 ± 1.2	0.235

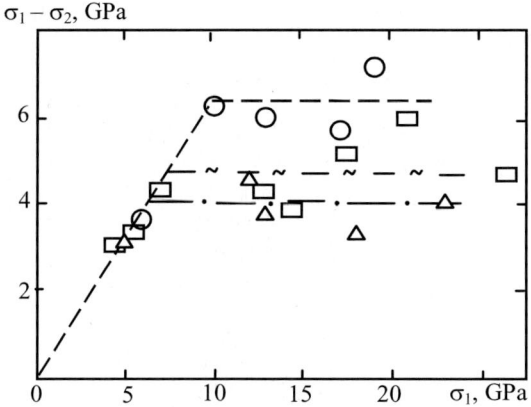

Fig. 6.15. Ceramics principal stress difference vs. shock wave amplitude: ∘ − cast KVPT (high density corundum with titanium); □ − pressed KVPT; △ − chylumin

Boron Carbide and Fused Quartz

Elastic precursor amplitudes σ_{HEL} of boron carbide with an initial density of $\rho_0 = 2.5$ g/cm^3 were recorded in [45] using the optical method for different sample thicknesses, h. The results: $\sigma_{HEL} = 16.2 \pm 2.9$ GPa, 15.8 ± 0.4 GPa, 14.9 ± 2.0 GPa, and 13.7 ± 0.4 GPa. The corresponding values of h: 3.2, 5.6, 7.2, and 9.6 mm. For $\nu = 0.188$ [48], the above values of σ_{HEL} give yield strengths of 12.5, 12.2, 11.54, and 10.5 GPa. The elastic sound speed is: $c_L = 13.78$ km/s.

For fused quartz (according to [48]), the experimental value of σ_{HEL} is 8.81 ± 0.11 GPa, corresponding to a dynamic yield strength Y_d of 7.0 GPa.

Polymers

Results of studies of the behavior of shock-compressed polymers using the principal stress measurement method are presented in Fig. 6.16 [49]. An examination of the data reveals that for shock front stresses up to $\sigma_1 \approx 3$ GPa, the principal stress difference $(\sigma_1 - \sigma_2)$ increases linearly with σ_1. The maximum in the difference $\sigma_1 - \sigma_2$ is reached at $\sigma_1 \approx 5 - 10$ GPa. With further increase in σ_1 (beyond the 5–10 GPa range), the principal stress difference tends to decrease. It is possible that this decrease is brought about by the onset of decomposition.

For paraffin, measurements using the principal stress method show that at $\sigma_1 \approx 2.3$ GPa, a hydrostatic state occurs behind the shock front [14].

High Explosives

In the stress range, where HE's behave as inert materials, the shear strength of solid HE may be studied using the principal stress measurement method [50].

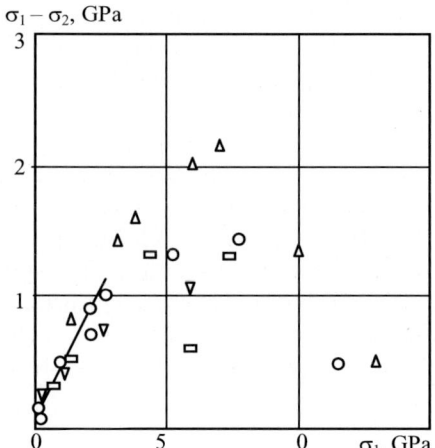

Fig. 6.16. Dynamic yield strength as principal stress difference vs. shock compression stress for polymers. Experimental points: ○ – PMMA; △ – polytetrafluorethylene; □ – polyethylene; ▽ – caprolon

For cast trinitrotoluene with initial density $\rho_0 = 1.6$ g/cm^3, within the range of σ_1 studied (0.3 GPa $\leq \sigma_1 \leq$ 1.9 GPa), it is seen that the difference in principal stresses increases linearly with $\sigma_1 : \sigma_1 - \sigma_2 \approx 0.08 + 0.155\sigma_1$ (GPa).

Similarly, for cast composition TG 50/50 with an initial density $\rho_0 = 1.65$ g/cm^3,

$$\sigma_1 - \sigma_2 \approx 0.12 + 0.20\sigma_1 \text{ (GPa) when } 0.36 \text{ GPa} \leq \sigma_1 \leq 1.9 \text{ GPa.}$$

For either type of HE, the difference $\sigma_1 - \sigma_2$ is significantly larger than the value of the static yield strength [51].

6.2.4 Taylor Method

The Taylor method is widely used to estimate material resistance to plastic strain in high-speed compression. The method has been used to test a range of materials used in engineering applications, including metals and alloys, brittle materials like metal glasses, and even solid propellants.

For a long time, the results of experiments with the Taylor method were used to evaluate, using Eqs. (6.5) and (6.6), the yield strength and relative strength of "same-name" materials (for example, steels) differing from each other in method of production, initial microstructure, presence of impurities or special additives.

Presently, with the availability of advanced computer simulation methods, experimental data obtained using the Taylor method, along with results of tests obtained using the widely known split Hopkinson bar method, are used for the construction and verification of plastic deformation models.

Results of several studies of the resistance to plastic strain using the Taylor method are presented in [20, 52] for several metals and their alloys. The experimental data and computational results obtained in [52] are summarized in Table 6.10.

The geometric dimensions of the samples tested in [52] were as follows: $L_0 = 120$ mm, $D_0 = 20$ mm, $L_0/D_0 = 5$. Numerical simulation of cylindrical sample deformation was performed in the context of the elastic-plastic material approximation. To obtain a satisfactory description of the final sample geometry, the values of dynamic yield strength, Y_d, specified in Table 6.10 must be included in the computation. Note that they are a factor of 1.4–1.6 smaller than those obtained using Eq. (6.5). This difference is attributable to significant approximations made by Taylor. It is apparent that the most

Table 6.10. Results of testing with the Taylor method [52]

Material	W_0,m/s	L_f,mm	L_f/L_0	ε	Y_d,GPa	$Y_d/\sigma_{0.2}$
Copper	110	90.0	0.90	0.10–0.48	0.4	8
	158	82.0	0.82			
	205	72.0	0.72			
	250*	62.0	0.62			
Steel St.3	120	93	0.93	0.07–0.35	0.7	3.3
	140	91	0.91			
	170	88	0.88			
	185	86	0.86			
	230	80	0.80			
	310*	70	–	–	–	–
Steel 30KhGSA (quenched) (annealed)	180	94	0.94	0.06–0.09	1.7	1.15
	215	92	0.92			
	230	91	0.91			
	220	92	0.92			
	270	89	0.89			
Aluminum AD1	200	76	0.76	0.3–0.5	0.14	1.4
	205	76	0.76			
	220	72	0.72			
	270	61	0.61			
	290	59	0.59			
Aluminum AMg6	185	89.5	0.895	0.11–0.29	0.35	2.2
	210	87.5	0.875			
	230	86.5	0.865			
	285	81	0.81			
	330	75	0.75			
Aluminum D16	200	91	0.91	0.09–0.21	0.50	1.7
	225	90	0.90			
	260	87	0.87			
	315	81	0.81			

* Experiments, in which material tearing was observed.

satisfactory approximation of reality results when one uses Y_d as a function of constitutive characteristics in the mathematical modeling.

Table 6.10 infers hardening for all tested materials in high-speed deformation: the yield strength in dynamic compressive strain is higher than the conventional quasi-static yield strength, $\sigma_{0.2}$. As in the case of the elastic precursor (see Table 6.4), the relative increase in yield strength tends to increase with decreasing static yield strength.

6.3 Experimental Methods for Spall Strength Estimation

An ability to resist failure under short-term intensive tensile loading is an important feature of solid media. In general, when a compressive load is rapidly applied to the sample under study, the result is the generation of unsteady compressive shock waves and expansion waves. A necessary condition for the production of tensile stresses is the interaction of two counter propagating expansion waves. This situation occurs most frequently when an unsteady shock wave moving through the material under study reaches an interface with a material of lower hardness (less dynamic impedance).

For tensile stresses (negative pressures in the hydrodynamic approximation) to form due to the interaction of counter propagating expansion waves, the wave amplitudes must be of sufficient magnitude. Thus, if two rarefaction waves of equal amplitudes move counter to each other across a planar sample compressed by a planar stationary shock wave to stress σ_1, it follows from the solution of the Riemann problem in the acoustic approximation that tensile stresses will form in the material since the amplitude of either wave will be larger than $0.5\sigma_1$. Under appropriate tensile stress conditions (amplitude, duration, etc.), a layer can be separated, whose thickness depends both on properties of the tested material and the characteristics of the tensile pulse. This particular type of failure is referred to as spall fracture, or spallation. Study of the spallation phenomenon seems at the present time to be the only practical avenue for research into the mechanics of material failure in the micro- and submicro-second time scales of tensile stress action.

All experimental methods for studying material spall strength that have been reported in the literature provide only indirect information regarding the damage processes taking place inside a solid and the tensile stresses that are acting on the material to produce that damage [2].

Spall fracture studies primarily employ two types of experiments: 1 – diagnostic recording of planar sample free surface velocity, or stress history at the interface between the planar sample of material under study and a medium of a lower dynamic impedance, and 2 – metallographic study of the spall fracture zone in samples that are recovered after loading.

Results from the first method provide the most accurate information about the stresses acting on the solid during spallation. In the second case, information is extracted from post-shot studies that shed light on the material's ability to resist short-term tensile stresses, the nature and extent of localized damage, and the effects of stress pulse characteristics on the material state in the failure region.

6.3.1 Methods for Generation of Pulsed Tensile Stresses

Intense short-duration tensile stresses are produced in deformed continuous media as the result of the wave processes induced in them by the application of high-energy sources. Solid HE, high-velocity impactors, electron beams, and laser radiation are used for this purpose [53]. Information about material failure under the action of pulsed tensile stresses has been acquired primarily from experiments involving conditions of high-rate uniaxial deformation. Laboratory methods for pulsed load generation (see Chap. 2) provide for the generation of a wide range of tensile stress pulses: from a few fractions of a GPa to several tens of GPa in amplitude, and from a few hundredths of a microsecond to a few microseconds in duration.

The generation of tensile stress in a solid target is often accomplished through the deceleration of a thin impactor driven by various energy sources onto a stationary target. Here the kinetic energy of the free-flying impactor is converted into shock compression energy in the target.

Consider cross-sectional dimensions to be infinite. When the thin impactor is decelerated, the Riemann problem results in a flow regime involving two shock waves, denoted S, moving to the left and to the right from the interface IF (Fig. 6.17).

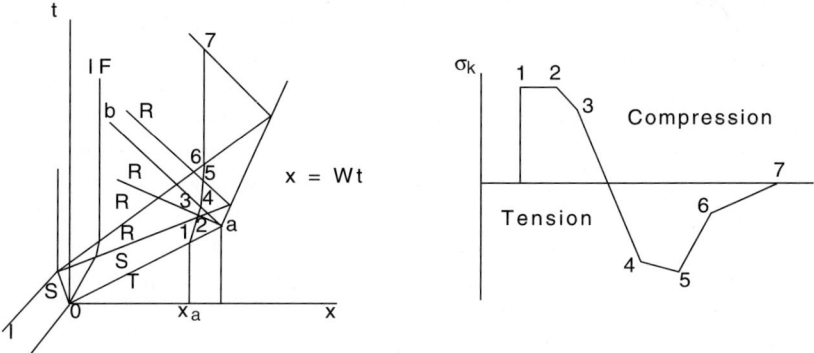

Fig. 6.17. Scheme of tensile stress formation in plate impact. I – impactor; T – target of material to be studied; S – shock wave; R – expansion wave; IF – interface

Their reflection from the free surfaces results in two centered rarefaction waves R moving toward each other across the impactor-target system. Their interaction leads to the development of tensile stresses. At any time, the maximum amplitude of the waves traversing the length of the sample is achieved

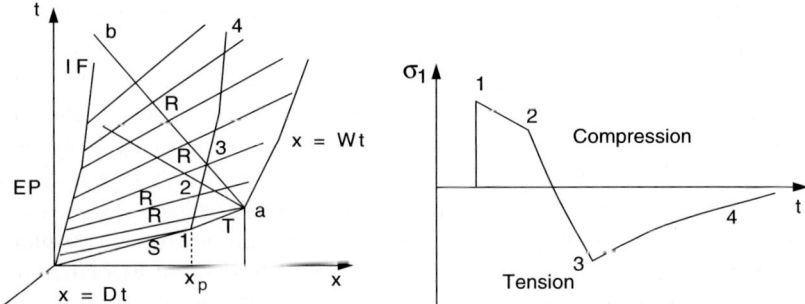

Fig. 6.18. Scheme of tensile stress formation for normally impinging detonation wave. EP – explosion products. The notation for T, S, R, IF is the same as in Fig. 6.17

at the extreme characteristic ab, and under the given loading conditions, this takes place at the intersection of characteristics of the counter-propagating expansion waves. The stress history at selected Lagrangian point x_A (see Fig. 6.17) is characterized by a compressive phase that is followed by transition to a tensile phase. In the acoustic approximation, given the same impactor and target materials, the tensile pulse is of rectangular shape and amplitude $|\sigma_t| = 0.5\rho_0 c_0 W_{imp}$, where W_{imp} is the impact velocity, and with duration of $\tau = 2h_{imp}/c_0$.

Tensile forces can also be produced in a solid through the application of a stress that decreases with time. When a detonation wave propagating normal to the target surface (Fig. 6.18a) impinges on the target, the stress across the EP-target interface is a decreasing function of time. Due to this unsteady decreasing stress profile, a decaying shock wave moves across the target. The maximum amplitude of the wave occurs at the EP-target interface at the time of impingement of the detonation wave onto the target, and is calculated through the solution of the Riemann problem. When the shock wave reflects from the free surface, a centered expansion wave moves into the target. The interaction of this expansion wave with the flow following the unsteady shock wave front results in the production of tensile stresses (Fig. 6.18b). Maximum tensile stress amplitude occurs on the extreme characteristic ab of the centered expansion wave, and increases along the characteristic toward point b. The impact process produces a stress history that is characterized by a compression phase, followed by an expansion phase.

The maximum tensile force amplitude for a specific type of HE and target material depends on the thickness ratio, $\alpha = l_T/l_{HE}$. When α is low, the stress amplitudes are small because of the small stress gradient through the target thickness. For this reason, relatively thin barriers experience no spall fracture, which agrees with the well-known observation of their spall-free acceleration by a detonation wave. When α is large, the tensile stress level is also low because of the decay of the shock wave as it moves across the target. Thus, tensile stresses reach their maximum at a certain $\alpha = \alpha_0$.

Methods for producing pulsed tensile loads using pulsed radiation energy sources (lasers, electron beams) have recently gained acceptance [54,55]. When heating of the material is fast, large stresses develop in the material due to inertial effects. When the heating is nonuniform, or when there are interfaces with a material of lower impedance, waves are generated. When a sample of finite size is heated promptly and uniformly over its mass, stresses develop in a direction x in the material ($\sigma_1^0 = \gamma_p \rho_0 \varepsilon_T$, where ε_T is the absorbed radiation energy concentration). In the simplest case of two free boundaries, in the acoustic approximation, the tensile stress amplitude is $\sigma_t = \sigma_1^0 = \gamma_p \rho_0 \varepsilon_T$ and tensile pulse duration is $\tau_t = h_0/c_0$, where h_0 is the sample thickness (Fig. 6.19a).

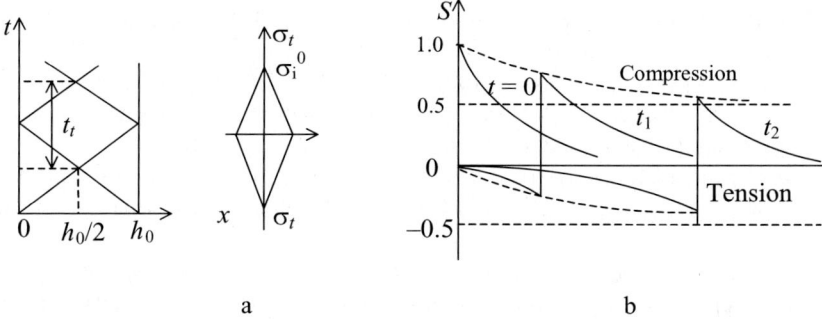

Fig. 6.19. Scheme of tensile stress formation in prompt uniform heating conditions. The exponential law of energy concentration distribution [53]

6.3.2 Spall Fracture Recording, Spall Stress Estimation

Methods used in the study of material resistance to short-term tensile stresses rely either on the recording of phenomena that are associated with failure, or on the detection of failure at some location within the sample [2,53]. We shall consider herein the methods used most frequently.

Measurement of Spall Layer Thickness

Thickness of the first separated spall layer is estimated either by pulsed X-radiography as in the experiments of [56], or by metallographic analysis of recovered samples. Time-dependence of the stresses in the spall plane may be calculated for the known loading conditions. The maximum tensile stress amplitude σ_t that the material can withstand without undergoing spall failure is typically regarded as the spall strength. Prompt failure is assumed whenever the tensile stress amplitude exceeds the spall strength.

The method is approximate since it includes no relaxation phenomena and it allows us to obtain only an approximate upper limit estimate for material resistance to tensile failure.

In experiments involving the deceleration of an impactor on a stationary target, tensile stress amplitude may be varied by varying the velocity of the impactor. By successively increasing the impact velocity over a relatively narrow range, the experimenter ensures a variety of loading levels in the material under study that are proportional to the impact velocity. In so doing, time parameters of the loading change slightly for fixed impactor and sample geometries. Time characteristics of the stress pulse may be changed by a transformation of the system (for example, by geometrical scaling of all sizes by a factor of n).

Metallographic analysis of recovered samples following the loading event reveals the level of material damage under given loading conditions and this may be compared to computational prediction [1, 2, 53].

It is thus possible to compare the degree of damage (from the generation of micro-cracks, to the development of larger cracks, to the formation of complete failure with spall layer rebound), as a function of the applied stress pulse parameters, and to examine the morphological features of spall fracture.

In cases involving a very high concentration of damage nucleation sites, such as in the case of a material discontinuity or upon the formation of a complete spall layer separation in the presence of relaxation phenomena, calculations of the critical failure stress under the assumption that no failure has occurred produces a significant overestimation of the spall strength. The best approximation of σ_t is made in the situation where microcracks have begun to develop in the spall plane but spall is still incomplete, i.e. the material has not completely lost its continuity, and the state of stress is still quite close to that in the continuous medium. Even in this case, the critical failure stress tends to be somewhat overestimated [2].

Taking into account the kinetics of damage initiation and evolution it is reasonable to characterize material resistance to pulsed tensile stresses with two critical levels of maximum tensile stress amplitude: the stress level σ_{1t} corresponding to micro-damage generation, and stress level σ_{2t} corresponding to complete macroscopic failure. The quantity σ_{1t} can be thought of as a physical strength property. The quantity σ_{2t} is a quantity useful for practical purposes. Its introduction is justified, as it derives from experimental data and allows us to predict complete spall fracture.

Measurement of Sample Free Surface Velocity as a Function of Time

The method of continuously recording free surface velocity of a sample loaded by a planar shock wave (see Fig. 6.20), using either a capacitive probe or

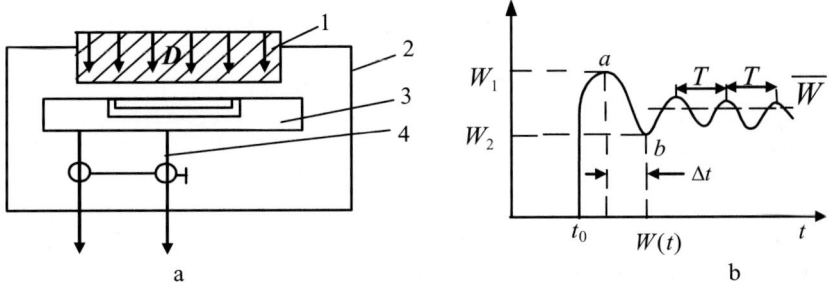

Fig. 6.20. Measurement of $W(t)$ with the capacitive probe method: **(a)** scheme of the experiment; **(b)** processed record of free surface velocity. 1 – material under study; 2 – metal base plate; 3 – measuring sensor; 4 – power source

laser interferometer is considered to be the most reliable and informative of available methods. The sample can be loaded either by plate impact or by a normal detonation wave. The failure stress is usually calculated as:

$$\sigma_t = 0.5\rho_0 c_0 (W_1 - W_2) ,   \tag{6.7}$$

where c_0 is typically assumed to be equal to the bulk sound speed, $(c_0 = c_B)$.

An alternate expression, one that includes the elastic-plastic properties of the material under study is given in [57]:

$$\sigma_t = \frac{\rho_0 c_L c_B}{c_L + c_B} \left[1 + \frac{\Delta_{spall}}{\Delta t} \left(\frac{1}{c_B} - \frac{1}{c_L} \right) \right] (W_1 - W_2) ,   \tag{6.8}$$

where c_L is the elastic sound speed, Δ_{spall} is the spall layer thickness, and Δt is specified in Fig. 6.20.

Use of a laser interferometer for the measurement of sample free surface velocity is discussed in detail in Chap. 9.

The value of σ_t can be approximately evaluated using the so-called "ready" spall method [1,58]. Using this method, the average velocity W_r of a thin layer pressed onto the sample (the thin layer is made from the same material as the sample), and the average velocity $\overline{W}$ of a natural spall layer are determined. The failure stress is then calculated from relationship: $\sigma_t = \rho_0 c_B (W_r - \overline{W})$. However, as shown in [59], this relationship is valid only for special cases of loading and specific types of EOS of the material under study. Under arbitrary loading conditions this relationship yields an erroneous result. At the present time, the "ready" spall method is not being used.

Measurement of the Shock Wave Profile in a "Soft" Medium

Information about spall fracture can be also acquired by continuously recording the stress history at a material-barrier interface with dielectric piezoresistive gauges (Fig. 6.21) [39, 60, 61]. The barrier material is chosen such that,

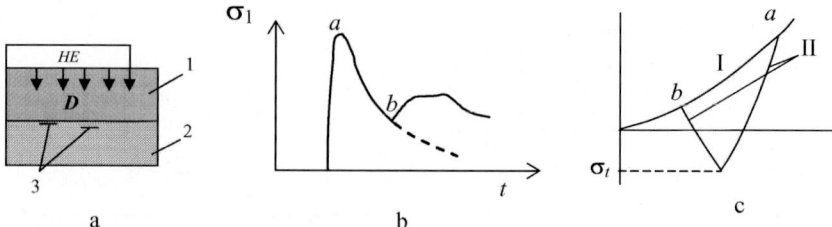

Fig. 6.21. Estimation of σ_t with the piezoresistive sensor: (**a**) scheme of the experiment; (**b**) processed record of stress history; (**c**) evaluation of σ_t. 1 – material to be studied; 2 – barrier; 3 – gauges; I – Hugoniot of barrier material; II – expansion isentropes of studied material

in unsteady shock wave impingement on the interface, an expansion wave is reflected into the material to be studied. For this purpose the barrier material has to be of lower dynamic impedance than the material under study.

The tensile stress σ_t, at which failure takes place in the sample under study, is evaluated by recording the maximum and minimum values of σ_1 (points a and b in Fig. 6.21b) produced in accord with the solution of the Riemann problem. In order to apply this method, the EOSs of the barrier and sample materials must be known. A minimum in the recorded profile of free surface velocity $W(t)$ or stress $\sigma_1(t)$ is inherent in either method under discussion. In either case, the decrease in velocity W (Fig. 6.20) or in stress σ_1 (Fig. 6.21b) on segment ab conforms to the stress distribution in the unsteady shock wave. The subsequent increase in σ_1 is due to the arrival of a shock wave emanating from the failure zone at the interface where the Lagrangian gauge is placed.

6.3.3 Metallographic Study of Spall Fracture Zone

Metallographic analysis of the spall fracture zone provides information about structural changes in metals and their alloys caused by shock compression and subsequent expansion, and what is particularly important, quantitative characterization of defects. The microstructure is often studied using an optical microscope with 50 to 2000 times magnification.

The microstructure is studied using specially prepared microsections produced by thorough polishing of the flat surface of a sample cut from a specimen that has been subjected to testing and then etching the polished surface with special reagents. A recording of the degree of damage within a zone of material discontinuity supplies information about the temporal evolution of failure, and about microstructural features that contribute to an estimation of the loading levels that correspond to the generation of damage on the microscopic level. A study of spall fracture zone photomicrographs suggests that different degrees of failure occur in the sample depending upon the loading parameters, from separated areas of local micro-damage to complete spall layer separation.

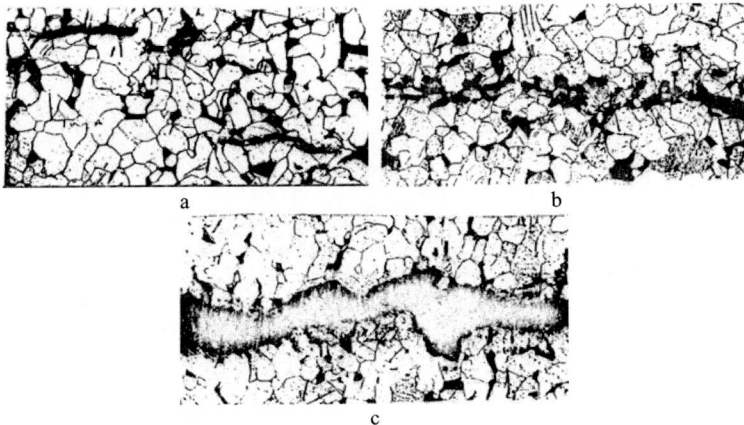

Fig. 6.22. Failure stages in steel St.3: (**a**) microfailure start; (**b**) microfailure and partial macrofailure; (**c**) complete spall fracture

The degree of failure can be characterized in accordance with some generally accepted quantitative measure of the level of damage. The authors of [62] propose five levels: 1 – complete spall fracture (presence of a main crack across the whole sample section), 2 – partial macroscopic failure (presence of separate macrocracks in the same section), 3 – intensive microfailure (presence of a large number of isolated or merged microcracks in the spall zone), 4 – weak microfailure (presence of separate microcracks), and 5 – microscopic total integrity of the sample (absence of any microcracks in the section at 1000 times magnification).

Sometimes three damage levels are introduced for a "rougher" estimation of the degree of damage, for example, as in [63] with the combination of levels 2, 3, and 4 mentioned above into a single level. Figure 6.22 presents photographs of micro-sections of different failure levels.

Bar charts showing the distribution of crack length versus distance from the loading plane and across the diameter of the sample provide a graphical representation of the degree of damage in each sample cross-section. The total crack length in the spall plane Σl is given as a function of pressure in Fig. 6.23 [64].

The spall fracture surface structure can be efficiently studied using an interactive image analysis system such as the one used in [55]. Spall surface images obtained with a microscope operated in reflected light and a TV camera are displayed and stored in memory. During data processing, average maximum surface roughness diameters, D_{max}, are estimated (Fig. 6.24).

Valuable information about changes in material microstructure and properties in shock wave loading can be acquired by measuring the microhardness of recovered intact samples [63]. Micro-X-ray fluorescence enables local determination of material chemical composition. The method is used for quanti-

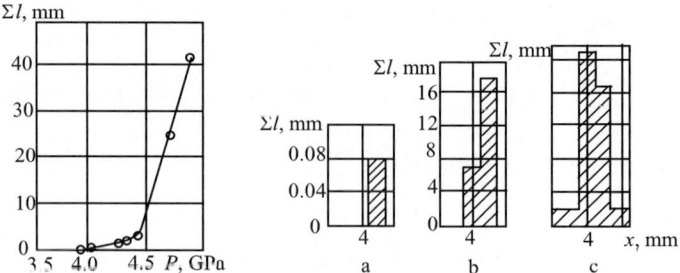

Fig. 6.23. Total crack length in titanium alloy VT-14 versus applied pressure and bar charts for different tensile stress amplitudes for the alloy VT-14: (**a**) $P = 4.0$ GPa; (**b**) $P = 4.7$ GPa; (**c**) $P = 4.9$ GPa

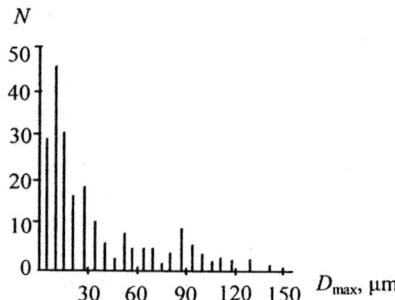

Fig. 6.24. Bar chart of spall surface roughness for lead sample $2 \cdot 10^{-4}$ m thick

tative chemical analysis of alloy microsections for elements within a 1–2 µm region. The method is highly sensitive: essentially any element in the periodic table can be detected in a 0.2–0.3 µm^3 volume, when their content ranges from 0.02 to 0.2 wt.% [65].

6.4 Spall Strength of Structural Materials

Since the publication by J. Rainhart of his pioneering paper on metal spall fracture under explosive loading [66], domestic and foreign investigators have accumulated extensive experimental data about spall strength and its relationship to the constitutive, physical, and mechanical properties of material.

Physical aspects of failure have been studied and general spall criteria (including identification of the parameters most significant for the process) have been developed for different classes of materials (from structural metals and their alloys to polymers, glasses, and heterogeneous solid high explosives (HE)).

The body of work that has been devoted to the study of spall fracture is of sufficient breadth, that when viewed as a whole, several general, material-type-independent trends that are directly related to the physical processes involved in spall fracture become apparent:

- failure zone localization in internal regions of the loaded sample;
- kinetic nature of the failure process;
- multipoint nature of failure initiation;
- statistical nature of defect size distribution;
- stress relaxation;
- individuality of the failure mechanism (brittle, viscous, mixed);
- dependence of failure initiation conditions on original material structure.

In this section we present some of the principal results from the body of research into the resistance of materials (primarily metals), to spall fracture. We shall consider issues such as the influence that material temperature, stress amplitude and duration, shock wave loading orientation, and initial material microstructure have on the process. To investigate these issues, the most important experimental data are recordings of the history of sample free surface velocity $W(t)$ and post-shot metallographic analysis. In the latter case, the spall strength σ_{spall} is taken to be the stress σ_{1t} that corresponds to the generation of microdamage. Spall strengths for a number of structural metals and alloys that are estimated from measurements of $W(t)$ only, are summarized in [2]. Despite the fact that the values of σ_{spall} are inherently approximate, general trends relating the effects of a given parameter on σ_{spall} can be considered to be quite realistic.

6.4.1 Spall Strength of Metals

Under the influence of shock wave loading, the material is compressed and heated by the shock wave prior to failure. Questions concerning the effect that compression (or shock front pressure) and a rise in temperature have on resistance to spall fracture arise naturally.

Temperature Dependence of σ_{spall}

The temperature effect can be investigated by conducting a series of experiments with different initial sample temperatures. This effect, as might be quantified through some functional relationship between spall strength and temperature, $\sigma_{spall}(T)$, has not been adequately studied. Having said this, results from the relatively small number of experiments that have been conducted for the purpose of investigating this issue enable us to discern some general trends (this is true in spite of the fact that these studies have not always used the same experimental methods). It may be stated that there is a general trend toward decreasing spall strength with increasing temperature (at least within the temperature range from room temperature ($T = 15-20°C$) to temperatures close to the melting point. However, the temperature effect

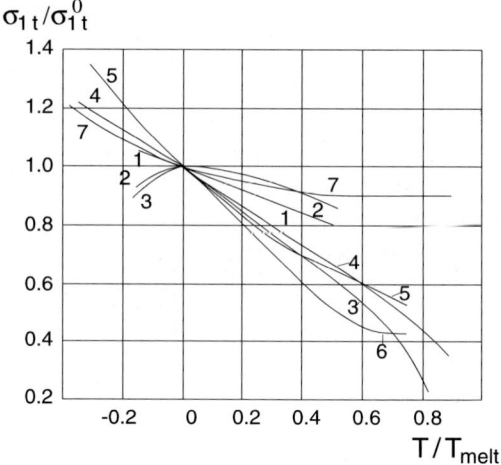

Fig. 6.25. Temperature dependence of spall strength σ_t: 1 – titanium VT-14, $\sigma_{1t}^0 = 4.81$ GPa; 2 – St.3, $\sigma_{1t}^0 = 3.89$ GPa; 3 – copper M1, $\sigma_{1t}^0 = 3.43$ GPa; 4 – AD1, $\sigma_{1t}^0 = 1.75$ GPa; 5 – alloy D16, $\sigma_{1t}^0 = 2.32$ GPa; 6 – alloy AMg6, $\sigma_{1t}^0 = 2.44$ GPa; 7 – Pb, $\sigma_{1t}^0 = 0.4$ GPa

under these conditions is relatively small, considerably smaller than that observed under conditions of quasi-static loading.

Figure 6.25 illustrates the temperature dependence of σ_{1t} for a number of metals subjected to a pulse loading of 1.3 µs duration [40]. As can be seen in the figure, for $T > 0°$C, a decrease in spall strength with increasing initial temperature is observed for all of the metals considered.

In [67–69], involving specimens subjected to explosive loading, the authors report a decrease in the spall strength of St.3 and Cu with increasing initial temperature (up to 500 °C).

In a study involving Al and Mg95, specimens loaded by impact (shock wave stresses of P = 5.8 GPa and P = 3.7 GPa, respectively, and initial temperature ranging up to (0.80–0.90 T_{melt}) [70], it is found that spall strength is only weakly influenced by temperature. As the temperature is increased further, however, spall strength rapidly decreases. Near the melting point, spall strength may approach the nonzero strength of liquid metal. One possible explanation for the threshold nature of the temperature effect may be the effect of local material melting due to local deformation at sites of microstructural stress concentration.

Dependence of σ_t on Shock Wave Amplitude

Presently, there is no unambiguous answer to the question of the effect of shock wave amplitude on critical tensile stresses σ_t [70–72]. The experimental data for five metals (Sn, Pb, Cu, Fe, U) in a wide range of shock wave amplitude [70]

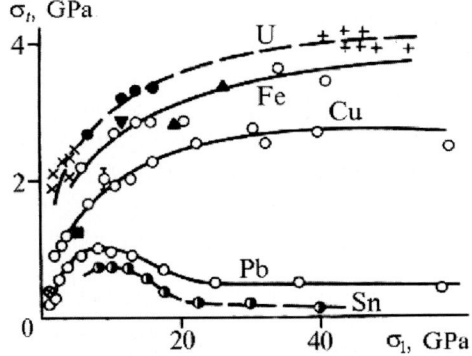

Fig. 6.26. Critical tensile stresses versus shock wave amplitude. $\times$, $\circ$, $\circ$, $\bullet$, $\blacktriangle$ – experimental points

are presented in Fig. 6.26. For Cu, Fe, and U, at shock wave amplitudes much below the magnitude required to produce melting, the value of σ_t increases monotonically until the applied shock amplitude, σ_1, approaches something on the order of 20–30 GPa; it remains essentially constant thereafter. For Sn and Pb, studied under conditions involving σ_1 ranging over a range of pressures that encompasses the melting pressure (stress), the curves $\sigma_t(\sigma_1)$ exhibit a peak.

According to [70], the ascending branch in the curve $\sigma_t(\sigma_1)$ can be attributed to metal hardening under the influence of plastic deformation brought about by the shock wave. This is confirmed by measurements of microhardness made on samples recovered after loading. For shock-wave heating in the temperature range $0.5T_r \leq T \leq T_{melt}$ (where T_r is the re-crystallization temperature and T_{melt} is the melting temperature), material re-crystallization begins, leading to a reduction in the strength of the metal (i.e. to softening [73]). The constancy of σ_t for U, Fe, and Cu in the range of high σ_1 may be the consequence of a competition between the above two processes (hardening and softening). For Sn and Pb at $\sigma_1 \geq 10$ GPa the principal contribution is made by softening. Behavior of the curves $\sigma_t(\sigma_1)$ for tin and lead at $\sigma_1 \geq 20$ GPa implies that a dramatic reduction in the strength of molten metal is unlikely [74]. The presence of the descending branch for Sn and Pb is corroborated by the results of [71].

Additional curves of $\sigma_t(\sigma_1)$ are presented in [72]. In this work no effect of shock-wave intensity on σ_t is observed for the titanium alloys (VT-5, VT-8), Al or its alloy AMg6, or for low-carbon steels. The experiments reported in [72] covered a wide range of σ_1: ($\sigma_1 \approx 5.6-77.0$ GPa for the titanium alloys, $\sigma_1 \leq 44$ GPa for Al, $\sigma_1 \leq 56$ GPa for AMg6, 6 GPa $\leq \sigma_1 \leq$ 90 GPa for steel). In the case of low-carbon steel, the polymorphic transition that occurs at $\sigma_1 = 13.0$ GPa does not affect σ_t.

Time Factor Influence on σ_t

Because there is only limited experimental data concerning the effect of tensile stress duration and tensile deformation rate on spall strength, only the general trends of this phenomenon can be ascertained. This trend consists of a relatively slight increase in the spall strength with decreasing tensile pulse duration t_t or with increasing tensile strain rate $\dot{\varepsilon}$. From Fig. 6.27, it is seen that with decreasing t_t, the critical stress of complete failure increases progressively [75,76].

Note that in the failure stress curve $\sigma(t_t)$ there is a kink at $t_t \approx 10^{-4}$ s [76], where, according to S.N. Zhurkov [77], the quasi-static durability branch transfers to the spall branch. According to [78], the difference between the dynamic and static branches of the temporal curve of strength is due to different mechanisms of irreversible deformation leading to failure: the thermally activated under-barrier dislocation slip on the static branch is replaced with the athermic above-barrier mechanism on the dynamic branch. According to [2], the dependence of spall strength on material deformation rate, $\dot{\varepsilon}$, in the unloading part of the pulse impinging on the free surface can be represented analytically as $\sigma_{spall} = A\,(\dot{\varepsilon})^{n}$, where A and n are constants. For some metals and alloys n ranges from 0.06 to 0.21, i.e. σ_{spall} weakly increases with increasing $\dot{\varepsilon}$.

Effect of Material Manufacture Technology and Thermal Processing on σ_{spall}

Experiments show that for a broad class of materials, critical tensile stress levels are material specific, depending on a number of factors, such as thermal processing, the rolling direction used in the manufacturing process, plastic pre-strain, and phase transformations [79]. Thermal preprocessing quite severely affects the conditions under which spall fracture will occur. A common trend is lower strength of annealed materials as compared to quenched materials. This

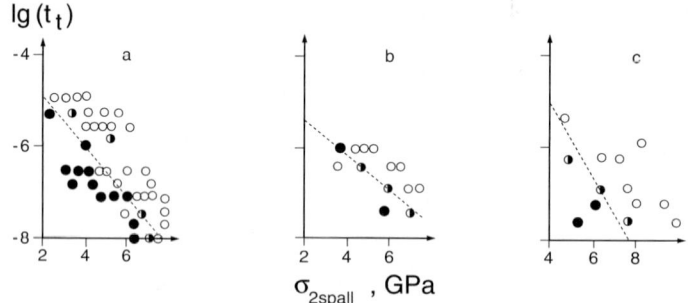

Fig. 6.27. Experimental curves $\sigma_{2spall}(t_t)$ characterizing metal durability in the submicrosecond loading range: Cu (**a**), Ni (**b**),Ti (**c**). $\bullet$ $-$ no failure; $\oplus$ $-$ presence of separate macrocracks; $\circ$ $-$ macroscopic failure

is illustrated by the fact that quenching steel 40Kh to HRC = 40 results in an increase in σ_{spall} of 40% over that of the material in the annealed state [67]. Fibrous microstructure and directedness in the location of inclusion clusters may be responsible for the generation and initial development of spall fracture in preferred directions. This implies that there may be an effect on spall strength that is attributable to the rolling direction used in the manufacturing process. This effect is seen in steels 30KhGSA and 12Kh18N10T where the spall strength is higher in the direction parallel to the rolling direction than in the direction perpendicular to it [79,80]. For the high-strength titanium alloy PT-3V, on the other hand, any anisotropy in strength is weak [80].

Dynamic plastic pre-strain ($\varepsilon = 0.5-5\%$) of the titanium alloy PT-3V does not significantly affect spall fracture characteristics for sample axis orientations either parallel or perpendicular to the rolling direction [80]. For steel 12Kh18N10T at $\varepsilon = 2-37\%$ the spall strength increases in the rolling direction and remains essentially constant in the perpendicular direction [80].

Spall Fracture of a Medium Experiencing a Phase Transition

Rarefaction shock waves can occur in materials possessing an anomalous Hugoniot (one with a section having upward convexity in the $P - V$ plane $\left(\partial^2 P/\partial V^2 < 0\right)$ [81–83]). Such an anomaly is possible in solids undergoing polymorphic phase transformation. The mechanism responsible for the formation of a rarefaction jump is theoretically considered in [81]. Atypical spall features that owe their existence to the aforementioned anomaly in the shock Hugoniot were observed in experiments conducted on iron and carbon steel that involved shock-wave loading followed by release [83]. These atypical features included the formation of two distinct types of spall: a typical spall producing a rough failure surface, and a spall with a very smooth failure surface (Fig. 6.28). The latter is formed in iron and steel during the interaction of rarefaction shock waves produced during unloading in the pressure range that includes the ($\varepsilon \to \alpha$) phase transition (and the states above it including the state of completion of the transition to the α phase).

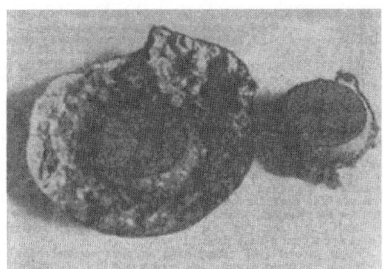

Fig. 6.28. "Smooth" spall in steel St.3 sample on explosive loading and unloading

This type of fracture is discussed in [82–84]. The formation of smooth-surface spall indicates that negative pressures (tensile stresses) exist over a very narrow zone during failure. In [82], the formation of smooth-surface spall is attributed to the existence of rarefaction shock waves. This explanation is supported by the fact that no spall fracture anomalies were detected in materials in which there were no polymorphic transformations in the studied pressure range.

6.4.2 Spall Strength of Polymers and HE

The results of a study of the critical tensile stresses σ_{1t} of plexiglas and poly-tetrafluorethylene conducted over a wide range of temperatures and involving a characteristic loading time of $\sim 1.3 \cdot 10^{-6}$ s [76] are presented in Fig. 6.29.

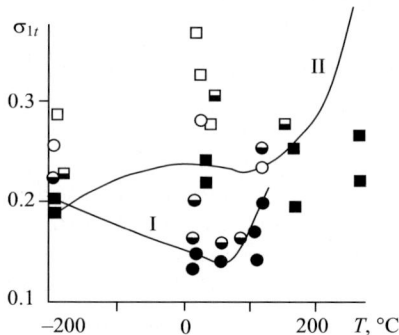

Fig. 6.29. Spall strength, σ_{spall}, versus temperature (——) for plexiglas ($\circ$) and polytetrafluorethylene ($\square$). $\blacksquare$, $\bullet$ – absence of visually observed failure; $\blacksquare$, $\bullet$, — partial spall fracture; $\square$, $\circ$— complete spall fracture. I – σ_{1t} for plexiglas, II – σ_{1t} for polytetrafluorethylene

An essential difference between the behavior observed here, and that which would be expected from metals or their alloys, is a sharp increase in σ_{1t} as initial temperature is increased above $\sim 110\,^{\circ}$C, which is near the softening temperature of either material. This feature can be surmised to be related to a change in the nature of the spall fracture, i.e. more viscous behavior.

For a characteristic tensile pulse time $\approx 1.5 \cdot 10^{-6}$ s, critical tensile stress amplitudes, σ_{cr}^{t}, resulting in crack generation, which are close to σ_{1t}, have been found for solid heterogeneous HE under conditions where they can be considered as inert materials. According to [85], the values of σ_{cr}^{t} are: 0.07 GPa for TNT, 0.12 GPa for cast composition TG 50/50, 0.26 GPa for composition GTK-70, 0.35 GPa for composition OFA-6.

Note that the failure of the above-mentioned HE proved to be close to quasi-brittle in its nature at room temperature ($T \approx 20\,^{\circ}$C).

6.5 Spall Fracture Criteria

6.5.1 Extreme State Criteria

Numerous experimental studies of material dynamic fracture (spall) have
shown it to be a complicated process that is influenced by many factors.
Factors that are important in material dynamic fracture include, among others, the constitutive state of the material during the tensile stress pulse and
numerous physical parameters such as temperature, original microstructure,
and the aggregative state of the material. This complicates the quantitative
description of spall fracture considerably. There is no theory of spall fracture
that accounts for all of the thermo-mechanical processes that may be involved.
The analysis and generalization of experimental data and theoretical concepts
of the failure mechanisms that are important, provide an insight into the many
details of this process (including those that are important at both the micro-
and macro-levels), and enable the prediction of spall fracture probability for
arbitrary pulsed force loads.

In early studies, material resistance to pulsed tensile stresses was characterized by a single quantity, the critical tensile stress amplitude σ_{cr} or the
critical failure tensile strain ε_{cr}, which were assumed to be material properties (or constants) for each specific material. Later studies, however, demonstrated that σ_{cr} and ε_{cr} are variables, dependent upon (among other things)
the loading condition of each specific material. For this reason, σ_{cr} or ε_{cr} are
now considered to be functions of a number of constitutive and other parame-
ters, such as the characteristic tensile stress time t_t, the tensile stress gradient
$d\sigma_t/dx$, the strain rate $\dot{\varepsilon}$, and the loading rate $d\sigma_t/dt$ [86].

The criteria alluded to above are evaluated from experiments that involve
failure in impact or explosive loading (mainly, through post-shot studies of
loaded samples). According to the criteria, spall fracture occurs in the region
where the criteria conditions, *viz* amplitude σ_{cr} (or ε_{cr}) and its relevant constitutive state parameters, are simultaneously first met. The criteria are hardly
general in nature, having their applicability strictly assured, only in the case
of loading scenarios that are the same as those used in the construction of the
various criteria. However, the parameters are useful for comparison between
different materials and arriving at engineering estimations of their durability.
It should be noted that attempts to apply linear fracture mechanics principles
to the description of spall fracture through mathematical analysis of the crack
instability condition have been unsuccessful. This may be because in the case
of spallation (in contrast to normal brittle fracture), the process is initiated
simultaneously at a large number of sites.

By considering the experimental data obtained under the simplest case
of loading states, high-velocity uniaxial deformation, domestic and foreign
investigators have constructed a number of spall fracture models that bear
clear-cut physical meaning and reflect the principal spall fracture features and
mechanisms. Models have been constructed using energy and kinetic concepts
of spall fracture.

6.5.2 Energy Concept of Spall Fracture

The energy criterion [87] is based on the assumption that in spall fracture, the work of material separation is performed at the expense of elastic energy stored in the tensile pulse. Thus, spallation is possible when the available elastic energy is sufficient for the process to proceed. Consider spallation in the acoustic approximation for tension which results from the simultaneous action of planar expansion waves. The necessary condition for unit surface failure is represented as:

$$\lambda = \int_0^{l_0} \frac{\sigma_t^2(x)\,dx}{AE} \quad \text{or} \quad \int_0^{l_0} \sigma_t^2(x)\,dx = \lambda AE \,, \tag{6.9}$$

where x is the path traveled by the expansion wave, l_0 is the spatial length of the wave, E is Young's modulus, σ_t is the tensile stress pulse amplitude, λ is the specific material fracture work per unit surface area, and $A = 2(1 - v)[(1 + v)(1 - 2v)]^{-1}$.

If we denote the spall layer thickness as δ, then from Eq. (6.9) it follows that:

$$\sigma_t^2 \delta = \beta \lambda E A \,. \tag{6.10}$$

The term β in (6.10) resides in the range from 1 to 3 depending on the shape of the tensile pulse. A triangular pulse corresponds to $\beta = 3$, and a rectangular pulse to $\beta = 1$. By noting that for a pulse that is triangular in shape, $dp/dt = E\varepsilon \approx (1/c_B)(\sigma_t/l_0)$, Eq. (6.9) can be expressed as [88]:

$$\sigma_t = (3\rho c_B E \lambda \dot\varepsilon)^{-1/3} \,, \tag{6.11}$$

which depends on the strain rate $\dot\varepsilon$. A similar equation was derived somewhat later in [89]. In [90], the recommendation is made for a single expression of both brittle and ductile materials:

$$\frac{\sigma_t^2}{\lambda} = \text{const.} \quad \text{or} \quad \frac{\sigma_t^3}{\lambda \rho \dot\varepsilon} = \text{const.} \tag{6.12}$$

It would be unjustified to assume that λ is a constant. Analysis of experimental data suggests that specific failure work is not a constant, but is a fairly strongly increasing function of tensile pulse time τ: $\lambda = \lambda(t)$. The failure zone is of a finite size. Consequently, an increase in λ is accounted for by an increase in the quantity of forming fracture surfaces.

According to [91], $\lambda(\tau)$ can be represented as:

$$\lambda = \lambda_0 (\tau/\tau_0)^m \,, \tag{6.13}$$

where $\lambda_0 = 2\gamma_0$ is the specific energy of surface formation, $\tau_0 = 10^{-13}$ s is the lattice oscillation period, and m is a constant. The exponent m varies in quite a narrow range, $m \approx 0.5$–0.7.

The energetic spallation theory is a finite-state theory and is not intended for the description of spall fracture transients. It imposes no limitations on the failure mechanism and supplies final results of the effect of the tensile stress.

6.5.3 Kinetic Models of Failure

Pulsed loading can result in significant tensile stresses within the body that lead to the formation and growth of microdefects (micropores and microcracks). The merger of these microdefects results in discontinuities and, hence, the formation of free surfaces. The damage evolves independently at different locations, but interacts with the stress wave in a complex manner, and some finite time interval is needed for its completion. Results of numerous experiments suggest that spall fracture is a process that evolves over time from the state of continuity to full or partial failure and that time is as important a factor as stress and strain. The time factor effect shows up in two important aspects: history of the applied stress, and time history of solid body damage. The kinetic nature of the spall fracture implies both limited tensile pulse duration and limited total failure time. Existing versions of the kinetic theory of solid body strength involve different physical concepts of failure, different methods of its description and different means of accounting for temporal effects.

In general, the extent of material damage ω depends on the combined influence of many variables and parameters that together characterize the perturbed state of the material and the physical properties of the material in that perturbed state. Mathematically, this complex relationship is expressed in terms of a kinetic (evolutionary) equation $d\omega/dt = f(\sigma, \sigma_c, \varepsilon, \dot{\varepsilon}, T, \omega, \eta...)$, where σ_c is the critical stress, ε is the strain, $\dot{\varepsilon}$ is the strain rate, T is the temperature, and η is a factor that accounts for structure. In the simplified versions of the kinetic equation used today, ω is expressed as a function of a limited number of variables.

In calculations, it is convenient to use the specific volume of defects V_T as a metric for damage and to study failure with respect this measure. Models, for which the mechanism for the accumulation of damage is independent of any previously achieved level of damage, are referred to as simple kinetic models.

Representative of this class of kinetic models is Zhurkov's strength model, which relates durability (lifetime), t_d, to the level of tensile stress σ_t and the current temperature T by the following relationship [77]:

$$t_d = t_k \exp\left(\frac{U_0 - \gamma\sigma_t}{kT}\right). \tag{6.14}$$

In (6.14), t_k is the lattice atomic oscillation period, U_0 is the activation energy, γ is the structure-sensitive factor, and k is the Boltzmann constant.

Many investigators have demonstrated that Eq. (6.14) is valid for various classes of materials in the durability range $t_d = 10^7$–10^{-3} s. However, in the durability range $t_d - 10^{-5}$ 10^{-8} s (inherent in shock-wave experiments), t_k, U_0, and γ have to be significantly modified from their static analogs if one is to accurately account for material behavior in the dynamic branch of the associated time dependence of strength.

A classic example of a model that falls into the simple kinetic model category is the Tuler-Bucher integral criterion [88]. This model has received wide acceptance. The criterion relates the damage level (damage integral) to the tensile stress history:

$$K = \int_{t_0}^{t} \left[\sigma_t \left(t \right) - \sigma_0 \right]^n dt ,$$
(6.15)

where n and σ_0 are constants selected to best describe the experiments. The integration with respect to t begins at time t_0, when $\sigma = \sigma_0$. When the condition $K = K_c$ (where K_c is an experimentally determined material constant), is met in some section, failure occurs.

As implied above, complex kinetic models are those for which the damage buildup mechanism depends on any previously achieved level of damage. Failure can therefore be either single-stage, or multistage.

In the characteristic single-stage kinetic model proposed in [92], the damage measure growth rate is given as a function of the acting stress σ_1 and current specific defect volume V_T.

$$\dot{V}_T = -\sigma_1 \frac{A_1 \exp \left(V_T / V_{T_0} \right) \exp \left(\sigma_1 / \sigma_0 \right)}{A_2 + \exp \left(\sigma_1 / \sigma_0 \right)} .$$
(6.16)

In Eq. (6.16), A_1, A_2, σ_0, and V_{T_0} are material constants, and σ_1 is positive in compression. The constants are selected so that the results of validation spallation experiments agree with those predicted by (6.16).

Presently, the concept that spall fracture involves a multistage evolutionary process is widely accepted [86]. The failure process can be broken down into several stages: microdefect generation, growth and evolution under the action of tensile stresses, merging of neighboring defects with the formation of macrocracks, material fracture and separation with the formation of one or more free surfaces. The duration of the various stages depends upon material properties and loading conditions, but in any case a finite time interval is required for the process to be completed. The question of stage characteristics and the conditions of transition from one stage to another remains open. Apparently, in certain perturbed states, several stages can proceed simultaneously. In this instance, each proceeding stage may exert a considerable influence on the dynamics of the other stages through the dependence of the perturbed state on the degree of damage.

Construction of a mathematical model for multistage failure that includes a full description of all failure stages represents at the present time, a formidable, if not insurmountable challenge. For this reason, wide attention has been paid to an approach in which spall fracture is considered to be a two-stage process with different stage interfaces, or without any interfaces [86].

In the two-stage kinetic model of spall fracture proposed in [93], the only structure parameter characterizing the degree of damage is relative void volume, $\omega = V_T / V$ ($0 \le \omega \le 1$). The evolutionary equation is given in the same form for both the stages:

$$\dot{\omega} = 0, \text{ if } \Delta P_s \leq 0; \quad \dot{\omega} = \frac{F(\omega)}{\eta} \Delta P_s \text{sign}(\Delta P_s), \text{ if } \Delta P_s > 0, \qquad (6.17)$$

with differences between the two stages being reflected through the function $F(\omega)$:

$$F(\omega) = \omega^{1/3} (1 - \omega)^{2/3} \quad \omega \leq 1/3 \text{ stage 1};$$
$$F(\omega) = \frac{\sqrt[3]{16}}{9} \omega^{-1/3} (1 - \omega)^{-2/3} \quad \omega > 1/3 \text{ stage 2}. \qquad (6.18)$$

The term ΔP_s is evaluated as:

$$\Delta P_s = \begin{cases} |P_s| - P_0, & P_s \leq 0; \\ |P_s| - \frac{P_0}{(1-\omega)^n}, & P_s > 0, \end{cases} \qquad (6.19)$$

where P_s is the pressure in the solid body (matrix) estimated as $P_s = K(1/V_s - 1) + 3K\alpha(T - T_0)$, α is the linear thermal expansion coefficient, P_0, n, and η are constants selected so that the best possible agreement with experimental results is achieved.

Stage 1 accounts for defect evolution up to the damage level $\omega \approx 0.3$; stage 2 for the merging of defects brought about by the rupture of partitions separating voids.

Significant progress in the development of kinetic spall fracture models has been brought about through the application of statistical methods [94, 95], used for the description of a representative ensemble of defects. A function of defect distribution, employing parameters that characterize size, shape and orientation of defects is introduced for this purpose. The approach combines macroscopic (continuum) and microscopic (micro-structural) failure concepts.

The kinetic spall fracture model known as the NAG model [94, 95], assumes that many defects discernible at magnification levels ~1000 exist in the material a priori.

For viscous failure, the initial stage involves the growth and gradual "closure" of pores that are roughly spherical in shape, while the final stage involving the fracture of inter-pore partitions resulting in the formation of cracks, is of relatively little significance.

Post-shot data for a great number of experiments on spall fracture of viscous materials under various time-temperature conditions of pulsed loading, suggests that the distribution of defects by size can be represented with the use of an exponential function:

$$N(R) = N_0 \exp(-R/R_m), \qquad (6.20)$$

where N_0 is the total number of pores per unit volume, N is the number of pores of a radius larger than R_m per unit volume, R is the pore radius, and R_m is a distribution parameter.

Defects are generated when the tensile stress resulting from the interaction of counter propagating rarefaction waves achieves some critical magnitude σ_{n_0}. If the tensile stress is greater than σ_{n_0}, that is $(\sigma > \sigma_{n_0})$, the pore quantity rise rate may be a function of stress.

Pore generation rate is described by the differential equation:

$$\dot{N} = \dot{N}_0 \exp\left(\frac{\dot{\sigma}_s - \sigma_{n_0}}{\sigma_1}\right) \Theta\left(\sigma_s - \sigma_{n_0}\right) , \qquad (6.21)$$

where σ_1, $\dot{N}_0$, and σ_{n_0} are constants, σ_{n_0} is a threshold stress below which no micropores are generated, Θ is the Heaviside step function, and σ_s is the stress in the solid.

The stage of pore growth up to final pore merger is described by the differential equation:

$$\dot{R} = \frac{\sigma_s - \sigma_{g_0}}{4\eta} R\Theta\left(\sigma_s - \sigma_{g_0}\right) , \qquad (6.22)$$

where σ_{g_0} and η are constants, R is the pore radius, σ_{g_0} is the threshold isolated micropore growth stress, and η is a constant similar in dimension and physical meaning to material viscosity.

Thus, in the NAG model failure is influenced by parameters $\dot{N}_0$, σ_{n_0}, σ_1, σ_{g_0}, η, and R_m, among which the first three control defect generation and the remainder control defect evolution under the action of tensile stresses. Methods for the evaluation of these parameters are discussed in [96]. Note that of the six parameters, the minimum size, R_m, of pores generated during the failure process remains the most uncertain.

In contrast to viscous failure, when constructing the model for brittle failure it is necessary to take into consideration both size and orientation of roughly disk-shaped cracks. In this case, the macroscopic parameter ω characterizing the degree of damage at the microlevel is a tensor quantity describing the anisotropic process of damage evolution [97]. Since no reliable experimental data and methods for studying anisotropic damage evolution are currently available, isotropy is assumed in most practical applications, where microdefects are distributed uniformly in all directions. In this case ω is a scalar.

According to [98], the volume distribution of microcracks by size, taking into account crack orientation, is an exponential and is described by an expression having the same form as that for the distribution of viscous failure [see Eq. (6.20)]

$$N(R) = N_0 \exp\left(-R/R_m\right) , \qquad (6.23)$$

where N is the number of cracks of radius larger than R_m per unit volume, N_0 is the total number of pores per unit volume, R is the crack radius, and R_m is a distribution parameter.

As the cracks grow under the action of stresses σ_n normal to their faces, the expressions for the crack generation rate and growth are

$$\dot{N} = \dot{N}_0 \exp\left(\frac{\dot{\sigma}_n - \sigma_{n_0}}{\sigma_1}\right) \Theta\left(\sigma_n - \sigma_{n_0}\right) \ ; \tag{6.24}$$

$$\dot{R} = \frac{\sigma_n - \sigma_{g_0}}{4\eta} R\Theta\left(\sigma_n - \sigma_{g_0}\right) \ . \tag{6.25}$$

In Eqs. (6.24) and (6.25), $\dot{N}_0$, σ_1, and η are constants, σ_{n_0} is the threshold stress below which no microcracks are generated in the medium, σ_{g_0} is the threshold stress below which no micropore growth occurs in the medium, η is a constant similar in dimension and physical meaning to material viscosity, Θ is the Heaviside function, and σ_n is the stress acting normal to the crack faces.

In the model, the crack volume is taken as the volume of an ellipsoid with semiaxes δ and R, where half of a maximum crack face opening is determined by the equality

$$\delta = \frac{4\left(1 - \nu^2\right)}{\pi E} R\sigma_n \ ,$$

where E is Young's modulus and ν is Poisson ratio.

As in the case of viscous failure, brittle failure kinetics is determined through the influenced of six parameters. The previous comments on the parameters for viscous failure apply here as well.

An important component of complex kinetic failure models is the constitutive equation for damaged materials that includes the damage effect on physical and mechanical characteristics of the material and, hence, on wave processes and spallation time history. At the present time, the construction of models of damaged materials is a subject of extensive research.

A dilation model of kinetic strength has been being recently developed [99]. In this model, fracture strength and test duration (time t required for both the loading and unloading phases) are related as follows:

$$\sigma = \frac{\varepsilon_{cr} E}{x}\left(1 - \frac{\alpha k T}{\varepsilon_{cr} C_0} \ln \frac{t}{t_k}\right) \ , \tag{6.26}$$

where k is the Boltzmann constant, T is temperature, ε_{cr} is the critical atomic bond strain, E is Young's modulus, C_0 is the atomic heat capacity, α is the linear thermal expansion coefficient, t_k is the lattice atomic oscillation period, and x is the overload factor. In Eq. (6.26), paramount is the relationship of the overload factor x to the actual defect structure in the stressed body. In [99] this relationship is determined from static strength theory and satisfactory agreement with experimental data is achieved for a number of metals.

Many examples using a kinetic model like NAG for the description of spall fracture of metals and their alloys (U, Cu, AMg6, armco iron, Ni) can be found in [86] and [100–103].

References

1. Glushak, B.L., Kuropatenko, V.F., and Novikov, S.A., *Study of Material Strength Under Dynamic Loading*, Nauka Publ., Siberian Branch, Novosibirsk, 1992.
2. Kanel, G.I., Razorenov, S.V., Utkin, A.V., and Fortov, V.E., *Shock Wave Phenomena in Condensed Matter*, Yanus-K Publ., Moscow, 1996, [see also Kanel, G.I., Razorenov, S.V., and Fortov, V.E., *Shock-Wave Phenomena and the Properties of Condensed Matter*, Springer-Verlag, New York, 2004].
3. Fortov, V.E., Altshuler, L.V., Trunin, R.F., and Funtikov, A.I., (eds.), *Shock Waves and Extreme States*, Nauka Publ., Moscow, 2000. [see also, *High-Pressure Shock Compression of Solids VII: Shock Waves and Extreme States*, Springer-Verlag, NY, 2004].
4. Batkov, Yu.V., Glushak, B.L., and Novikov, S.A., Material Resistance to Plastic Deformation in Shock Waves (Review), TsNIIAtominform, Moscow, 1990.
5. Yaziv, D., Rosenberg, Z., and Partom, Y., "Determination of the Pressure Dependence of Yield Strength in 2024 Al Under Dynamic Loading with In-Material Manganin Gauges," *Proc., 3rd Conf. on the Mechanical Properties of Materials at High Rates of Strain*, Oxford, UK, 1984, pp. 105–109.
6. Altshuler, L.V., Brazhnik, M.I., and Telegin, G.S., "Strength and Elasticity of Iron and Copper at High Shock-Wave Compression Pressures," *Zhurnal Prikladnoi Mekhaniki i Tecknicheskoi Fiziki*, 1971, No. 6, pp. 159–166, [English trans., *Journal of Applied Mechanics and Technical Physics*, Vol. 12, No. 6, 1971, pp. 921–926].
7. Curran, D.R., "Nonhydrodynamic Attenuation of Shock Waves in Aluminum," *Journal of Applied Physics*, Vol. 34, No. 9, 1963, pp. 2677–2685.
8. Bakhrakh, S.M., Ivanov, A.G., Kovalev, N.P., et al., "Aluminum Strength with Elastic-Plastic Compression in Shock Waves," *in* Detonation. Critical Phenomena. Physical and Chemical Transformations in Shock Waves, OChKhF, Chernogolovka, 1978, pp. 94–100.
9. Rosenberg, Z., Partom, Y., and Yaziv, D., "The use of In-Material Stress Gauges for Estimating the Dynamic Yield Strength of Shock-Loaded Solids," *Journal of Applied Physics*, Vol. 56, No. 1, 1984, pp. 143–146.
10. Dremin, A.N., and Kanel, G.I., "Compression and Rarefaction Waves in Shock-Compressed Metals," *Zhurnal Prikladnoi Mekhaniki i Tekhnicheskoi Fiziki*, 1976, No. 2, pp. 146–153, [English trans., *Journal of Applied Mechanics and Technical Physics*, Vol. 17, No. 2, 1976, pp. 263–267].
11. Batkov, Yu.V., Novikov, S.A., Sinitsyna, L.M., and Chernov, A.V., "Study of Shear Stresses in Metals at a Shock Front," *Problemy Prochnosti*, 1981, No. 5, pp. 56–59, [English trans., *Strength of Materials*, Vol. 13, No. 5, 1981, pp. 601–605].
12. Chartagnac, P.F., "Determination of Mean and Deviatoric Stresses in Shock Loaded Solids," *Journal of Applied Physics*, Vol. 53, No. 2, 1982, pp. 948–953.
13. Batkov, Yu.V., Glushak, B.L., and Novikov, S.A., "Strength of Aluminum, Copper and Steel at Front of Shock Wave," *Fizika Goreniya i Vzryva*, Vol. 25, No. 5, 1989, pp. 126–132, [English trans., *Combustion, Explosion, and Shock Waves*, Vol. 25, No. 5, 1989, pp. 635–640].
14. Batkov, Yu.V., Glushak, A.B., Glushak, B.L., Novikov, S.A., and Fishman, N.D., "Study of the Stressed State of Shock-Compressed Solids by the Method

of Principal Stresses," *Fizika Goreniya i Vzryva*, Vol. 31, No. 5, 1995, pp. 114–121, [English trans., *Combustion, Explosion, and Shock Waves*, Vol. 31, No. 5, 1995, pp. 605–611].

15. Batkov, Yu.V., Novikov, S.A., and Timonin, L.M., "Strength Behind Shock Wave in Ceramic Materials Based on Alumina," *Proc., 10th Int. Conf. High Energy Rate Fabrication*, Ljubljana, Yugoslavia, 1989, pp. 256–262.

16. Lipkin, J., and Asay, J.R., "Reshock and Release of Shock-Compressed 6061–T6 Aluminum," *Journal of Applied Physics*, Vol. 48, No. 1, 1977, pp. 182–189.

17. Asay, J.R., and Chhabildas, L.C., "Estimating Shear Strength of Aluminum Alloy 6061-T6 Subjected to Compression in Shock Loading," *in* Shock Waves and Metal Rapid-Rate Deformation Phenomena, Metallurgiya Publ., Moscow, 1984, pp. 110–120 [*see also*, Asay, J.R., and Chhabildas, L.C., "Determination of the Shear Strength of Shock Compressed 6061-T6 Aluminum," *in* Shock Waves and High Strain-Rate Phenomena in Metals: Concepts and Applications, Meyers, M.A., and Murr, L.E., eds., Plenum Publ., New York, 1981, pp. 417–431].

18. Asay, J.R., Chhabildas, L.C., and Dandekar, D.P., "Shear Strength of Shock-Loaded Polycrystalline Tungsten," *Journal of Applied Physics*, Vol. 51, No. 9, 1980, pp. 4774–4783.

19. Jones, S.E., Gillis, P.P., and Foster, J.C., Jr., "On the Equation of Motion of the Undeformed Section of a Taylor Impact Specimen," *Journal of Applied Physics*, Vol. 61, No. 2, 1987, pp. 499–502.

20. Wilkins, M.L., and Guinan, M.W., "Impact of Cylinders on a Rigid Boundary," *Journal of Applied Physics*, Vol. 44, No. 3, 1973, pp. 1200–1206.

21. Ivanov, A.G., and Novikov, S.A., "Capacitive Data Transmitter Method for Recording the Instantaneous Velocity of Moving Surfaces," *Pribory i Tekhnika Eksperimenta*, 1963, No. 1, pp. 135–138, [English trans., *Instruments and Experimental Techniques*, 1963, No. 1, pp. 128–131].

22. Taylor, J.W., and Rice, M.H., "Elastic-Plastic Properties of Iron," *Journal of Applied Physics*, Vol. 34, No. 2, 1963, pp. 364–371.

23. Zhugin, Yu.N., and Krupnikov, K.K., "Induction Method of Continuous Recording of the Velocity of a Condensed Medium in Shock-Wave Processes," *Zhurnal Prikladnoi Mekhaniki i Tekhnicheskoi Fiziki*, 1983, No. 1, pp. 102–108, [English trans., *Journal of Applied Mechanics and Technical Physics*, Vol. 24, No. 1, 1983, pp. 88–93].

24. Novikov, S.A., Kashintsov, V.I., Fedotkin, A.S., Sinitsyn, V.A., Bodrenko, S.I., and Koltunov, O.I., "Measurement of the Velocities of Current-Conducting Shells with a Sensor of the Electromagnetic Type," *Fizika Goreniya i Vzryva*, Vol. 22, No. 1, 1986, pp. 71–74, [English trans., *Combustion, Explosion, and Shock Waves*, Vol. 22, No. 1, 1986, pp. 67–70].

25. Fowles, G.R., "Shock Wave Compression of Hardened and Annealed 2024 Aluminum," *Journal of Applied Physics*, Vol. 32, No. 8, 1961, pp. 1475–1487.

26. Kozlov, E.A., Muzyrya, A.K., Chinkova, R.Kh., and Shorokhov, E.V., "A Cleavage in the Process of Forming in Copper," *Fizika Goreniya i Vzryva*, Vol. 20, No. 4, 1984, pp. 123–126, [English trans., *Combustion, Explosion, and Shock Waves*, Vol. 20, No. 4, 1984, pp. 471–474].

27. Zlatin, N.A., Mochalov, S.M., Pugachev, G.S., and Bragov, A.M., "Laser Differential Interferometer," *Zhurnal Tekhnicheskoi Fiziki*, Vol. 43, No. 9, 1973, pp. 1961–1964, [English trans., *Soviet Physics Technical Physics*, Vol. 18, No. 9, 1974, pp. 1235–1237].

28. Davison, L., and Graham, R.A., "Shock Compression of Solids," *Physics Reports*, Vol. 55, No. 4, 1979, pp. 255–379.

29. Ivanov, A.G., Novikov, S.A., and Sinitsyn, V.A., "Investigation of Elastic-Plastic Waves in Explosively Loaded Iron and Steel," *Fizika Tverdogo Tela*, Vol. 5, No. 1, 1963, pp. 269–278, [English trans., *Soviet Physics – Solid State*, Vol. 5, No. 1, 1963, pp. 196–202].

30. Novikov, S.A., Sinitsyn, V.A., Ivanov, A.G., and Vasilyev, L.V., "Elastoplastic Properties of a Number of Metals Exposed to Explosive Shock Loading," *Fizika Metallov i Metallovedenie*, Vol. 21, No. 3, 1966, pp. 452–460, [English trans., *The Physics of Metals and Metallography*, Vol. 21, No. 3, 1966, pp. 135–144].

31. Altshuler, L.V., "Use of Shock Waves in High-Pressure Physics," *Uspekhi Fizicheskikh Nauk*, Vol. 85, No. 2, 1965, pp. 197–258, [English trans., *Soviet Physics Uspekhi*, Vol. 8, No. 1, 1965, pp. 52–91].

32. Kleshchevnikov, O.A., Tyunyaev, Yu.N., Sofronov, V.N., Ogorodnikov, V.A., Ivanov, A.G., and Mineev, V.N., "Dynamic Yield Point and Specific Work for Breakage During Spalling of a Number of Structural Steels," *Fizika Goreniya i Vzryva*, Vol. 22, No. 4, 1986, pp. 102–106, [English trans., *Combustion, Explosion, and Shock Waves*, Vol. 22, No. 4, 1986, pp. 482–485].

33. Novikov, S.A., and Sinitsyn, V.A., "Effect of Temperature on Elastoplastic Properties of Steel in Explosive Loading," *Problemy Prochnosti*, 1976, No. 12, pp. 104–106, [English trans., *Strength of Materials*, Vol. 8, No. 12, 1976, pp. 1482–1484].

34. Kanel, G.I., "Model of the Kinetics of Metal Plastic Deformation Under Shock-Wave Loading Conditions," *Zhurnal Prikladnoi Mekhaniki i Tekhnicheskoi Fiziki*, 1982, No. 2, pp. 105–110, [English trans., *Journal of Applied Mechanics and Technical Physics*, Vol. 23, No. 2, 1982, pp. 256–260].

35. Dremin, A.N., Kanel, G.I., and Chernikova, O.B., "Resistance of Aluminum AD-1 and Duraluminum D-16 to Plastic Deformation Under Shock Compression Conditions," *Zhurnal Prikladnoi Mekhaniki i Tekhnicheskoi Fiziki*, 1981, No. 4, pp. 132–138, [English trans., *Journal of Applied Mechanics and Technical Physics*, Vol. 22, No. 4, 1981, pp. 558–562].

36. Glushak, B.L., Novikov, S.A., and Batkov, Yu.V., "Constitutive Equation for Describing High Strain Rates of Al and Mg in a Shock Wave," *Fizika Goreniya i Vzryva*, Vol. 28, No. 1, 1992, pp. 84–89, [English trans., *Combustion, Explosion, and Shock Waves*, Vol. 28, No. 1, 1992, pp. 79–83].

37. Bolshakov, A.P., Novikov, S.A., and Sinitsyn, V.A., "Dynamic Uniaxial Tension and Compression Curves of Copper and Alloy AMg6," *Problemy Prochnosti*, 1979, No. 10, pp. 87–88, [English trans., *Strength of Materials*, Vol. 11, No. 10, 1979, pp. 1159–1161].

38. Vereshchagin, L.F., Synthetic Diamonds and Hydrostatic Extrusion: Selected Works, Nauka Publ., Moscow, 1982.

39. Stepanov, G.V., Elastic-Plastic Deformation of Metals Under Action of Pulsed Loads, Naukova Dumka Publ., Kiev, 1979.

40. Novikov, S.A., "Shear Stress and Spall Strength of Materials Under Shock Loads (Review)," *Zhurnal Prikladnoi Mekhaniki i Tekhnicheskoi Fiziki*, 1981, No. 3, pp. 109–120, [English trans., *Journal of Applied Mechanics and Technical Physics*, Vol. 22, No. 3, 1981, pp. 385–394].

41. Novikov, S.A., and Sinitsyna, L.M., "Effect of the Pressure of Shock Compression on the Critical Shear Stresses in Metals," *Zhurnal Prikladnoi Mekhaniki*

i Tekhnicheskoi Fiziki, 1970, No. 6, pp. 107–110, [English trans., *Journal of Applied Mechanics and Technical Physics*, Vol. 11, No. 6, 1970, pp. 983–986].

42. Batkov, Yu.V., Novikov, S.A., and Chernov, A.V., "Shear Strength of Solids and its Effect on Plane Shock Wave Propagation," *Fizika Goreniya i Vzryva*, Vol. 22, No. 2, 1986, pp. 114–120, [English trans., *Combustion, Explosion, and Shock Waves*, Vol. 22, No. 2, 1986, pp. 238–244].

43. Riecker, R.E., and Towle, L.C., "Shear Strength of Grossly Deformed Cu, Ag, and Au at High Pressures and Temperatures," *Journal of Applied Physics*, Vol. 38, No. 13, 1967, pp. 5189–5194.

44. Simonov, I.V., and Chekin, B.S., "High-Velocity Collisions of Iron Plates," *Fizika Goreniya i Vzryva*, Vol. 11, No. 2, 1975, pp. 274–281, [English trans., *Combustion, Explosion, and Shock Waves*, Vol. 11, No. 2, 1975, pp. 237–242].

45. Gust, W.H., and Royce, E.B., "Dynamic Yield Strengths of B$_4$C, BeO, and Al$_2$O$_3$ Ceramics," *Journal of Applied Physics*, Vol. 42, No. 1, 1971, pp. 276–295.

46. Rosenberg, Z., Yaziv, D., Yeshurun, Y., and Bless, S.J., "Shear Strength of Shock-Loaded Alumina as Determined with Longitudinal and Transverse Manganin Gauges," *Journal of Applied Physics*, Vol. 62, No. 3, 1987, pp. 1120–1122.

47. Mashimo, T., Hanaoka, Y., and Nagayama, K., "Elastoplastic Properties under Shock Compression of Al$_2$O$_3$ Single Crystal and Polycrystal," *Journal of Applied Physics*, Vol. 63, No. 2, 1988, pp. 327–336.

48. Sugiura, H., Kondo, K., and Sawaoka, A., "Dynamic Response of Fused Quartz in the Permanent Densification Region," *Journal of Applied Physics*, Vol. 52, No. 5, 1981, pp. 3375–3382.

49. Batkov, Yu.V., Novikov, S.A., and Fishman, N.D., "Shear Stresses in Polymers Under Shock Compression," Shock Compression of Condensed Matter – 1995, Schmidt, S.C., and Tao, W.C., eds., AIP Press, Woodbury, NY, 1996.

50. Batkov, Yu.V., Novikov, S.A., and Fishman, N.D., "Investigation of Shear Stress on a Shock Front in Solid High Explosives (HE)," *Fizika Goreniya i Vzryva*, Vol. 19, No. 3, 1983, pp. 120–122, [English trans., *Combustion, Explosion, and Shock Waves*, Vol. 19, No. 3, 1983, pp. 357–359].

51. Afanasyev, G.G., and Bobolev, V.K., *Shock Initiation of Solid High Explosives*, Nauka Publ., Moscow, 1968.

52. Glushak, A.B., and Novikov, S.A., "Metal Resistance to Plastic Deformation in High-Velocity Compression," *Kimecheskaya Fizika*, Vol. 19, No. 2, 2000, pp. 65–69.

53. Glushak, B.L., Novikov, S.A., Ruzanov, A.I., and Sadyrin, A.I., Deformed Media Failure Under Pulsed Loading, NNSU Publishing House, Nizhni Novgorod, Russia, 1992.

54. Khokhlov, N.P., Sviridov, V.A., Glushak, B.L., Novikov, S.A., and Ivanov, A.G., "Damping of Stress Waves of Short Duration in Aluminum and Copper," *Pisma v Zhurnal Tekhnicheskoi Fiziki*, Vol. 6, 1980, pp. 1427–1430, [English trans., *Soviet Technical Physics Letters*, Vol. 6, No. 12, 1980, pp. 616–617].

55. Bonyushkin, E.K., Zhukov, I.V., Zavada, N.I., et al., "Features of Structural Material Spall Fracture in Conditions of Rapid Volume Heating and in Explosive Loading," *Voprosy Atomnoi Nauki i Tekhnini. Seriya: Impulsniye Reaktory i Prostiye Kriticheskiye Sborki*, 1988, No. 1, pp. 53–61.

56. Breed, B.R., Mader, C.L., and Venable, D., "Technique for the Determination of Dynamic-Tensile-Strength Characteristics," *Journal of Applied Physics*, Vol. 38, No. 8, 1967, pp. 3271–3275.

57. Romanchenko, V.I., and Stepanov, G.V., "Dependence of the Critical Stresses on the Loading Time Parameters During Spall in Copper, Aluminum, and Steel," *Zhurnal Prikladnoi Mekhaniki i Tekhnicheskoi Fiziki*, 1980, No. 4, pp. 141–147, [English trans., *Journal of Applied Mechanics and Technical Physics*, Vol. 21, No. 4, 1980, pp. 555–561].

58. Altshuler, L.V., Novikov, S.A., and Divnov, I.I., "The Relationship Between the Critical Breaking Stresses and the Time of Failure as a Result of the Explosive Stressing of Metals," *Doklady Akademii Nauk SSSR*, Vol. 166, No. 1, 1966, pp. 67–70, [English trans., *Soviet Physics – Doklady*, Vol. 11, No. 1, 1966, pp. 79–82].

59. Novikov, S.A., and Chernov, A.V., "Determination of the Spall Strength From Measured Values of the Specimen Free-Surface Velocity," *Zhurnal Prikladnoi Mekhaniki i Tekhnicheskoi Fiziki*, 1982, No. 5, pp. 126–129, [English trans., *Journal of Applied Mechanics and Technical Physics*, Vol. 23, No. 5, 1982, 703–705].

60. Batkov, Yu.V., Glushak, A.B., and Novikov, S.A., "Using Manganin Gauge to Study Spall Phenomena in Metals During Explosive Loading," *Proc., 4th All-Union Workshop on Detonation*, Vol. 1, 1988, pp. 154–157.

61. Batkov, Yu.V., Golubev, V.K., Novikov, S.A., Sobolev, Yu.S., and Trunin, I.R., "Recording the Spalling Failure of Copper and Lead with Explosive Loading," *Fizika Goreniya i Vzryva*, Vol. 24, No. 1, 1988, pp. 89–92, [English trans., *Combustion, Explosion, and Shock Waves*, Vol. 24, No. 1, 1988, pp. 82–85].

62. Golubev, V.K., Novikov, S.A., Sobolev, Yu.S., and Yukina, N.A., "Effect of the Temperature and of the Loading Time on the Strength and Fracture of Mild Steel and Steels St.3 and 12Kh18N10T in Spalling," *Problemy Prochnosti*, 1985, No. 6, pp. 28–34, [English trans., *Strength of Materials*, Vol. 17, No. 6, 1985, pp. 763–769].

63. Glazov, V.M., Vigdarovich, V.N., Microhardness of Metals, Metallurgizdat Publ., Moscow, 1962.

64. Novikov, S.A., Sobolev, Yu.S., and Yukina, N.A., "Study of Microdamage Accumulation with Spalling in Titanium Alloy VT14," *Zhurnal Prikladnoi Mekhaniki i Tekhnicheskoi Fiziki*, 1988, No. 2, pp. 128–131, [English trans., *Journal of Applied Mechanics and Technical Physics*, Vol. 29, No. 2, 1988, pp. 281–283].

65. Rid, S., Electron Probe Microanalysis, Mir Publ., Moscow, 1979.

66. Rainhart, J., "Some Quantitative Data on Spallation of Metal Subjected to Explosive Loading," *in* Mekhanika, No. 3(19), Izdatelstvo Inostrannoy Literatury Publ., Moscow, 1953, pp. 96–103 [*see also*, Rinehart, J.S., "Some Quantitative Data Bearing on the Scabbing of Metals under Explosive Attack," *Journal of Applied Physics*, Vol. 22, No. 5, 1951, pp. 555–560].

67. Novikov, S.A., Divnov, I.I., and Ivanov, A.G., "Failure of Steel, Aluminum and Copper Under Explosive Shock Loading," *Fizika Metallov i Metallovedenie*, Vol. 24, No. 4, 1966, pp. 607–615, [English trans., *The Physics of Metals and Metallography*, Vol. 21, No. 4, 1966, pp. 122–128].

68. Novikov, S.A., Sobolev, Yu.S., Glushak, B.L., Sinitsyn, V.A., and Chernov, A.V., "Effect of Temperature on Breaking Strength on Spallation in Copper,"

Problemy Prochnosti, 1977, No. 3, pp. 96–98, [English trans., *Strength of Materials*, Vol. 9, No. 3, 1977, pp. 345–347].

69. Razorenov, S.V., Bogatch, A.A., Kanel, G.I., et al., "Elastic-Plastic Deformation and Spall Fracture of Metals at High Temperatures," Shock Compression of Condensed Matter – 1997, Schmidt, S., Dandekar, D., and Forbes, J., eds., AIP Press, Woodbury, NY, pp. 447–450.

70. Ogorodnikov, V.A., Ivanov, A.G., Tyunkin, E.S., Grivorev, V.A., and Khokhlov, A.A., "Dependence of Spall Strength of Metals on the Amplitude of a Shock-Wave Load," *Fizika Goreniya i Vzryva*, Vol. 28, No. 1, 1992, pp. 94–98, [English trans., *Combustion, Explosion, and Shock Waves*, Vol. 28, No. 1, 1992, pp. 88–92].

71. Kanel, G.I., Razorenov, S.V., and Utkin, A.V., "The Spall Strength of Metals at Elevated Temperatures," Shock Compression of Condensed Matter – 1995, Schmidt, S.C., and Tao, W.C., eds., AIP Press, Woodbury, NY, 1996.

72. Kanel, G.I., Razorenov, S.V., and Fortov, V.E., "Cleavage Strength of Metals Over a Wide Range of Shock-Load Amplitudes," *Doklady Akademii Nauk SSSR*, Vol. 294, No. 2, 1987, pp. 350–352, [English trans., *Soviet Physics Doklady*, Vol. 32, No. 5, 1987, pp. 413–414].

73. Deribas, A.A., Nesterenko, V.F., and Teslenko, T.S., "Universal Dependence of Metal Hardening on Shock Density," *Fizika Goreniya i Vzryva*, 1982, No. 6, pp. 68–74.

74. Ivanov, A.G., "Phenomenology of Fracture and Spalling," *Fizika Goreniya i Vzryva*, Vol. 21, No. 2, 1985, pp. 97–104, [English trans., *Combustion, Explosion, and Shock Waves*," Vol. 21, No. 2, 1985, pp. 218–224].

75. Borin, I.P., Novikov, S.A., Pogorelov, A.P., et al., "On Metal Failure Kinetics in Submicrosecond Durability Range," *Doklady Akademii Nauk SSSR*, Vol. 266, No. 6, 1982, pp. 1377–1380.

76. Novikov, S.A., "Strength with Quasistatic and Shock-Wave Loading," *Fizika Goreniya i Vzryva*, Vol. 21, No. 6, 1985, pp. 77–85, [English trans., Combustion, Explosion, and Shock Waves, Vol. 21, No. 6, 1985, pp. 722–729].

77. Regel, V.R., Slutsker, A.I., and Tomashevsky, E.I., *Kinetic Nature of Solid Strength*, Nauka Publ., Moscow, 1974.

78. Merzhievskii, L.A., and Titov, V.M., "Criterion for Metal Durability in the Microsecond Range," *Doklady Akademii Nauk SSSR*, Vol. 286, No. 1, 1986, pp. 109–113, [English trans., *Soviet Physics Doklady*, Vol. 31, No. 1, 1986, pp. 73–75].

79. Golubev, V.K., Material Strength and Failure Under Intensive Pulsed Loading. Metals and Alloys (Review), TsNIIAtominform Publ., Moscow, 1983.

80. Ogorodnikov, V.A., Ivanov, A.G., Luchinin, V.I., Khokhlov, A.A., and Tsoi, A.P., "Effect of Size, Manufacturing and Prestrain on High-Rate Failure (Spall) of PT-3V Titanium Alloy and 12Kh18N10T Steel," *Fizika Goreniya i Vzryva*, Vol. 31, No. 6, 1995, pp. 130–139, [English trans., *Combustion, Explosion, and Shock Waves*, Vol. 31, No. 6, 1995, pp. 726–733].

81. Zeldovich, Ya.B., and Raizer, Yu.P., *Physics of Shock Waves and High-Temperature Hydrodynamic Phenomena*, Nauka Publ., Moscow, 1966, [English trans., Academic Press, NY, Vol. 1 (1966), Vol. 2 (1967); Reprinted in a single volume by Dover Publ., Mineola, NY 2002].

82. Ivanov, A.G., and Novikov, S.A., "Rarefaction Shock Waves in Iron and Steel," *Zhurnal Eksperimentalnoi i Teoreticheskoi Fiziki*, Vol. 40, No. 6, 1961,

pp. 1880–1882, [English trans., *Soviet Physics JETP*, Vol. 13, No. 6, 1961, pp. 1321–1323].

83. Ivanov, A.G., Novikov, S.A., and Tarasov, Yu.I., "Fragmentation Phenomena in Iron and Steel Caused by Explosive Shock Wave Interactions," *Fizika Tverdogo Tela*, Vol. 4, No. 1, 1962, pp. 249–260, [English trans., *Soviet Physics – Solid State*, Vol. 4, No. 1, 1962, pp. 177–185].

84. Batkov, Yu.V., Ivanov, A.G., and Novikov, S.A., "Study of the Unloading of Steel Shock-Compressed Above the Phase-Transition Point," *Zhurnal Prikladnoi Mekhaniki i Tekhnicheskoi Fiziki*, 1985, No. 6, pp. 142–144, [English trans., *Journal of Applied Mechanics and Technical Physics*, Vol. 26, No. 6, 1985, pp. 890–893].

85. Novikov, S.A., "Mechanical Properties of Solid High Explosives in Intense Shock Loading," *Khimicheskaya Fizika*, Vol. 18, No. 10, 1999, pp. 95–102.

86. Glushak, B.L., Trunin, I.R., Novikov, S.A., and Ruzanov, A.I., "Numerical Simulation of Metal Spall Fracture," *in*, Fractals in Applied Physics, RFNC-VNIIEF, Arzamas-16, 1995, pp. 59–122.

87. Ivanov, A.G., "Cleavages in a Quasiacoustic Approximation," *Fizika Goreniya i Vzryva*, Vol. 11, No. 3, 1975, pp. 475–480, [English trans., *Combustion, Explosion, and Shock Waves*, Vol. 11, No. 3, 1975, pp. 401–405].

88. Butcher, B.M., Barker, L.M., Munson, D.E., and Lundergan, C.D., "Influence of Stress History on Time-Dependent Spall in Metals," *AIAA Journal*, Vol. 2, No. 6, 1964, pp. 977–990, [Russian trans., *Raketnaya Tekhnika i Kosmonavtika*, Vol. 2, No. 6, 1964, pp. 3–18].

89. Grady, D.E., "The Spall Strength of Condensed Matter," *Journal of the Mechanics and Physics of Solids*, Vol. 36, No. 3, 1988, pp. 353–384.

90. Ivanov, A.G., and Ogorodnikov, V.A., "Do Brittle and Plastic Materials Differ When Spalling?" *Prikladnaya Mekhanika i Tekhnicheskaya Fizika*, 1992, No. 1, pp. 102–06, [English trans., *Journal of Applied Mechanics and Technical Physics*, Vol. 33, No. 1, 1992, pp. 91–95].

91. Ogorodnikov, V.A., Ivanov, A.G., and Luchinin, V.I., "Scale Effect During Dynamic Destruction of Brittle Viscous Materials," *Book of Abstr., Int. Conf. on High Pressure Science and Technology*, Honolulu, Hawaii, 1999, p. 400.

92. Kanel, G.I., and Shcherban, V.V., "Plastic Deformation and Cleavage Rupture of Armco Iron in a Shock Wave," *Fizika Goreniya i Vzryva*, Vol. 16, No. 4, 1980, pp. 93–103, [English trans., *Combustion, Explosion, and Shock Waves*, Vol. 16, No. 4, 1980, pp. 439–446].

93. Volkov, I.A., "Numerical Modeling of the Spallation of Elastoplastic Solids," *Problemy Prochnosti*, 1991, No. 1, pp. 63–67, [English trans., *Strength of Materials*, Vol. 23, No. 1, 1991, pp. 73–79].

94. Seaman, L., Curran, D.R., and Shockey, D.A., "Computational Models for Ductile and Brittle Fracture," *Journal of Applied Physics*, Vol. 47, No. 11, 1976, pp. 4814–4826.

95. Seaman, L., Curan, D.R., and Shockey, D.A., "Development of a Microfracture Model for High Rate Tensile Damage," *Proc. of the Int. Conf. on Creep and Fracture of Engineering Materials and Structures*, Swansea, Wales, UK, 1981, pp. 345–364.

96. Novikov, S.A., Ruzanov, A.I., Trunin, I.R., and Yukina, N.A., "Microdamage Accumulation in Nickel," *Problemy Prochnosti*, 1993, No. 10, pp. 41–46, [English trans., *Strength of Materials*, Vol. 25, No. 10, 1993, pp. 735–739].

97. Kachanov, L.M., *Fundamentals of Failure Mechanics*, Nauka Publ., Moscow, 1974.

98. Seaman, L., Curran, D.R., and Crewdson, R.C., "Transformation of Observed Crack Traces on a Section to True Crack Density for Fracture Calculations," *Journal of Applied Physics*, Vol. 49, No. 10, 1978, pp. 5221–5229.

99. Sanin, I.V., Vorobev, A.I., and Gornovoi, A.A., "Kinetic-Statistical Model of the Cleavage of Metals," *Fizika Goreniya i Vzryva*, Vol. 23, No. 1, 1987, pp. 67–70 [English trans., *Combustion, Explosion, and Shock Waves*, Vol. 23, No. 1, 1987, pp. 61–64].

100. Ruzanov, A.I., "Numerical Investigation of Splitting Strength with Alowance for Microdamage," *Izv. AN SSSR Mekhanika Tverdogo Tela*, Vol. 19, No. 5, 1984, pp. 109–115, [English trans., *Mechanics of Solids*, Vol. 19, No. 5, 1984, pp. 105–111].

101. Glushak, B.L., Novikov, S.A., Trunin, I.R., and Ruzanov, A.I., "Numerical Simulation of Spall Fracture in One-Dimensional Expansion Waves," *Prikladnyje Problemy Prochnosti i Plastichnosti*, No. 53, 1995, pp. 56–62.

102. Glushak B.L., Trunin, I.R., and Uvarova, O.A., "Spall Fracture Kinetics in α Phase of Armco Iron," *Khimicheskaya Fizika*, Vol. 17, No. 2, 1998, pp. 80–85.

103. Trunin, I.R., Glushak, B.L., and Novikov, S.A., "Simulation of Stress Waves and Spall Fracture of Copper in Volume Heating Conditions," *Proc., 4th Zababakhin Scientific Talks*, RFNC-VNIITF, Snezhinsk, Russia, 1995, pp. 287–294.

Estimation of Temperatures
in Shock-compressed Transparent Material

M.A. Mochalov

In thermodynamics, temperature T is a value characterizing the direction of heat exchange among bodies. In equilibrium, temperature is the same for all bodies comprising the system. A change in temperature of a body entails a change in almost all of its physical properties: length, volume, density, elastic properties, electrical conductivity, etc. The temperature measurement can be based on a change in any of the properties of the body, providing the temperature function of the property is known.

The material shock compression temperature can be used to validate the EOS of a material at high temperatures and pressures by imposing a limitation on two parameters: specific heat, C_V, and Grüneisen factor, $\gamma = V\,(\partial P/\partial E)_V$, which relates heat values to pressure in a shock-compressed material. Knowledge of temperature is critical in studies of such phenomena as dielectric-metal transition, melting, dissociation, and ionization. In trying to understand these phenomena, temperature proves to be a much more accurate metric (compared to the Hugoniot) for the validation of theoretical models.

A simultaneous experimental measurement of compressibility, pressure, and temperature provides a complete set of thermodynamic data about the state of the shock-compressed material that is important for the validation of an EOS.

7.1 Optical Radiation and Its Characteristics

7.1.1 Optical Range

Optics is the branch of physics that deals with the phenomena of electromagnetic radiation. Foundational to this discipline is the development and understanding of the laws that govern the generation of electromagnetic radiation, its propagation, and the nature of its interaction with materials. Various categories of electromagnetic radiation may be established according to wavelength, λ. Wavelength is the distance corresponding to one oscillation period, or the distance between successive minima or maxima. The following units are used for wavelength characterization: $1\ \mu m = 10^{-6}\ m = 10^{-4}\ cm = 10000\ \text{Å}$; $1\ nm = 10^{-9}\ m = 10^{-7}\ cm = 10\ \text{Å}$; $1\ \text{Å} = 10^{-10}\ m = 10^{-8}\ cm$.

Besides wavelength, radiation may also be characterized in terms of frequency, ν. Frequency ν is the number of oscillations per unit time. The wavelength and the frequency are interrelated: $\nu = \upsilon/\lambda = c/\lambda$, where υ is the electromagnetic oscillation propagation velocity and c is sound speed in vacuum ($c = 2.9979 \cdot 10^8\ m/s$). In all systems the unit of measurement for frequency is hertz (cycles/s).

Laws describing the propagation of light in a transparent medium are based on the concept of light as a set of light beams that are treated in terms of geometrical optics. In geometrical optics, the light beam is a line along which electromagnetic radiation propagates. Geometrical optics does not account for light wave properties and relevant diffraction effects.

The optical range is a part of the electromagnetic radiation spectrum covering the visible, ultraviolet and infrared spectra. Presently, the following wavelength values are accepted for the optical range boundaries: the left boundary at 1 nm and the right at 1 mm. The region adjacent to it on the left ($\lambda < 1$ nm) is X-ray radiation (brought about by rapid electron deceleration), and on the right ($\lambda > 1$ mm) radio waves.

The wavelength of 380 nm is considered to be the short-wavelength boundary of the visible range and 780 nm to be the long-wavelength boundary. Visible spectrum boundaries and the wavelengths for various colors appear in Fig. 7.1.

Violet		Blue	Green	Yellow	Orange		Red

380   455   490   575   595   620   780 λ, nm

Fig. 7.1. Visible spectrum boundaries

On the short-wavelength side ($\lambda < 380$ nm) adjacent to the violet spectral region is the ultraviolet region; on the long-wavelength side adjacent to the red spectral region is the infrared region. This partitioning of the electromagnetic spectrum, into regions separated by inter-spectral boundaries is conventional. Wavelengths near the boundaries (on either side of a boundary) may be excited by various methods.

Radiation characterized by one wavelength (or frequency) is called monochromatic. Radiation composed of a combination of the monochromatic radiations is called complex radiation. The spectrum of the radiation can include both several separate wavelengths and their continuous sequence. In the former case the line spectrum is dealt with, in the latter the continuous spectrum.

Strictly monochromatic radiation does not exist in nature. Monochromatic radiation is radiation, which contains a very narrow wavelength band which is described by a single wavelength.

The wavelength scale for a continuous radiation spectrum can occupy any interval, both small and large. For thermal radiation, for example, the continuous spectrum can stretch from zero to infinity.

7.1.2 Principal Properties of Optical Radiation

Refractive Index

A medium through which radiation propagates is called homogeneous if its refractive index is identical everywhere. In such a medium light beams are straight and light propagates along a straight-line. The ratio of light speed in vacuum to light speed in a given medium is referred to as the absolute refractive index of the medium: $n = c/\upsilon = (\varepsilon\mu)^{1/2} \approx (\varepsilon)^{1/2}$. Here ε and μ are relative medium dielectric and magnetic permeabilities, respectively. This expression is valid for non-ferromagnetic media ($\mu \approx 1$). For any medium other than a vacuum, $n > 1$; this value depends on the light frequency and medium characteristics, i.e. temperature and density. For gases at standard conditions, $n \approx 1$.

The ratio of light speed υ_1 in the first medium to light speed υ_2 in the second medium is called the relative refractive index n_{21} of the second medium relative to the first: $n_{21} = \upsilon_1/\upsilon_2 = n_2/n_1$. Here n_1 and n_2 are absolute refractive indices of the first and second media. If $n_{21} > 1$, then the second medium is denser than the first.

Light Dispersion

The dependency of the material absolute refractory index n on frequency ν of radiation impinging on material (or on wavelength λ_0 in vacuum) is called light dispersion: $n = f(\nu) = \varphi(\lambda_0)$.

From the relation $n = c/\upsilon$ it follows that the wavelength of light can be determined from the dependence of the light wave propagation velocity on the light wave frequency: $\upsilon = f_1(\nu)$.

Dispersion is called normal, if the refractory index increases with increasing frequency; otherwise dispersion is abnormal. Dispersion leads to a decomposition of white light to monochromatic components having certain frequencies. This property is frequently used in various monochromators to separate a monochromatic wave for various physical studies.

Light Refraction and Reflection

The phenomena of light refraction and reflection can be observed on a perfectly planar interface separating two media. When crossing the interface, the direction of light propagation changes. The angle i between the incident beam and the perpendicular to the interface is called angle of incidence (Fig. 7.2). The angle of reflection i' is defined as the angle between the reflected beam and the perpendicular. The angle of refraction r is defined as the angle between the refracted beam and the perpendicular.

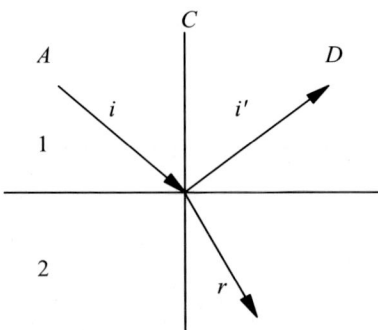

Fig. 7.2. Graphic representation of light reflection and refraction

Laws of Reflection

1. Incident beam A, reflected beam D, and perpendicular C erected to the interface of two media at the beam incidence point lie in one plane.
2. Angle of reflection i' equals the angle of incidence i.
3. The laws of reflection are valid at the inverse direction of the light beam path (reversibility of the light beam path).

Light reflection meeting these laws is called specular. If specular reflection does not hold, then the laws of reflection are invalid. The light reflection can be diffuse or mixed diffuse-specular.

Laws of Refraction

1. The incident and refracted beams, combined with the perpendicular erected from the interface of the two media form a plane.
2. The ratio of the sines of the incident angle and the refraction angle is a constant, which is equal to the relative refractive index of the two media: $\sin i / \sin r = (n_2/n_1) = n_{21}$.
3. The incident and refracted beams are mutually invertible.
4. The laws of light reflection and refraction are valid for isotropic media in the absence of light absorption.

Figure 7.3 serves to illustrates four possible outcomes when a light beam traveling through medium (1) reaches an interface with medium (2). For purposes of illustration, we assume that the optical density of medium (1) exceeds that of medium (2), i.e. $n1 > n2$. The first possibility (illustrated in the figure) is the case wherein the angle of incidence and refraction are both zero. By Snell's law, it is obvious that for a non-zero angle of incidence, the angle of refraction must exceed the angle of incidence (as is depicted in the illustration). Moreover, since $n1 > n2$, there must be some angle of incidence (less than $\pi/2$) for which the angle of refraction is precisely $\pi/2$ (not explicitly illustrated in Fig. 7.3, but easily imagined). The value of incidence angle for

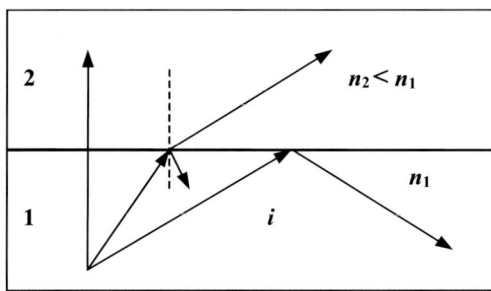

Fig. 7.3. Graphic representation of the complete internal reflection

which the angle of refraction is $\pi/2$, is referred to as the critical angle of incidence. If the angle of incidence exceeds the critical angle, then the light beam is completely reflected (as illustrated by the right-most ray-trace in Fig. 7.3. This phenomenon is referred to as complete internal reflection. To illustrate, if light is traveling through material (1) with refractive index n, and reaches an interface with material (2) possessing a refractive index of 1, the sine of the critical angle of incidence is $1/n$.

Light Absorption

An electromagnetic wave passing through a material is associated with a loss of energy due to its conversion to various internal energy forms, or to secondary radiation energy. Secondary radiation differs from the primary, both in spectral composition and in direction of propagation.

Light absorption is defined as a decrease in the energy of light as it travels through a material. If light flow Φ passes through a medium, then as it travels the distance of l to $l + dl$ it will be attenuated by:

$$-d\Phi = \Phi\mu dl ,\tag{7.1}$$

where μ is an absorption parameter, dependent upon wavelength and the chemistry and state of the material. From Eq. (7.1), it follows that:

$$\Phi = \Phi_0 \left[\exp\left(-\mu l\right)\right] ,\tag{7.2}$$

where Φ_0 is the incident light flux, and Φ is the flux after passing through a material layer of thickness l.

The law of absorption written in this form is called the Bouger-Lambert law. It is valid for linear optics, i.e., those radiation intensities, which do not change medium properties.

Quantity $\mu = (I/l)\ln\left(\Phi_0/\Phi\right)$ is called the linear optical absorption constant. Its numerical value determines thickness of a material layer, through which the light intensity decreases by a factor of e.

If χ (a parameter of light absorption per unit of material concentration), is introduced instead of μ, then Eq. (7.2) becomes:

$$\Phi = \Phi_0 \left[\exp \left(-\chi K'l \right) \right] \ , \tag{7.3}$$

where K' is material concentration. This formula form is referred to as the Bouger-Lambert-Beer law.

Light Scattering

In addition to absorption, scattering also attenuates electromagnetic radiation as it passes through a medium. Therefore, the absorption parameter in the Bouger-Lambert-Beer formula must account for both effects (true absorption, K_{true}, plus scattering factor, K_{scat}): $K = \chi K' = K_{true} + K_{scat}$.

The light flux in the propagation direction decreases faster with scattering than with absorption alone. Scattering causes the light to alter its direction of propagation, and the scattering material begins to emit light. The scattering level depends on the radiation wavelength and scattering particle size. When there are coarse particles, light impinging on them, through geometrical optics laws, is reflected at different angles. If the size of the particles is comparable to the wavelength, diffraction effects begin to play a decisive role. This is called diffraction scattering. When the scattering particles are smaller than the light wavelength, the scattering intensity is:

$$I \sim \nu^4 \sim \lambda^{-4} \ . \tag{7.4}$$

This relation was found by Rayleigh and is referred to as Rayleigh's law.

Further information about the material discussed in this section can be found in the relevant literature (see, [1–5]).

7.2 Thermal Radiation

7.2.1 Kirchhoff Law

The radiation process is attended with an energy loss. For the process to continue, it is necessary to compensate for the energy decrease. Energy sources absorbed by the radiating body can be different. This section discusses a process, whereby the thermal energy of the body is converted to radiation.

Each body's temperature corresponds to a certain energy level of the emitting particles – atoms and molecules (the Boltzmann distribution). The higher the temperature of the body, the particles within the body exist at high energy levels. Energy is emitted due to spontaneous transitions of excited particles to lower energy levels. To excite the emitting particles – raising their energy levels – occurs from thermal motion and (in a general case) the quantum absorption from the surrounding electromagnetic field. Thermal energy can be replenished through various processes: chemical reactions, electric current flow, etc. If the energy has had time to become distributed uniformly before radiation takes place, the radiation is thermal in its nature.

When the radiating bodies and the radiation in an insulated body reach a steady state, the thermal radiation is considered to be in equilibrium. Imagine that a number of bodies are placed in a vacuum inside an impermeable, perfectly reflecting shell. After some time the bodies will be at identical temperatures as a result of the energy exchange through emission and absorption. When equilibrium is established, each body releases and absorbs identical quantities of energy. This pivotal feature of thermal radiation follows the second law of thermodynamics: steady thermal equilibrium of an insulated system cannot be broken through any heat exchanges within the system. If an insulated system is in equilibrium the temperature is identical for all the bodies, the released and absorbed energies will be equal for each body. Therefore it is clear that bodies with a lower absorptivity should emit less per surface area.

Using the second law of thermodynamics, in 1859 R. Kirchhoff proved the ratio of radiant intensity to absorptivity at each wavelength for a given temperature is constant for all bodies and depends on the wavelength and temperature only:

$$M_\lambda(\lambda, T)/\alpha(\lambda, T) = M_{\lambda S}(\lambda, T) , \tag{7.5}$$

where $M_\lambda(\lambda, T)$ is the radiant excitation; $\alpha(\lambda, T)$ is emmisivity of the body; $M_{\lambda S}(\lambda, T)$ is a universal wavelength and temperature function identical for all bodies. The physical meaning of this function becomes clear, if the notion of a blackbody which completely absorbs its radiation is introduced. Absorptivity of this body is equal to one at any temperature and at any wavelength. If $\alpha(\lambda, T) = 1$ is substituted to Eq. (7.5), then function $M_{\lambda S}(\lambda, T)$ will be always equal to the blackbody radiance.

Thus, the Kirchhoff law can be formulated as follows: the ratio of the spectral radiance to the absorptivity is a wavelength and temperature function universal for all bodies, with this function being blackbody radiance. From this it follows, in particular, that both for integral (by spectrum) radiant emittance and absorptivity

$$M(T)/\alpha(T) = M_S(T) , \tag{7.6}$$

where $M(T)$ and $\alpha(T)$ refer to any body and $M_S(T)$ is blackbody radiant emittance at a given temperature.

Although the surface absorptivity of any real body is always less than one, a blackbody model can be constructed. It can be made from a cavity with a small hole, whose area is small compared to the internal surface area of the cavity (Fig. 7.4). A beam going through the hole, can only escape after multiple reflections, the absorptivity for the hole can be made close to one.

This condition, in particular, is met by hollow radiator constructions used as radiation reference sources.

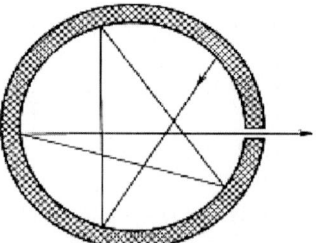

Fig. 7.4. Blackbody Model

Particularly high absorptivities of real bodies are usually attributed to porous surface structure, from which a beam incident to the surface experiences absorption several times before leaving the surface (i.e. soot, black velvet). This phenomenon accounts for the deep colors in porous-structured colored surfaces.

From the Kirchhoff law, it follows that the blackbody is a limiting case for any thermal radiator, knowledge of the radiation laws is important for solution of both theoretical and practical problems.

7.2.2 Blackbody Radiation

Experimental studies indicate that the blackbody spectral radiance increases at all wavelengths with increasing temperature, with its peak shifted toward shorter wavelengths. The assumption made by M. Planck was that energy is released and absorbed in certain portions, i.e. quanta, which led him in 1900 to an accurate description for the blackbody radiation spectral distribution as a function of temperature. This law is given by equation:

$$M_\lambda(\lambda, T) = c_1 \lambda^{-5} \left(e^{c_2/\lambda T} - 1 \right)^{-1},   \quad (7.7)$$

where c_1 and c_2 are the first and second radiation constants: $c_1 = 2\pi hc^2 = (3.7417749 \pm 0.0000022) \cdot 10^{-16}$ W $\cdot$ m^2; $c_2 = hc/r = (1.438769 \pm 0.000012) \times 10^{-2}$ m $\cdot$ K; c is light speed in vacuum; r is Boltzmann's constant; h is Planck's constant.

Spectral radiance vs. radiation wavelength can be represented as radiation isotherms (as shown in Fig. 7.5).

Planck's Eq. (7.7) is written for radiant emittance of full (nonpolarized) radiation. For the radiance of full radiation, c_1/π should appear instead of c_1. For a single polarization plane these coefficients are divided by 2. From Planck's equation, it follows that at any temperature the blackbody radiates the entire wavelength range, from zero to infinity, but for wavelengths quite far from the maximum, spectral radiant intensity becomes as small as one desires. At room temperature, for example, blackbody full radiation is small (which

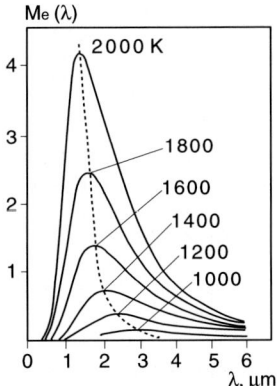

Fig. 7.5. Spectral radiance $M_e(\lambda)$ vs. wavelength at various temperatures T. - - - - positions of maximums

is clear from the Stefan-Boltzmann law alone), and its fraction accounted for by the visible spectrum is negligible as compared to the full radiation.

The distribution of the relative spectral flux of a Planck radiator is presented in Fig. 7.6.

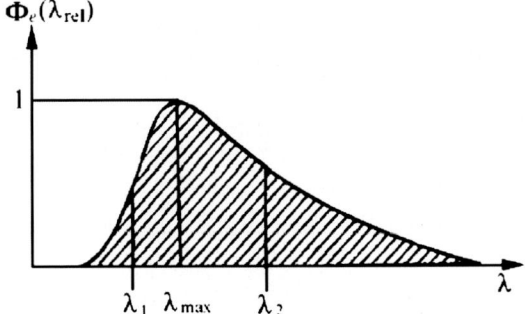

Fig. 7.6. Distribution by wavelengths λ of the relative spectral radiation flux Φ_e (λ_{rel}) of the Planck radiator

If $\lambda T/c_2 \ll 1$, the Planck equation reduces to the Wien equation (by expansion of the multiplier $(e^{c_2/\lambda T} - 1)^{-1}$ in series while neglecting higher-order terms):

$$M_\lambda(\lambda, T) = c_1 \lambda^{-5} e^{c_2/\lambda T}. \tag{7.8}$$

If $\lambda T/c_2 \gg 1$, we arrive at the Rayleigh-Jeans equation (by the series expansion of the function $e^{c_2/\lambda T}$, while neglecting higher-order terms):

$$M_\lambda(\lambda, T) = 2\pi c r \lambda^{-4} T. \tag{7.9}$$

These equations are used as convenient approximations for some cases. For example, when using Eq. (7.8) instead of Eq. (7.7) for temperatures below 2856 K, the error becomes 0.1% for wavelengths < 730 nm (for $\lambda = 400$ nm the error is $< 3 \cdot 10^{-4}\%$).

Experimental studies show that the blackbody spectral radiance increases with increasing temperature for all wavelengths, with its peak shifted toward short wavelengths (see Fig. 7.5).

The peak position determined by the Wien law is

$$\lambda_{\max}T = b , \tag{7.10}$$

where b is the Wien constant equal to $(0.2897756 \pm 0.0000024) \cdot 10^{-2}$m $\cdot$ K.

Wien's displacement law illustrates why radiation for heated bodies shifts toward the long-wavelength spectrum with decreasing temperature (Fig. 7.7).

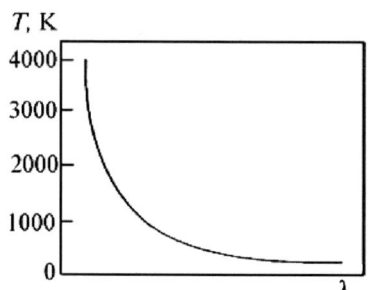

Fig. 7.7. Graphic representation of the Wien displacement law

In 1894 experimental studies by I. Stefan and theoretical reasoning by L. Boltzmann led to the following law for blackbody radiation: the blackbody radiation (radiant emittance) is proportional to temperature raised to the fourth power:

$$M = \sigma T^4 , \tag{7.11}$$

where σ is the Stefan-Boltzmann constant, $b = (5.67051 \pm 0.00019) \cdot 10^{-8}$ W $\cdot$ m$^2 \cdot$ K^{-4}.

Fundamentals of modern pyrometry measuring real temperatures of irradiating solid bodies are discussed thoroughly in the literature (see, e.g., [6]).

7.2.3 Laws of Blackbody Radiation

Lambert Law

In the general case, luminance of a surface element of the source depends not only on the surface element location, but also on the radiation direction.

However, there are surfaces which are irradiating where their luminance can be with some level of accuracy considered direction-independent:

$$L(\theta) = \text{const} , \qquad (7.12)$$

where θ is an angle between the radiation direction under consideration and the normal to the surface element. (In the general case, for an asymmetric radiator, the luminance depends not only on polar angle θ, but also on azimuthal angle ϑ.)

Sources radiating in accordance with this law are called Lambert radiators. The Lambert law is strictly obeyed by a blackbody radiator; however, there are several radiator types which approximately obey it, namely: incandescent bodies, luminescent layers, as well as some light-scattering media. Deviations from the Lambert law become significant, as a rule, at large angles θ, i.e. those directions nearly tangential to the surface.

Radiance of the Lambert radiator surface areas is independent of the radiation direction, i.e. $L_0 = L_\theta$. The luminance vector locus is a hemisphere with its center on the surface element (Fig. 7.8).

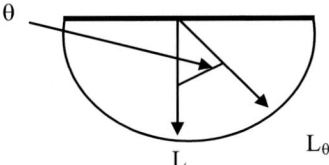

Fig. 7.8. Graphic representation of the first Lambert Law $L(\theta) = \text{const}$

If light of intensity I_θ strikes a surface at an angle θ to a surfaces normal the intensity is given by $I_\theta = I_0 \cos \theta$, where I_0 is radiation intensity in the direction normal to the surface. For a surface element radiating according to Lambert's law a sphere describes the light intensity distribution about every direction (see Fig. 7.9).

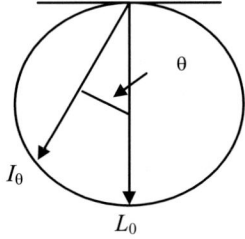

Fig. 7.9. Curve of light intensity according to Lambert's law

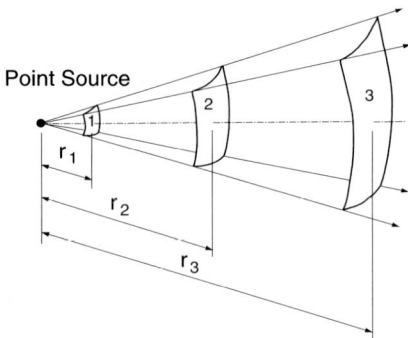

Fig. 7.10. Graphic representation of the law of inverse squares

Irradiances E_i of surface areas irradiated by a point source at the same angles are inversely proportional to the squared distance from the source to the surface (Fig. 7.10): $E_1/E_2 = r_2^2/r_1^2$.

7.2.4 Radiation of Real Bodies

Thermal radiation of real bodies always differs from blackbody radiation, as it not only absorbs and radiates, but also reflects and transmits radiation. In the most general case, following an energy balance equation which holds for any body:

$$\varepsilon\left(\lambda, T\right) + R\left(\lambda, T\right) + \tau\left(\lambda, T\right) = 1, \tag{7.13}$$

where $\varepsilon\left(\lambda, T\right)$ is body absorptivity; $R\left(\lambda, T\right)$ is body reflectivity; and $\tau(\lambda,\ T)$ is transparency of the body. For a blackbody at any temperature, $\varepsilon(\lambda, T) = \varepsilon_0(\lambda, T) = 1$ and $R\left(\lambda, T\right) = \tau\left(\lambda, T\right) = 0$. $I_0\left(\lambda, T\right)$ and $\varepsilon\left(\lambda, T\right)$ are radiating body characteristics and their ratio, according to Kirchhoff's law, is universal, with a body independent value for equilibrium radiation at given temperature T and wavelength λ:

$$I_0(\lambda, T)/\varepsilon_0\left(\lambda, T\right) = I\left(\lambda, T\right)/\varepsilon\left(\lambda, T\right), \tag{7.14}$$

where $I_0\left(\lambda, T\right)$ and $\varepsilon_0\left(\lambda, T\right)$ refer to blackbody radiation. In real bodies the surface cannot be completely black. An important conclusion follows from this is that real body radiation, both integral (by spectrum) and in any spectral range can be only less than the blackbody radiation at the same temperature.

The radiating body is gray, if its absorptivity is constant, which depends only on temperature, material, and body surface state for the entire wavelength range. As $\varepsilon_0\left(\lambda, T\right) = 1$, from Eq. (7.14),

$$I\left(\lambda, T\right) = \varepsilon\left(\lambda, T\right) I_0\left(\lambda, T\right) \tag{7.15}$$

follows, which relates the real body radiation with the blackbody radiation. If the body is not black, then the measured temperature does not coincide with the true temperature.

7.2.5 Conventional Temperatures

In optical pyrometry, the notions of luminance and color temperatures are introduced for real body characterization. Luminance temperature (T_{lum}) is the temperature for the blackbody, which has the same monochromatic luminance for a selected wavelength, as the body with temperature T, i.e.

$$I_0(\lambda, T_{lum}) = I(\lambda, T) . \qquad (7.16)$$

In practice, when measuring temperature with the luminance method, absolute body emissivity is not measured, but is compared with the emissivity of a reference black radiator having a known luminance temperature T_r. Then from relation

$$R(\lambda, T, T_r) = \varepsilon(\lambda, T) I_0(\lambda, T)/I_0(\lambda, T_r), \qquad (7.17)$$

assuming $\varepsilon(\lambda, T) = 1$, the luminance temperature of the material can be determined.

Up to temperatures $T \sim 6000$ K using Wien's formula Eq. (7.8) is certainly justified in the visible spectrum. It is easy to show [6] that the luminance and true temperatures of the radiating body are related as follows:

$$\ln \varepsilon(\lambda, T) = c_2/\lambda (1/T - 1/T_{lum}), \qquad (7.18)$$

where $c_2 = 1.438$ cm $\cdot$ deg and λ is radiation wavelength.

The systematic error of a luminance temperature measurement is determined from the difference between the inverse luminance values and true temperatures. Taking into account Eq. (7.18), the luminance temperature measurement error [6] can be written as

$$\Delta T_{lum}^{-1} = T_{lum}^{-1} - T^{-1} = \frac{\lambda}{c_2} \ln \varepsilon(\lambda, T) . \qquad (7.19)$$

Then we obtain the following expression for the relative methodic luminance temperature measurement error $\delta T = (T - T_{lum})/T$:

$$|\delta T_{lum}| = (T - T_{lum})/T = (\lambda T_{lum}/c_2) \ln \varepsilon(\lambda, T), \qquad (7.20)$$

from which it is seen that the relative error of the luminance is lower for shorter wavelengths and lower temperatures.

Color temperature is another conventional pyrometric technique used for characterizing radiating bodies. The color temperature of a non-blackbody radiator having a temperature T is defined as the temperature T_{color} where the ratio of the spectral luminances at wavelengths λ_1 and λ_2 is the same as the object of interest. To find the color temperature, radiation in two spectral ranges,

$$I_1(\lambda_1, T) = \varepsilon(\lambda_1, T) I_0(\lambda_1, T) \quad \text{and} \quad I_2(\lambda_2, T) = \varepsilon(\lambda_2, T) I_0(\lambda_2, T).$$

is experimentally measured. Then from the relation

$$R\left(\lambda_1, \lambda_2, T\right) = \left[\varepsilon\left(\lambda_1, T\right)/\varepsilon\left(\lambda_2, T\right)\right]\left[I_0\left(\lambda_1, T\right)/I_0\left(\lambda_2, T\right)\right] \tag{7.21}$$

with $\left[\varepsilon\left(\lambda_1, T\right)/\varepsilon\left(\lambda_2, T\right)\right] = 1$ one can find the second conventional temperature, i.e. the color temperature:

$$R\left(\lambda_1, \lambda_2, T\right) = I_0\left(\lambda_1, T_{color}\right)/I_0\left(\lambda_2, T_{color}\right). \tag{7.22}$$

As in the case of the luminance temperature, the absolute value of the color temperature measurement error can be written as follows:

$$\Delta T_{color}^{-1} = T_{color}^{-1} - T^{-1} = -\frac{\Lambda}{c_2}\ln\frac{\varepsilon(\lambda_1, T)}{\varepsilon(\lambda_2, T)}. \tag{7.23}$$

Then, from analogy with the luminance method, at the spectral ratio temperature the relative systematic error in the measurement is:

$$\delta T_{color} = \frac{(T - T_{color})}{T} = \frac{\Lambda T_{color}}{c_2}\ln\frac{\varepsilon(\lambda_1, T)}{\varepsilon(\lambda_2, T)} \tag{7.24}$$

Here Λ is the equivalent spectral ratio pyrometer wavelength determined as $\Lambda = \lambda_1^{-1} - \lambda_2^{-1}$. From Eq. (7.24) it is seen that at an identical ratio $\varepsilon(\lambda_1, T)/\varepsilon(\lambda_2, T)$ relative error δT_{color} is lower for shorter wavelengths and lower temperatures.

To compare the pyrometry systematic errors of the luminance Eq. (7.19) and color Eq. (7.23) methods, consider relation [6]:

$$\frac{\Delta T_{color}^{-1}}{\Delta T_{lum}^{-1}} = \frac{\Lambda\left[\ln\varepsilon(\lambda_1, T) - \ln\varepsilon(\lambda_2, T)\right]}{\lambda_r \ln\varepsilon(\lambda_r, T)}. \tag{7.25}$$

If the luminance temperature is measured over the luminance interval at an effective wavelength $\lambda = \lambda_2$, then, having introduced notations $\lambda_2/\lambda_1 = S$ and $\left[\ln\varepsilon\left(\lambda_1, T\right)/\ln\varepsilon\left(\lambda_2, T\right)\right] = M$, we obtain:

$$\frac{\Delta T_{color}^{-1}}{\Delta T_{lum}^{-1}} = \frac{M - 1}{S - 1}. \tag{7.26}$$

From Eq. (7.26) it is seen that if $M > S$ ($\lambda = \lambda_2$) the luminance method is preferable and if $M < S$ the color temperature is closer to the true temperature. Thus, it is clear that the choice of pyrometric technique should depend on the specific experimental conditions.

7.3 Optical Materials for Physical Studies

Optical materials used in physical studies include lenses, mirrors, beam splitters (plane-parallel and wedge-shaped), prisms, windows of various types, and filters for radiation reduction and spectral range separation. The materials have to be highly transparent, strong, immune to various effects, readily machinable, and relatively cheap.

The most important characteristic for glasses is spectral radiation transmission. Passage of radiation of wavelength λ through a homogeneous absorbing (non-scattering) material layer is described with Bouger-Lambert law: $I = I_0 \exp(-kl)$. Here I_0 and I are light intensities at the entry into and exit from the material layer of thickness l. This law assumes that the absorbing radiation fraction is radiation flux-independent and the flux absorbed in the elementary layer is proportional to its thickness. From this assumption it follows that this form of the absorption law is valid with linear optics, i.e. for radiation intensity values at which the medium properties remain unchanged.

The transmission factor of a layer of thickness l for monochromatic radiation, by definition, is $\tau_\lambda = I/I_0 = \exp(-k_\lambda^* l)$, where $k_\lambda^* = -(\ln \tau_\lambda / l)$ is the natural exponent of absorption.

To calculate optical material transmissions, not the natural, but the decimal logarithm is used most frequently. In this case glass absorptivity k_λ for light of wavelength λ is determined from expression $k_\lambda = -(\log \tau_\lambda / l)$, where τ_λ is the transmission factor of a layer of thickness l for monochromatic radiation of wavelength λ.

Glass optical density d_λ for monochromatic light of wavelength λ is related to absorptivity k_λ and the transmission factor τ_λ as $d_\lambda = -\log \tau_\lambda = k_\lambda l$. When calculating glass optical density, it is necessary to take this into account, in addition to light absorption, losses due to reflection from two glass surfaces and apply other corrections.

7.3.1 Glazier Materials

Glazier materials are used in many experimental devices for radiation transfer from one medium to another without any contact between them. When selecting a material for the windows, it is necessary to consider its strength, if a high-pressure gas medium is used in the apparatus, the linear expansion factor for cryogenic equipment, its transparency, and its immunity to environmental effects. Various types of optical glass, fused quartz, artificial sapphire (Al_2O_3), and various other transparent crystals are used in fabrication of the windows. Here we consider the spectral characteristics for widely used optical materials in physical shock wave research.

Fused Quartz

This glass is produced by grinding and melting natural quartz. Its initial density is $\rho_0 = 2.2$ g/cm^3 and refractive index is $n = 1.46$ (for $\lambda = 589$ nm). The refractive indices for laser radiation wavelengths of type KU quartz glass appear in Fig. 7.11. Spectral transmission of type KU fused quartz in the 190–1100 nm spectrum range is presented in Fig. 7.12.

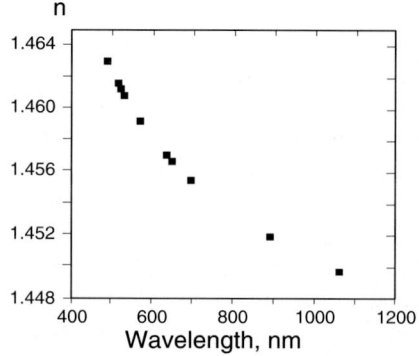

Fig. 7.11. Refractive index of KU glass vs. wavelength [7]

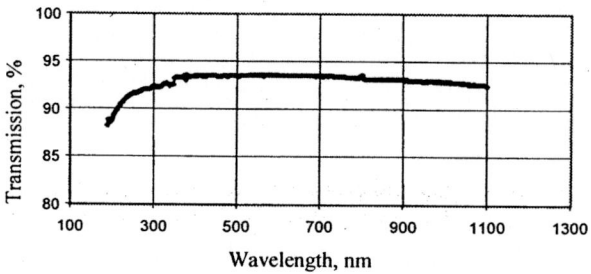

Fig. 7.12. Spectral transmission of type KU-1 fused quartz 1.95 mm thick

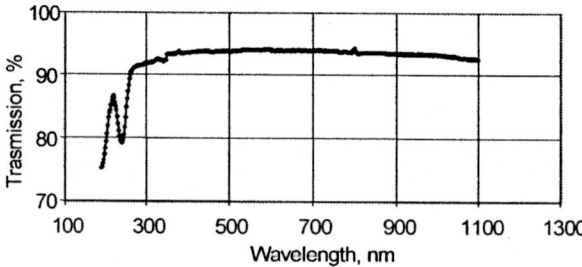

Fig. 7.13. Spectral transmission of type KI fused quartz 10 mm thick

As seen from the figure, this glass has a high transmission factor for radiation up to 200 nm and, therefore, it is widely used in studies of radiation over the entire visible spectrum, including the ultraviolet. Figure 7.13 presents the spectrum transmission of KI type (potassium iodide) quartz glass. The ultraviolet radiation transmission at 200 nm wavelength is seen to be lower by 15% in this glass than in KU type ultraviolet quartz glass. Also, at ~250 nm wavelength, an absorption band is seen in KI type quartz, which should be taken into account when setting up experiments.

Synthetic Sapphire

Synthetic sapphire (Al_2O_3) is an excellent material on account of its hardness, strength, radiation transmittance, and resistance to various external effects. Spectral transmission of a synthetic sapphire specimen is plotted in Fig. 7.14.

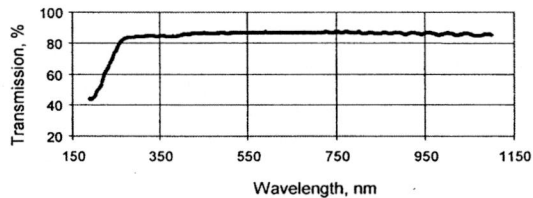

Fig. 7.14. Spectral transmission of synthetic sapphire 2 mm thick

Type K-8 Optical Glass (Crown)

Note that K-8, like KI is a quartz glass. As compared to fused quartz and synthetic sapphire, it is a cheaper glass having less strength and resistance to external effects. However, because of its high optical transmission in the visible and near infrared, this glass finds wide utility for the fabrication of optical parts of objectives: simple lenses, wedge-shaped and plane-parallel plates, prisms, and in combination with other types of glass. This material is widely used in experiments to study laser and thermal radiations. The refractive index for K-8 vs. laser radiation wavelength appears in Fig. 7.15. Spectral transmission of a K-8 specimen is plotted in Fig. 7.16.

As seen from the figure, at wavelengths shorter than 400 nm the K-8 transmission drops from 90% to zero, as a result this glass unsuitable for experiments in the ultraviolet wavelengths.

Polymethylmethacrylate (PMMA)

Owing to its availability, this glass is frequently used as windows in experiments with inert gases in the visible and near-infrared regions. The spectral transmission of plexiglas appears in Fig. 7.17. The abrupt drop in the radiation transmission for wavelengths shorter than 300 nm makes the material unsuitable for experiments in the ultraviolet region. A serious disadvantage of this organic glass is its low mechanical strength.

In experiments studying optical radiation, wide application is also found for other materials, among which are the metal halides, such as LiF, NaCl, KCl and others, more than can be mentioned here.

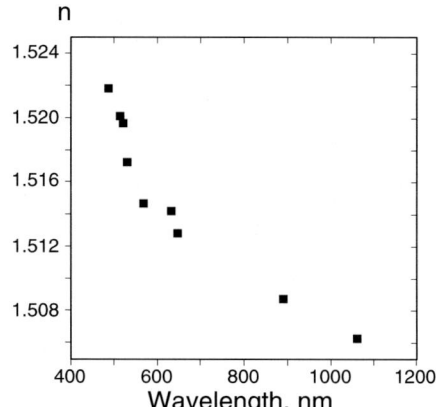

Fig. 7.15. Refractive index for K-8 vs. wavelength [7]

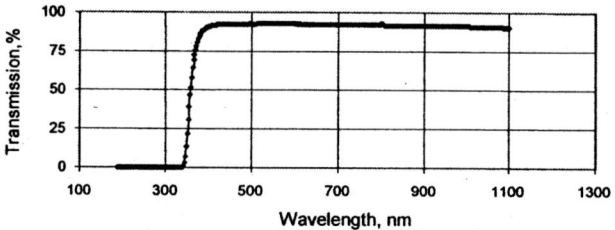

Fig. 7.16. Spectral transmission of crown K-8 3 mm thick

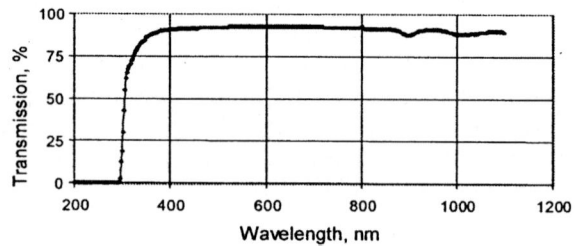

Fig. 7.17. Spectral transmission of PMMA 8 mm thick

Lithium Fluoride (LiF)

The ionic crystal LiF is used in optical studies over a wide spectral region. The spectral transmission of a LiF specimen appears in Fig. 7.18.

Substrates made of LiF are widely used for recording thermal radiation at the shock wave front in experiments with cryogenic temperatures ≤77 K.

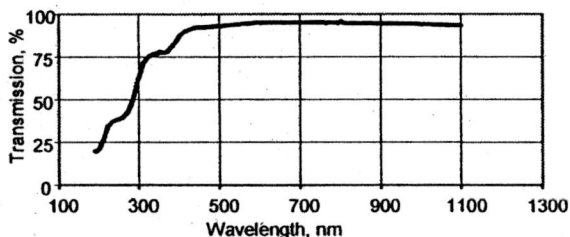

Fig. 7.18. Spectral transmission of LiF 12 mm thick

7.3.2 Filters

A filter by definition is an optical element designed for changing radiation spectral composition or radiant flux. Spectral and compensation filters can be distinguished by their purpose. The first type serves to separate the desired spectral region, the second type is for bringing the spectral characteristic to a required form.

Neutral Light Filters

The neutral filters are compensation type filters that attenuate radiation without any change in the spectral composition.

A wide application is found for neutral glass filters with a characteristic size of 40×40 mm or 80×80 mm and up to 3 mm in thickness in experimental measurements for a shock wave's thermal radiation front. To fabricate glass neutral filters, high-quality colored glass is used, which has stable characteristics in time and uniform absorption in area. Their transmission weakly depends on the incidence angle of the radiation onto the glass surface.

These filters are quite strong mechanically and the filter surfaces are finished and polished so that the roughness is no greater than 0.05 µm.

The spectral transmission for three typical neutral filters in the 300–700 nm spectral range appears in Fig. 7.19.

As seen from the figure, for the visible spectrum (450–650 nm) the transmission of the filters presented in the figure is essentially constant.

Interference Filters

These filters are used, as a rule, for separation of narrow spectral regions and for radiation suppression outside those regions. In the simplest case, the filter is a dielectric film covered on either side with a semi-transparent metallic film. The transmission band of the filter is formed by interference of incident radiation and light reflected from the interfaces of the thin dielectric and metal films (Fig. 7.20).

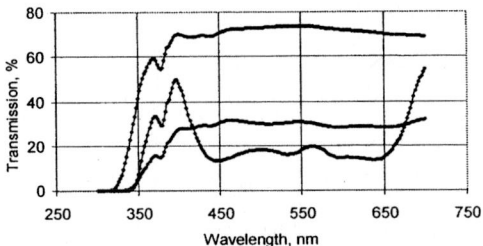

Fig. 7.19. Spectral transmission of neutral light filters

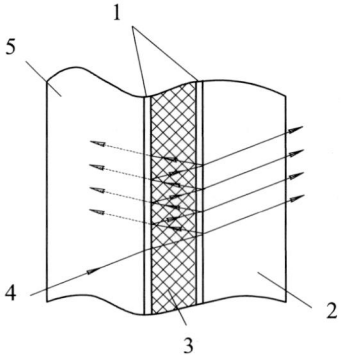

Fig. 7.20. Schematic view of transmitting interference filter: 1 – semitransparent metal layers; 2, 5 – glass filter; 3 – dielectric layer; 4 – incident beam

Numerous reflections from the metal layers result in a system of beams, which are passed through or reflected, interfering with each other. When the dielectric layer thickness is a multiple of $\lambda/2$, the propagation difference between two interfering neighboring beams is λ and electromagnetic oscillation composition results in passing radiation gain.

The wavelength changes with changing dielectric thickness, but the bandwidth increases. To reduce bandwidth, it is necessary to increase the metal film reflectivity, which leads to reduction in the filter transmission. Thus, in interference filters, there is a trade-off between the bandwidth and radiation transmission factor.

Characteristics of four visible spectrum interference light filters are presented in Fig. 7.21. As seen from the figure, the filters transmit from 38 to 56% in the wavelength range from 406 to 600 nm at bandwidth $\sim$10 nm of each (at a level of half the maximum transmission).

A feature of the multi-layered interference filters is the displacement of the transmission level and spectrum for radiation deviation from normal incidence angle. This effect should not be neglected when measuring thermal radiation. In pyrometers, where input radiation is collected by a focusing lens, it is

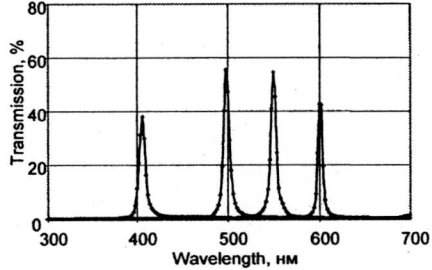

Fig. 7.21. Spectral transmission of visible-spectrum interference light filters

necessary to use a condenser to form a parallel light beam, in which the
interference filters are placed.

Figure 7.22 shows the spectral transmission of an interference light filter.
The figure shows the transmission peak shifting toward shorter wavelengths,
a decrease in the transmission coefficient, and the appearance of an additional
transmission at wavelengths longer than 500 nm.

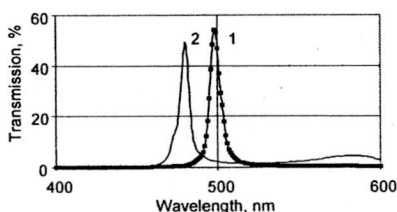

Fig. 7.22. Spectral transmission of an interference light filter at normal incidence
(1) and at an angle of 30° (2)

Colored Glass

Colored light filters are fabricated from colored high-quality optical glass.
Industry produces filters of 40 × 40 mm and 80 × 80 mm in size. Spectral
characteristics of these light filters are uniform over the entire surface and
are essentially insensitive to the radiation incidence angle. As for the neutral
filters, the colored light filter surfaces have an optical finish and wedging no
greater than 2″.

The spectral transmission of two colored light filter types, blue-green and
orange, 3 mm thick, is presented in Fig. 7.23 and Fig. 7.24, respectively [8].

The figures show that the colored light filters can be used both for separation of a certain transmission band and for suppression of undesirable
radiation.

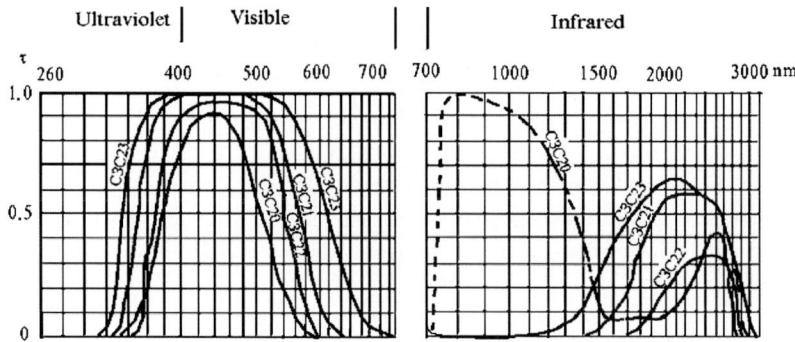

Fig. 7.23. Spectral transmission of blue-green light filters

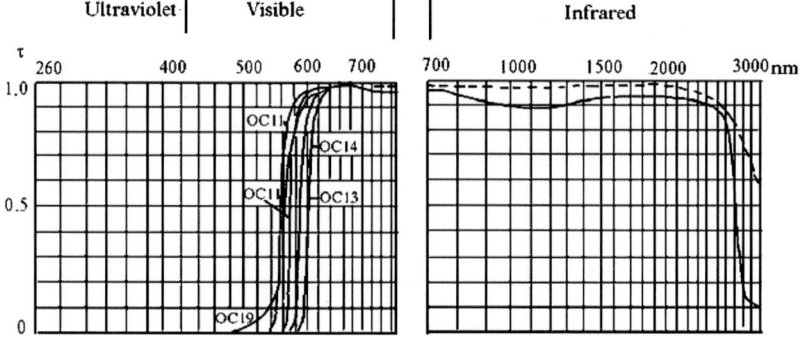

Fig. 7.24. Spectral transmission of orange light filters

Many colored glasses change spectral transmission under the action of high temperature. In particular, spectral transmission of yellow, orange and red light filters shifts by 10–15 nm upon heating at 100°C increments [8]. This should be taken into account, when the colored light filters are located near powerful thermal radiation sources.

7.4 Fundamentals of Optical Temperature Measurement

Under conditions of rapid material compression and heating ($t \sim 10^{-6}$–10^{-7} s) during a shock wave, highly-sensitive and quick-response optical methods provide temperature measurement with good time resolution. A set of temperature measurement methods using temperature-dependent and blackbody emissivities is called pyrometry. The devices used for this purpose are referred to as pyrometers. Principal components of any pyrometer are: a light-sensitive receiver; an optical path for radiation going to the receiver; light filters (or a monochromator) for separating the needed spectral range; recording devices and a timing system.

The optical path consists either of a system of lenses and diaphragms cutting off extraneous radiation or light guides. Using light guides in pyrometer design is a promising approach for it reduces radiation losses and, thus, enhances the pyrometer sensitivity. Principal requirements for the recording devices are fast operation and amplitude linearity.

Photographic and electro-optical recording methods are distinguished by the type of pyrometer light-sensitive element used. In the first case, the radiation blackens a photographic film, which can later be evaluated. In the second case, a photomultiplier used in combination with a fast oscilloscope records photoelectric current as a result of the radiation flux.

Depending on a receiver type, the following two methods for radiation recording can be mentioned: photographic and photoelectric.

7.4.1 Photographic Method

In order to estimate the radiance temperature, the photographic method [9, 10] consists of measuring the irradiance of a light-sensitive layer produced by the light source under study within a selected spectral region. In order to accurately estimate the radiance temperature, monochromatic light should be used. Monochromatic radiation is radiation lying within a very narrow wavelength range, from λ to $\lambda + d\lambda$. However, radiation is quite hard to record on photographic film for narrow spectral regions ($d\lambda \sim 10$ nm). So colored glass filters are used to provide a broader spectral range ($d\lambda \sim 100$ nm) thus providing sufficient accuracy to measure the radiance temperature [8].

The photographic emulsion is a selective radiant energy receiver, i.e. a receiver whose absorptivity depends on incident radiation wavelength. Therefore, radiation of different wavelengths produces different efficiency in terms of how the emulsion responds to the radiation. A radiation flux affecting the photographic film within wavelength range λ and $d\lambda$ is determined as

$$dA = E_\lambda S_\lambda d\lambda \qquad (7.27)$$

where S_λ is spectral sensitivity of the photographic emulsion and E_λ is irradiance on the photographic emulsion surface. From photographic image theory it is known that the relation between spectral irradiance E_λ on a light-sensitive layer positioned in the device focal plane and spectral radiance I_λ of the photographic light source can be expressed as

$$E_\lambda = \pi/4\tau_0 I_\lambda \left(2a/f\right)^2 F\left(\varphi\right) , \qquad (7.28)$$

where τ_0 is the objective transmission factor; f is the objective focal length; $2a/f$ is the device aperture ratio; $F(\varphi)$ is the photometric field error of the device; and I_λ is the Planck function Eq. (7.7). When a narrow spectral region, $d\lambda = \lambda_2 - \lambda_1$, is separated with a light filter whose transmission factor is τ_λ, the flux acting on the photographic layer can be represented as:

$$A = \int\limits_{\lambda_1}^{\lambda_2} I_\lambda S_\lambda \tau_\lambda d\lambda \ . \tag{7.29}$$

Radiation acting on the photographic film blackens the film, after photochemical processing, the film can be evaluated with opacity measurements, i.e. optical density d (see, e.g., [11]). Because the radiation is recorded on photographic film over some spectral range, $d\lambda$, on account of the multispectral nature of the radiation and receiver's selectivity, the photographic emulsion optical density d is expressed by formula

$$d - d_0 = \gamma \log (tA) + B \ , \tag{7.30}$$

where d is image optical density; d_0 is fog density; γ is photographic emulsion contrast factor; t is exposure time; and B is a constant including design features of the optical system and the distance to the radiation source. Value A is given by Eq. (7.29). Integration limits λ_1 and λ_2 are the wavelength limits for selective filter transmission and photographic film spectral sensitivity.

The potential for quantitative measurement using the photographic layer as a tool is valid only for the linear segment of the characteristic curve [11], where values γ, d, B and the flux in Eq. (7.30) can be considered constant under certain photochemical processing conditions. The logarithm of the flux ratio for the studied and reference sources can be estimated from the expression

$$\log (A/A_r) = (d - d_r)/\gamma + \log (V/V_r) + \log (\tau_r/\tau) + \log Q \ . \tag{7.31}$$

Here d and d_r are photographic layer optical densities for the studied and reference radiations, V is the scanning speed, and τ is the neutral light filter transmission in the recording of the radiation under study, and the subscript "r" refers to recording of the reference source. Q is a correction factor relating to reflections at the interface between the material under study and air, which is to be estimated for the experimental conditions.

For the selected region $d\lambda$, integral $A = \int_{\lambda_1}^{\lambda_2} I_\lambda S_\lambda \tau_\lambda d\lambda$ is determined by numerical integration for temperatures from 10^3 K to 10^5 K. The computations include the neutral filters' effective transmission as a function of temperature for the spectral region under study. The problem of radiance temperature estimation with this method is solved by the following. Fluxes A are determined for a number of temperatures from Eq. (7.29) in the working spectral region and a calibration chart $\log A = f(1/T)$ is constructed. A typical calibration chart covering temperatures up to 8000 K is presented in Fig. 7.25. The known reference light source temperature is entered on the calibration chart (point A). The logarithm ratio between the radiation fluxes A_{exp} under study and reference A_{ref} of light sources are plotted from point A parallel to the x-axis. Temperature at point b corresponds to the luminance temperature for the source studied. Then the obtained results are assigned to the spectral region characterized with effective wavelength λ_{ef}.

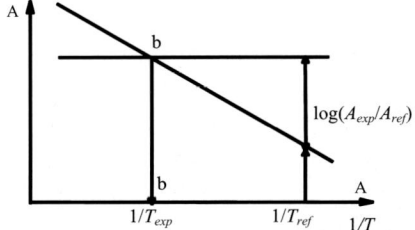

Fig. 7.25. Radiation flux vs. temperature

Based on the assumption that the recorded radiation is of thermal nature and described with the Planck function, the effective wavelength is determined from the expression:

$$\lambda_{ef} = \int_\infty^\infty I_0\left(\lambda, T\right) S(\lambda) \lambda d\lambda \left/ \int_\infty^\infty I_0\left(\lambda, T\right) S(\lambda) d\lambda \right. .$$ (7.32)

Effective wavelengths λ_{ef} are meaningful only for the radiating body temperature T, for which they are defined. When temperature changes, the values of λ_{ef} also change.

7.4.2 Photoelectric Method

Radiance temperature estimation with the photoelectric method [6, 10, 12–19] consists of measuring the light flux, from the light source under study, impinging on a photoelectric cathode of a photomultiplier for a given spectral region. The light flux is transformed by the photomultiplier into an electric signal, which is recorded by an oscilloscope. The photoelectric cathode is a selective receiver of radiant energy, therefore the intensities of the studied and reference light sources, as in the photographic method, must be compared to their radiation fluxes by the formula

$$A = \int_{\lambda_1}^{\lambda_2} I_\lambda \tau_\lambda V_\lambda d\lambda ,$$ (7.33)

where τ_λ is the interference filter transmission; V_λ is the photoelectric multiplier cathode sensitivity; I_λ is the radiance estimated by Planck function Eq. (7.7).

It is necessary that the measuring components operate in the linear region of their characteristics; i.e. the photomultiplier response is in the linear region and the oscilloscope operates in the linear region for the amplitude characteristics. In this case the signal amplitude on the oscilloscope display can be represented as

$$h = \alpha KSAL^{-2} \tag{7.34}$$

where α is the coefficient of proportionality between the signal amplitude and light flux; K is the neutral filter transmission; S is the luminous object area; A is radiation flux Eq. (7.29); L is distance from the luminous body to the facility.

At constant coefficient α the radiant flux of the light source under study is determined from relation

$$A = (h_{exp}K_{ref})/(h_{ref}K_{exp}) \, A_{ref}, \tag{7.35}$$

where h_{ref} is the signal amplitude on the oscilloscope from the reference light source; h_{exp} is the signal amplitude on the oscilloscope from the light source under study; K_{exp} is the neutral filter transmission for radiation from the light source under study; K_{ref} is the neutral filter transmission for radiation from the reference light source.

A constant coefficient α implies that the reference and studied light source radiances are recorded with identical sensitivities and amplification factors for the photomultiplier and the oscilloscope. For the measurements the reference and studied light sources are positioned equidistant from the measuring device and their radiation passes through the same diaphragm thereby eliminating inaccuracies relating to the dependence of the radiation flux on distance and luminous object area. The radiance temperatures are estimated in the same manner as the photographic method.

7.5 Pyrometers for Radiation Recording

7.5.1 Single-channel Pyrometer

A typical experimental scheme for single-channel pyrometry using the photographic method is presented in Fig. 7.26. Radiation from central specimen part (1) is released through diaphragm (2) and reflector (3) to input objective (4) of SFR. The image of the diaphragm is formed by the objective in the plane of slit (5), which removes a narrow strip. The shock wave front radiation removed by the slit is projected by objective (6) through rotating mirror (7) onto the photographic film (8) and exposes it. If necessary to separate the spectral range and attenuate the radiation, colored (10) and neutral (11) light filters of a known transmission factor and wavelength relation are used.

When the sensitivity is $T_{min} \sim 3000$ K, the method offers a number of benefits. The streak-camera image allows the estimation of the radiance distribution over the specimen's surface. This is especially important, as high-temperature luminosity regions relating to shock wave reflection from structural elements can be detected and eliminated from the processing. Significant distortions can exist at high impact velocities upon the target, i.e. at high temperatures. Besides, the streak-camera image strictly fixes the state of the specimen at the moment of shock compression.

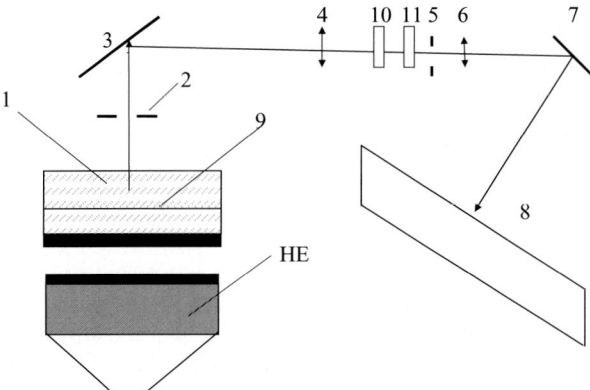

Fig. 7.26. The photographic method for measuring luminance temperatures: 1 – material under study; 2 – diaphragm; 3 – external reflector; 4 – input objective; 5 – slit, 6 – objective; 7 – rotating mirror; 8 – photographic film; 9 – shock front; 10 – color filter; 11 – neutral density filter

With currently available equipment the photographic method for temperature measurement is no more labor-intensive than the photoelectric method.

7.5.2 Photoelectric Pyrometer

The basic pyrometer design used in studying optical properties of shock-compressed transparent dielectrics with the photoelectric method is discussed in [12]. The radiance is recorded in the blue ($\lambda = 470$ nm) and red ($\lambda = 625$ nm) spectral regions (Fig. 7.27).

First objective (1) of focal length $F_1 = 750$ mm projects the light source image onto diaphragm (2) which limits extraneous radiation. The diaphragm

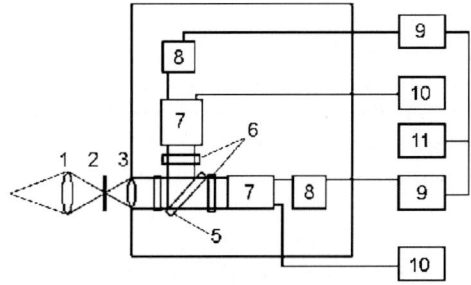

Fig. 7.27. Flow chart of the photoelectric pyrometer: 1, 3 – objectives; 2 – diaphragm; 4 – neutral filters; 5 – beam splitter; 6 – spectral filters; 7 – photoelectric multiplier; 8 – cathode follower; 9 – oscilloscope; 10 – power supply; 11 – signal generator

is placed at the focus of second objective (3) of $F_2 = 210\,\text{mm}$, which ensures a parallel light beam following it.

Photomultipliers (7) are positioned such that, there is constant image size on the first objective input lens on the photoelectric cathodes and a uniform exposure to the light. This prevents the photomultiplier's band characteristics impacting the results. A colorless optical glass plate (5) positioned at an angle of 60° to the light flux divides the light flux about equally between two photomultipliers. The photomultipliers considerably reduce the measured temperature threshold as compared with the photographic method limits. The requirements of the explosive experiment with the measurement being far (∼10 m) from the specimen and the use of light filters with a narrow bandwidth (100 Å) results in the lower limit of recorded temperatures being 1500–2000 K. The signal arriving from the photoelectric multiplier is recorded with oscilloscope (9) ensuring a rise time of $5 \cdot 10^{-8}$ s.

To separate narrow spectral ranges in the blue and red visible spectral regions, interference filters were used that had a maximum transmission corresponding to 4780 and 6250 Å, respectively. To improve wavelength characteristics, the interference light filters were combined with glass ones. If necessary, the intensity of the light flux impinging on the photoelectric multiplier was adjusted with neutral filters of a known spectral transmission.

For the reference light source, an IFK type lamp (impulse Xe lamp of 50 W power) was used [20]. The IFK-50 lamp radiance temperature measured on the same facility was 6900 K at $\lambda = 4780$ Å and 6200 K at $\lambda = 6250$ Å.

Conditions for the temperature measurement were set such that the operation of the photomultipliers was in the linear response region and that the oscilloscope was in the linear amplitude region. The method's accuracy was estimated by Eq. (7.34) reasoning that the IFK-50's lamp reference temperature was estimated to be within 5% and that of signal amplitude h, neutral filter transmission factor K, and luminous object area S was within 2%.

The above-described pyrometer has been widely used to study optical characteristics of transparent dielectrics [12, 13–15]. It has detected luminescence, melting, and, at high pressures ($P > 100$ GPa), nonequilibrium radiation of shock-compressed ionic crystals – NaCl, KCl, CsBr, LiF. The facility has also been used to estimate the absorptivity in NaCl, CsBr and other halogenides as well as in CCl_4 under shock compression conditions [21–23]. Modern pyrometers measure temperature with 4 to 6 spectral ranges at once [24–27].

7.5.3 Electro-optical Pyrometer

Recently, widespread acceptance has been gained by electro-optical pyrometers for radiation recording, where the optical path consists of light guides (Fig. 7.28).

The device is composed of four identical channels. Optical radiation from the specimen is transmitted through light guide 2 to the inlet of one of the channels. The outlet end of the light guide is located in the focal plane of an

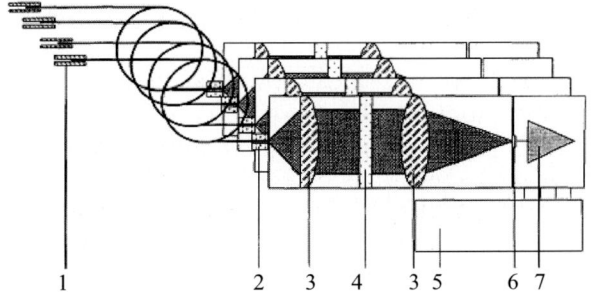

Fig. 7.28. Four-channel electro-optical pyrometer

optical system composed of two lenses 3. Optical components (interference, neutral or color filters) are positioned in the parallel light beam between the lenses. The optical spectrum radiation selected by the filters goes to optical receiver 6. The low photo diode current is amplified by the electronic circuit 7 and enters a recording device (oscilloscope, digitizer).

Light guides with a small spectral attenuation factor reduce optical losses, enhancing the pyrometer sensitivity (making the minimum measurable temperature lower).

To conduct temperature measurements with any pyrometer, it is necessary to have the pyrometer optical channel calibrated with a reference radiation source. Such reference light sources can be tungsten-ribbon filament lamps, pulsed xenon lamps of the IFK-50 type or any other radiation source with known characteristics.

7.5.4 Reference Radiation Sources

Temperature measurement in shock-wave experiments involves recording the shock front radiation and comparing it with the radiation from a reference source of known temperature. Tungsten lamps with ribbon or helical filaments are the reference sources most widely used.

Tungsten lamps with a filament in the form of ribbon (with the supply voltage 6–8 kV and power from 40 to 200 W) placed in a bulb made of quartz glass or sapphire are very convenient for spectral measurements. Also, lamps with a filament in the form of a helix (with the supply voltage 6–8 kV and power from 7.5 to 20 W) are widely used in optical devices. The filament of the lamps forms something like a cavity, inside which multiple reflections occur, and radiation going out of the inner-coil spaces proves to be more like blackbody radiation than radiation from an ordinary tungsten filament or ribbon. While ribbon lamps require high-current regulated power supplies rated at 6–30 A, the operating current of the helical lamps is usually 1–3 A. Standard B5-series regulated power supplies can be used for these lamps.

The radiance temperature of an incandescent body surface is measured with a pyrometer with a disappearing filament, for example, the visible

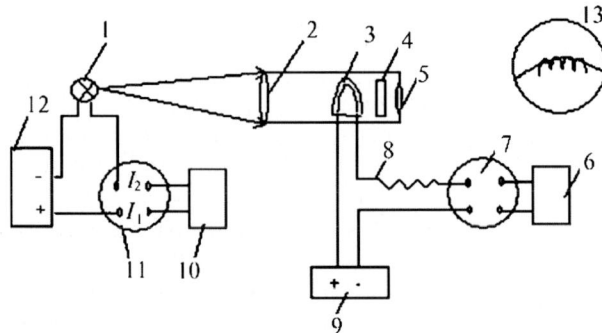

Fig. 7.29. Scheme of pyrometer tungsten lamp calibration: 1 – lamp to be calibrated; 2 – objective of pyrometer EOP-66; 3 – pyrometer lamp filament; 4 – filter; 5 – ocular lens; 6, 10 – voltmeters; 7 – standard resistor (accuracy class 0.01) $R_H = 0.1\ \Omega$; 8 – potentiometer; 9 – pyrometer power unit; 11 – standard resistor (accuracy class 0.01) $R_H = 0.001\ \Omega$; 12 – regulated power supply; 13 – filaments of the lamp to be calibrated and the reference pyrometer lamp

spectrum precision pyrometer EOP-66 (electro-optical converter). The pyrometer arrangement and the scheme for measuring radiance temperatures of the helical or ribbon tungsten lamps are presented in Fig. 7.29.

The pyrometer is composed of input objective (2), comparison lamp (3), light filter (4), and ocular lens (5). To prepare for the temperature measurement, the ocular lens is moved until the pyrometer lamp filament image becomes sharp. Then, without any change in the ocular lens position, the objective is moved to bring the filament image of lamp (1) coincident with that of the comparison lamp (3). After heating the lamp to be calibrated for ~15 minutes, the comparison lamp is turned on and the filament radiances of the lamp being calibrated and the pyrometer lamp are visually equalized by varying the current flowing through the comparison lamp. In doing so, the pyrometer lamp filament image disappears on the background of the calibrated lamp image, which implies their radiance temperatures coincide. The radiance is compared with monochromatic light at a wavelength $\lambda = 0.655 \pm 0.01\ \mu$m separated by the light filter which is a component of the pyrometer assembly. 10 to 15 current measurements within 10^{-5}A accuracy are taken for each lamp; the measurements are averaged. The calibrated lamp temperature is estimated by temperature dependence on current flowing through the pyrometer lamp filament supplied by the calibration certificate.

To extend the temperature measurement range, the pyrometer assembly provides for several neutral filters attenuating the studied item image radiance. This enables temperature measurement (for example, with pyrometer EOP-66) up to 10000°C without an increasing the comparison lamp filament temperature greater than 1400°C.

Using the tabulated data for the dependence of tungsten's true temperature for the EOP-66 pyrometer's operating wavelength and the experimental

luminance temperature, it is easy to calculate the true temperature for pyrometer lamps as

$$T_{lum} = \frac{c_2}{\lambda} \frac{1}{\frac{c_2}{\lambda T} - \ln \varepsilon_\lambda} \; .$$

Experimental studies show that luminosity of tungsten-filament lamps of
~2000–3000 K temperature is insufficient for radiation recording onto photographic film. So, in shock-wave experiments, experimental facility calibration
requires reference light sources of greater luminance temperatures. A suitable
light source may be pulsed xenon lamp IFK-50.

The lamp IFK-50 radiance temperature is estimated by comparing the
radiance of the lamp with some reference source, for example, a ribbon tungsten lamp pre-calibrated by radiance temperature using precision pyrometer
EOP-66 or by the State Metrology Service. Lamp IFK-50 can be calibrated
for radiance temperature using any of the pyrometers, whose schemes are
presented in Sect. 7.5. Typical oscillograms of radiation pulses for the IFK-50
lamp at wavelengths of 500 and 550 nm appear in Fig. 7.30.

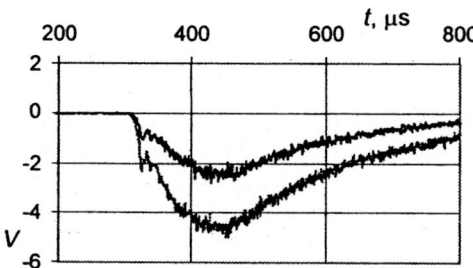

Fig. 7.30. Typical oscillograms of radiation pulses of lamp IFK-50. Duration
– 50 µs/div

7.6 Temperature of Shock-compressed Materials

7.6.1 Ionic Crystals

The characteristics of shock front radiation in NaCl, KCl, KBr, LiF and CsBr
ionic crystals and their pressure functions were studied for the first time in
[12–14, 21–23]. Temperature was measured with two methods: the electro-
optical pyrometer method for pressures up to 100 GPa (see Fig. 7.28) and
the photographic method for higher pressures (see Fig. 7.26). References [27,
28] measured ionic crystal temperatures for pressures up to 50 GPa using a
six-channel optical pyrometer.

The typical experimental scheme for temperature measurement of shock-compressed ionic crystals appears in Fig. 7.31 [12]. The shock front radiation was recorded for wavelengths of 478 nm and 625 nm. The ionic crystal specimens were single crystals 40 × 20 mm in size. There was no less than 95% light transmission by the crystals. The shock wave in the crystals was produced by impacting a metal plate accelerated by EP up to velocity W.

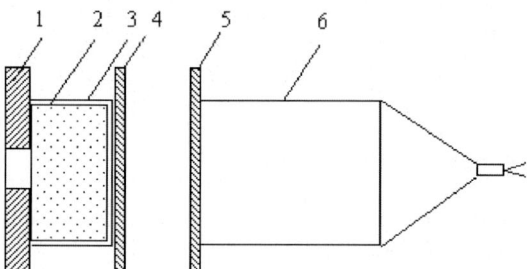

Fig. 7.31. Experimental scheme for temperature measurement

Various shock wave intensities were generated in crystal 2 by varying the power of HE 6, materials and thickness of accelerated plate 5 and shielding plate 4. A black dye layer 3 served to remove air bubbles from the gap between the base plate and crystal.

The shock front radiation was reduced by diaphragm (1) for luminosity separation of the central crystal region. A typical shock front luminosity oscillogram obtained from an ionic crystal temperature measurement is given in Fig. 7.32.

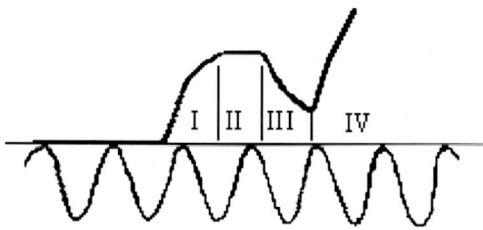

Fig. 7.32. Shock front radiation oscillogram (sinusoid $f = 2$ MHz)

The oscillogram shows four periods, which can be clearly distinguished in the recorded luminosity. Period I: radiation radiance grows due to the increasing thickness of the layer of material compressed by the shock wave. The period is mainly determined by the compressed material transparency. The duration of constant luminance period (II) is due to both the compressed material transparency and the time the rarefaction wave enters the crystal.

When the distance between the shock wave and the rarefaction wave is about equal to the nontransparent compressed material layer thickness Δ, the luminous intensity begins to decay (period III). Period IV is where the rarefaction wave reaches the crystal's free surface causing a shock wave in the air with a high temperature at its front.

The radiance temperature was estimated from the constant (or maximum) amplitude luminonsity segment. The constant radiance of segment II means that the material layer following the front is essentially opaque and the measured radiance temperature at this state coincides with the true equilibrium temperature of the compressed material. Indeed, their difference is solely due to light reflection from the wave front. The studies conducted in [12–14] showed that the front reflectivity was no higher than 2–3% even for most intensive shock waves up to 100 GPa in NaCl, KCl, KBr, and CsBr, and at temperatures up to 5500 K contributes an error less than 1% to the measured temperature. Therefore, no correction for reflection was introduced.

Temperatures in NaCl and KCl measured in [12] are summarized in Table 7.1. The table gives averaged (over four or five measurements) radiance temperatures. In all the cases, the temperatures measured in the blue spectral region proved greater than in the red region. This difference, according to [12], characterizes the measurement accuracy and has nothing to do with the non-blackbody character of the radiation from the studied ionic crystals. Experiments [27, 28] used a six-channel optical pyrometer, which allowed emissivity of shock-compressed ionic crystals NaCl, KCl and LiF to be estimated too. The results of these papers are presented in Table 7.2.

Table 7.1. Radiance temperature in shock-compressed ionic crystals NaCl and KCl

D, km/s	P, GPa	ρ/ρ_0	T_{rad}, K	
			$\lambda = 478$ nm	$\lambda = 625$ nm
NaCl ($\rho_0 = 2.165$ g/cm^3)				
7.40	46.5	1.65	2400	2250
7.59	50.0	1.67	2950	2750
7.85	54.7	1.70	3450	3200
8.13	61.5	1.75	3450	3250
8.52	70.5	1.81	3850	3600
8.91	79.0	1.85	4800	4400
KCl ($\rho_0 = 1.194$ g/cm^3)				
5.64	23.7	1.60	2350	2150
5.75	25.0	1.62	2500	2300
6.10	29.0	1.65	3500	3200
6.53	35.3	1.71	3900	3750
7.00	42.7	1.78	3900	3700
7.36	48.5	1.81	4250	3900
7.50	50.8	1.83	4650	4250
7.77	56.0	1.86	5600	5300

Table 7.2. Emissivity of shock-compressed ionic crystals NaCl, KCl, and LiF

Crystal	P, GPa	T, K	E	References
NaCl	30.9±0.3	3010	0.017	[26]
		1930	1.0	
NaCl	35.3±0.6	3190	0.007	
		1850	1.0	
NaCl	21.7±0.5	2216	0.319	[27]
	25.0±0.5	3429	0.009	
	28.0±0.7	3409	0.082	
	30.0±0.9	4313	0.029	
	31.4±0.8	2520	0.213	
KCl	14.6±0.3	3641	0.002	
	19.9±0.5	3128	0.07	
	30.0±0.6	3928	0.073	
LiF	42.0±1.0	2715	0.003	
	47.1±1.2	3152	0.005	

Phase diagrams for NaCl and KCl phase diagrams are presented in Figs. 7.33 and 7.34, which also plot the calculated melting curves and Hugoniots for the crystals [13]. The phase diagrams for KBr and CsBr are of a similar form [10].

As seen from the figures, on shock compression a rise in radiance temperature is initially observed in the crystals up to 30–40 GPa pressures, then the temperature remains essentially constant, while at pressures greater than 50 GPa a temperature rise is observed again. Reference [12] attributes the constant temperature segment to the ionic crystal melting behind the shock

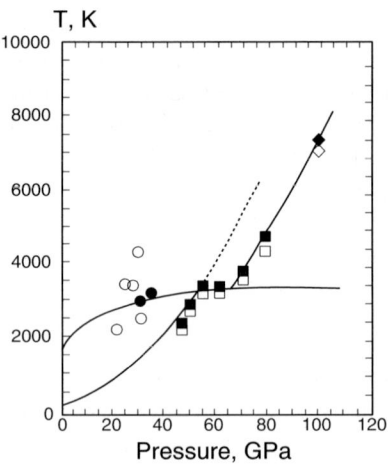

Fig. 7.33. Phase diagram of NaCl. Experiment [12]: ■ $- \lambda = 478$ nm; □ $- \lambda = 625$ nm; experiment [14]: ◆ $- \lambda = 440$ nm; ◇ $- \lambda = 650$ nm; • $-$ from [26]; ○ $-$ from [27]

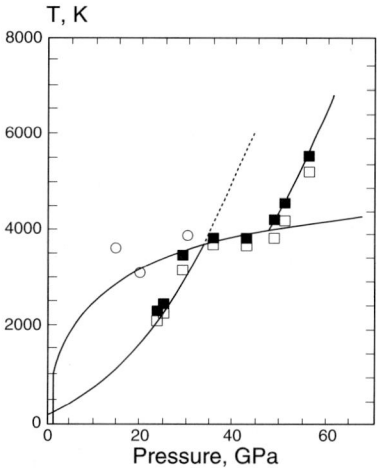

Fig. 7.34. Phase diagram of KCl. Experiment [12]: ■ – $\lambda = 478$ nm; □ – $\lambda = 625$ nm; ○ – from [27]

wave front. As compared to melting at atmospheric pressure, in the region where two phases exist, shock compression leads to a slight rise in radiance temperature, as for the normal melting curves $dT/dP > 0$. For many materials, melting in shock compression does not essentially affect the kinematic parameter variation [12], so temperature measurements prove to be an accurate tool for studying properties of materials under shock compression.

A similar method was used to measure radiance temperatures in the shock compression of KBr, CsBr and LiF crystals [10, 14]. Reference [27] measures 2715 K and 3152 K at 42 GPa and 47.1 GPa pressures in LiF. Analysis of the results presented in Figs. 7.33 and 7.34 suggests that up to ~25 GPa in KCl and ~45 GPa in NaCl the recorded light fluxes are higher than for the temperatures calculated assuming that the radiation is an equilibrium radiation, which obeys the Planck and Kirchhoff laws. The most significant difference has been recorded for LiF at pressures up to 100 GPa. The radiance temperature versus shock compression in LiF is presented in Fig. 7.35.

In this case the measured temperature is a conventional characteristic of the recorded fluxes and the radiation itself is of a definitely non-thermal nature. Thus, on LiF compression to 34 GPa the shock compression energy is 1.5 kJ/g. About half of the energy (~0.65 kJ/g) is consumed to overcome the elastic repulsive forces. In order to heat LiF to 2080 K at 34 GPa, which assumes that the temperature corresponds to thermal radiation, requires energy of 4 kJ/g, i.e. about 5 times greater than the shock compression energy.

Further study of radiation in shock-compressed ionic crystals in [14] shows that at 50–500 GPa the measured radiance temperatures fall well below the calculated values. The greater the pressure in the shock-compressed crystal, the larger the difference is. For example, Fig. 7.36 plots radiance

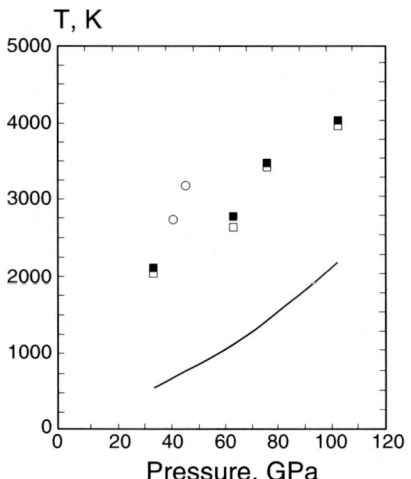

Fig. 7.35. Radiance temperature of shock-compressed LiF. Experiment [10]: ■ − λ = 478 nm, □ − λ = 625 nm; ○ − from [27]; — − calculation [10]

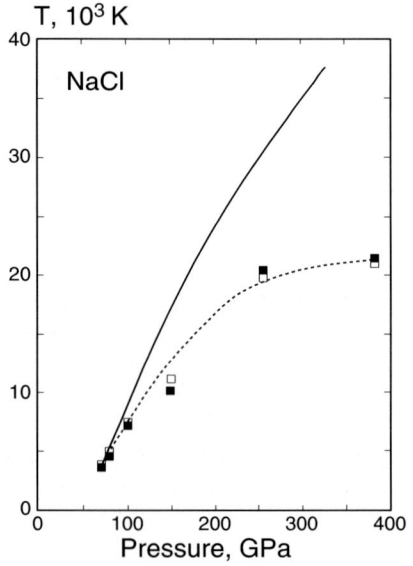

Fig. 7.36. Luminance temperature of shock-compressed NaCl. Experiment [14]: ■ − λ =   = 478 nm, □ − λ = 625 nm. ——— − calculation [14], - - - - approximation

temperature in NaCl vs. pressure. The figure shows that at a pressure, for example, of 200 GPa, the measured temperature differs from the expected value by ∼ 10000 K. The largest difference observed is for CsBr [10], in which a drop in temperature at pressures higher than 150 GPa is seen.

The work of [14] demonstrates that neither the heating layer nor shock front reflectivity can cause a decrease in radiance of the experimentally recorded radiation. A semi-quantitative theory explaining this discrepancy is presented in [28]. Reference [28] shows that behind the shock wave density increase there is a layer with nonequilibrium electron temperature, whose radiation is responsible for the observed shock front temperature. Radiation from deeper layers, where the electron temperature coincides with the equilibrium temperature, is completely shielded by the layer.

7.6.2 Liquefied Inert Gases: Argon, Krypton, Xenon

Temperature versus pressure for shock-compressed liquid noble gases was measured for liquid argon in [15, 29], liquid xenon in [18, 30] and liquid krypton in [31]. A typical radiance temperature in shock-compressed liquid argon versus pressure [29] appears in Fig. 7.37. As seen from Fig. 7.37, the equilibrium temperatures in argon calculated from the EOS do not describe the experimental data, even taking into account the electron thermal excitation.

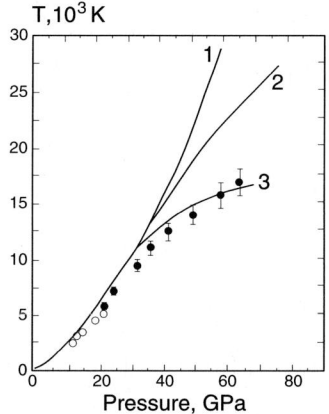

Fig. 7.37. Radiance temperature versus pressure in liquid argon. Calculation from [29]: 1 – equilibrium temperature without account for electron excitation; 2 – equilibrium temperature with account for electron excitation; 3 – effective temperature in red region. Experiment: • is the data from [29], ○ is that from [15]

Satisfactory agreement of the computation (curve 3) with the experiment was obtained by considering the kinetic theory of electron heating by lattice photons. Because of the finite time for the electron transfer from the bound state to the intermediate state and for electron heating, the equilibrium temperature between the electrons and the lattice is established later than a layer of optical density ~1 forms at the shock front. As a result, the recorded radiation from the layer corresponds to a lower temperature than the equilibrium temperature. The fact that the shock compression temperature in liquid xenon measured with a multi-channel pyrometer proved lower than the calculated equilibrium temperature even while taking into account emissivity [18], also supports this interpretation.

7.6.3 Liquid Nitrogen

Liquid nitrogen shock compression temperature as a function of pressure was studied in [15–17]. The results are presented in Fig. 7.38. References [16, 17] attribute the unusual decrease in rate of temperature increase around the 30 GPa pressure range to the onset of molecular nitrogen dissociation, where shock compression energy is absorbed which leads to the observed behavior.

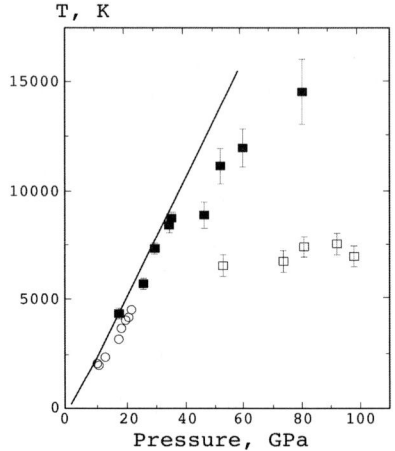

Fig. 7.38. Temperature of shock-compressed nitrogen versus pressure. Experiment: ○ – data from [15]; ■ – single compression data from [17]; □ – double compression data from [17]; — – computation [32]

Reference [17] mentions a most interesting feature of the liquid nitrogen radiation on the second-shock Hugoniot (Fig. 7.38). When the shock wave is reflected from sapphire (Al_2O_3) at pressures as low as ∼50 GPa, the light flux from the reflected shock front proved lower than the light flux from the first shock wave at the same pressure. In [17] this feature is referred to as shock-cooling of liquid nitrogen, with the effect associated with molecular nitrogen dissociation.

Figure 7.38 also plots the calculated $T - P$ function from [32] which agrees satisfactorily with the experimental data up to ∼35 GPa pressures.

7.6.4 Methane Halogen Derivatives

The temperature dependence for shock-compressed carbon tetrachloride was obtained for the first time in [11]. Further experiments were reported in [33, 34]. The compiled experimental results are presented in Fig. 7.39.

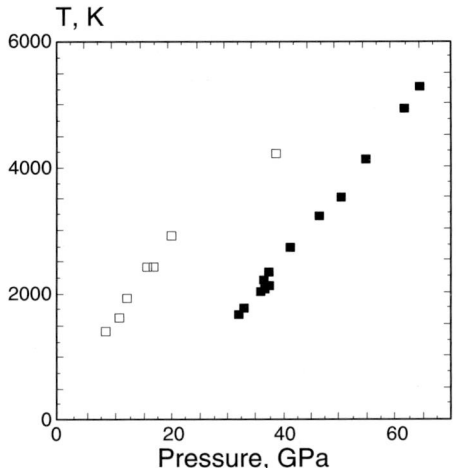

Fig. 7.39. Shock compression temperatures in carbon tetrachloride and bromoform. $\square$ – CCl$_4$ [33]; $\blacksquare$ – CHB$_3$ ($P+$ 20 GPa) [33]

As for ionic crystals, a characteristic segment with a relatively slow temperature rise in the 20–30 GPa pressure range was recorded in carbon tetrachloride [10].

At higher pressures, the slope of the $T-P$ curve changed again. The change is evidence of energy absorption related transformations at these ranges. According to [10], the weak temperature rise at 20–30 GPa is due to melting of CCl$_4$, which is in the solid phase between 0.2–20 GPa. The figure also plots radiance temperature versus pressure in bromoform.

The interest in the methane halogen derivatives relates to their wide application as indicator liquids in studying shock and detonation waves [34]. In this case the shock front radiation plays the role of a sensor having high amplitude and time characteristics. The strong dependence for radiation radiance based on the shock wave pressures allows slight changes occurring at the wavefront to be recorded. Recently, indicator liquids have found wide utility in studying sound speeds of various materials in the "overtake" release method [34, 35]. Which indicator liquid is chosen is problem-specific. When studying sound speeds in metals, denser liquids are preferred. (For example, bromoform's initial density is $\sim$3 g/cm^3.)

Here we give a linear approximation for the $T - P$ dependence based on experimental data for carbon tetrachloride over 8.5–20 GPa [34]: $T = (343 \pm 112) + (121 \pm 7) \, P$, where P is in GPa and T is in K.

7.7 Study of Shock-compressed Material Properties with Optical Methods

7.7.1 Luminous Reflectivity

A number of shock-wave measurements require optical measurements on the specimen side, which has not yet been traversed by the shock wave. If thermal self-radiation is excluded, information about the shock-compressed material's optical properties can be acquired by recording reflected light. To study optical properties under explosive experimental conditions, S.B. Kormer and associates developed a method of light reflection from moving optical boundaries [36]. The boundaries can be:

- shock front in a transparent material;
- interface of two compressed materials with various refractivity;
- moving free surface of material the shock wave has passed through.

In the idealized representation, the shock front can be considered as an infinitely thin discontinuity surface between resting and moving materials. Actually, the shock front is of a finite width because of viscosity and other relaxation processes. For the shock front to reflect light, its thickness δ must be much less than incident light wavelength λ. This allows resolution of front smearing which is on the order of hundreds of angstroms, which is inaccessible to any other methods.

Thanks to the features of the method, it was possible to determine for the first time that the shock front for condensed material is highly reflective [37]. Next, it turned out that the shock front smoothness is stable, as artificial perturbations lead only to a short-time distortions and rapid smoothing of the shock front [38]. The method of light reflection detected the water-ice phase transformation occurring as a result of a two-step pressure application with a first wave amplitude of 20–40 kbar [39]. Features of the phase transformation are loss of transparency and a diffuse scattering following the shock front. The detected phase transition has a pronounced relaxation nature. The method of light reflection also allowed studying elastic-plastic deformation of shock-compressed glass and front structure in it depending on the applied pressure [40].

The detection of the detonation front smoothness in liquid HE, chemical decomposition following the shock front in CCl_4, and shock front smearing for the shock wave propagating through material that has been already compressed complement the range of problems which is now possible to investigate as a result of the reflection method development.

Experimental Method

The experimental scheme for recording light reflection from a shock front appears in Fig. 7.40.

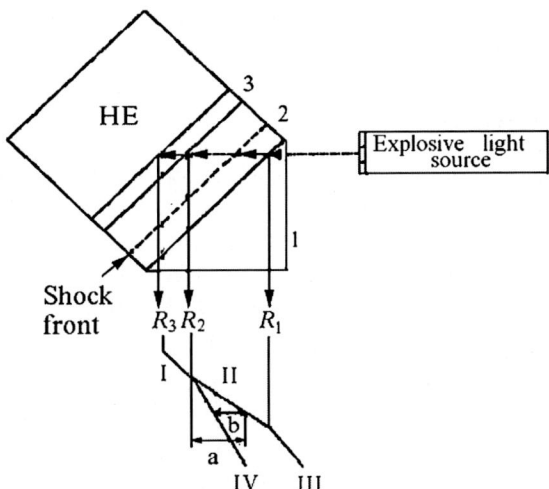

Fig. 7.40. Experimental scheme for measuring light reflectivity from a shock front and transparent material refractive index

Material (2), which is to be studied, is placed between plate (3) (in contact with HE) and glass prism (1). The shock front radiation from the external explosive accent light source penetrates the material under study through prism (1). As a rule, shock front luminosity in gaseous argon ($P_0 = 1$ atm) is used as the accent lighting source. The experimental timing is such that the wave in argon begins to emit light somewhat earlier than the shock wave begins propagating through the specimen under study. In so doing, the streak-camera image will initially record straight lines implying light reflection from unmoving boundaries (1–2), (2–3) and the internal surface of plate (3), providing the plate is optically transparent and refractivities of materials (1), (2), and (3) are different. From shock wave arrival time at the plate and then at the specimen examined the light will be reflected from the plane shock front and, since it moves, the accent lighting source slit image reflected from the front will move across the film. When the shock wave arrives at the external specimen surface, the reflection from the front will merge with that from boundary (1–2).

If, upon their shock wave compression, materials (2) and (3) remain transparent, then the streak-camera image will also record the light reflection from moving optical boundaries (1–2) and (2–3). Figure 7.40 plots reflected beams, which can be recorded for this setup.

To evaluate the light reflectivity from the shock front, it is necessary either to know the ratio of reflected and incident light flux intensities or simultaneously record the light reflected from the shock front and the same light beam reflected from an unmoving optical boundary (point of reference), whose reflectivity is known.

We now turn our attention to the scheme presented in Fig. 7.40. In this setup, light reflection from any of the unmoving boundaries R_1, R_2, and R_3 can be considered as a reference reflection. If a photomicrograph corresponding to some time t is taken, then it is possible to measure, for example, the light flux ratio I_I/I_{II} and, hence, reflectivity from the shock front. Then, if light reflectivity R_1 from reference boundary (1–2) is known, the light reflectivity from shock front R_{sw} is calculated as

$$R_{sw} = \frac{I_{II}}{I_I} \frac{R_I}{(1 - R_I)^2}. \tag{7.36}$$

Figure 7.41 presents a typical streak-camera image obtained when measuring the shock front reflectivity in a transparent material. The streak-camera image corresponds to the above-described experimental setup.

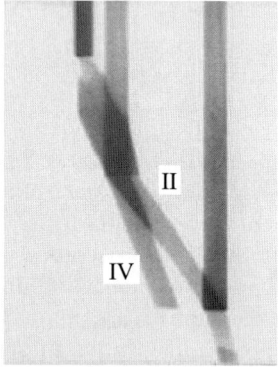

Fig. 7.41. Streak-camera image of the experiment

This method has been used to study the shock front reflectivity in many materials, both liquid and solid, which are initially optically transparent [10, 41, 42].

Under explosive experimental conditions, according to [10], the method's resolution is $R \geq 10^{-3}$. The reflectivity error of separate measurements is $\Delta R/R \approx (5\text{--}10)\%$.

At high pressures and front temperatures of ~ 10000 K it is essentially impossible to measure reflection using external accent lighting, since its radiance becomes comparable with the shock front radiance. In such cases front self-luminosity can be used to measure reflection. By Zeldovich's proposal,

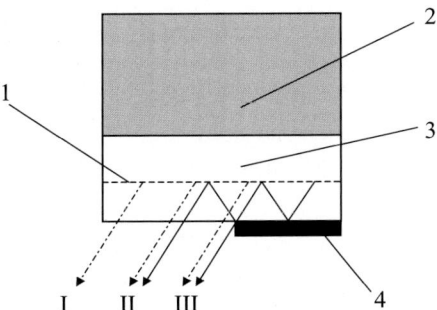

Fig. 7.42. Scheme of light reflectivity measurement using shock front self-radiation: 1 – shock front; 2 – HE; 3 – material under study; 4 – reflector

this method was used in [14]. The experimental setup measuring reflectivity of the shock front using its self-luminosity appears in Fig. 7.42.

For this experimental setup the specimen is in contact with HE on one side and is shielded with a reflector having reflectivity ξ on the other side. Light reflectivity η from the shock front can be determined by comparing the front radiance at separate locations.

As seen from the figure, the shock front self-luminosity is recorded at location I – II and the front self-luminosity and light reflected from the reflector at location II – III. Let shock front radiance be I_0, then the light which has been reflected from the reflector of reflectivity ξ and returned to the front will have radiance $I_0\xi(1-r)^2$, where r is light reflectivity from the specimen's free boundary. Then the total radiance of the front area reflecting light running from the reflector with reflectivity η will be:

$$I = I_0 \left[1 + \xi\eta(1-r)^2\right] . \qquad (7.37)$$

Having determined the ratio of shock front self-luminosity radiance to luminosity of the area reflecting light from the reflector, Eq. (7.37) can be used to determine light reflectivity η from the shock front.

From this scheme, [31] obtains $\eta = 13\%$ from the shock front in liquid krypton at $P = 76.1$ GPa.

7.7.2 Refractive Index of Shock-Compressed Optically Transparent Materials

Analysis of the streak-camera image presented in Fig. 7.41 allows the following to be inferred:

(a) The material parameters change in a jump, with the transitional region width, i.e. shock front thickness, being less than the incident light wavelength. This is confirmed by reflection of the recorded light from the shock front.

(b) The shock front is smooth and the light reflection is specular, which is confirmed by absence of any diffuse background in the streak-camera image and absence of any slit image smearing reflected from the shock front.

This means that the reflectivity unambiguously depends on a change in the refractive index at the shock front, and this fact can be used to measure the refractive index.

The streak-camera image of a typical experiment presented in Fig. 7.41 renders a time history of the process. The distance between beams II and IV (see Fig. 7.40) corresponding to the reflections from the shock front and from the boundary (2–3) at some time t depends on the thickness of the material layer compressed in the shock wave and on the angle φ at which the light passes through the layer, i.e. on the refractive index of the compressed specimen. Hence, having measured distance $(a–b)$ between the beams in the streak-camera image, if the compression ratio in the shock wave is known, the compressed material refractive index can be calculated.

Write the relation for the calculation of the refractive index for the shock-compressed material [10]:

$$\frac{n_3^2 - n_1^2 \sin^2 \varphi}{n_2^2 - n_1^2 \sin^2 \varphi} = \left(\frac{\mu}{\eta}\right)^2, \tag{7.38}$$

where $\mu = a/b$; φ is the angle of light incidence onto the prism base; n_1 is the refractive index of the prism material; n_2 is the refractive index of the material under study before the shock front; n_3 is the refractive index of the shock-compressed material; and η is compression.

The measured refractive index along the beam path is unaffected by the transitional layer at the shock front, whose width is much less than the compressed material layer thickness.

Thus, for materials remaining transparent during compression the refractive index in the compressed state can be estimated immediately by recording the paths of the beams reflected from the compressed material boundaries. This experimental scheme also allows determination of whether the material remains transparent after the shock compression.

The developed method measured the refractive index of shock-compressed materials with accuracy $\Delta n/n \cong \pm 0.5\%$. The following materials were studied: water, liquid high explosive, benzene, toluene, ethyl alcohol, acetone, carbon tetrachloride, ionic crystals LiF, NaCl, CsBr, CsJ, KCl, KBr, and optical glass TF-5 [36–42].

Liquid Dielectrics

The refractive index of liquid dielectrics versus compression ratio is plotted in Fig. 7.43 [42]. From the figure it follows that all experimental data can be described by a unique function of the form

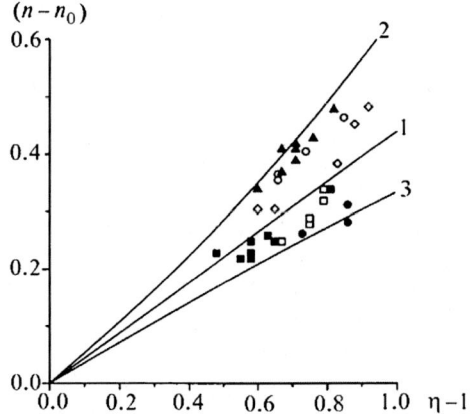

Fig. 7.43. Refractive index of liquid dielectrics versus compression ratio. Experiment [41]: ■ – glycerin, □ – ethyl alcohol, • – acetone, ○ – benzene, ◇ – carbon tetrachloride, ▲ – toluene. Calculation for average $n_0 = 1.44$: 1 – from Eq. (7.40), 2 – from Eq. (7.41), 3 – from Eq. (7.42)

$$n = n_0 + (dn/d\eta)(\eta - 1) \tag{7.39}$$

where $\eta = \rho/\rho_0$ is the compression, n_0 is the refractive index under standard conditions, and the derivative $(dn/d\eta)$ is determined for each liquid from static measurements at low pressures. For all liquids studied, $(dn/d\eta) = (n_0 - 1)$ to a good enough approximation. As readily seen, in this case the Gladstone-Dale empirical dependence

$$(n - 1)/\rho = \text{const} \tag{7.40}$$

provides a better description of the experimental refractive index as a function of density than Lorentz-Lorentz dependence

$$\left(n^2 - 1\right)/\left(n^2 + 2\right)\rho = \text{const} \tag{7.41}$$

or Drude dependence

$$n^2 - 1/\rho = \text{const} \tag{7.42}$$

This discrepancy may be due to decreasing material specific polarizability in compression. Also note that the above dependences $n(\eta)$ do not include the refractive index's temperature dependence.

Ionic Crystals

The above-discussed methods were used to study optical properties of shock-compressed alkali-halide compounds. Figure 7.44 presents the experimental curves for the refractive index as a function of compression ratio for single crystals CsBr (1) and LiF (2) which do not undergo a polymorphic phase transformations within the studied pressure range [41]. The vertical lines separate the region of CsBr single crystal melting in shock wave that was determined from calculations [41].

Of a similar form are dependences $n(\eta)$ for other crystals: NaCl, KCl, KBr, CsJ. It turns out that the density dependence of refractive index can be written in the form of Eq. (7.39) with constant slope $(dn/d\eta) = \text{const}$ for each crystal. Unlike liquids, for ionic crystals $(dn/d\eta) < (n_0 - 1)$.

The experimental studies [41] covered the density ranges not only of solid, but also liquid phase, when ionic crystal melting occurred due to heating at the shock front. The measurements indicated that both the refractive index and its derivative increased abruptly during melting in compressed state (see Fig. 7.44). Nevertheless, unique dependences of the form $n = n_0 + (dn/d\eta)\,(\eta - 1)$, where $(dn/d\eta)_1 > (dn/d\eta)_s$, which have a common point at $\sigma \approx 1$, can be adopted for either phase over a wide density range. This accounts for the seeming contradiction that melting leads to a decreased refractive index at atmospheric pressure and to increasing one under compression.

The refractive index change $(dn/d\eta)$ with increasing density is presented in Fig. 7.45. As seen from the figure, the values of $(dn/d\eta)$ for ionic crystals proved significantly less than what might be expected from the Gladstone-Dale, Lorentz-Lorentz or Drude relations.

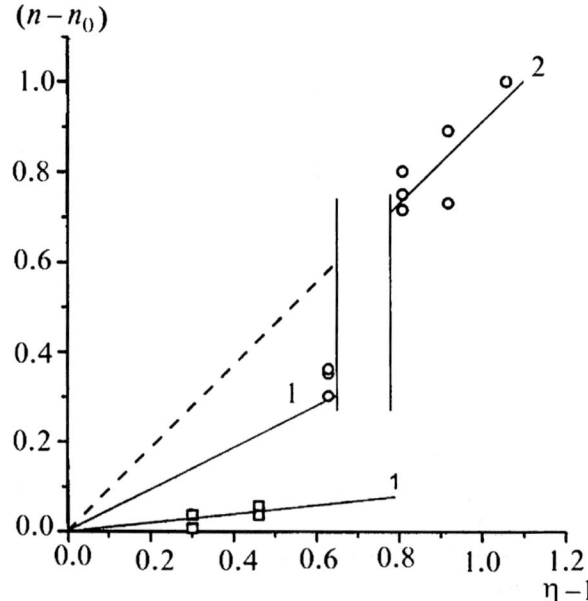

Fig. 7.44. Refractive index of CsBr (○) and LiF (□) versus compression ratio: 1 – solid phase, 2 – liquid phase. The verticallines separate the region of CsBr melting in shock wave

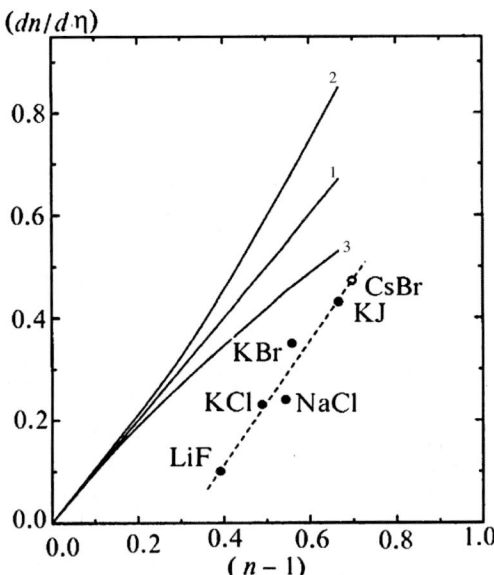

Fig. 7.45. Comparison of the experimental values of $(dn/d\eta)$ for shock-compressed ionic crystals to those calculated by Eqs. (7.40) – 1, (7.41) – 2, and (7.42) – 3:- - - - experimental data interpolation

The experimental fact that under shock compression the refractive index of ionic crystals increase less than the theoretical representations suggest, is attributed to decreasing electronic polarizability of atoms in [41].

7.7.3 Optical Absorption Constant in Shock-compressed Material

Studies of ionic crystal optical properties [10] show that, during shock compression up to 70 GPa, essentially all crystals, except LiF, lose their transparency. The absorptivities of KBr, KCl, NaCl and CsBr crystals were evaluated in [13, 21–23].

The authors of [12] have experimentally recorded the radiance increase with time during the shock wave propagation through a NaCl specimen, which is attributed to the increase in thickness of the shock-compressed material layer and its transparency.

The radiation intensity of plane layer in the normal direction is described by expression

$$I = I_0 \left[1 - \exp\left(-\alpha l \right) \right] = I_0 \left[1 - \exp\left(-\alpha \left(D - u \right) t \right) \right]. \qquad (7.43)$$

Here I_0 is radiation intensity of the optically dense layer; $l = (D - u)t$ is shock-compressed material layer thickness; t is time of shock wave's motion through the material. From Eq. (7.39) it is easy to obtain an expression evaluating the average optical absorption constant α:

$$\alpha = -\left[1/(D-u)\,t\right]\ln\left(1 - I/I_0\right) , \qquad (7.44)$$

where intensity variation I/I_0 with time determined from the oscillograms, whose characteristic form is shown in Fig. 7.32, D and u were taken from compressibility experiments [43–45]. The radiance temperature of the shock-compressed specimen was simultaneously measured by the amplitude of the signal (I_0) in each experiment.

For this type of experiment the primary attention is given to the initial segment of the radiation radiance increase. As noted in [21], the light flux initially increases slowly, as concentration N_e of electrons is low and $\alpha \sim N_e$. Therefore, the signal rise does not correspond to Eq. (7.39). Only with increasing N_e and a relevant increase in α does the light flux increase and does the behavior of this increase agree with Eq. (7.39) for a constant α. The first time this method was used in [9] for the measuring of the optical absorption constant of shock-compressed noble gases.

In the experiments [21–24] at wavelengths 4780 Å and 6250 Å the measured absorptivities for NaCl, KCl, KBr and CsBr crystals proved close, but higher by more than a factor of $\sim$100 than the absorptivity at standard conditions ($\sim$0.05 cm^{-1}). The averaged data from the measurements over two spectral ranges is presented in Fig. 7.46.

The shock front's radiation radiance increase over the initial segment prior to equilibrium between the electrons and the lattice depends on various factors, such as non-uniform shock wave ingress into the specimen, and initial

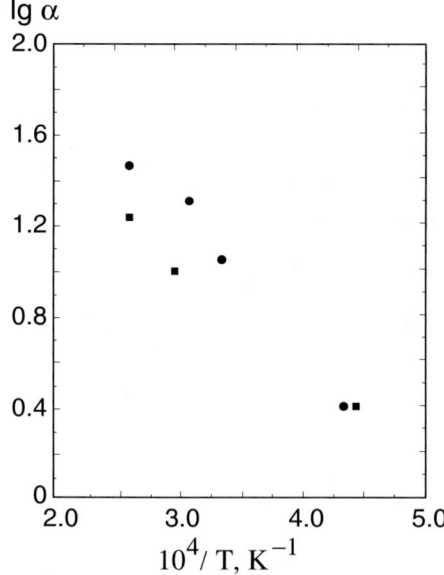

Fig. 7.46. Optical absorption constant versus temperature in shock-compressed ionic crystals: ■ – NaCl; • – KCl

density fluctuations of the HE from one experiment to another. Hence, up to 10 experiments are needed to evaluate α accurately [9].

References

1. Gurevich, M.M., *Photometry* (Theory, Methods, Devices), 2nd Improved and Supplemented Ed., Energoatomizdat Publ., Leningrad, 1983.
2. Rabek, J.F., *Experimental Methods in Photochemistry and Photophysics* (in 2 Volumes), Vol. 1, Mir Publ., Moscow, 1985 (in Russian), [see also Rabek, J.F., Photochemistry and Photophysics (in 6 Volumes), CRC Press, Boca Raton, FL, 1992 (in English)].
3. Kozelkin, V.V., and Usoltsev, I.F., *Fundamentals of Infrared Engineering*, Mashinostroyeniye Publ., Moscow, 1967.
4. Epshtein, M.I., *Optical Radiation Measurements in Electornics*, Energoatomizdat Publ., Moscow, 1990.
5. Tsykulin, M.A., and Popov, E.G., *Radiating Properties of Shock Waves in Gases*, Nauka Publ., Moscow, 1977.
6. Svet, D.Ya., *Optical Methods for Measuring True Temperatures*, Nauka Publ., Moscow, 1982.
7. *Catalog of Optical Glasses*, Mashinpriborintorg Publ. House, 1980.
8. *Catalog of Colored Glass*, Masinostroyeniye Publ., Moscow, 1967.
9. Model, I.Sh., "Measurement of High Temperatures in Strong Shock Waves in Gases," *Zhurnal Eksperimentalnoi i Teoreticheskoi Fiziki*, Vol. 32, 1957, pp. 714–726, [English trans., *Soviet Physics JETP*, Vol. 5, No. 4, 1957, pp. 589–601].
10. Kormer, S.B., "Optical Study of the Characteristics of Shock-Compressed Condensed Dielectrics," *Uspekhi Fizicheskikh Nauk*, Vol. 94, 1964, pp. 641–687, [English trans., *Soviet Physics Uspekhi*, Vol. 11, No. 2, 1968, pp. 229–254].
11. Shashlov, B.A., *Theory of Photographic Process*, Kniga Publ., Moscow, 1971.
12. Kormer, S.B., Sinitsyn, M.V., Kirillov, G.A., and Urlin, V.D., "Experimental Determination of Temperature in Shock-Compressed NaCl and KCl and of their Melting Curves at Pressures up to 700 kbar," *Zhurnal Eksperimentalnoi i Teoreticheskoi Fiziki*, Vol. 48, 1965, pp. 1033–1049, [English trans., *Soviet Physics JETP*, Vol. 21, No. 4, 1965, pp. 689–700].
13. Kirillov, G.A., Kormer, S.B., and Sinitsyn, M.V., "Equilibrium Radiation of Shock-Compressed Ionic Crystals," *Pisma v Zhurnal Eksperimentalnoi i Teoreticheskoi Fiziki*, Vol. 7, No. 10, 1968, pp. 368–370, [English trans., *JETP Letters*, Vol. 7, No. 10, 1968, pp. 290–291].
14. Kormer, S.B., Sinitsyn, M.V., and Kuryapin, A.I., "Nonequilibrium Radiation from Shock Compressed Ionic Crystals at Temperatures above 1 eV, I," *Zhurnal Eksperimentalnoi i Teoreticheskoi Fiziki*, Vol. 55, 1968, pp. 1626–1630, [English trans., *Soviet Physics JETP*, Vol. 28, No. 5, 1969, pp. 852–854].
15. Voskoboinikov, I.M., Gogulya, M.F., and Dolgoborodov, Yu.A., "Temperatures of Shock Compression of Liquid Nitrogen and Argon," *Doklady Akademii Nauk SSSR*, Vol. 246, 1979, No. 3, pp. 579–581, [English trans., *Soviet Physics Doklady*, Vol. 24, No. 5, 1979, pp. 375–376].

16. Radousky, H.B., Nellis, W.J., Ross, M., Hamilton, D.C., and Mitchell, A.C., "Molecular Dissociation and Shock-Induced Cooling in Fluid Nitrogen at High Densities and Temperatures," *Physical Review Letters*, Vol. 57, No. 19, 1986, pp. 2419–2422].

17. Nellis, W.J., Radousky, H.B., Hamilton, D.C., Mitchell, A.C., Holmes, N.C., Christianson, K.B., and van Thiel, M., "Equation-of-State, Shock-Temperature, and Electrical-Conductivity Data of Dense Fluid Nitrogen in the Region of the Dissociative Phase Transition," *Journal of Chemical Physics*, Vol. 94, No. 3, 1991, pp. 2244–2257.

18. Radousky, H.B., and Ross, M., "Shock Temperature Measurements in High Density Fluid Xenon," *Physics Letters A*, Vol. 129, No. 1, 1988, pp. 43–46.

19. Nellis, W.J., Ross, M., and Holmes, N.C., "Temperature Measurements of Shock-Compressed Liquid Hydrogen: Implications for the Interior of Jupiter, *Science*, Vol. 269, No. 5228, 1995, pp. 1249–1252.

20. Voitenko, A.E., Kuznetsov, F.O., and Model, I.Sh., "The Use of an IFK-50 Lamp as a High-Intensity, Pulsed, Standard Source," *Pribory i Tekhniki Eksperimenta*, No. 6, 1962, pp. 121–123, [English trans., *Instruments and Experimental Techniques*, No. 6, 1962, pp. 1184–1186].

21. Kormer, S.B., Sinitsyn, M.V., Kirillov, G.A., and Popova, L.T., "An Experimental Determination of the Light Absorption Coefficient in Shock-Compressed NaCl. The Absorption and Conductivity Mechanism," *Zhurnal Eksperimentalnoi i Teoreticheskoi Fiziki*, Vol. 49, 1965, pp. 135–147, [English trans., *Soviet Physics JETP*, Vol. 22, No. 1, 1966, pp. 97–105].

22. Grogorev, F.V., Kirillov, G.A., Kormer, S.B., et al., "Study of Electrical Conductivity and Absorption of Shock-Compressed Ionic Crystals in 150–800 kbar Pressure Range," *proc., 2nd All-Union Symposium on Combustion and Detonation*, Yerevan, 1965, pp. 259–262.

23. Grigorev, F.V., Kirillov, G.A., Kormer, S.B., et al., "Compressibility, Temperature, Electrical Conductivity, and Absorptivity of Shock-Compressed Carbon Tetrachloride," *proc., 2nd All-Union Symposium on Combustion and Detonation*, Yerevan, 1965, pp. 255–257.

24. Radousky, H.B., and Mitchell, A.C., "A Fast UV/Visible Pyrometer for Shock Temperature Measurements to 20,000 K," *Review of Scientific Instruments*, Vol. 60, No. 12, 1989, pp. 3707–3710.

25. Lyzenga, G.A., and Arens, T.J., "Multiwavelength Optical Pyrometer for Shock Compression Experiments," *Review of Scientific Instruments*, Vol. 50, No. 11, 1979, pp. 1421-1424, [Russian trans., *Probory dlya Nauchnykh Issledovanii*, No. 11, 1979, pp. 101–105].

26. Kondo, K., and Ahrens, T.J., "Heterogeneous Shock-Induced Thermal Radiation in Minerals." *Physics and Chemistry of Minerals*, Vol. 9, No. 3–4, 1983, pp. 173–181.

27. Schmitt, D., Svendsen, B., and Ahrens, T.J., "Shock Induced Radiation From Minerals," Shock Waves in Condensed Matter – 1985, Gupta, Y.M., ed., Plenum Press, NY, 1986.

28. Zeldovich, Ya.B., Kormer, S.B., and Urlin, V.D., "Nonequilibrium Radiation From Shock Compressed Ionic Crystals at T > 1 eV. II," *Zhurnal Eksperimentalnoi i Teoreticheskoi Fiziki*, Vol. 55, 1968, pp. 1631–1639, [English trans., *Soviet Physics JETP*, Vol. 28, No. 5, 1969, pp. 855–859].

29. Grigoryev, F.V., Kormer, S.B., Mikhailova, O.L., Mochalov, M.A., and Urlin, V.D., "Shock Compression and Brightness Temperature of a Shock Wave Front

in Argon. Electron Screening of Radiation," *Zhurnal Eksperimentalnoi i Teoreticheskoi Fiziki*, Vol. 88, No. 4, 1985, pp. 1271–1280, [English trans., *Soviet Physics JETP*, Vol. 61, No. 4, 1985, pp. 751–757].

30. Urlin, V.D., Mochalov, M.A., and Mikhailova, O.L., "Liquid Xenon Study Under Shock and Quasi-Isentropic Compression," *High Pressure Research*, Vol. 8, 1992, pp. 595–605.

31. Glukhodedov, V.D., Kirshanov, S.I., Lebedeva, T.S., and Mochalov, M.A., "Properties of Shock-Compressed Liquid Krypton at Pressures of up to 90 GPa," *Zhurnal Eksperimentalnoi i Teoreticheskoi Fiziki*, Vol. 116, No. 2, 1999, pp. 551–562, [English trans., *Journal of Experimental and Theoretical Physics*, Vol. 89, No. 2, 1999, pp. 292–298].

32. Zubarev, V.N., and Telegin, G.S., "The Impact Compressibility of Liquid Nitrogen and Solid Carbon Dioxide," *Doklady Akademii Nauk SSSR*, Vol. 142, No. 2, 1962, pp. 309–312, [English trans., *Soviet Physics – Doklady*, Vol. 7, No. 1, 1962, pp. 34–36].

33. Gogulya, M.F., Condensed Media Shock Compression Temperatures, MIFI Publishing House, Moscow, 1988.

34. Gogulya, M.F., and Dolgoborodov, A.Yu., "Indicator Method for Studying Shock and Detonation Waves," *Khimicheskaya Fizika*, No. 12, 1994, pp. 118–127.

35. Holmes, N.C., Nellis, W.J., and Ross, M., "Sound Velocities in Shocked Liquid Deuterium," Shock Compression of Condensed Matter – 1997, Schmidt, S.C., Dandecar, D.R., and Forbes, J.W., eds., AIP Press, Woodbury, NY, 1998.

36. Zeldovich, Ya.B., Kormer, S.B., Sinitsyn, M.V., and Yushko, K.B., "A Study of the Optical Properties of Transparent Materials Under High Pressure," *Doklady Akademii Nauk SSSR*, Vol. 138, No. 6, 1961, pp. 1333–1336, [English trans., *Soviet Physics – Doklady*, Vol. 6, No. 6, 1961, pp. 494–496].

37. Zeldovich, Ya.B., Kormer, S.B., Krishkevich, G.V., and Yushko, K.B., "Investigation of the Smoothness of the Detonation Front in a Liquid Explosive," *Doklady Akademii Nauk SSSR*, Vol. 158, No. 5, 1964, pp. 1051–1053, [English trans., *Soviet Physics – Doklady*, Vol. 9, No. 10, 1965, pp. 851–853].

38. Zeldovich, Ya.B., Kormer, S.B., Krishkevich, G.V., and Yushko, K.B., "Smoothness of the Detonation Front in a Liquid Explosive," *Doklady Akademii Nauk SSSR*, Vol. 171, No. 1, 1966, pp. 65–68, [English trans., *Soviet Physics – Doklady*, Vol. 11, No. 11, 1967, pp. 936–939].

39. Kormer, S.B., Yushko, K.B., and Krishkevich, G.V., "Phase Transformation of Water Into Ice VII by Shock Compression," *Zhurnal Eksperimentalnoi i Teoreticheskoi Fiziki*, Vol. 54, 1968, pp. 1640–1645, [English trans., *Soviet Physics JETP*, Vol. 27, No. 6, 1968, pp. 879–881].

40. Kormer, S.B., Krushkevich, G.V., and Yushko, K.B., "Investivation of Optical Properties of Shock Compressed Lead Glass (Elastic-Plastic Wave in Glass)," *Zhurnal Eksperimentalnoi i Teoreticheskoi Fiziki*, Vol. 52, 1967, pp. 1478–1484, [English trans., *Soviet Physics JETP*, Vol. 25, No. 6, 1967, pp. 980–985].

41. Kormer, S.B., Yushko, K.B., and Kirshkevich, G.V., "Dependence of the Refractive Index on the Density of the Solid and Liquid Phases of Shock-Compressed Ionic Crystals. Relaxation Time of Phase Transformation Under Shock Compression," *Pisma v Zhurnal Eksperimentalnoi i Teoreticheskoi Fiziki*, Vol. 3, No. 2, 1966, pp. 64–69, [English trans., *JETP Letters*, Vol. 3, No. 2, 1966, pp. 39–42].

42. Yushko, K.B., Krishkevich, G.V., and Kormer, S.B., "Change in the Refractive Index of a Liquid Compressed by a Shock Wave. Anomalous Optical Properties of Carbon Tetrachloride," *Pisma v Zhurnal Eksperimentalnoi i Teoreticheskoi Fiziki*, Vol. 7, No. 1, 1968, pp. 12-16, [English trans., *JETP Letters*, Vol. 7, No. 1, 1968, pp. 7–10].

43. Altshuler, L.V., Kuleshova, L.V., and Pavlovskii, M.N., "The Dynamic Compressibility, Equation of State, and Electrical Conductivity of Sodium Chloride at High Pressures," *Zhurnal Eksperimentalnoi i Teoreticheskoi Fiziki*, Vol. 39, 1960, pp. 16–24, [English trans., Soviet Physics JETP, Vol. 12, No. 1, 1961, pp. 10–15].

44. Altshuler, V.L., Pavlovskii, M.N., Kuleshova, L.V., and Simakov, G.V., "Investigation of Alkali-Metal Halides at High Pressures and Temperatures Produced by Shock Compression," *Fizika Tverdogo Tela*, Vol. 5, No. 1, 1962, pp. 279–290, [English trans., *Soviet Physics – Solid State*, Vol. 5, No. 1, 1963, pp. 203–211].

45. Kormer, S.B., Sinitsyn, M.V., Funtikov, A.I., Urlin, V.D., and Blinov, A.V., "Investigation of the Compressibility of Five Ionic Compounds at Pressures up to 5 Mbar," *Zhurnal Eksperimentalnoi i Teoreticheskoi Fiziki*, Vol. 47, 1964, pp. 1202–1213, [English trans., *Soviet Physics JETP*, Vol. 20, No. 4, 1965, pp. 811–819]

8

Determination of Detonation Parameters and Efficiency of Solid HE Explosion Products

V.M. Belsky and M.V. Zhernokletov

Problems related to the design of HE containment vessels, an estimation of their security and ability to withstand a variety of external effects, requires versatile experimental research into high explosive properties, the explosion itself, and its thermomechanical effect on the environment. Knowledge pertaining to the mechanisms involved in the initiation and propagation of a detonation wave is needed for effective work in a number of important practical areas, such as:

- the operational reliability and stability of initiating devices;
- HE charge resistance to different power effects;
- critical conditions for detonation propagation in thin layers.

To estimate the thermomechanical effects that the detonation of an explosive device will have on the environment, it is necessary to know the detonation wave parameters of the HE used, and equations of state of its explosion products.

8.1 Study of the Detonation Initiation

The result of the application of an impulsive load to HE depends on the spatial and temporal characteristics of the applied load and the specific properties of the HE. In general, there are two basic results: one involving detonation, and the other devoid of any of the chemical reactions that are associated with detonation. For relatively low levels of loading, the second of these two possible results takes place. In this case the shock wave propagates through the HE as though it were an inert material. With a sufficiently high level of level of loading, however, a fast exothermal chemical decomposition takes place, the energy of which is deposited into an amplification of the shock front. Given sufficient geometrical freedom, this amplification will eventually lead to the appearance of a normal detonation wave, "deep" in the interior of the charge (at some distance from the point of load application). The process just described, leading to normal detonation under the action of an impulsive load applied to the HE charge, is referred to as shock initiation. The critical level of the applied load that is necessary for detonation to occur, characterizes the shock-wave sensitivity of the HE. Knowledge of this sensitivity is required for a clear-cut understanding of the level of permissible insult to high explosives, and to predict their behavior under given conditions of use.

The principal purposes behind the study of shock initiation are: determination of the mechanisms for the formation of a normal detonation regime, construction of a physically sound model for detonation, and an estimation of shock-wave sensitivity.

The formation of a normal detonation regime is a complex process. The nature of the transition from an initiating shock wave (ISW) to a detonation wave, and the depth within the HE charge that this transition takes place, depend on the loading parameters (amplitude, pulse duration and shape), structural features of the HE charge (density, fabrication processes, particle size of HE), and on the initial temperature.

A number of experimental methods using shock waves and various techniques for recording the response of HE to an applied load are being employed to study the issues of detonation regime formation in heterogeneous high explosives and to determine the characteristics of their shock-wave sensitivity.

8.1.1 Comparison Methods

The simplest methods for studying HE shock-wave sensitivity are those that compare different high explosives under identical conditions of initiation.

Experiments of this type, first reported in the literature in the 1930s [1], did not attempt a quantification of the strength of the initiating shock wave penetrating the HE, rather, they characterized the sensitivity of the HE to the shock wave in terms of the thickness of an inert base plate, through which a stationary detonation could be initiated (Fig. 8.1). It was believed that the thicker the base plate through which a stationary wave could be initiated, the more sensitive the high explosive. A witness plate was used to register the occurrence of detonation initiation.

This method is meaningful so long as the geometry, mass of the active charge and its HE type, and the material of the base plate separating the active charge from the HE under study are invariable. The results obtained using this method are problem specific – change the geometry, or the mass, or the active HE charge type, and the results of the experiment are changed. In the Russian literature, this method is referred to as the method for determining HE shock-wave sensitivity (SWS) by base plate thickness. In the foreign literature, this method is known as the "gap-test" method [2].

Later, shock-wave sensitivity came to be determined in terms of the pressure of the shock front at the time of transmission from the base plate to the HE under study (this pressure is in part a function of base plate thickness). Thus, SWS came to be quantified in terms of the critical pressure,

BC

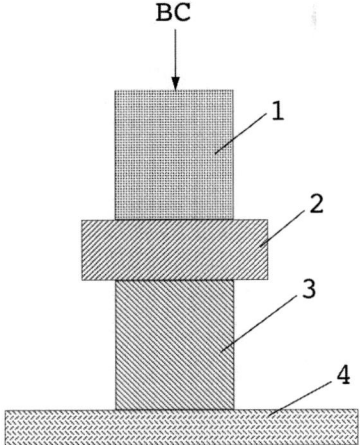

Fig. 8.1. Typical scheme of the experiment for determination of HE shock-wave sensitivity by base plate thickness: 1 – active HE charge; 2 – base plate; 3 – charge of HE to be studied; 4 – witnessing plate; BC stands for blasting cap

P_{cr}, required for the initiation of the HE under specified experimental conditions [3] (the method for SWS estimation by determining the critical pressure of initiation).

Another HE shock wave sensitivity parameter, detonation delay, τ_{del}, proved to be a more suitable measure in practice [4]. It is estimated using the loading system of the previous method, including the constant-thickness base plate, but with the witness plate replaced by a pair of electric contacts located at the ends of the charge under study.

The time required for the wave front (initially an ISW, later transitioning into a detonation wave) to travel through the HE under study is recorded directly in the experiment. Detonation delay (Fig. 8.2), is calculated as the difference between the experimentally measured time, and the time it would take for a stationary wave to travel the length L of the charge length if propagating at velocity D:

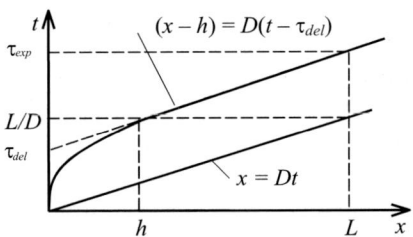

Fig. 8.2. The $x - t$ diagram of the initiation process: h is the depth of normal detonation actuation

$$\tau_{del} = \tau_{exp} - \frac{L}{D}. \tag{8.1}$$

The shorter the value of τ_{del}, the higher the shock-wave sensitivity.

In the graph of Fig. 8.2, the region from $x = 0$ to $x = h$ (a region of ISW acceleration) is referred to as the detonation formation region, and the distance h, at which the ISW velocity becomes equal to that of normal detonation, is called the depth of detonation actuation.

Presently, the above-described methods are the principal tools for obtaining qualitative and semi-quantitative data pertaining to degree to which the shock wave sensitivity of HE is affected by various factors (density, HE particle size, HE reprocessing technology, etc). These methods require a minimal quantity of HE and a minimal expenditure of time to acquire the needed information.

8.1.2 Optical Wedge-test Method

This method records the time of shock wave arrival at the inclined surface of a wedge-shaped charge under study (Fig. 8.3) [5]. During the experiment, the plane of the shock wave, together with the inclined plane of the charge under study, form a wedge with a given vertex angle. The $x - t$ diagram of the passage of the ISW through the charge of HE under study, is obtained by geometric transformation of the axial coordinates to the coordinates of points on the inclined surface of the wedge.

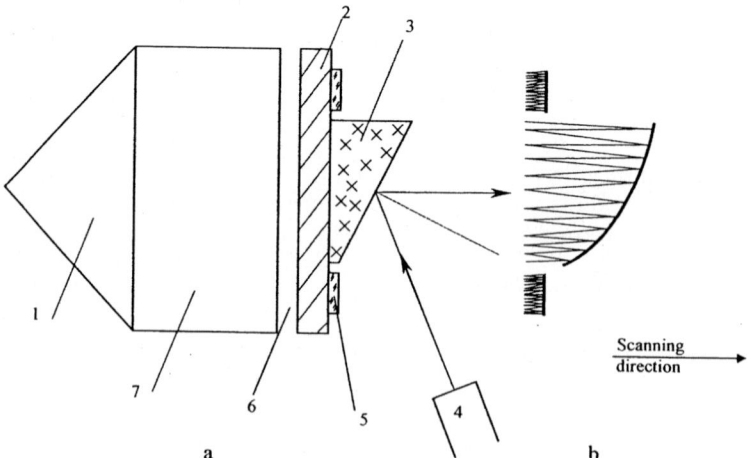

Fig. 8.3. Typical scheme of the experiment on recording the ISW front movement across the wedge-shaped HE charge (**a**) and typical streak-camera image of the process (**b**): 1 – plane detonation wave generator; 2 – base plate; 3 – charge of HE to be studied; 4 – explosive (pulse) accent lighting; 5 – time mark of the shock wave penetration into the HE under study; 6 – air gap; 7 – HE charge

To enable visualization of the passage of the shock wave through the charge of HE under study, a thin (∼0.01 mm) film with a specular surface is placed on the inclined surface and this film is exposed to the beam of an external pulsed light source. The reflected light is directed to a streak camera operated in the slit scanning mode.

A change in the angle that light is reflected from the mirrored film occurs upon shock front arrival. This ensures an abrupt cutoff of light arriving at the film and a precise recording of the passage of the shock front.

Differentiation of the resulting $x - t$ diagram provides the shock wave velocity. The distance at which a normal detonation wave is generated, is given by function $D(x)$. This distance is frequently used as a characterization of the sensitivity of the HE to rectangular-profile shock waves of given amplitude. The wedge-test method and its variations have found wide use in experiments aimed at determining the depth of detonation actuation and the time history of an ISW in the HE charge. It is also possible, in the same experiment, to determine the nonreactive ("cold") HE shock compression parameters by extrapolating the shock-wave velocity to the base plate-HE interface.

8.1.3 Method of Interface Deceleration

This method, suggested in [6], represents an application of the electromagnetic method for recording the mass velocity profile following the passage of an ISW front. An electromagnetic sensor is positioned at the interface between a base plate and the HE under study. In this setup, information concerning the release of energy in shock-compressed HE is obtained by recording the base plate-HE interface deceleration. Detonation of the HE leads to an increase in pressure on the base plate, resulting in its deceleration – hence, the name of the method. The interfacial deceleration results in a deviation in what would be a rectangular velocity profile were it not for the occurrence of detonation in the HE. The $P - u$ diagram for the method, elucidating the interface deceleration, appears in Fig. 8.4 [6].

Since the recorded mass velocity profile depends on the energy release rate, a change in the velocity profile corresponding to a change in any single characterizing feature of the HE under study (particle size, manufacturing process (casting, pressing), nature of the gas in charge pores, etc.), is evidence of the effect of that feature on the release of energy following passage of the ISW front (all other factors, such as HE density, and initial pressure being held constant).

As with the previous method, the method of interface deceleration allows us to obtain nonreactive HE shock compression parameters in the same experiment.

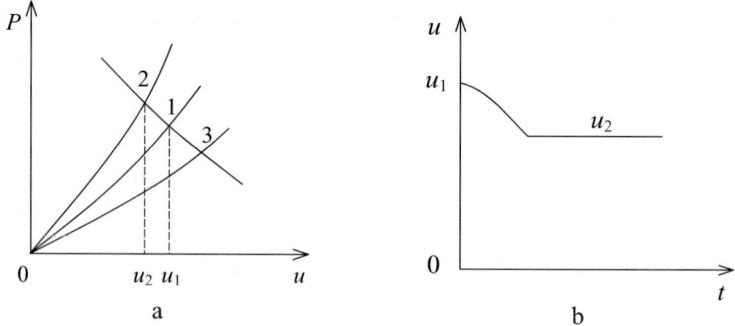

Fig. 8.4. The $P - u$ diagram (**a**) and mass velocity profile during base plate-HE interface deceleration (**b**): 1, 2, 3 – Hugoniots of starting HE, detonation products, base plate, respectively; u_1 – initial mass velocity at the base plate-HE interface, u_2 – decomposition product mass velocity

8.2 Formation of the Detonation Regime

8.2.1 Effect of Initiating Pulse Amplitude and Profile

The investigations reported in [7, 8], on the effect of the initial amplitude of the initiating pulse pressure, show that the transition of the ISW to a detonation wave involves a gradual, monotonically increasing acceleration in shock speed from that of the initial pulse to that of the stationary detonation wave (Fig. 8.5).

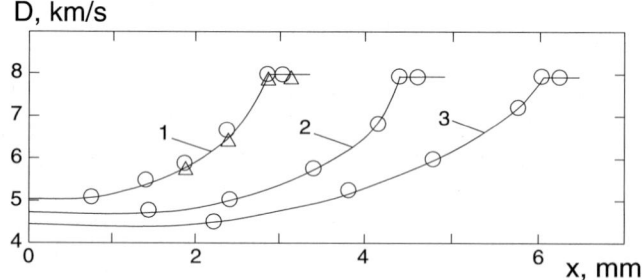

Fig. 8.5. ISW velocity vs. distance in the charge of cast composition B (trinitro-toluene/hexogen 40/60) of 1.71 g/cm^3 initial density. Initiation pressures: 1 – 9.5 GPa, 2 – 7.7 GPa, 3 – 6.7 GPa

Increasing the initiation pressure reduces the region of transition to detonation. For the same amplitude of the initiating shock-wave incident on the HE, the detonation actuation depth depends on density, HE particle size, temperature, and manufacturing processes. When the density and HE particle size

Table 8.1. Values of coefficients A and B in equation (8.2) for some HEs [9]

HE	ρ, g/cm^3	A	B	Pressure range, GPa
Octogen (HMX)	1.891	1.18	0.59	$4.4 < P < 9.6$
PBX-9404	1.840	1.12	0.67	$2 < P < 25$
Petn	1.72	0.65	0.59	$2 < P < 4.2$
TATB	1.876	1.417	0.40	$11 < P < 16$
Tetryl	1.70	0.79	0.42	$2.2 < P < 8.5$

decrease, the detonation actuation depth also decreases. ISW acceleration in cast HE is greater than that in pressed HE of the same density. Studies reported in [4], revealed a simple empirical relationship between the detonation actuation depth in an HE sample (applicable to HEs with an initial density $\geq$ 0.95 of the theoretical value) and the pressure of the ISW. That relationship is expressed as:

$$\ln P = A - B \ln h \qquad (8.2)$$

where P is the initiating shock wave pressure (GPa) and h is the detonation actuation depth (mm).

The values of the coefficients A and B for several HEs and explosive compositions are presented in Table 8.1.

The gradual transition of an ISW to a detonation wave is not always observed. For example, a nonmonotonic transformation to the detonation regime is observed in cast trinitrotoluene ($\rho_0 = 1.62$ g/cm^3 initiated by a rectangular-profile shock wave of $\sim$5 GPa and 7.5 GPa) [8,10]. The shock front velocities recorded in [8] are plotted in Fig. 8.6. The plot reveals a weak acceleration in shock velocity from about 18 mm depth ($P \sim 6$ GPa) to about 40 mm depth, becoming considerably steeper beyond 40 mm as the ISW transitions to detonation.

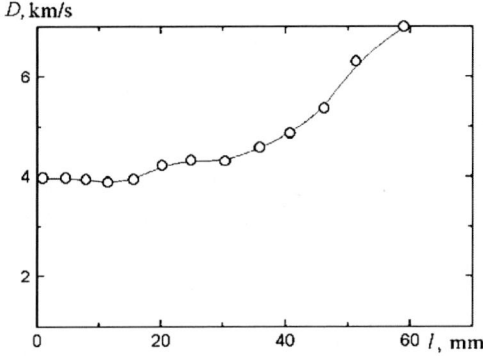

Fig. 8.6. Initiating shock wave velocity vs. distance in cast trinitrotoluene charge. The initial shock wave amplitude is 5 GPa

A similar trend was detected in the shock-wave initiation of cast TNT subjected to an initiation pressure of 7.5 GPa [10]. A change in the nature of $u(t)$ at $P \sim 113$ GPa has been recorded with the method of interface deceleration in cast fine-crystal trinitrotoluene ($\rho_0 = 1.62$ g/cm^3) [11] (see Fig. 8.7). The observed features are associated with the release of energy behind the ISW front and its transition to a detonation shock wave.

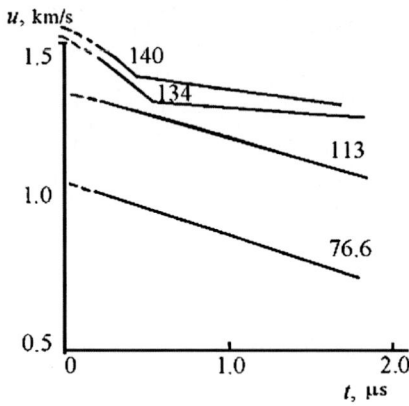

Fig. 8.7. Inert barrier-cast trinitrotoluene interface velocity versus time at different pressures. Initial HE pressures (GPa) are specified on the lines

Oscillograms of mass velocity profile taken at different depths in porous, pressed, and cast trinitrotoluene charges initiated by triangular-profile shock waves [12] are presented in Fig. 8.8. It is seen that for porous HE, the profile of the mass velocity following the ISW front continues to have a negative slope, even as the ISW develops into a detonation wave. Such a profile is characteristic of any low density HE ($\rho_0 \leq 1$ g/cm^3), irrespective of the ISW amplitude and HE particle size [12]. It may therefore be concluded that in low-density HE, the energy released immediately following the shock front is always more intense than any subsequent energy release.

In pressed and cast TNT, the ISW that starts out with a triangular profile upon entry into the HE charge, gradually transforms into a profile possessing an increasing mass velocity, evidence of increased pressure at some distance from the shock front. At the final stage of transformation into the detonation regime, the mass velocity profile takes on a negative slope.

The evolution of the pressure wave profiles recorded by manganin gages in cast trinitrotoluene under initiation by a rectangular-profile shock wave is presented in Fig. 8.9 [13]. It is seen that the pressure immediately following the shock front (compression wave) increases with increasing depth into the charge. This increase in pressure proceeds relatively quickly and results in a steepening of the shock pressure profile. The growth in the ISW amplitude abruptly increases when the compression wave overtakes the shock front.

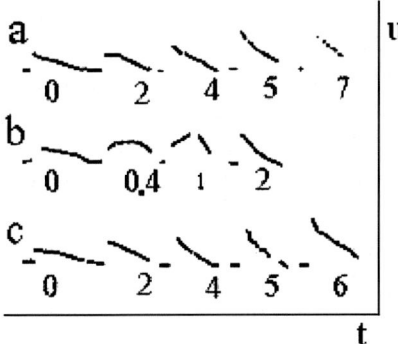

Fig. 8.8. Oscillograms of mass velocity profiles in TNT: **a** – pressed; **b** – cast; **c** – porous ($\rho \leq 1$ g/cm^3) TNT. The figures near the profiles denote the distance from the base plate to the sensor (in mm)

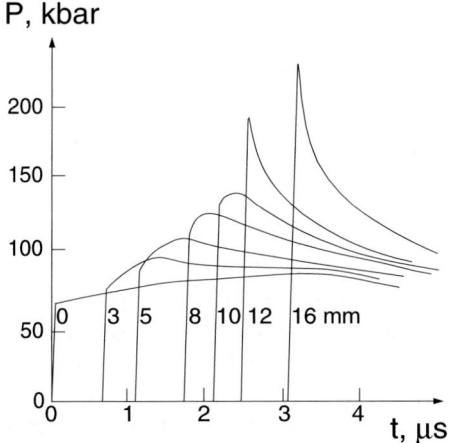

Fig. 8.9. Typical pressure profiles in cast TNT of initial density 1.62 g/cm^3. The figures specify the distances from the base plate to the sensor

In the final stage of detonation regime formation, the pressure profile has a negative slope. Shock amplitude is not the only parameter affecting detonation regime formation. Other factors that may be important include, but are not limited to, the rate of increase in the shock wave pressure pulse, and the pulse duration [14, 15].

Results from the studies on detonation initiation in TNT/RDX 50/50 by shock waves produced by the impact of a thin steel plate [14] are presented in Fig. 8.10.

An interesting feature of the process (see Fig. 8.10) is a region of decreasing pressure at the ISW front. The source of this feature is the effect of the rear-

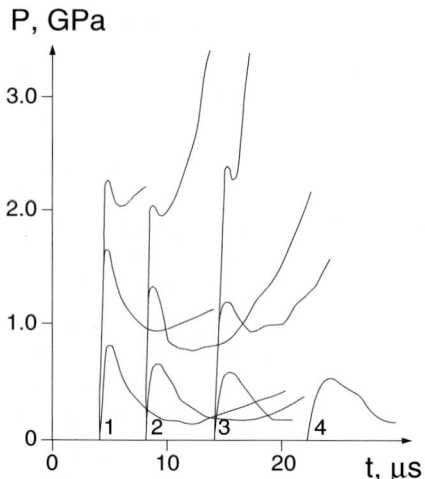

Fig. 8.10. Time history of the plane nonstationary shock wave parameters in TNT/RDX 50/50 sample: $1 - h = 4$ mm; $2 - h = 8$ mm; $3 - h = 14$ mm; $4 - h = 22$ mm; h is the distance to the surface of the impactor impact on the HE sample

ward rarefaction wave formed during the early stages of ISW passage through the sample under study. The region of the decreasing pressure following the ISW front is reduced with increasing initial amplitude of the initiating pulse. It is also reduced as the ISW propagates. The magnitude of the energy release-related pressure increase that follows depends largely on the initial shock wave amplitude. With increasing initial amplitude, the rate of pressure increase in a given region increases abruptly (see Fig. 8.10).

The experimental results presented in Figs. 8.9 and 8.10 indicate that the bulk of the energy release in HE occurs in a wide zone following the shock front. During the stage of detonation regime formation, the initiation process depends on the energy released in the entire volume of shock-compressed HE, while the pressure increase is determined by the transfer of energy from material already traversed by the wave. Spatial and temporal characteristics of the flow is determined, to a large extent, by processes occurring immediately behind the shock front and depends both on pressure and its gradient. Experiments [5, 8, 15–18] suggest that the HE decomposition reaction rate is very sensitive to the pressure gradient following the front of the ISW, up to its complete cessation.

The effect of the shock wave pulse pressure increase on the initiation of solid heterogeneous HE is illustrated in Fig. 8.11, which compares the wave profiles during the initiation of PBX-9404 by a shock wave and by compression waves with different rates of pressure increase [19].

The experiments show no pressure increase that would be indicative of a release of energy associated with initiation prior to the time that the compression waves transition into shock waves. Obviously the time required for

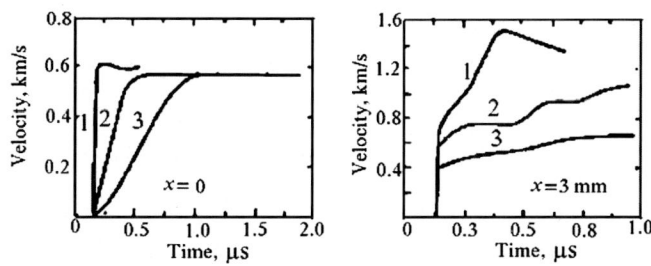

Fig. 8.11. Effect of loading rate on PBX-9404 initiation: **a** – initiating pulse profiles; **b** – velocity profiles at a distance of 3 mm from the sample surface

detonation regime formation will be longer with compression waves than will be the case when the initiating wave is a shock wave.

Similar results have been obtained for pressed TNT ($\rho_0 = 1.56$ g/cm^3 loaded by a pressure pulse of 1–3 GPa/μs rise rate and 2.0–4.2 GPa amplitude) [20].

The study of detonation initiation in HE subjected to single and double shock wave loading (second shock stronger than the first) suggests that the second shock wave suppresses the explosive transformation reaction that appears following the front of the first. This effect is referred to as shock desensitization, and it manifests itself both on pressed and cast HE (B-3, TNT/RDX 50/50, TNT) [21, 22]. One of the most notable features of the desensitization effect is deeper detonation actuation in the double-shocked HE. This is brought about as a consequence of the explosive transformation reaction rate being noticeably slower behind the second shock than it is behind the front of a single shock wave of the same amplitude.

No exhaustive account of this effect has yet been found. However, since any "smearing" of the loading (either monotonic or stepped) leads to a decrease in the HE shock heating temperature, the difference in the detonation regime formation times may be due to the difference, both in reaction foci concentration and in the intensity of HE decomposition processes on exposure to pulses of different front rates.

8.2.2 Recording Initiation Processes with the Flash Radiography Method

Flash radiography is used significantly less frequently in the study of detonation initiation than is the case for other methods. This is the case, even though it offers a number of indisputable advantages: such as process visualization, the absence of any distortions of the flow under study (since there are no measuring elements that interact with the environment), a recording of the material state upon passage of peak pressures (i.e., following the shock and detonation wave fronts and on multiple loading with possible intermediate unloading phases. Finally, flash radiography offers a unique opportunity,

unavailable with other techniques, to render the inner structure of the sample
(by its X-ray image) at any phase of the experimental process.

With the advent at VNIIEF of pulsed X-ray facilities such as ERIDAN (see
Chap. 3), X-ray experiments came to be used in the study of HE detonation
initiation processes on a much larger scale. The apparatuses produce a soft
radiation spectrum of short gamma pulse duration.

In studies of detonation activation, some of the most informative experiments are those involving the experimental device sketched in Fig. 8.12.

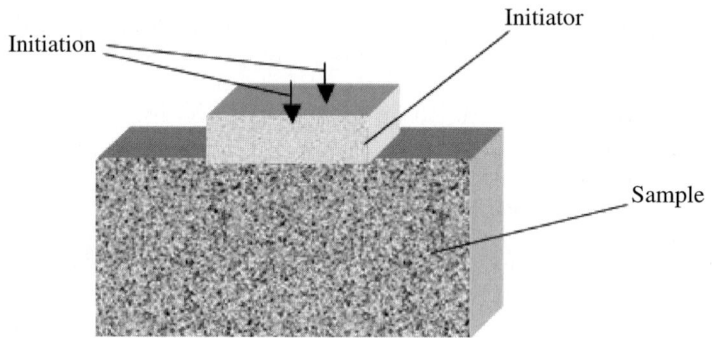

Fig. 8.12. The experimental device for the detonation activation studies

Detonation is actuated along an axis lying in the planar interface between
the initiator and sample by a diverging shock wave that is produced by the
parallelepiped-shaped initiator. The resultant shock (detonation) wave front
has the shape of a cylindrical surface and is projected onto the X-ray image
plane as a curved line. This expanding cylinder, with dimensions aligned with
the X-ray line of sight, provides excellent image contrast. X-ray experiments
designed for the study of plane-wave loading of HE samples employ a planar shock wave generator instead of blasting cartridges. The X-ray images of
Fig. 8.13 illustrate the ability of the method to differentiate between regions of
unperturbed HE, shock-compressed HE, EPs, and the shock and detonation
wave fronts.

Presently, RFNC–VNIIEF has the capability of studying HE detonation
initiation processes using flash radiography method in either single- or double-
frame X-ray modes. In order to acquire additional experimental data pertaining to the initiation processes, diagnostics have been refined so as to combine
the capabilities of flash radiography, with the electrocontact and manganin
techniques and with methods for the determination of the distribution of material density.

The method for determining the distribution of material density involves
a determination of X-ray film photometric density d as a function of optical
depth $Z = \rho L$ of the sample under study. This $d(z)$ function is then compared
to the known $d(z)$ of a reference sample ("optical" wedge). For this purpose

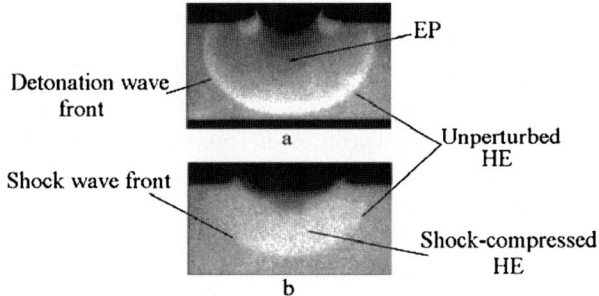

Fig. 8.13. Detonation (**a**) and shock (**b**) waves in the X-ray image

the sample to be studied and the reference sample (of a known geometry fabricated from the same explosive composition) are X-rayed simultaneously on the same X-ray film.

Knowledge of the material density following the ISW front enables us to obtain information about the explosive transformation kinetics as well as new quantitative data on the evolution of the density profile.

By way of example, Fig. 8.14 plots the experimental and computed curves of the material density distribution following the front of a diverging shock wave with partial HE decomposition, and following a detonation wave. The coordinate $x = 0$ (mm) is referenced to the plane of HE charge initiation.

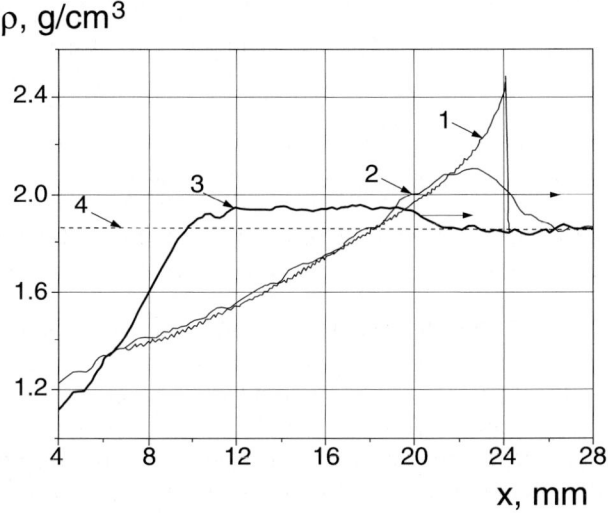

Fig. 8.14. Material density versus distance following the front of the shock wave with partial HE decomposition and detonation wave: 1 – computation for the detonation wave. Experiment: 2 – detonation wave; 3 – shock wave; 4 – initial density

8.2.3 Nature of Chemical Reaction Run Following Shock Front

Profiles of the experimentally recorded gas-dynamic variables (mass velocity and pressure) depend largely on the kinetics of the detonation process. Therefore, empirically obtained profiles of these variables (recorded at various locations within the sample under study), can be used to draw certain conclusions regarding the kinetics of the process.

The Lagrangian form of the 1D field equations (conservation of mass and momentum) is:

$$(\partial V / \partial t)_h = V_0 \, (\partial u / \partial h)_t \; ; \tag{8.3}$$

$$V_0 \, (\partial P / \partial h)_t = - \, (\partial u / \partial t)_h \; , \tag{8.4}$$

where h is the Lagrangian coordinate of the particle (the current Lagrangian coordinate of a particle is a function of time and the initial coordinate of that particle at time = zero), V is the specific volume, V_0 is the initial specific volume, P is the pressure, and u is the mass velocity. These equations can be used to calculate current values of the specific volume $V(t)$ and mass velocity $u(t)$ at various Lagrangian points if V_0 and the pressure field $P(h,t)$ are known from experimental observation. Alternatively, these equations can be used to calculate current values of pressure $P(t)$ and specific volume V(t) at various Lagrangian points if V_0 and the mass velocity field $u(h,t)$ are known from experimental observation. In either case, the unknown quantities are calculated through integration of the equations of motion.

The equations of motion (8.3) and (8.4) may be recast in terms of the phase velocities. Let phase velocities c_u and c_p be defined as follows [23]:

$$c_u = \left(\frac{\partial h}{\partial t} \right)_u = - \frac{\left(\frac{\partial u}{\partial t} \right)_u}{\left(\frac{\partial u}{\partial h} \right)_t} \; ; \tag{8.5}$$

$$c_P = \left(\frac{\partial h}{\partial t} \right)_P = - \frac{\left(\frac{\partial P}{\partial t} \right)_u}{\left(\frac{\partial P}{\partial h} \right)_t} \; . \tag{8.6}$$

Using these relationships, equations (8.3) and (8.4) may be recast as:

$$\left(\frac{du}{dt} \right) = \frac{1}{\rho_0 c_P} \left(\frac{\partial P}{\partial t} \right), \quad h = \text{const} \; ; \tag{8.7}$$

$$\left(\frac{dV}{dt} \right) = - \frac{1}{\rho_0 c_u} \left(\frac{du}{dt} \right), \quad h = \text{const} \; . \tag{8.8}$$

Recognizing that (8.7) and (8.8) are actually expressions of the classical 1D wave equation, it follows that the physical significance of c_u and c_P is that of phase velocities of propagation (at fixed levels of mass velocity and pressure, respectively).

Equations (8.7) and (8.8) may be combined in such a way as to eliminate mass velocity, resulting in the following pressure-specific volume relationship, valid along the phase path pertaining to some given coordinate [23]:

$$\left(\frac{\partial V}{\partial P}\right)_h = -\frac{1}{\rho_0^2 c_P c_u} . \tag{8.9}$$

The phase path reflects a change in volume (and as a consequence of continuity, in its velocity as well) due to pressure and the nonequilibrium process under consideration (detonation). The challenge, is to isolate from this combined effect, only the part contributed by the nonequilibrium transformation, and express it through experimentally measured values. The computational scheme to do this is the following:

– the shock front path and lines of constant pressure (isobars) are constructed from the experimental data in t-h coordinates;
– the isobars are differentiated to determine $c(P)$ for various Lagrangian coordinates;
– equation (8.4) is integrated and mass velocity profiles $u(t)$ are determined from the dependencies of $c(P)$ and the profiles $P(t)$;
– the procedure is repeated to calculate profiles $V(t)$ from knowledge of the $u(t)$ profiles;
– the profiles of $P(t)$ and $V(t)$ for the same coordinate yield values of P and V at different times, which may be used to determine the path of state change.

All advanced calculations assume that the chemical composition of the medium can be described by a single scalar variable α, a measure of the degree of HE decomposition (fractional part of the HE that has decomposed). The thermal and chemical nonequilibrium in heterogeneous HE, which complicates the construction of an EOS for the reacting mixture, is replaced with individual equations of state for each constituent. Pressure equilibrium is assumed (within a given microvolume the pressure in the explosion products is assumed equal to the pressure in the unreacted HE) [24]. Under these assumptions, the specific volume of the initial and final reaction product mixture is assumed to be addative:

$$V_{mix} = \alpha V_{EP} + (1 - \alpha)V_{HE} . \tag{8.10}$$

Although the error associated with estimating the portion of HE that is decomposed can be significant, having an approximate functional relationship $\alpha(t)$ enables us to make some conclusions regarding the general laws governing the process.

The h-t diagram of the process presented in Fig. 8.15 [13] has been obtained by processing a series of the experimental pressure profiles (see Fig. 8.9) obtained in a study of cast TNT.

The illustration clearly shows that in the initial phase, the compression waves that appear as a result of material decomposition (initiated by the shock wave) overtake the shock front and increase its amplitude and velocity. The change in the EP mass fraction with time for cast TNT, according to the results of the same experimental series, is plotted in Fig. 8.16. The EP mass

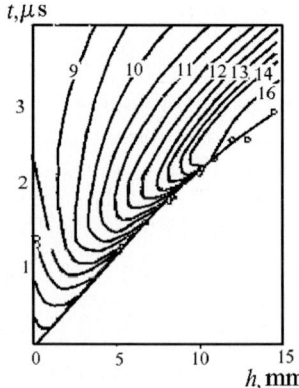

Fig. 8.15. h-t diagram of the shock wave evolution in cast trinitrotoluene. The figures in the curves specify pressure (GPa)

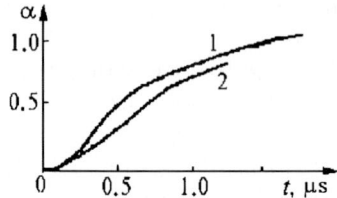

Fig. 8.16. EP mass fraction versus time for layers with coordinates 5 (2) and 8 (1) mm

fraction following the shock front is observed to monotonically increase with time.

For a given time t, the deeper into the HE (e.g., 8 mm versus 5 mm), the greater the concentration of EP. The decomposition rate of cast trinitrotoluene found by differentiation of the curves $\alpha(t)$ presented in Fig. 8.16 reaches a maximum $\approx 1.4 \cdot 10^6$ s^{-1} at $\alpha \approx 0.3$.

The resultant curves of such a calculation for the decomposition rate, $d\alpha/dt$, of cast trinitrotoluene as a function of EP mass fraction, α, for locations with coordinates 5 and 8 mm into the HE is plotted in Fig. 8.17. It should be emphasized that the decomposition rate of pressed TNT at the same pressures is observed to be 5–7 times higher than that of cast TNT [25].

The large spatial extent of the chemical decomposition zone and the smoothness of energy release following the shock front are demonstrated in [15, 17, 18]. In an investigation of $\alpha(t)$, the authors of [26] observed a transition in reaction rate for an HE composition consisting of 46.6% TATB, 49.3% octogen, and 4.1% inert binding material. The transition involves the transformation from a single-stage HE decomposition reaction following the shock front, to a two-stage reaction as the detonation proceeds. This transition from

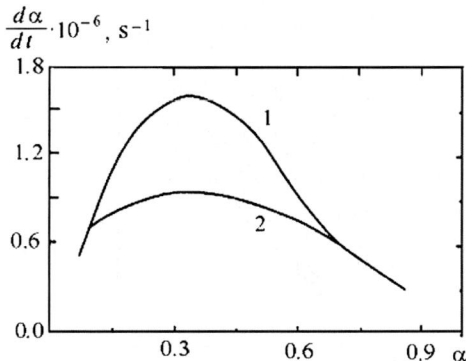

Fig. 8.17. Cast trinitrotoluene decomposition rate versus EP concentration for the layers with coordinates 5 (1) and 8 (2) mm

a single-stage to two-stage reaction is due to the different shock-wave sensitivities of TATB and octogen.

The patterns of behavior that are seen in investigations of detonation wave formation, provide strong evidence that the release of energy in detonating HE and shock front intensity are interrelated. This is in agreement with the concepts of a focusing mechanism of reaction initiation and energy release in HE following the shock front (the combustion mechanism) [27].

8.3 Experimental Estimation of Normal Detonation Wave Parameters

The principal variables describing HE detonation are the stationary detonation rate D and the three Jouguet point variables: pressure P_J, mass velocity u_J, and sound speed c_J. These variables serve to characterize the state of reacted HE at the end of the chemical reaction zone. An additional important parameter is the reaction zone width. Experimental study of the stationary detonation process employs a variety of methods for studying material properties under dynamic pressures. Most of these are discussed in Chaps. 3 and 9. For this reason, this section focuses not on a general description of methods themselves, but rather on particular features of these method, as may be applied to the study of the very fast processes that must be understood if one is to understand the behavior of detonating HE.

It should be noted that recently developed experimental methods estimate, not only the values of the pertinent state variables at the detonation wave front, but also pressure, mass velocity and density distributions quite far from the detonation front. This experimental data can be used both for the straightforward determination of EP expansion adiabats, and for the verification and validation of mathematical models for the equation of state of EP.

8.3.1 Measurement of Detonation Rate

Detonation rate D is a major HE charge characteristic. There is a method for the measurement of detonation rate that requires no measuring devices. It is known as Dautriche's method. The method consists of making a comparison between the detonation rate to be estimated, and the known constant-detonation-rate of demolition cord (DC). Measurements of D by this method are taken as follows. The ends of the piece of DC are inserted into the HE charge so that they are separated by a distance l (measured along the axis of the HE charge). Only the ends of the DC are inserted into the HE charge; the remainder of the DC length forms a "loop" that is not confined by the attachment to the HE. The middle portion of the loop is made straight and laid upon a lead or brass plate. The DC is positioned on the plate so that the mid-point of the length of cord rests atop a line scratched into the plate. The straight middle portion of the DC is aligned perpendicular to the line scratched in the plate. The HE is detonated at one end. As the detonation wave travels down the length of the HE charge, it encounters one end of the DC and ignites it, initiating the run of a detonation wave through the DC. The distance from the point of detonation in the HE charge to the first DC end piece is unimportant so long as it is far enough removed from the detonator that steady detonation in the HE charge is achieved prior to the interaction of the wave with the first DC end piece. After igniting the first end piece, the detonation wave traveling through the HE charge will ignite the second end piece at a time l/D later. Detonation is now traveling from the two ends of the DC loop toward a collision. At the point of collision of the detonation waves traveling through the DC toward each other a clear imprint is made in the plate marking the point of collision. Let us note the distance from this point to the light scratched in the plate as h. Then the relation $l/D = 2h/D_c$ is valid, from whence the detonation rate expression $D = lD_c/2h$ follows. The accuracy of D measurement with the method is ~3–5%. The method is not actually used currently because of its low accuracy; it has been briefly described here as a historical fact.

Advanced methods ensure significantly better accuracy in the measurement of D. These include, for example, the discrete method using electrocontact sensors, the method of continuous measurement using rheostat sensors, and optical methods using photorecorders (either with image mirror scanning, or frame-by-frame filming, or slit scanning).

Electrocontact sensors positioned at discrete depths into the HE charge register the passage of a detonation wave. Arrival of the detonation wave triggers the closing of an electric circuit (details of specific sensor types are discussed in Chap. 3), and pulses in current are produced. Electronic oscilloscopes record these current pulses. The spread of sensor actuation time depends on sensor design, but typically is from 1 to 10 ns. Measured time intervals and known distances between sensors are used to estimate the detonation rate. The accuracy of the measurement of detonation rate using electrocontact sensors can be better than 1%.

Optical methods are used widely in measurements of the kinematic characteristics of explosive processes. Detonation rate is usually estimated using a slit photorecording method in which a narrow strip is selected from the whole of the item being filmed, and the motion of luminous material within this strip is recorded. Slit photorecording produces a path-time diagram such as that illustrated in Fig. 8.18 [28], with the role of the time coordinate being played by the trace across the focal surface of the photorecorder, which moves at constant scanning rate V_{sc}. When the detonation rate vector is perpendicular to the scanning direction, the following expression is valid for the detonation rate: $D = kV_{sc} \tan \varphi$, where k is the optical system reduction ratio (equal to the ratio of the item height to the item image height), and φ is the angle of inclination of the tangent to the path-time curve.

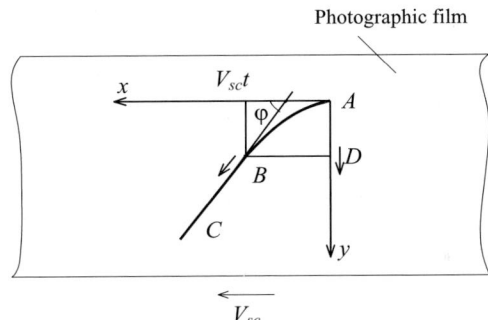

Fig. 8.18. Processing of the slit streak-camera image for detonation rate estimation

Experience with this method leads us to conclude that the greatest contribution to total error arises from estimation of the angle φ. To minimize this error, the photorecording should be performed at $\varphi = 45°$. If experiments employing this method are properly conducted, the maximum relative error in measurements of the constant detonation rate is no greater than 1% [28].

Additionally, photorecording methods are used to study the shape of the stationary detonation front in transparent and opaque HE charges, as well as detonation front structure in transparent HE [29].

8.3.2 Method of Spallation

The first method to estimate condensed HE detonation wave properties by recording the shock wave evolution in a metal plate, i.e. barrier, was the method of spallation (see Chap. 4). Using this method, a shock wave is set into motion through a plate by the arrival of a detonation wave at the interface between a charge of HE and the plate. Such experiments are conducted with plates of various thicknesses. The nature of the shock front is determined by measuring the free surface velocity W of the plate. The free surface velocity can be measured using electrocontact sensors, optical methods, or by employing capacitive velocity sensor techniques.

Mass velocity, u_{pl}, following the shock front in the plate is evaluated by the rule of doubled velocities, according to which $u_{pl} = W/2$. Since these experiments are conducted with plates constructed from materials for which the Hugoniot is well known, u_{pl} can be used to estimate shock pressure $P(u)$. Thus, the pressure profile at the HE-barrier interface may be qualitatively determined from the free surface velocity W as a function of plate thickness x. The shape of the decay curve reflects the sequential effect of the reaction zone and Taylor rarefaction wave on the plate (Fig. 8.19). The region of abrupt increase in W for small x is a consequence of the effect that the chemical reaction zone in the detonation wave has on the strength of the shock wave in thin plates.

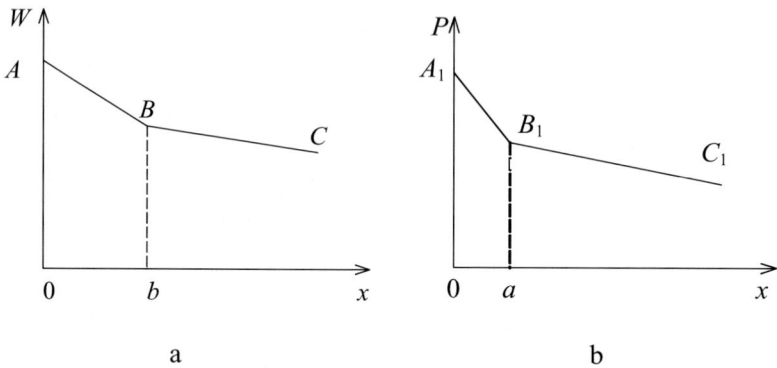

Fig. 8.19. Plate free surface velocity versus thickness (**a**) and the relevant pressure in the detonation wave (**b**)

Point A_1 corresponds to the peak pressure in the detonation wave chemical reaction. It is estimated by $u_a = W_a/2$, where W_a is the metal plate free surface velocity extrapolated to zero thickness. At point B ($x = b$) the chemical reaction effect has ceased. Having experimentally determined the kink position ($k = b$), the relationship between the width of the chemical reaction zone a, and the value of parameter b, can be found from an analysis of the motion of the waves in the plate and in the EP. The wave set in x-t coordinates is depicted in Fig. 8.20.

The chemical transformation zone width is calculated by the following expression, readily derived using the $x - t$ diagram:

$$a = \frac{(D - u_b)(u_b + c_b - D_b)}{D_b(u_b + c_b - u_c)} b \ . \tag{8.11}$$

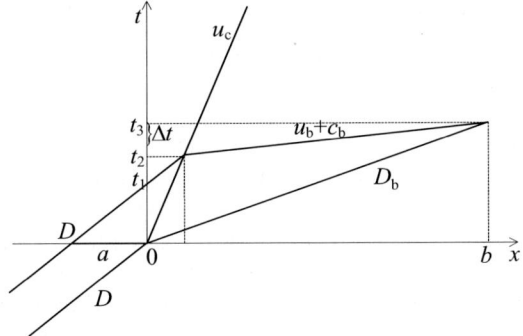

Fig. 8.20. The $x - t$ diagram of flow in the method of spallation: to the *right* of the ordinate axis – metal, to the *left* – HE

In Eq. (8.11) u_b, c_b, and D_b are the mass velocity, sound speed, and shock wave velocity in the metal (averaged over width b), respectively, u_c is the average velocity of the EP-metal interface, and D is the detonation rate. Note that the calculation of the reaction zone width does not include the shock wave reflected in the EP.

When measuring the free surface velocity, the gauge length S is taken such that the motion recording time is shorter than the time required for reflections from the free surface to transit the plate (of thickness Δ) twice: $S/W \leq 2\Delta/D_b$; otherwise the plate free surface will be driven by a second compression wave.

The average relative error in the measurement of W depends on the thickness of the metal plate. For plates 3 or more mm thick, the average relative error in the measurement is $\approx 1\%$.

To avoid any ambiguity that may arise from difficulties associated with the interpretation of experimental data for a determination the Chapman-Jouguet point on the chemical reaction zone boundary, experiments are conducted with HE charges of different length. In taking this approach, the decay curves $u(x)$ are compared, and the Chapman-Jouguet parameters are determined from their intersection (Fig. 8.21).

Let us examine the relationship between the field variables associated with HE detonation and those of the shock wave in the metal plate. The $P - u$ diagram of the processes that occur upon reflection of the detonation wave from the plate is presented in Fig. 8.22. From the $P - u$ diagram and the law of conservation of momentum, it follows that: $\Delta P_{refl} = P_b - P_J$; $\Delta P_{refl} = \rho_0 D_{refl} \Delta u_{refl}$; $\Delta u_{refl} = u_b - u_J$; $P_J = \rho_0 D u_J$; and $P_b = \rho_0 D_b u_b$.

We can determine the slope of the EP deceleration curve at the point corresponding to normal detonation, i.e., at the Jouguet point, using Eqs. (1.52) and (1.53):

$$(dP/du)_J = - (\rho c)_J = -\rho_0 \frac{n+1}{n} D \frac{n}{n+1} = \rho_0 D .$$

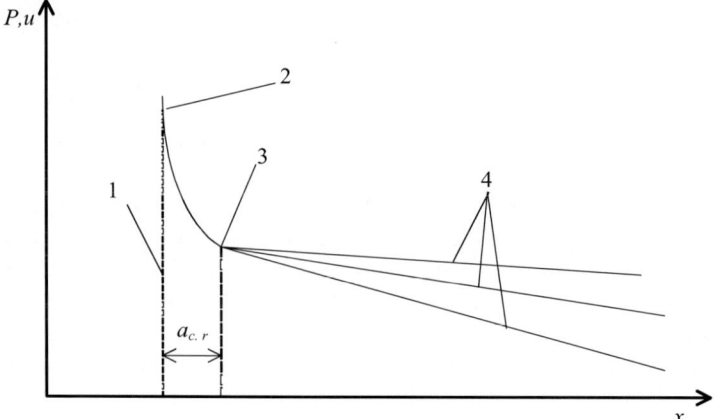

Fig. 8.21. On estimation of Chapman-Jouguet state by unloading wave intersection: 1 – shock front; 2 – chemical peak pressure; 3 – Chapman-Jouguet state; 4 – Taylor unloading waves for different HE lengths; $a_{c.r}$ – chemical reaction zone width

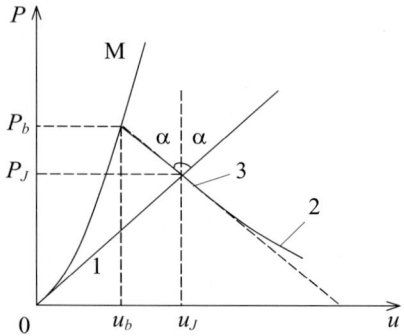

Fig. 8.22. The $P-u$ diagram of the detonation wave reflection from a metal barrier: 1 – detonation beam $P = \rho u D$; 2 – EP expansion curve; 3 – specular reflection of the detonation beam; $0M$ – metal shock Hugoniot

Observe that the expansion initiates as a specular reflection from the line segment $P = \rho_0 D u$, representing possible states at the wave front for given ρ_0 and D in the $P - u$ plane. Note that this result is general, independent of the form of the equation of state.

In the acoustic approximation, the slopes of the reflected wave and the deceleration curve will be the same, i.e., $-\rho_J D_{refl} = -\rho_0 D$. In view of this,

$$P_b - P_J = -\rho_0 D \left(u_b - u_J \right) = \rho_0 D \left(\frac{P_J}{\rho_0 D} - \frac{P_b}{\rho_{0M} D_b} \right) ,$$

and on reduction,

$$P_J = \frac{P_b}{2} \left(1 + \frac{\rho_0 D}{\rho_{0M} D_b} \right) . \tag{8.12}$$

Next the relations $P_J = \rho_0 D^2/(n+1)$ and $u_J = D/(n+1)$ can be used to determine the EP polytropic exponent n.

The method of spallation is not free of difficulties. For example, the field variables are affected by plate strength, time dependence of the resistance to failure, and the shock wave reflected from the plate into the chemical reaction zone.

The strength effect on the detonation wave field variables is assessed in experiments involving compound and continuous plates of equal thickness made of the same material. The compound plate is composed of two parts having thicknesses l_1 and l_2 and the continuous plate has thickness $l = l_1 + l_2$. In compound plates, thickness of the external part is constant, l_2, and thickness of the part in contact with the HE charge is variable, l_1. The velocity of the plate of thickness l_2 is measured in the experiments. The artificial spall thickness, l_2, is smaller than the natural spall thickness, l_3, for a given HE for all thicknesses $l = l_1 + l_2$ used. For thin continuous plates, in which $l < l_3$, the plate velocity gradually decreases with increasing l.

When $l > l_3$, a kink appears on the curve $u = f(x)$. In this case, the velocity of the spalled part of the plate of thickness l_3 will depend on the portion of the wave pulse that the material at the location of failure has experienced at the time of fracture. Spall fracture, in general, involves a process of damage initiation and evolution leading to fracture. For fracture to occur, it is therefore necessary that the tensile stresses act for a period of time. During this time, a portion of the rarefaction wave passes the point at which spallation will occur, leading to a decrease in the velocity of the spalled portion of the plate (l_3 thick) as compared to the velocity of the compound plate at $l_2 = l_3$. When $l_2 < l_3$, the rebound velocity of the compound plates will be always greater than that of the continuous plate.

The graph in Fig. 8.23 plots the results of experiments [30] for a nitromethane-acetone mixture (75/25 by volume) with continuous and compound aluminum plates ($l_2 = 0.3$ mm). As seen in the figure, natural spallation produces two kinks in the curve $W = W(x)$, whereas artificial spallation produces only one kink, which corresponds to the Chapman-Jouguet point. The results of Chapman-Jouguet pressure measurements [31] in trinitrotoluene and composition B using the method of artificial spallation are presented in Table 8.2.

Table 8.2. Detonation wave parameters for pressed trinitrotoluene and composition B estimated by the method of artificial spallation

HE	Barrier material	ρ_0, g/cm^3	D, km/s	P_J, GPa
	Al	1.63	6.86	19.8
	Plexiglas	1.63	6.86	19.7
Pressed trinitrotoluene	Brass	1.63	6.86	19.6
	Mg	1.63	6.86	19.4
Composition B (TNT/RDX 36/64)	Plexiglas	1.68	7.50	23.9

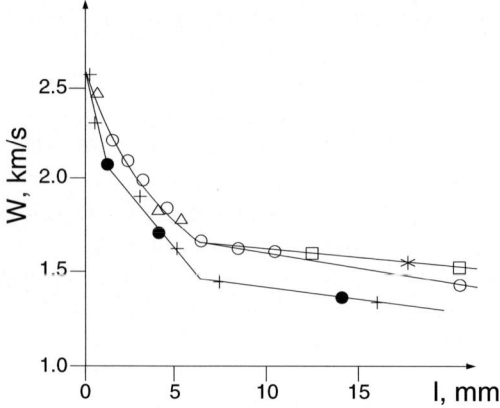

Fig. 8.23. Function $W = W(x)$ for continuous ($+$, $L/d = 1.3$; $\bullet$, $L/d = 2.5$) and composite ($\triangle$, $L/d = 1.3$; $\circ$, $L/d = 2.5$; $\square$, $L/d = 5.0$; $*$, $L/d = 10$) plates at different ratios of charge length L to diameter d ;1 – natural spallation; 2 – artificial spallation

8.3.3 Method of Closed Contacts

When measuring mass and wave velocities in barriers made of various materials, open electric contacts separated by a dielectric layer are sometimes used. A pulse in current appears at the time of their closing, when the insulation layer ruptures. The sluggishness of such contacts, by which is meant the time interval from the time of arrival of the shock wave at the contact to the time of a noticeable drop in voltage across it, and its dependence on the pressure in the insulation layer, can lead to some systematic error when recording unsteady shock-wave motion.

The method of closed contacts, a discrete electrical method for the estimation of shock wave velocity, is largely free of these problems. This method involves recording the times of shock front arrival at closed electric contacts positioned at given distances from each other along the shock wave path. The principle of operation of the closed contacts in the shock wave is discussed in detail in Chap. 3 (Fig. 3.17 presents schematic views of the closed contact sensor and a typical oscillogram produced by the passage of a shock wave through it).

Thickness and time of shock wave motion in each foil are used to calculate average velocities of shock wave propagation through the barrier and to determine the function $D = f(x)$, where x is the barrier thickness. The transformation to Jouguet pressure is then accomplished using Eq. (8.12).

In terms of accuracy, the method of closed contacts is comparable to the technically more complex method of laser wave velocity measurement [32], which can be used only in optically transparent media. The method is particularly effective for measurements of the velocity of a relatively weak shock wave ($P \sim 1$ GPa in amplitude), and for measurements of an unsteady shock wave.

8.3.4 Magnetoelectric Method for Recording Mass Velocity

The magnetoelectric method discussed in detail in Chap. 3 is used widely in studies of detonation waves. It is attractive because it enables tracking of the mass velocity profile both inside the chemical reaction zone, and outside it – in the rarefaction wave. A conceptual diagram of the magnetoelectric method for the measurement of mass velocity in EP appears in Fig. 8.24.

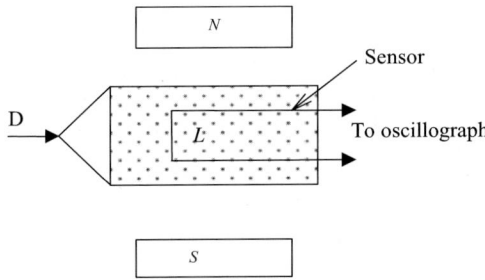

Fig. 8.24. Scheme of recording mass velocity profile behind the detonation wave front with the electromagnetic method. D is a detonator. N, S are the magnet poles. Magnetic field lines H are perpendicular to the working part of the sensor

A sensor, fabricated in the form of an aluminum foil crossbar (0.05–0.1 mm thick and 12–15 mm long) is positioned inside the charge of HE to be studied. The entire system is placed in a homogeneous magnetic field. In this way, the working plane of the sensor should cross field lines during its motion. When the detonation wave passes through the charge, the sensor gets caught up in the motion of EP. Electromagnetic Force (EMF) induced on the sensor is recorded on an oscilloscope. It must be assumed that the presence of the sensor in the HE does not significantly distort the detonation wave profile.

The EP velocity at the Chapman-Jouguet point and the chemical reaction time are estimated by a characteristic kink in the $E(t)$ record (such as that presented in Fig. 8.25 [33]). The pressure at the Chapman-Jouguet point is estimated by the relationship $P_J = \rho_0 u_J D$.

Among the internal methods, the electromagnetic method provides the best time resolution t_{res}. The value of the t_{res} depends on the time it takes the sensor to attain EP velocity. The best resolution that has been achieved thus far with this method is ~ 0.01 μs, and the accuracy in the measurement of velocity is $\sim 2\%$.

Chapman-Jouguet values estimated using the magnetoelectric method [34] for some HE are presented in Table 8.3.

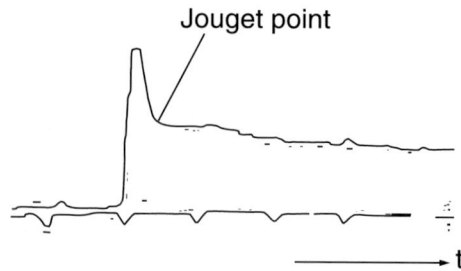

Fig. 8.25. Typical oscillogram of signal recorded by the mass velocity sensor in detonation wave for pressed TNT, $\rho_0 = 1.35$ g/cm^3

Table 8.3. Chapman-Jouguet parameters of some individual and mix HE

HE	ρ_0, g/cm^3	D, km/s	u_J, km/s	P_J, GPa
Pressed trinitrotolene	1.59	6.94	1.6	17.7
Cast trinitrotoluene	1.62	6.85	1.61	17.9
Hexogen (RDX)	1.72	8.50	2.12	31.0
Hexogen (RDX)	1.59	8.14	2.00	25.9
Tetryl	1.68	7.50	1.87	23.6
Petn	1.66	8.10	1.83	24.6

The magnetoelectric sensor can be used to record the reflected rarefaction wave propagating in the EP. The recording method is discussed in [35]. A block of paraffin possessing a lower dynamic hardness than the EP is attached to the end of the HE charge. The rarefaction wave resulting from the detonation front reflection from the paraffin moves in the same direction as the EP flow. The arrival of the reflected wave at the sensor increases its velocity, and a characteristic rise appears in the $u(t)$ profile. The experimentally measured motion of the reflected wave enables us to calculate the sound speed in the rarefaction wave, and by using an appropriate EOS (such as the Landau-Stanyukovich EOS), we can determine the density and pressure distribution behind the detonation front.

8.3.5 X-ray Method

In addition to the electromagnetic method, the X-ray method [36, 37] is also a direct method for studying the motion of EP. In the X-ray method, the EP state is determined by experimentally recording the position and motion of the DW front using thin metal foils placed inside HE charge (Figs. 8.26a, b).

In the first scheme, which is called the "zebra" scheme, thin (0.02 mm) lead foils, which act as Lagrangian movement sensors, are positioned between HE layers comprising the charge to be studied. In the second scheme, a narrow strip of the same foil is positioned at an angle to the detonation wave front, enabling continuous tracking of its displacement under the action of

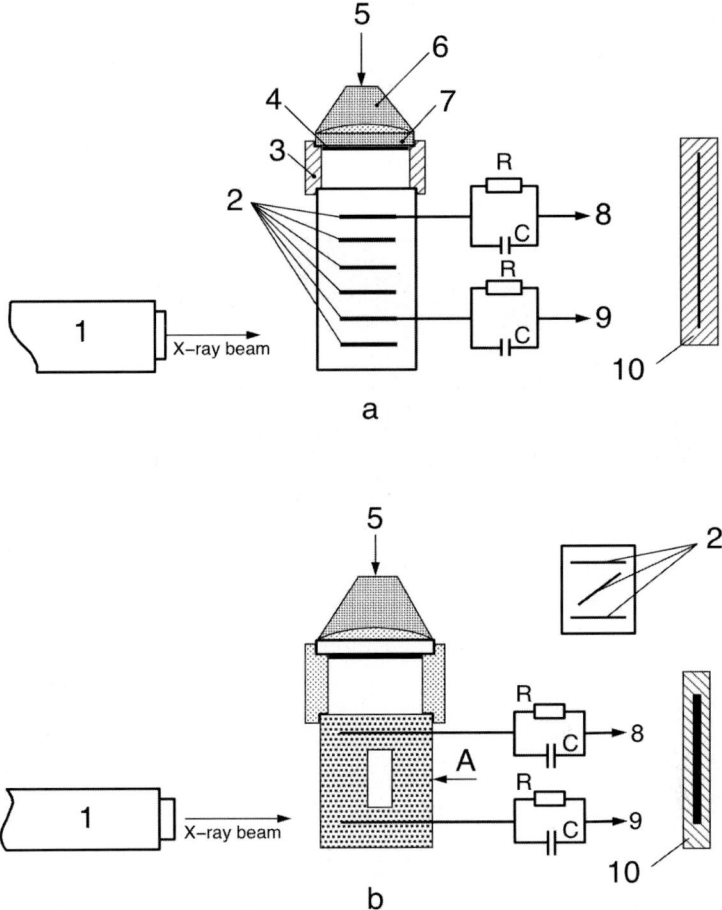

Fig. 8.26. Schemes of detonation parameter estimation with the X-ray method:
a – with the "zebra" charge; **b** – with the inclined foil: 1 – X-ray apparatus; 2 –
lead foil; 3 – plastic foam ring; 4 – Al foil (0.1 mm thick); 5 – blasting cap; 6 –
plane detonation wave generator; 7 – HE charge; 8, 9 – contacts for detonation rate
measurement; 10 – film pack

detonation products and a determination of the flow structure in the vicinity
of the detonation front.

In the first scheme, at X-ray pulse time the detonation wave front is located, as a rule, in the space between the two foils farthest from the initiation
plane. In the second scheme, the detonation wave front is normally located
at a distance from the initiation plane such that a portion of the foil has not
yet been traversed by the detonation wave. In either case, the pulse from the
X-ray source is timed so that the detonation wave will be a given distance
from the initiation plane.

The X-ray source is a pulsed X-ray apparatus with a pulse duration at half-height ~0.1 μs at full maximum discharge voltage ~1 MV and ~20 kJ of stored energy, enabling reliable X-ray radiography of charges up to 120 mm in diameter. The accuracy in estimates of displacement with these pulse parameters is ~0.4 mm. A characteristic photographic print of an experimental X-ray image is presented in Fig. 8.27.

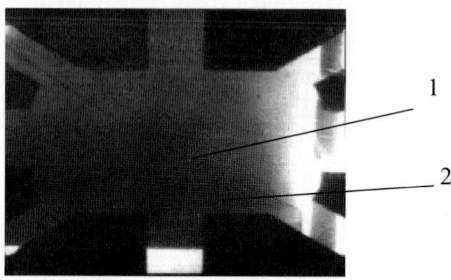

Fig. 8.27. A typical X-ray image from the inclined-foil experiment: 1 – position of the lead foil at the shot moment; 2 – HE charge end

In experiments aimed at investigating detonation wave structure, initiation can be performed in a number of different ways: by plane-wave generators, by simultaneous initiation across the end plane using exploding grids, and by thin foil impact on the HE.

The experimental results can be processed and used to estimate the detonation field variables and to construct an EOS of the EP, only when the motion of the EP behind the front is self-similar. Consequently, initiation with lens systems is undesirable since they exert a rather severe influence on the motion of the EP behind the detonation wave front. Use of these lens systems leads to misrepresentation of the results and difficulties in their interpretation.

Perhaps the most acceptable method of initiation for the experiments just described is initiation by means of a short pulse generated by impact on the HE of a thin (0.1 mm) aluminum foil traveling at a velocity of 5–6 km/s. According to reliable estimates, a pressure pulse ~50 GPa amplitude and ~0.02 μs duration occurs in the HE during such an impact, which is quite sufficient for the production of steady-state detonation in the HE from the very beginning. Moreover, any potential supracompression that may occur will not significantly affect the motion of the EP behind the front, since this is a fairly short-time event, comparable in duration to the chemical reaction zone. Experimental evidence shows that normal detonation conditions are established very quickly. At distances of 3–5 mm from the impact plane, the detonation rate differs from the normal rate by no more than 1.5%.

Since this method is Lagrangian, the relations needed for processing the experimental results should be expressed in terms of a Lagrangian coordinate system.

In the Lagrangian coordinate system, the law of conservation of mass enables immediate computation of EP density:

$$\rho = \rho_0 \frac{dx_0}{dx} \, . \tag{8.13}$$

Here x is the coordinate measurement from the detonation wave front to the foil in the preliminary image, and x_0 is the coordinate measurement from the detonation wave front to the same foil, but moving with the EP.

The following well-known relationship for isentropic flow [38],

$$du = \frac{cd\rho}{\rho}; \qquad \frac{x_0}{t} = \frac{\rho}{\rho_0} c \tag{8.14}$$

may be combined with the following relationships, which follow from the self-similarity condition,

$$\frac{x_i}{t} = u_i + c_i; \qquad \frac{x_i}{L/D} = u_i + c_i \tag{8.15}$$

to derive expressions for the mass velocity and sound speed:

$$u = D \left(\frac{x}{L} - \frac{x_0}{L} \frac{dx_0}{dx} \right); \qquad c = D \frac{x_0}{L} \frac{dx_0}{dx} \tag{8.16}$$

The Jouguet parameters are estimated in accordance with the laws of conservation of mass and momentum at the detonation jump

$$\rho_J = \rho_0 \frac{dx_0}{dx}; \quad P_J = \rho_0 D^2 \left(\frac{x}{L} - \frac{x_0}{L} \frac{dx}{dx_0} \right) \quad \text{with} \quad \frac{x}{x_0} = 1 \, . \tag{8.17}$$

In the above, L is the detonation wave front coordinate, D is the detonation rate, and u is the mass velocity following the detonation wave front.

In terms of an equation of state, it is of interest to specify either $P(u)$ or $P(V)$ ($V = 1/\rho$ is specific volume), whose isentropic derivatives are related to impedance ρc by the following well-known relations [38]

$$\rho c = \left| \frac{dP}{du} \right| = \left| \frac{du}{dV} \right| = \left| -\frac{dP}{dV} \right|^{1/2} \, . \tag{8.18}$$

Thus, in addition to determining u or c from the experimental data, the above relationships, can be used to determine derivatives of the flow variables. This in turn provides the means for a simple solution to the problem posed: finding the material expansion adiabat from the experimental profile of centered rarefaction waves.

Once $x\,(x_0, t_1)$ at known time t_1 in the moving medium has been determined from the X-ray experiment, we use the law of conservation of mass $V = V_0 dx/dx_0$ to find $V = V_0 f(x_0)$ and the inverse function $x_0(V)$. From Eqs. (8.14) and (8.18) we obtain:

$$P\left(V\right) = P_j - \frac{\rho_0^2}{t_1^2} \int_{V_j}^{V} x_0^2\left(V\right) dV \qquad (8.19)$$

and the parametric form of $P(u)$ (with V being the parameter) by Eqs. (8.19) and (8.20),

$$u\left(V\right) = -\frac{\rho_0}{t_1} \int_{V_j}^{V} x_0\left(V\right) dV . \qquad (8.20)$$

It should be noted that the flow variables of the EP at and near the detonation front are more accurately determined from experiments with the inclined foil than in experiments with foils placed parallel to the planar detonation front. In this case, the displacements are recorded "continuously", which allows for measurement of the displacement to be taken significantly closer to the front. Consequently, the distance of extrapolation to the detonation front is smaller using the inclined foil method, thereby enabling a more accurate computation. Another feature of the inclined foil method that makes it a more attractive method is that it requires less inert pad mass (one inclined foil instead of several positioned parallel to the detonation wave front). Fewer pads means fewer density discontinuities in the HE charge (due to the presence of the pads). These features: less pad mass, fewer density discontinuities, and the ability of detecting the flow structure in the vicinity of the detonation front all testify to the attractiveness of using this method.

For purposes of comparison, Table 8.4 presents Chapman-Jouguet variables for a 50/50 mixture of TNT/RDX estimated using the X-ray method, electromagnetic method, and the method of barriers.

Table 8.4. Chapman-Jouguet parameters for mixture TNT/RDX 50/50

Method	ρ_0, g/cm^3	D, km/s	u_J, km/s	P_J, GPa
X-ray	1.66	7.56	2.04	25.6
Electromagnetic	1.68	7.65	2.07	26.6
Method of barriers	1.66	7.55	2.02	25.3

From Table 8.4 it is seen that all three methods provide essentially the same results. The maximum difference in the Jouguet pressure is no greater than 5%.

8.3.6 Jouguet Parameter Estimation Using Laser Measuring Systems

Laser methods ensure high resolution in time when applied to investigations of the structure of the detonation front. A detailed description of laser measuring systems and their application to dynamic experiments can be found in

Chap. 9. Features of the laser method, as specifically applied to estimation of the Jouguet state are discussed below.

Laser Wave Velocity Measurement (LWVM) Method

The LWVM method [32] involves the measurement of shock wave velocity in a package of transparent plates that are in intimate contact with the end surface of the high explosive under study. The time required for the shock to wave travel across each plate is measured. Knowing the plate thicknesses, average shock wave velocity is easily estimated. Accuracy in this measurement is typically 0.5–1.0% on plates with a thickness ≥ 1 mm, and 1.5–2.0% on plates with a thickness ~ 0.2 mm. The first step in determining the Jouguet state involves an assessment of the high-pressure zone of the detonation front. This assessment is made in terms of the decay of the shock wave that occurs in barriers, measured over a short distance. The Jouguet pressures are then calculated using Eq. (8.12). A complication with this method is that it is discrete, and requires a fine barrier surface finish. Experimental results [32] obtained using this method are summarized in Table 8.5.

Table 8.5. Chapman-Jouguet parameters for individual and mix HE estimated with the LWVM method

HE	ρ_0, g/cm^3	D, km/s	Barrier material	P_J, GPa
Pressed TNT	1.56	6.84	Plexiglas	18.6 ± 0.3
			Glass TF-5	17.7 ± 0.4
	1.58	–	Plexiglas	18.6
Phlegmatized petn	1.65	8.11	–"–	24.6 ± 0.4
			Glass TF-5	24.1 ± 0.4
Phlegmatized HMX	1.77	8.69	Plexiglas	31.8 ± 0.4
Agated RDX	1.78	–	–"–	35.2
HMX	1.87	–	–"–	36.9

Laser Interferometry Method

When using laser interferometry, the nature of the high-pressure zone is determined from measurements of the mass velocity profile at the interface between the HE under study and an inert barrier. To facilitate this measurement, light-reflecting metal pads are placed between the HE charge and the barrier. Aluminum or copper foil [39] 0.012–0.025 mm thick is typically used. The foil is pressed directly onto the end surface of the charge to be studied. Water is the most frequently used barrier material [40], although LiF crystals are also sometimes used for this purpose [41]. In the latter case, an aluminum layer (one micrometer thick) is applied directly onto the crystal surface. Results of experimental measurements [40] are summarized in Table 8.6.

Table 8.6. Chapman-Jouguet parameters estimated with the laser interferometry method

HE	ρ_0, g/cm^3	D, km/s	u_J, km/s	P_J, GPa
	1.41		2.0	–
Phlegmatized RDX	1.60	6.84	2.1	–
	1.67		2.2	–
PBX-9502	1.87		-	28.5

Photoelectric Method

In the photoelectric method, the nature of the high-pressure zone is recorded by shock front luminance in an indicator liquid in contact with the charge of HE being studied (Fig. 8.28).

A container holding the indicator liquid (bromoform, carbon tetrachloride, other methane halogen derivatives) is placed on the end of the HE to be studied. A shock wave enters the indicator liquid at the interface between HE and indicator. To block detonation product radiation, a thin ($\sim$0.05 mm) aluminum foil is placed on the end surface of the charge. Diaphragms possessing a hole through which light may pass are used to clip the visible part of the shock front to the required size. Shock front radiation is transmitted to a photoelectric multiplier through an optical fiber and recorded with an electrooptical pyrometer.

Indicator liquid pressure is estimated by luminance as a gage function of pressure. Because this dependence is strong (as noted in Chap. 4), the accuracy of the pressure estimation is quite high. Examples of recorded shock

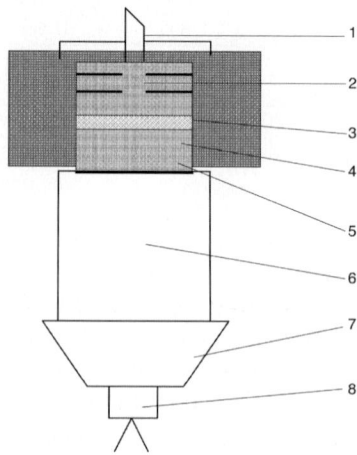

Fig. 8.28. Scheme of the experiment on Jouguet pressure estimation with indicator liquid: 1 – light guide; 2 – diaphragms; 3 – glass plate; 4 – indicator liquid; 5 – Al foil; 6 – charge of HE under study; 7 – plane-wave lens; 8 – detonator

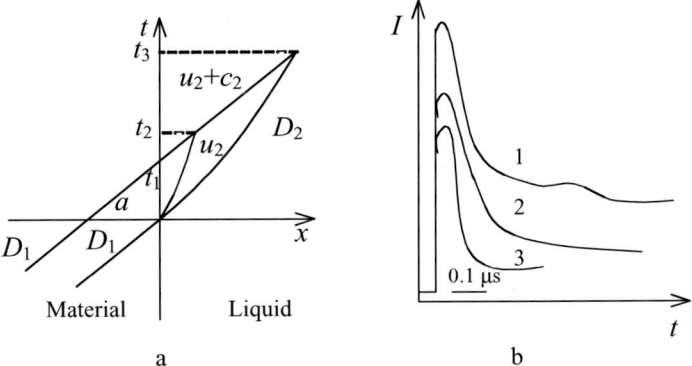

Fig. 8.29. The $x - t$ diagram of the flow (**a**) and typical records of shock front radiation luminance in carbon tetrachloride (**b**) in contact with the charges of: 1 – pressed TNT/RDX 50/50 ($\rho_0 = 1.65$ g/cm^3); 2 – cast trinitrotoluene ($\rho_0 = 1.60$ g/cm^3); 3 – pressed trinitrotoluene ($\rho_0 = 1.60$ g/cm^3)

front radiation luminance [42] in carbon tetrachloride situated at the end of detonating TNT and 50/50 TNT/RDX charges are presented in Fig. 8.29b. In each case, is there evidence of the nature of the high-pressure zone. Jouguet pressures are found using Eq. (8.12).

The photoelectric method can be used to measure any change in mass velocity at the interface. Conversion, from the experimentally determined pressure at the shock front in the indicator liquid to the state of the field variables at the interface, is performed using the $x-t$ diagram of wave propagation. The $x-t$ diagram presented in Fig. 8.29a is constructed using the experimentally determined pressure-time history and known Hugoniot of the indicator liquid.

Times t_2 and t_3 (corresponding to the occurrence of identical velocities u_2 at the HE-indicator liquid interface and shock front in the liquid) are related as follows:

$$\int_0^{t_2} u_2 \, dt + (u_2 + c_2)(t_3 - t_2) = \int_0^{t_3} D_2 \, dt \, ,$$

where D_2 and c_2 are, respectively, shock wave velocity and sound speed behind the shock front with mass velocity u_2. Knowing the relationship between pressure and sound speed in the indicator liquid ($c_2(P)$), and the shock wave velocity in the liquid ($D_2(t)$), the solution of the above equation yields the change in mass velocity at the interface. The times of the pressure change in the chemical reaction zone may then be found from the following relations:

$$t_1 = a/D_1; \qquad t_2 = D_1 t_1/(D_1 - u_{ave}),$$

where a is the chemical reaction zone width, D_1 is the detonation rate, and u_{ave} is the average mass velocity at the interface. The time resolution of the method is ~ 5 ns [42].

Pressures for cast and pressed TNT and a 50/50 mixture of TNT/RDX at the Jouguet point that have been found using the above-discussed method are presented in Table 8.7.

Table 8.7. Chapman-Jouguet parameters estimated with the photoelectric method

HE	ρ_0, g/cm^3	P_J, GPa
Pressed TNT	1.60	20.86
		20.27
		20.27
Cast TNT	1.60	20.95
		20.45
		20.29
Pressed TNT/RDX 50/50	1.65	26.07
		25.21
		24.95

8.3.7 Method of Barriers

This method, discussed in detail in Chap. 4, is used to measure the shock wave field variables in an inert material (barrier) in contact with the HE under study. Since the velocity and pressure of EP adjacent to the barrier are the same as those in the barrier at the time of shock reflection, the variable to be measured experimentally in the method of barriers is the wave velocity D_{bar} in the barrier.

When the detonation wave reaches the barrier, this precipitates the propagation of a reflected wave through EP. If the dynamic impedance of the barrier is greater than that of the EP ($\rho_0 D_{bar} > \rho_{0HE} D$), the EP compress (decelerate) and their pressure increases. If the dynamic impedance of the barrier is less than that of the EP, the EP expand and their pressure drops. Having measured the wave velocity in a barrier with a known Hugoniot, we can determine the EP expansion or deceleration point from that curve.

Because shock waves decay in barriers, in order to find the true reflection state, wave velocities are measured for different thicknesses of barriers and the results are extrapolated to the interface.

Strictly speaking, to obtain state variables on the EP deceleration curve, functions $D(x)$ or $u(x)$ (where x is barrier thickness), have to be extrapolated to a barrier thickness that corresponds to the boundary of the peak chemical high-pressure zone. Estimates for a number of materials (Mg, Al, teflon) suggest that this thickness is ~1 mm. Differences in the estimated values of D and u arrived at by extrapolating to $x = 0$ versus $x = 1$ mm are on the order of 1%, which is within the measurement error.

The experimental results are plotted on a graph of P versus u, and the straight line $P = \rho_0 D u$, which represents the locus of points for a given deto-

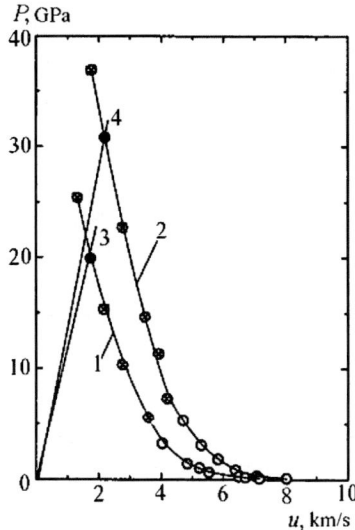

Fig. 8.30. EP expansion isentropes: 1 – TNT [43]; 2 – TNT/RDX 25/75 [44]; 03, 04 – detonation beams; ● – Jouguet point. The experimental data on EP expansion to condensed ⊗ and gaseous ○ barriers

nation rate D, is drawn. The intersection point of that straight line with the experimental curve $P(u)$ determines the pressure and mass velocity at the Chapman-Jouguet point (Fig. 8.30).

This construction is used not only to determine the Jouguet point, but also the isentropic derivative, which is common to the isentrope, detonation adiabat, and re-compression adiabat at the Jouguet point. Consequently, when the pressures are very close to the Jouguet state pressure, the shock Hugoniot and isentrope can be assumed to coincide and are therefore described by a single curve. The accuracy of this assumption has proven to be adequate in many cases. The above-described method was used in [43, 44] to find EP unloading isentropes of TNT, a 50/50 composition of TNT/RDX, and a 25/75 composition of TNT/RDX. It should be noted that by using inert gases such as argon or xenon (initially compressed to 1–100 atm) for the barrier material, the run of the expansion waves can be carefully tracked, enabling determination of the expansion isentrope at pressures below 1 GPa. Moreover, doing so reveals noticeable disagreement between the computational prediction produced by a cubic EOS for the EP ($P = A\rho^3$) and the experimental evidence.

8.3.8 Equations of State of EP

Monographs [29, 45] and review papers [46, 47] on condensed HE detonation inevitably touch on the question of the EOS of EP. This is natural since the EOS of the EP is required, both for the solution of the system of differential equations of motion, and for computation of the Chapman-Jouguet state.

Accurate theoretical calculation of the EOS of condensed HE EP requires the use of interaction potentials and a calculation of the statistical sum, which is difficult. Therefore, when seeking the EOS of EP, one tries to combine the theoretical concepts of EP behavior at high densities, pressures, and temperatures with experimental data from detonation wave studies.

Two approaches to the construction of the EOS of EP have been recently established. In one case, the contribution of each individual component comprising the EP is accounted for and summed to arrive at a total; the other approach does not consider the EP composition but rather focuses on an averaged description. The second approach includes the widely used Mie-Grüeneisen form of the EOS of EP, which includes pressure and energy on a reference curve (for example, Eq. (1.2)). The EP expansion isentrope passing through the Jouguet point serves as a convenient reference curve, and may be derived in a rather straightforward manner from the experimental data.

Many versions of the EOS for EP have been developed. This section treats the simplest forms of EP EOS that have been included in RFNC–VNIIEF computer programs.

For description of the motion of EP, wide acceptance has been gained by a very simple EOS: $P = A\rho^n$ (Eq. (1.49)) and its extension $P = (\gamma - 1)\rho E + A\rho^n$ (Eq. (1.50)). Derivations of the principal relations needed for determination of the Jouguet state for the polytropic EOS of EP are provided in Chap. 1.

Most EP expansion related problems are significantly simpler to solve when $n = 3$, i.e., when the EP EOS has the form $P = A\rho^3$.

L.D. Landau and K.P. Stanyukovich were the first to note good agreement of that equation with experimental data and the simplicity of problem solution that it affords. The simplification is related to the fact that in this case, the Riemann invariants are $u \pm 2/(n - 1) = u \pm c$, i.e., they coincide with the slope of the characteristics $x = (u \pm c)t$. Therefore the characteristics of the opposite families do not interact with each other, and turn out to be straight lines.

Recent analysis of many Russian and American studies suggests that the detonation variables for the most widely used HE have been estimated to an accuracy of 2–3% [48]. The appropriate value of the isentropic exponent n at the Jouguet point is to a good approximation, 3, for many condensed HE. Deviations toward higher or lower values are in general no greater than 5%.

The polytropic exponent n is therefore taken to be 3 in the solution of many gas-dynamic problems (other HE parameters may be modified in order to minimize the error in the computation). A typical problem involves calculation of the high-pressure interaction that takes place when a detonation wave reaches a barrier (reflection from a metal barrier or plate acceleration). In this case it is desirable that the simplified EOS provide the correct pressure of the wave that is reflected into the EP.

In deceleration on a perfectly rigid wall, $u = 0$ and pressure in the reflected wave is determined by Eq. (1.57), which would be reasonably written in this case as:

$$P_{refl} = \frac{\rho_0 D^2}{n+1} \left(\frac{3n-1}{2n} \right)^{\frac{2n}{n-1}} .$$

According to E.I. Zababakhin [49], the above calculation of P_{refl} is retained unchanged if we replace n with 3 and introduce an effective value of $\rho_0 D^2$ in place of the true value:

$$\frac{\rho_0 D^2}{n+1} \left(\frac{3n-1}{2n} \right)^{\frac{2n}{n-1}} = \left(\rho_0 D^2 \right)_{eff} \frac{16}{27} . \tag{8.21}$$

It is usually convenient to introduce an effective detonation rate, assuming the HE density to be equal to the true value. In this case the following expression for the effective detonation rate applies:

$$\frac{D_{eff}}{D} = \frac{3}{4} \sqrt{\frac{3}{n+1}} \left(\frac{3n-1}{2n} \right)^{\frac{n}{n-1}} . \tag{8.22}$$

$P - u$ diagrams for both the true and effective HE are shown in Fig. 8.31.

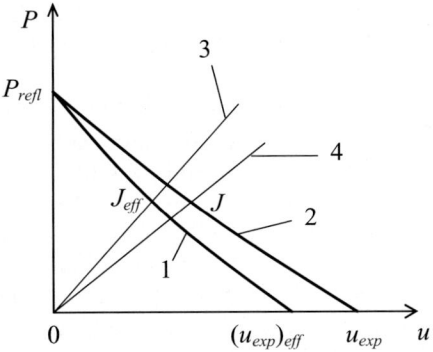

Fig. 8.31. EP deceleration curves (1, 2) and wave beams for the effective (3) and true (4) HE

The deceleration curves intersect at $u = 0$ and then diverge. Points P_J and $P_{J(eff)}$, which correspond to the normal wave are also different. The extent of divergence between the two deceleration curves can be quantified by calculating the difference between the EP expansion rates for either HE at zero pressure $u_{exp}(P = 0)$:

$$\frac{(u_{exp})_{eff}}{u_{exp}} = \frac{D_{eff}}{D} \frac{n^2 - 1}{3n - 1} . \tag{8.23}$$

The values of D_{eff}/D and $(u_{exp})_{eff}/u_{exp}$ for different values of n as calculated by Eqs. (8.22) and (8.23) are summarized in Table 8.8.

Table 8.8. Ratio between detonation rates and EP expansion rates for effective and true HE with different values of n

n	D_{eff}/D	$(u_{exp})_{eff}/u_{exp}$
2.0	1.175	0.703
2.5	1.074	0.868
3.0	1.000	1.000
3.5	0.940	1.130

From the table it is seen that for $n \approx 3 \pm 0.3$ the value of $(u_{exp})_{eff}/u_{exp}$ evaluated at $P = 0$ differs by no more than 10%, i.e., the true and effective HE deceleration curves not only intersect at the upper point, but remain close to one another throughout.

To enhance the accuracy of the solution, primarily for problems where the degree of expansion of the explosion products relative to the Jouguet state is significant, RFNC–VNIIEF utilizes an EOS for the EP of the form $P = P_S(\rho) + \Gamma\rho(E - E_S)$, in which the isentrope dependence suggested by V.N. Zubarev [43] is expressed as $P_S = A^{-k/\rho} + B\rho^\gamma$, where A, k, B, and γ are constants. At VNIIEF, this EOS has been called Zubarev's simplified form of EP EOS. It removes the above-mentioned disadvantages of the polytropic EOS (Eq. (1.48)), and is used for the numerical solution of problems that require an accurate description of EP motion over a wide range of specific volume.

An experimental basis, necessary for an evaluation of the five constants appearing in the EOS of EP is: determination of state variables at the Jouguet point, the expansion isentrope, and detonation rate D as a function of initial density ρ_0. The requirement that the isentrope pass through the Jouguet point, and do so in such a way that Michelson-Rayleigh straight-line tangency is satisfied, supplies two equations for estimation of the constants. The parameters that remain free are selected so that the experimental data on the unloading curve and $D(\rho_0)$ are best approximated. Table 8.9 summarizes the parameters at the Jouguet point and constants in the EOS of the EP of three HE according to [43, 44].

For purposes of illustration, Fig. 8.32 compares the detonation rate as a function of initial TNT density, as calculated using the above discussed EOS with the constants from Table 8.9, with Camlet's linear empirical description, and the experimental data of [43].

Table 8.9. Constants in the EOS of EP for three HE

HE	ρ_0, g/cm³	D, km/s	u_J, km/s	P, GPa	A, GPa	B, GPa	k	γ	Γ
Trinitrotoluene	1.63	7.00	1.74	19.9	521.7	1.762	7.876	1.6	0.9
TNT/RDX 50/50	1.67	7.61	2.03	25.8	442	1.679	7.113	1.6	0.9
TNT/RDX 25/75	1.72	8.02	2.20	30.5	459.3	1.755	7.037	1.6	0.65

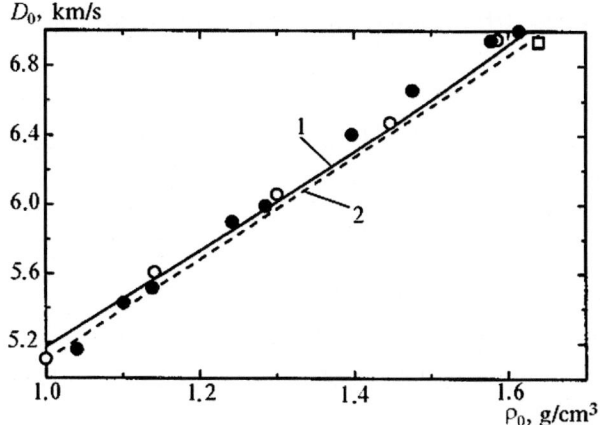

Fig. 8.32. Detonation rate versus initial trinitrotoluene density: 1 and 2 – calculations by EOS with the constants from Table 8.9 and by Camlet's empirical relation; □, •, ○ – experiment

For purposes of demonstrating the validity of using the above-discussed EOS for EP in the solution of initial-boundary value problems, we present Fig. 8.33, which shows a comparison between the velocity predicted, and that experimentally measured, of a steel shell driven by the expansion of EP. The explosive was a 25/75 composition of TNT/RDX, and the expansion of its EP was modeled using the parameters of Table 8.9. The steel shell had a thickness of 3.08 mm and an outer radius of 77 mm. Such shells are used as impactors in the hemispherical shock wave generators SC-3, SC-4, and SC-7 discussed in Chap. 2. The HE charge had an outer radius of 220 mm. A Mie-Gruneisen EOS was used in the calculation for steel.

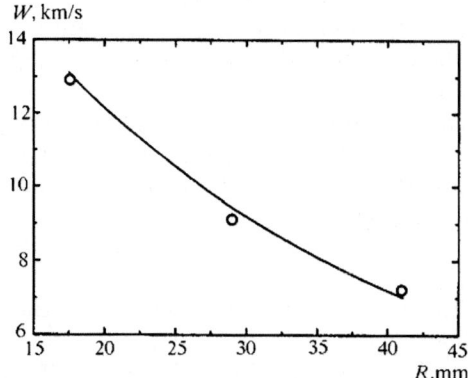

Fig. 8.33. 3.08-mm-thick steel shell velocity versus measurement radius: *solid line* – computation; *open circles* – experiment

8.4 Estimation of Detonation Cutoff Diameter

The term cutoff diameter refers to the diameter of a cylindrical charge of HE, below which a self-sustained detonation is impossible regardless of initiating pulse power. For any cylindrical charge with a diameter less than the cutoff diameter, a detonation wave will cease to propagate at some distance from the initiation point. The existence of detonation cutoff diameter is associated with energy losses in the detonation chemical reaction zone. Since the chemical transformation zone in the detonation wave is of finite dimension, a rarefaction wave enters the reaction zone from the side surface of the charge. The material mass that has been swept by the rarefaction wave ceases to participate in the chemical transformation, and this leads to heat losses in the reaction zone and to a slower detonation rate.

For a charge of sufficient size, the relative energy losses in the detonation wave are negligible. With decreasing charge diameter, the relative energy losses increase and at some point the energy release and losses in the chemical reaction zone become unbalanced. This imbalance occurs at the cutoff diameter and the detonation wave ceases to propagate; a stable steady detonation becomes impossible in the HE charge. Reasoning from these concepts, Yu.B. Khariton formulated the principle of the critical condition for detonation propagation: detonation can propagate steadily through the charge if the chemical reaction duration (τ_x) is shorter than the time of the material expansion in the radial direction $\tau = d/2c$. The detonation cutoff diameter is determined by $d_{cutoff} \approx 2c\tau_x$, where c is the average sound speed in the detonation wave.

The cutoff diameter is an important characteristic of HE. It allows estimation of the minimum initiator force needed for actuation of a stable HE detonation, and together with the detonation rate, estimation of the reaction zone width, the time of chemical reaction, and the degree of its completion in the detonation wave. The most widespread schemes for the estimation of detonation cutoff diameter are presented in Figs. 8.34a and 8.34b [50, 51].

In the conical charge, the detonation is initiated near the base of the cone so that the detonation front propagates in the direction of decreasing charge diameter. The detonation cesation point is sometimes determined from photographic records of the detonation propagation, but more often from the imprint left on a witness plate from the detonation wave shattering effect. Due to the supracompression effect (a consequence of detonation propagation from larger to smaller diameter) the detonation wave will cease to propagate at diameters somewhat smaller than the actual cutoff diameter. To reduce this effect to a minimum, it is necessary to strive for the minimum possible apex angle in the cone.

A more accurate, but also more labor-intensive method for the estimation of cutoff diameter uses stepped cylindrical charges (see Fig. 8.34b). In this case the length l of each cylinder has to be sufficient for all charge transients re-

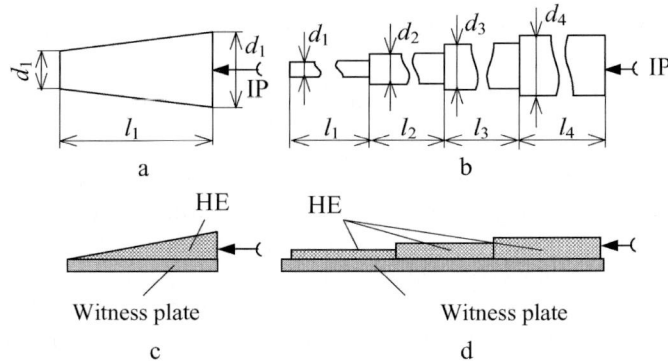

Fig. 8.34. Experimental methods for estimation of detonation cutoff diameter (d_{cutoff}) and critical detonation thickness (h_{cr}): **a** – conical charge method; **b** – cylindrical charge method; **c** – wedge method; **d** – stepped charge method; IP – initiating pulse

lating to the "supporting" action of the EP from the previous larger-diameter cylinder to be completed. If $d \geq 5d_{cutoff}$, the length of the cylinder must be $\geq 3d$. If $d \cong d_{cutoff}$, then the required length of the cylinder increases to 10–$12d_{cutoff}$.

Using either of the aforementioned methods, measurement of the cutoff diameter of high-density HE charges (not confined by a shell) to be accurate, it is necessary that cylindrical charges of perhaps very small diameter be fabricated. This fabrication must be done very carefully and involves significant manufacturing difficulty. It is this fact that led to the development of the wedge method and stepped charge method (Figs. 8.34c and 8.34d) for measurements of "critical thickness." In these methods, initiation is also performed near the charge base and the detonation wave propagates in the direction of decreasing thickness. The point of detonation extinction is determined from the imprint left on a metal witness plate. However, it must be recognized that the witness plate, whose dynamic hardness is greater than that of HE, undoubtedly affects the HE chemical decomposition reaction rate in the detonation wave.

In response to the above-mentioned shortcoming with the wedge method, the method has been modified [52] to avoid the influence of the witness plate material on critical material detonation depth. In this modified method, the wedge-shaped HE charge is placed with one side surface on the witness plate and is initiated by a plane detonation wave generator over the entirety of the other side surface (Fig. 8.35). The detonation wave extinction point is registered by the imprint left on a witness plate.

Experiments show that the detonation cutoff diameter depends on the nature of the HE: charge structure (HE density, grain size), method of fabrication, component content, and temperature [53–55]. The general trend is such

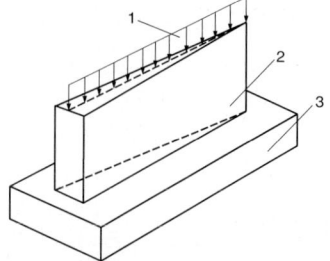

Fig. 8.35. Modified method for estimation of HE charge cutoff depth: 1 – linear detonation generator; 2 – wedge of HE to be studied; 3 – witnessing plate

that the detonation cutoff diameter increases with increasing HE particle size, and decreases with increasing density and temperature [54]. A strong dependence of the cut-off diameter on initial temperature is evidenced by the data for molten trinitrotoluene and nitroglycerine [53], reproduced in Fig. 8.36.

The effect that method of manufacture can have on the detonation cutoff diameter is demonstrated in Table 8.10 for the example of cast versus pressed TNT of the same density. The cutoff diameter is noticeably larger in cast TNT. The cutoff diameter of cast trinitrotoluene also depends on the charge cooling rate: the faster the cooling rate, the larger the cutoff diameter. According to US investigators [54], it ranges from 14 mm to 38 mm. Such a strong effect

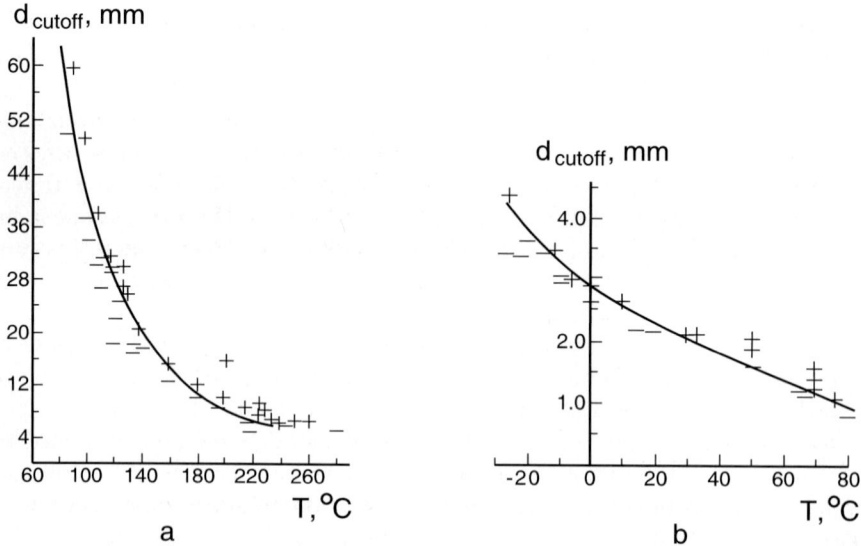

Fig. 8.36. Liquid trinitrotoluene (**a**) and nitroglycerine (**b**) detonation cutoff diameter versus temperature. + – stationary detonation; — – failure

Table 8.10. Trinitrotoluene cutoff diameters

ρ_0, g/cm^3	d_{cutoff}, mm	Particle size, μm
0.8	10	
1.00	9.0	70–200
1.46	4.3	
1.55	3.3	
1.0	9	60

is attributed to a significant change in the microstructure of the charge for different rates of cooling.

Cutoff diameters of several high explosive single crystals are summarized in Table 8.11 [55].

Table 8.11. HE single crystal cutoff diameters

HE	ρ_0, g/cm^3	d_{cutoff}, mm
Petn	1.78	5.0
Hexogen (RDX)	1.80	7.0
Octogen (HMX)	1.904	18
TNT	1.663	110

8.5 He Brisance, Efficiency (Fugacity), and Launching Ability

Work done by expanding EP is given by $A = \int_{V_1}^{V_2} P dV$ (where V_1 and V_2 are the EP volumes at the beginning and at the end of the work event, and P is the EP pressure), and is equal to the area under the process curve in $P - V$ space, as illustrated in Fig. 8.37. In general, the amount of work done depends on several factors: the initial pressure of the EP, the density ($\rho = 1/V$) relative to the density of the condensed HE, the chemical composition, and the expansion ratio. In turn, the expansion ratio depends on both the intensity and duration of the interaction between the EP and the medium upon which work is being conducted. Hence, explosion efficiency is affected by properties both of the HE itself, and of the loaded medium. Thus, when solving practical problems, common practice is to use a relative measure of HE efficiency as opposed to some absolute measure. The relative measure compares the action of the "new" HE to that of HE that has been tested under the same conditions and is used as a reference.

Two types of explosion work are distinguished: brisant forms, directly related to the effects of the detonation wave front, and fugacious forms, which are for the most part related to the work done by EP expansion. The term efficiency is often used instead of fugacity.

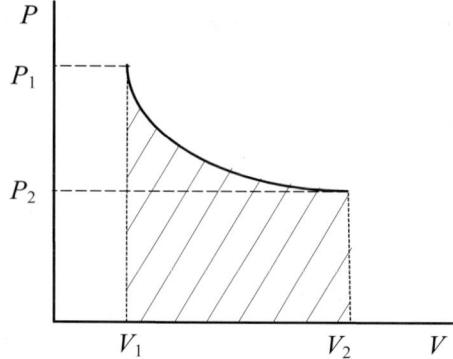

Fig. 8.37. Work of expanding EP

Brisance (shattering effect) is the ability of the HE to do damage as a result of an abrupt EP impact (sudden rise in pressure) in regions in near proximity to the HE . The brisant form of work is done in a few microseconds, and the fraction of energy expended is a few to several tens of percent. The HE shattering effect is most pronounced in armor-piercing EP action, rock fragmentation, and different forms of explosive metal working.

An example of fugacious action is the ejection of soil brought about by the explosion of a charge placed in a borehole. In fugacious EP work, up to 85% of total explosion energy is released over a few milliseconds. The part of the explosion energy that is not converted to mechanical work is termed thermodynamic losses. There are also intermediate forms of explosion work, where the work time (time of energy extraction from the EP) is longer than that of the explosion shattering action, but shorter than that of the fugacious action. The work of launching is an example of an intermediate form of HE work.

The variety of forms and factors affecting explosion work means that in principle, it is impossible to develop a unique method for determining the efficiency of an explosive. For this reason, numerous experimental methods are used in the estimation of HE efficiency, and these methods enlist the use of a number of materials in the role of receiving media (e.g. air, water, hard rocks, earth, or metals (steel, lead)).

8.5.1 Shattering Action of Explosion

Hess's Test

A very simple and widely used method for testing the shattering effect of HE is Hess's test, a diagram of which is presented in Fig. 8.38 [56]. In Hess's test, a lead cylinder 60 mm in height and 40 mm in diameter is positioned vertically on a massive steel support. A steel plate 10 mm in thickness and 41 mm in diameter is placed atop the lead cylinder. A 50-g charge of HE (surrounded by

ED - 8

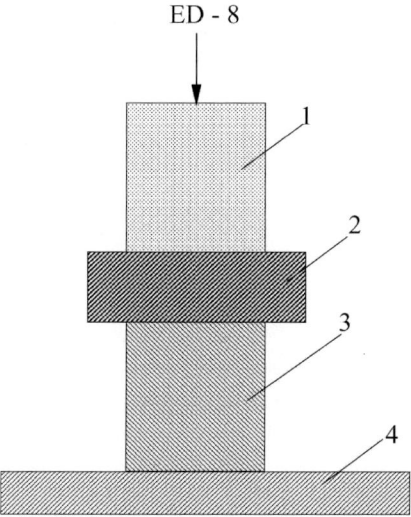

Fig. 8.38. Diagram of HE testing for the shattering effect by the Hess's test: 1 – HE charge; 2 – steel base plate; 3 – lead cylinder; 4 – massive steel support

a 40-mm-diameter paper shell) is placed onto the steel plate. In the standard test the charge density is 1.0 g/cm^3. Under action of the charge explosion, the lead cylinder is deformed. The difference between the average height (in mm) of the lead cylinder before and after the explosion is taken as a measure of brisance. The brisance of a few HE as measured using Hess's test [56] are as follows: 20–22 mm for tetryl, 16 mm for trinitrotoluene, and 14.8 mm for amatol 80/20.

Because of its extreme simplicity, Hess's test is widely used as an acceptance test for industrial HE. However, it has significant shortcomings as a method for the assessment of brisance:

– the final result is not in units of energy or work, but in units of length;
– since the resistance to deformation increases as the test proceeds, the relationship between the length of the cylinder and the work of deformation is nonlinear;
– sometimes the reduction in cylinder height is largely dependent on the HE detonability and cutoff diameter;
– the standard version of Hess's test is only possible for HE with low brisance, since powerful HE (petn, hexogen, etc.) result in essentially complete destruction of the lead cylinder.

For HE with low detonability, which can not achieve steady detonation in charges confined within a 40-mm-diameter paper shell, and which require a distance for detonation acceleration that is greater than that afforded in the standard test, Hess's test is modified somewhat: the HE is placed in steel rings, the charge is made longer, and its mass is increased (to 100 g). When the sensitivity of the HE is low (low enough that a blasting cap may prove to be insufficient), 5-g blasting cartridges of tetryl or other powerful HE are used to actuate detonation.

Cast-Bardin Impulse Meter

The brisance of more powerful HE is estimated using the Cast-Bardin impulse meter [55], a diagram of which is presented in Fig. 8.39. Using this method, a cylindrical charge of the HE under study (20 mm in diameter) is placed onto an impulse meter composed of changeable plates, dutchman, and piston. The piston enters into a guide cylinder and compresses a copper crusher that is positioned on a support that is screwed into the base of the impulse meter. A relative measure of explosion impulse is made by the degree of deformation experienced by the copper crusher. As the crusher's resistance to deformation increases with increasing reduction in crusher height, the principle of using crusher height reduction as a comparative measure of brisance becomes more complex, and less accurate.

An attempt to consider the crusher deformation physically was made by Sadovsky and Pokhil [56]. According to [54], the explosion impulse can be represented as $I = (2M\sigma\omega\delta H/(H - \delta H))^{1/2} = (2M\sigma\omega\alpha)^{1/2}$, where σ is the saturation stress (maximum on a plot of true stress versus true strain) of copper, ω is the crusher volume, H is the crusher height, δH is the crusher height reduction, M is the mass involved in motion, I is the explosion impulse, and $\alpha = \delta H/(H - \delta H)$.

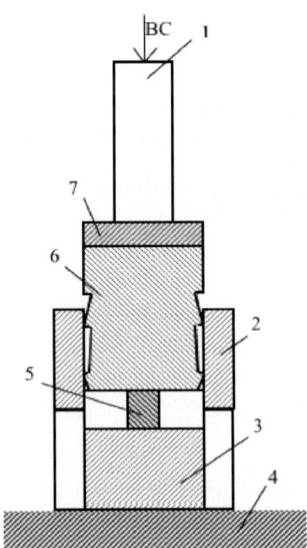

Fig. 8.39. Diagram of Cast's -Bardin's impulse meter: 1 – HE; 2 – guide cylinder; 3 – support; 4 – base; 5 – copper crusher; 6 – piston; 7 – lead Dutchman

Since σ, ω, and M are constants, we have for the relative impulse: $I_{rel} = I/I_{ref} = (\alpha/\alpha_{ref})^{1/2}$.

Table 8.12 [57] presents the relative impulses of three explosives measured by the height reduction of a copper crusher. In this table, the 50/50 composition of TNT/RDX (initial density $\rho_0 = 1.68$ g/cm^3) is taken as the reference material.

Table 8.12. Relative impulses measured by copper crusher reduction

HE	ρ_0, g/cm^3	I_{rel}, %
TNT/RDX 50/50	1.68	100
Hexogen (RDX)	1.80	119.3
TNT	1.60	83.6
PETN	1.77	117.4
Tetryl	1.70	102.3
Nitroglycerine	1.60	91.6

8.5.2 Estimation of Total Explosion Work

Trautzle Bomb

A test designed to measure the efficiency of HE was suggested by Trautzel, and standardized at the 2nd International Congress on Applied Chemistry in 1903 [56]. The Trautzle bomb (Fig. 8.40) is a massive lead "cylinder" 200 mm in height and diameter. On one side it has a 25-mm-diameter blind hole going 125 mm in depth along the axis of the cylinder. A 10-g charge of the HE under study (contained within in a paper sleeve), is placed at the bottom of the hole. The free part of the channel is then filled with dry quartz sand. The HE charge is initiated with primer ED-8-E. A characteristic swelling results from the explosion in the bomb.

A relative measure of HE efficiency is provided by the expansion of the bomb. The measure is based on the change in volume ΔV (in cubic centimeters) of the "cylinder," minus the initial hole volume and the portion of expansion that is considered to be produced by the electric detonator (30 cm^3). Bomb expansions for typical HE's are [58]: 465–520 cm^3 for hexogen; 480–500 cm^3 for PETN; 285 cm^3 for trinitrotoluene; 370 cm^3 for amatol 80/20; 520 cm^3 for nitroglycerine; and 195–220 cm^3 for ammonium nitrate.

Note that in this simple and widespread method the efficiency is expressed not in energy units, but in units of volume.

Ballistic Pendulum

The ballistic pendulum is a method for estimating the efficiency of HE that is based on the energy criterion. Using this method, the explosion work is

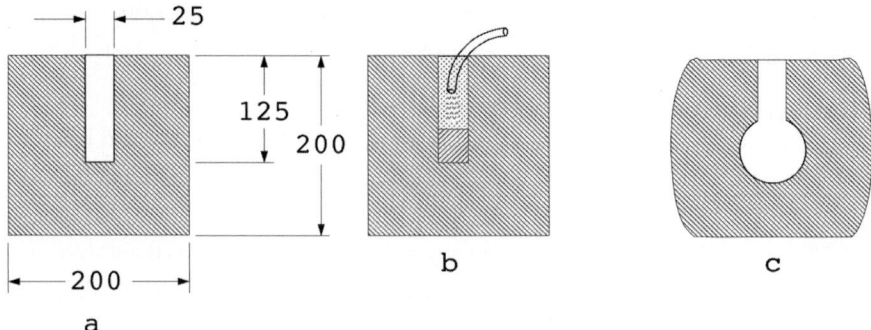

Fig. 8.40. Trautzle bomb: **a** – before rigging; **b** – rigged; **c** – after the explosion

measured by the momentum (or energy of motion) that is attained by an inert body under the action of expanding EP. Ballistic pendulum designs are diverse. One of them appears in Fig. 8.41.

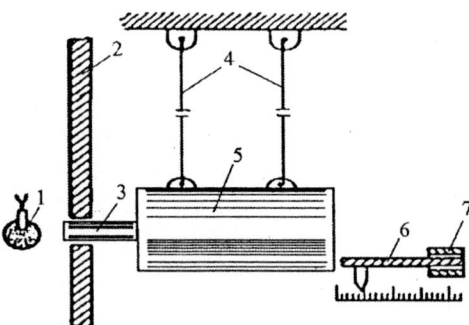

Fig. 8.41. Ballistic pendulum: 1 – HE charge; 2 – shield; 3 – pendulum nose; 4 – suspensions; 5 – pendulum body; 6 – deviation meter; 7 – machine yoke

The basic parts of any ballistic pendulum are an inertial mass M that gains some momentum MW under the action of the explosion, and a diagnostic that records either the position of mass M before and after the explosion, or estimates the velocity that is gained by it. The mass M is fixed by rigid guys to an immovable support. The ballistic pendulum suspension length ranges from 1.5 m to 2.5 m and the mass from 50 kg to 1000 kg. The HE charge is placed at some distance from the nose of the pendulum, or immediately at its end.

The pendulum swings through angle φ under the action of the explosion. The impulse accepted by the pendulum is calculated as $I = M \left(2gl \left(1 - \cos\varphi\right)\right)^{1/2}$, where M is the pendulum mass, g is the gravitational acceleration, φ is the angle of pendulum motion, and l is the suspension length.

This relation is valid for angles no larger than $20°$. If the horizontal pendulum displacement x is measured, then for small angles the impulse is calculated as $I = M 2\pi x/T$, where T is the pendulum period of swing.

The explosion work is calculated as $A = Mgh = Mgl(1 - \cos\varphi) = I^2/2M$, where h is the rise height of the pendulum center of gravity. The accuracy of the ballistic pendulum is 2–5%.

Ballistic Mortar

The ballistic mortar (Fig. 8.42) is a variation on the ballistic pendulum, with some differences being that in this case, a blasting chamber (2) is fabricated in the body of the movable mass M_1 into which the charge of HE under study is placed, and there is an expansion chamber into which a massive projectile piston (1) is placed before firing. When the charge is fired, the EP (initially occupying the space of chamber (2)), push the projectile piston (1) out of its chamber. Work is done by the EP in the mortar-projectile system as they expand from volume V_2 to volume $(V_1 + V_2)$. The expansion ratio is precisely defined as $(V_2 + V_1)/V_2$, where the initial volume is that of blasting chamber V_2, and the final volume is the total volume of the blasting and expansion chambers $(V_2 + V_1)$.

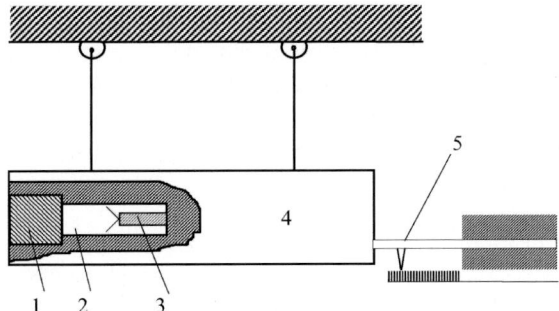

Fig. 8.42. Diagram of the ballistic mortar: 1 – projectile; 2 – blasting chamber; 3 – HE charge; 4 – mortar; 5 – measuring ruler

The amount of work done to deflect the mortar is $A_1 = Mgh = Mgl(1 - \cos\varphi)$, where M is the mortar mass, $h = l(1 - \cos\varphi)$ is the gain in elevation of the mortar center of gravity, and l is the mortar suspension length. The work done to launch the projectile is $A_2 = mW^2/2$, where m and W are the projectile piston mass and velocity, respectively. The total work can calculated under the assumption of equality of momenta gained by the mortar and the projectile: $Mu = mW$, where u is the mortar velocity which can be found from the relation $Mgl(1 - \cos\varphi) = Mu^2/2$. Thus, $u^2 = 2gl(1 - \cos\varphi)$. Since $W = Mu/m$, $W^2 = (M^2/m^2) 2gl(1 - \cos\varphi)$. Substituting into the above expression for A_2, the work done to launch the projectile is: $A_2 = M^2gl(1 - \cos\varphi)/m$.

The total work of the explosion is $A = A_1 + A_2 = Mgl\,(1 + M/m)\,(1 - \cos\varphi)$. If we denote $E_0 = Mgl(1 + M/m)$, we obtain $A = E_0(1 - \cos\varphi)$. Note that for a mortar of a given design E_0 is a constant.

To facilitate cleaning of the blasting chamber, to reduce EP heat losses (through conduction), and to reduce plastic deformation of the chamber walls, a strong steel insert $\approx$15-mm thick is placed into the chamber. To prevent gas leaks, the gap between the projectile and its chamber is eliminated using an annular seal. When the experimental conditions are strictly met, the spread in experimental data is 2–5%.

Trinitrotoluene Equivalent

There are two basic options for comparison testing of HE efficiency. One option is to compare the efficiency of two HE charges of the same mass, the other is to compare the mass required to achieve the same efficiency as that of a given mass of reference HE. Trinitrotoluene is typically taken as the reference HE. The ratio of mass of the trinitrotoluene charge to that of the HE under study is called the trinitrotoluene equivalent. This ratio serves as a measure of efficiency. The estimation of HE efficiency in terms of a trinitrotoluene efficiency parameter has gained wide acceptance. Trinitrotoluene with a density of 1.5 g/cm^3, which produces exothermic reaction energy of 4186 kJ/kg (1000 kcal/kg), is taken as the reference. In accordance with the foregoing discussion, the trinitrotoluene equivalent can be expressed as $\alpha = A/A_{TNT}$, where A and A_{TNT} are efficiencies of the HE under study and of TNT, respectively. Or alternatively, the trinitrotoluene equivalent can be expressed as the ratio of trinitrotoluene mass to the mass of the other HE under the condition that the efficiencies are the same (A = constant). In this case, $\alpha = M/M_{TNT}$.

One of the methods for determining a trinitrotoluene equivalent involves measuring the intensity of shock waves generated by the expansion of EP into air when charges of trinitrotoluene and the HE under study are detonated. It is assumed that the shock wave that is generated in air propagates with minimal thermodynamic losses. Therefore, the efficiency of the HE as a shock wave generator can be estimated by measuring the total momentum of the shock wave (or overpressure) at its front. In the experimental determination of such a trinitrotoluene equivalent, a determination is made of the mass of trinitrotoluene required to generate an air shock wave of equivalent intensity as that generated by a unit mass of the HE under study.

The trinitrotoluene equivalent is typically calculated as follows. First the shock front overpressure ΔP that is generated in air by the explosion of a spherical charge of mass M of the HE under study is experimentally determined. The measurement is made at some specified distance R from the mass center, and the measured overpressure is substituted into Sadovsky's relation [57]:

$$\Delta P = 0.84 \frac{M_{TNT}^{1/3}}{R} + 2.7 \frac{M_{TNT}^{1/3}}{R^2} + 7 \frac{M_{TNT}}{R^3}.$$

Sadovsky's equation is then solved for M_{TNT}. This yields the trinitro-toluene charge mass that would give the same shock front overpressure at distance R as a charge of mass M of the HE under study. If the charge is detonated on the surface of the ground, another Sadovsky's relation is used:

$$\Delta P = 1.06 \frac{M_{TNT}^{1/3}}{R} + 4.3 \frac{M_{TNT}^{1/3}}{R^2} + 14 \frac{M_{TNT}}{R^3}.$$

It should be emphasized that in the experimental estimation of the trinitro-toluene equivalent in terms of the shock wave parameters, the measurements should be taken at distances where the shock wave has formed completely, detached from the EP and begun to propagate independently in the environment. The zone before detachment of the shock wave from the EP is called the short-range explosion zone. Its radius depends on the kinetics of energy release in the EP. The slower the energy release, the longer is the short-range zone radius. A large short range zone radius is characteristic of HE containing metal (e.g., alumotol, ammonal). The typical HE near zone radius is about ten times the charge radius.

The trinitrotoluene equivalent for some HE [59] is as follows: 1.0 for TNT; 1.37 for hexogen; 1.37 for PETN; 1.39 for octogen; and 0.57 for ammonite.

8.5.3 Estimation of HE Launching Action

The phrase "HE launching action" refers to the ability of the HE to transform the chemical energy of the explosive into kinetic energy of the body being launched. The measure of work performed is the kinetic energy gained by the launched body.

Rather than using an absolute measure of launching action, common practice is to use a relative measure, (η), which compares the launching action of the HE under study to that of a comprehensively verified reference under the same conditions: $\eta = m_x W_x^2 / m_{ref} W_{ref}^2$. Masses of the launched plates need not be the same.

The pattern of the flow of EP and plate motion for 1D launch is illustrated in the $x - t$ diagram of Fig. 8.43.

When the adiabatic exponent of the EP has the value $n = 3$, the equations of gas dynamics can be solved in closed form yielding a simple analytical expression for the limiting $(t \to \infty)$ velocity $W_{\lim}$ of a plate of mass m (constructed from an incompressible material) launched by the HE. The limiting velocity is expressed in terms of HE charge parameters (charge density ρ_0, length l, and detonation rate D) as shown below [49]:

$$W_{\lim} = \frac{27}{16} \frac{m}{\rho_0 \ell} D \left\{ 1 + \frac{16}{27} \frac{\rho_0 \ell}{m} - \sqrt{1 + \frac{32}{27} \frac{\rho_0 \ell}{m}} \right\}.$$

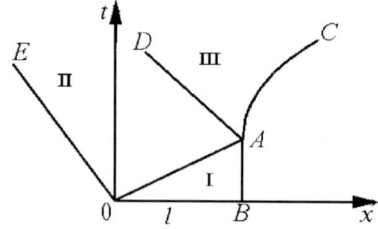

Fig. 8.43. The $x - t$ diagram of EP flow and plate motion for 1D launching: $0A$ – the detonation wave whose propagation velocity is equal to detonation rate D; AC – the plate path; AD – the shock wave reflected from the plate; $0E$ – EP expansion boundary. I – the region of original HE; II – the region of the centered expansion wave; III – the region of interaction of the centered expansion wave with the flow following the front of the shock wave reflected in the EP

Let us denote the ratio of HE charge mass to plate mass per unit area as $\rho_0 \ell / m = k$. Using k in the above, yields:

$$\frac{W_{\lim}}{D} = \frac{27}{16k}\left(1 + \frac{16}{27}k - \sqrt{1 + \frac{32}{27}k}\right) . \tag{8.24}$$

Examination of Eq. (8.24) reveals that as $k \to 0$ (small charge, heavy plate), $W_{\lim} \to 0$, and as $k \to \infty$, $W_{\lim} \to D$ (i.e., a very thin plate attains the same velocity as that of EP freely expanding into vacuum).

Let us introduce φ, the ratio of plate energy to the energy of the HE (i.e., the energy extracted by the plate from the HE):

$$\varphi = \frac{E_{pl}}{E_{HE}} = \frac{mW_{\lim}^2}{D^2} = \frac{8}{k}\left(\frac{W_{\lim}}{D}\right)^2,$$

$$\text{where} \quad E_{pl} = \frac{mW_{\lim}^2}{2}, \quad E_{HE} = \rho_0 \ell \frac{D^2}{16} .$$

Using Eq. (8.24) in the above yields:

$$\frac{27^2}{32k^3}\left\{1 + \frac{16}{27}k - \sqrt{1 + \frac{32}{27}k}\right\}^2 ; \tag{8.25}$$

$\varphi(k)$ has a maximum, occurring at $k = 81/32$, with $\varphi_{\max} = 256/729 = 0.351$ and $W_{\lim}/D = 1/3$.

The maximum energy transfer quantity (35.1%) is acquired by a plate whose mass is less than that of the HE mass by as little as a factor of 2.53. For values of k other than that which produces the maximal transfer of energy, the energy extraction factor decreases, but remains quite high over a wide range, as is seen in Table 8.13.

Table 8.13. Calculated energy extraction factors [49]

k	1.0	2.0	2.5	3.0
φ	0.296	0.347	0.351	0.349
k	4.0	10	20	30
φ	0.338	0.225	0.176	0.137

The actual acceleration of the launched plate is stepped because of its finite compressibility. The $P - u$ diagram for compressible plate acceleration is presented in Fig. 1.5b, Chap. 1.

Limiting plate launch velocity is determined by the intersection of the HE expansion isentrope with the Hugoniot of air. Higher launching velocities can be obtained by ensuring conditions of overdriven detonation, in which the pressure and velocity are higher than those under standard conditions.

For real launchers involving HE charges of finite mass, the launching velocities are always lower than the limiting value, even for thin (0.1–0.3 mm) metal plates. This is due to the drop in pressure at the EP-launched body interface caused by the arrival of rear and lateral rarefaction waves. The acceleration process essentially terminates after 3 or 4 reverberations in the plate. Once this occurs, plate launching velocity can no be longer estimated from a $P - u$ diagram based on the analytical solution. Experimental data must be employed.

The efficiency of converting HE charge energy into plate kinetic energy depends on a number of factors: the nature of the particular HE, the mass (both of charge and of plate), the charge design, the method of initiation, plate thickness and plate material, and a number of other parameters. Hence, the number of parameters that can influence the launching ability of a particular HE is great; so great as to render an absolute measure of launching ability impractical, thereby forcing the adoption of relative measures, expressed in terms of the launching ability of a reference. In practice this is the approach taken: the efficiency of a "new" HE is estimated by testing the launching action of the "new" HE using a "standard" charge design. The relative launching action is estimated by comparing (in terms of a ratio) the square of free surface velocities of launched bodies of identical material, mass, and shape launched by the "new" HE and by a reference HE with the charges being identical in volume and shape. The measurement of free surface velocity must of course be made at the same point in the flight path of the two launched bodies. Since the HE expansion ratio determines the work done by specific HE explosion products on the launched body, and this is in turn influenced by the geometry of the HE charge, it would be most accurate to study and compare the launching ability of different HE and compounds in experiments involving the actual intended application (i.e., full-scale design testing). But full-scale tests are labor-intensive and expensive. They often require large quantities of HE and complex (expensive) experimental setups. Therefore, such experiments are conducted only in the final stage of HE study as control (verification) tests.

Sets of experimental devices that are fairly simple and inexpensive have been developed for the purpose of estimating launching action. These devices are intended to provide a metric for judging the appropriateness of a given HE in some specific application. The judgment is often based on observations of some parameter directly related to launching action (such as flight distance), taking into account the ratio between the mass of the launched plate and HE charge mass. These methods provide a fast and reliable estimation of the potential for a particular HE in a specific application.

Each of these devices may be categorized according to class of spatial-dimensionality: 1D devices, involving the launching of a plate from the end of a cylindrical charge, and 2D devices, involving the launching of a cylindrical shell from the surface of a cylindrical charge.

With respect to diagnostics, the experimental methods can be subdivided into three groups: those using X-ray, optical, or electrocontact methods. The choice of one or another technique is problem-specific. If the investigators are more interested in a qualitative assessment of the event (flying plate state, shape, degree of fragmentation, etc.), then frame-by-frame high-speed photography of the process using a high-speed framing camera is generally a good choice. To provide the most comprehensive pattern of plate deformation or destruction, pulsed X-radiography is a good choice. Neither of these methods ensures a precise measurement of the spatial coordinates of the flying plate or expanding shells (or of their velocities). When taking X-ray images, it is necessary that the radiation pulse be normal to the axis of the cylindrical charge. Otherwise, the X-ray image will produce overestimated measures of shell expansion. In X-radiography, the error of the estimated normal component of the expansion velocity of a shell launched by detonation products propagating from an HE charge is 2–3%.

In the electocontact methods, several electrical contacts are positioned at various fixed distances from the plate surface and are used for recording the $x - t$ diagram of plate motion. When the plate (or plate fragments) comes into contact with one of these electrcontacts, it closes the contact (takes to zero ground). An oscilloscope is used to record the times at which the various contacts go to ground. To determine the launching action, plate velocity is obtained by differentiation of the resulting $x - t$ diagram.

Method of Rebounding Indicator

In perhaps the simplest method [60] (illustrated in Fig. 8.44), the velocity of equal-thickness steel plates (one driven by the reference HE and the other by the HE under study), are determined at a given flight distance and kinetic energies are compared. The HE charges are initiated over the entire surface. At the end of its flight path, steel plate (2) is decelerated against steel base

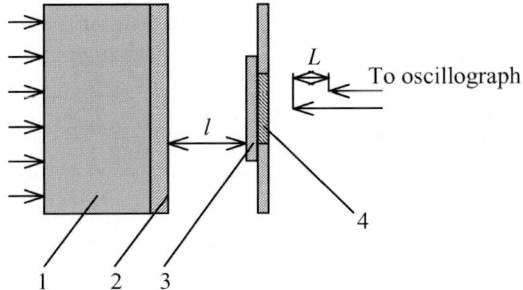

Fig. 8.44. Method of rebounding indicator: 1 – high explosive charge; 2 – steel plate; 3 – steel base plate; 4 – rebounding indicator; l – plate's flight length; L – indicator's flight length

plate (3), which is in intimate contact with aluminum rebounding indicator (4). The indicator is pressed into a steel ring to avoid side unloading effects.

When the shock wave has arrived at the surface of the indicator, the indicator rebounds at a velocity associated with that of the steel plate at the time of its impact on the base plate. The time of indicator arrival is measured at two specified distances using electric contacts and an oscilloscope in real time. Thicknesses of the steel base plate and the rebounding indicator are chosen such that the rear unloading wave does not influence the motion of the indicator.

Indicator flight velocity, W_{Al}, is calculated from the measured time and known flight distance. Transformation from the aluminum indicator velocity to the steel plate velocity at the time of the plate impact onto the steel base plate is performed through solution of the Riemann problem (see Chap. 1):

$$\rho_0 \left(a_0 + a_0 \left(W_{Fe} - \frac{W_{Al}}{2} \right) \right) \left(W_{Fe} - \frac{W_{Al}}{2} \right) = \rho_1 \left(a_2 + a_3 \frac{W_{Al}}{2} \right) \frac{W_{Al}}{2},$$

where ρ_0, a_0, and a_1 are the density and fitting coefficients in the $D - u$ relationship for steel; ρ_1, a_2, and a_3 are density and fitting coefficients in the linear $D - u$ relationship for aluminum; W_{Al} is the measured aluminum indicator velocity; and W_{Fe} is the desired steel plate velocity. The relevant $P - u$ diagram for the process appears in Fig. 8.45.

The total error of the method (which includes errors in the measurement of flight distance and time, deviations from the nominal Hugoniots of the base plate and rebounding indicator materials, and shock wave nonplanarity), is no higher than 2.5%.

Method M-60

This method [61], whose scheme appears in Fig. 8.46, records the $x - t$ diagram of the motion of a steel plate driven by EP of the HE charge under

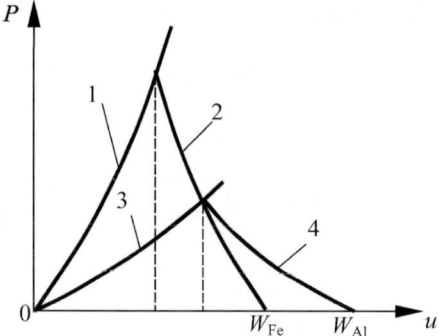

Fig. 8.45. The $P - u$ diagram of the interaction between the steel plate and rebounding indicator: 1, 3 and 2, 4 — Hugoniots and isentropes of steel and aluminum, respectively

study. The recording is performed using electrocontact sensors positioned at different distances from the end surface of the plate. To reduce the effect of side rarefaction waves, the charge is put into a thick-walled steel shell. Differentiation of the $x - t$ diagram produces plate velocity, which characterizes the HE launching action in this design. Relative launching action is determined by comparing the maximum kinetic energy $(E_{kin})_{\max}$ of a plate launched by the HE under study to that attained by a plate launched by the reference HE. The procedure for making this comparison is multi-step. First, velocities of plates of different thickness are obtained experimentally and a plot of E_{kin} versus mass is constructed. The maximum kinetic energy $(E_{kin})_{\max}$ is determined from the plot. This procedure is followed for each HE (the one under study and the reference) and the results are compared.

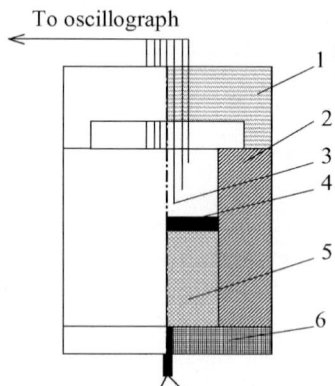

Fig. 8.46. Scheme of method M-60: 1 – organic glass unit; 2 – casing; 3 – electrocontact sensors; 4 – plate; 5 – HE charge; 6 – lid

Method of Cylindrical Shell Launching

This method is known as method T-20 at VNIIEF [61,62], and as the "cylinder test" abroad. The scheme of the experiment is presented in Fig. 8.47a. The charge under study is placed inside a cylindrical tube (180–200 mm in length, with an inner diameter of 20 mm and a wall thickness of 2 mm) and is initiated by a special plane wave detonation generator.

At a distance equal to several charge diameters from the initiation point, the detonation propagation process becomes steady-state in the coordinate system related to the detonation wave. As the detonation wave propagates, a point will eventually be reached (at some distance L from the initiation surface, at time $t = L/D$), when the cylinder will begin to expand. The onset of cylinder expansion will be marked by an increase in radial velocity of the cylindrical shell. At arbitrary cross section x_0, shell time of flight (a measure of time commencing with the arrival of the detonation wave at x_0) is related to the total time of the process counted from the time of initiation as: $\tau = t - x_0/D$. The shell launching diagram is presented in Fig. 8.48.

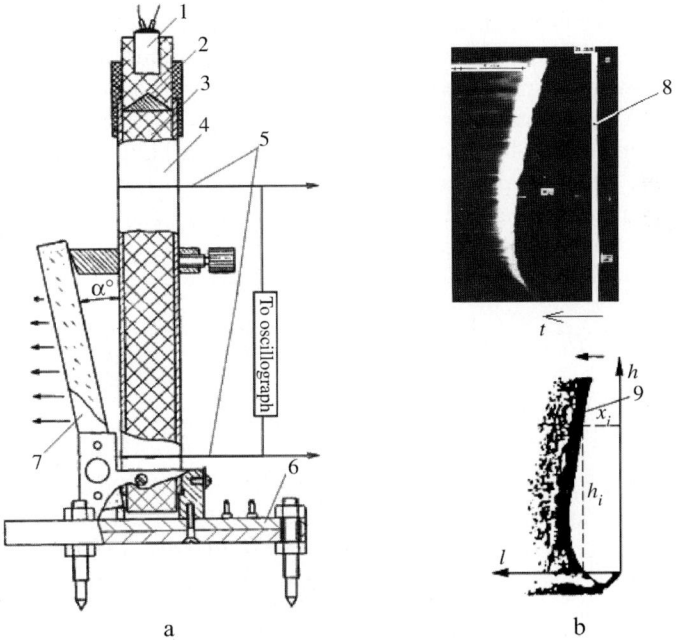

a   b

Fig. 8.47. Method T-20. Scheme of the tube rigging and recording of the shell motion across the organic glass wedge cutoff (**a**) and streak camera image of the shell impact on the wedge (**b**): 1 – detonator; 2 – plane-wave generator; 3 – HE charge; 4 – tube; 5 – velocity sensors; 6 – support; 7 – organic glass cutoff; 8 – slit image; 9 – line of impact

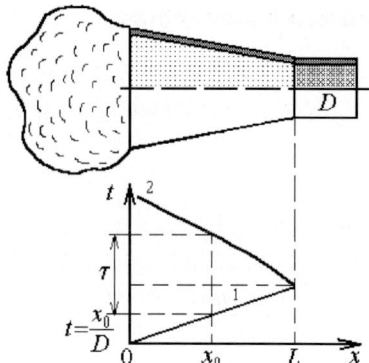

Fig. 8.48. The $x-t$ diagram of the cylindrical shell launching: 1 – detonation wave path; 2 – position of the external shell boundary

During photography, impact of the expanding cylindrical shell onto a transparent organic glass plate, positioned at angle α to the cylinder's axis of symmetry, is recorded continuously through the streak camera slit. At impact time, the air layer between the shell and the transparent plate becomes luminous; the streak camera records this luminosity. The luminosity ceases when the shock wave penetrates the plate, at which time the plate loses its transparency and thus plays the role of a light shutter. A typical streak camera image of cylindrical shell impact on the organic glass light shutter is presented in Fig. 8.47b. The streak camera images can be processed either from films or from magnified photographic prints. The scheme of the calculation using a photographic print is also shown in Fig. 8.47b.

The expansion parameters are determined from:

$$R_i - R_0 = M h_i \sin(\alpha); \qquad W_i = \frac{D\, tg\,(\alpha)}{1 + \frac{D(dx/dh)_i}{V_{sc} M \cos(\alpha)}};$$

$$t_i = \frac{x_i}{V_{sc}} + \frac{h_i M \cos(\alpha)}{D} - \frac{z_0}{D_1},$$

where W_i and t_i are the radial expansion velocity and time of shell expansion corresponding to arrival of the shell at radial position R_i; R_i–R_0 is the distance from the shell surface at radial position R_i and the initial position of the external shell surface; D is the detonation rate of the HE; α is the angle between the plane of the surface of the organic glass plate and the axis of symmetry of the cylindrical shell; M is a scale factor; V_{sc} is the linear rate of image scanning provided by the streak camera; x_i and h_i are the coordinates describing points on the line (appearing in the experimental streak-camera image) representing impact of the shell onto the glass plate; D_1 is the velocity of the shock wave in the shell material; and z_0 is the shell thickness. The point of detonation luminosity release from the tube end opposite the end where initiation occurs is taken as the reference point. Simultaneous to the recording of the shell expansion, electrocontact sensors measure the time of detonation wave propagation through the charge on a $\sim$100-mm gauge length.

The relative launching ability v of some HE determined by cylindrical shell expansion (with 7-fold expansion of the EP) is summarized in Table 8.14 [61].

Table 8.14. Launching ability of some HE with 7-fold expansion of EP

HE	ρ, g/cm^3	v
Trinitrotoluene	1.60	1.0
Nitromethane	1.14	1.120
XTX-8003	1.554	1.265
TATB	1.854	1.349
TNT/RDX 40/60	1.728	1.492
PETN	1.765	1.629
HMX/trinitrotoluene 78/22	1.813	1.606
Octogen (HMX)	1.894	1.715
Hexogen (RDX)	1.80	1.641
BTP	1.859	1.682

The EP expansion ratio Ω, is determined as the squared ratio of the current cylindrical shell inner radius r_x to initial radius r_0: $\Omega = (r_x/r_0)^2$. Error in the estimation of relative launching action of propagating detonation EP when the factor of expansion is from 1.5 to 10, is no greater than 0.5%.

Experimental data of cylindrical copper shell velocities obtained with technique T-20 not only serve to characterize HE efficiency, but may also be used as the basis for the determination of coefficients appearing in EOS's that describe expanding EP isentropes.

References

1. Byurlo, E., *Detonation Through Effect*, Art. Academy Publishing House, Moscow, 1934.
2. Cachia, G.P., and Whitbread, E.G., "The Initiation of Explosives by Shock," *Proc., Royal Society of London, Series A*, Vol. 246, No. 1245, 1958, pp. 268–273.
3. Roth, J., "Shock Sensitivity and Shock Hugoniots of High-Density Granular Explosives," *Proc., 5th International Symposium on Detonation*, Aug. 18–21, 1970, Pasadena, CA, pp. 219–230.
4. Ramsay, J.B., and Popolato, A., "Analysis of Shock Wave and Initiation Data for Solid Explisives," *Proc., 4th International Symposium on Detonation*, Oct. 12–15, 1965, White Oak, Maryland, pp. 233–238.
5. Campbell, A.W., Davis, W.C., Ramsay, J.B., and Travis, J.R., "Shock Initiation of Solid Explosives," *Physics of Fluids*, Vol. 4, No. 4, 1961, pp. 511–521.
6. Koldunov, S.A., Shvedov, K.K., and Dremin, A.N., "Decomposition of Porous Explosives Under the Effect of Shock Waves," *Fizika Goreniya i Vzryva*, Vol. 9,

No. 2, 1973, pp. 295–304, [English trans., *Combustion, Explosion, and Shock Waves*, Vol. 9, No. 2, 1973, pp. 255–262].

7. Campbell, A.W., Davis, W.C., and Travis, J.R., "Shock Initiation of Detonation in Liquid Explosives," *Physics of Fluids*, Vol. 4, No. 4, 1961, pp. 498–510.

8. Jacobs, S.J., Liddiard, T.P., Jr., and Drimmer, B.E., "The Shock-to-Detonation Transition in Solid Explosives," *Proc., 9th International Symposium on Combustion*, Aug. 27–Sep 1, 1962, Ithaca, NY, pp. 517–529.

9. Dobratz, B.M., *LLNL Explosives Handbook: Proberties of Chemical Explosives and Explosive Simulants*, Lawrence Livermore National Laboratory, Livermore, CA, 1981.

10. Dremin, A.N. and Koldunov, S.A., "Detonation Initiation by Shock Waves in Cast and Pressed Trinitrotoluene," *in* Explosive Practice 63/20, Nedra Publ., Moscow, 1967, pp. 37–50.

11. Shvedov, K.K., and Dremin, A.N., "Effect of Charge Aggregate State and Structure on Trinitrotoluene Decomposition in Shock Waves," *in* Combustion and Detonation, *Proc., 4th All-Union Symposium on Combustion and Detonation*, Nauka Publ., Moscow, 1977, pp. 440–446.

12. Dremin, A.N., Savrov, S.D., Trofimov, V.S., and Shvedov, K.K., *Detonation Waves in Condensed Media*, Nauka Publ., Moscow, 1970, p. 164.

13. Kanel, G.I., and Dremin, A.N., "Decomposition of Cast Trotyl in Shock Waves," *Fizika Goreniya i Vzryva*, Vol. 13, No. 1, 1977, pp. 85–92, [English trans., *Combustion, Explosion, and Shock Waves*, Vol. 13, No. 1, 1977, pp. 71–77].

14. Batkov, Yu.V., Novikov, S.A., Pogorelov, A.P., and Sinitsyn, V.A., "Investigation of the Process of Explosive Transformation of the Composite TG 50/50 Behind a Nonstationary Shock Front," *Fizika Goreniya i Vzryva*, Vol. 15, No. 5, 1979, pp. 139–141, [English trans., *Combustion, Explosion, and Shock Waves*, Vol. 15, No. 5, 1979, pp. 676–678].

15. Uokerli, D., et al., "Studying Shock-Wave Initiation of PBX-9404," *in* Detonation and High Explosives, Borisov, A.A., ed., Mir Publ., Moscow, 1981, pp. 269–290.

16. Lobanov, V.F., "Initiating-Wave Parameter Determination for TG 50/50," *Fizika Goreniya i Vzryva*, Vol. 22, No. 5, 1986, pp. 104–111, [English trans., *Combustion, Explosion, and Shock Waves*, Vol. 22, No. 5, 1986, pp. 589–594].

17. Glushak, B.L., Novikov, S.A., and Pogorelov, A.P., "Shock-Wave Initiation of Solid Heterogeneous Explosives," *Fizika Goreniya i Vzryva*, Vol. 20, No. 4, 1984, pp. 77–85, [English trans., *Combustion, Explosion, and Shock Waves*, Vol. 20, No. 4, 1984, pp. 429–436].

18. Grin, L., Nidik, E., Li, E., and Tarver, C., "PBX-9404 Chemical Decomposition Initiation by Weak Shock Waves," *in* Detonation and High Explosives, Mir Publ., Moscow, 1981, pp. 107–122.

19. Setchell, R.E., Ramp-Wave Initiation of Granular Explosives," *Combustion and Flame*, Vol. 43, 1981, pp. 255–264.

20. Doronin, G.S., Yermolovich, E.I., and Rabotinsky, A.N., "Pressed Trinitrotoluene Decomposition Kinetics During Smeared-Front Pulse Initiation," *Proc., 1st All-Union Symposium on Macroscopic Kinetics and Chemical Gas Dynamics*, Chernogolovka, 1984, Vol. 1, Part 1, pp. 30–31.

21. Campbell, A.W., and Travis, J.R., "The Shock Densensitization of PBX-9404 and Composition B-3," *Proc., 8th International Symposium on Detonation*, Jul. 15–19, Albuquerque, NM, 1985, pp. 1057–1068.

22. Batkov, Yu.V., Glushak, B.L., and Novikov, S.A., "Desensitization of Pressed Explosive Compositions Based on TNT, RDX, and HMX Under Double Shock-Wave Loading," *Fizika Goreniya i Vzryva*, Vol. 31, No. 4, 1995, pp. 89–92, [English trans., *Combustion, Explosion, and Shock Waves*, Vol. 31, No. 4, 1995, pp. 482–485].

23. Fowles, R., and Williams, R.F., "Plane Stress Wave Propagation in Solids," *Journal of Applied Physics*, Vol. 41, No. 1, 1970, pp. 360–363.

24. Batalova, M.V., Bakhrakh, S.M., and Zubarev, V.N., "Excitation of a Detonation in Heterogeneous Explosives by Shock Waves," *Fizika Goreniya i Vzryva*, Vol. 16, No. 2, 1980, pp. 105–109, [English trans., *Combustion, Explosion, and Shock Waves*, Vol. 16, No. 2, 1980, pp. 227–231].

25. Belinets, Yu.M., Dremin, A.N., and Kanel, G.I., "Kinetics of Pressed-TNT Decomposition Behind a Shock Front," *Fizika Goreniya i Vzryva*, Vol. 14, No. 3, 1978, pp. 111–116, [English trans., *Combustion, Exposion, and Shock Waves*, Vol. 14, No. 3, 1978, pp. 361–365].

26. Nutt, G.L., and Erickson, L.M., "Reactive Flow Lagrange Analysis in RX-26-AF," Shock Waves in Condensed Matter – 1983, Asay, J.R., Graham, R.A., and Straub, G.K., eds., Elsevier, Amsterdam, 1984, pp. 605–608.

27. Belyaev, A.F., Bobolev, V.K., and Sulimov, A.A., *Condensed System Deflagration-to-Detonation Transition*, Nauka Publ., Moscow, 1973, p. 292.

28. Dubovik, A.S., *Photographic Recording of Fast Processes*, Nauka Publ., Moscow, 1964, p. 467.

29. Orlenko, L.P., ed., *Physics of Explosion*, 3rd Edition, Vol. 1, Fizmatlit Publ., Moscow, 2002.

30. Veretennikov, V.A., Dremin, A.N., Rozanov, O.K., and Shvedov, K.K., "Applicability of Hydrodynamic Theory to the Detonation of Condensed Explosives," *Fizika Goreniya i Vzryva*, Vol. 3, No. 1, 1967, pp. 3–10, [English trans., *Combustion, Explosion, and Shock Waves*, Vol. 3, No. 1, 1967, pp. 1–5].

31. Jameson, R.L., and Hawkins, A., "Shock Velocity Measurements in Inert Monitors Placed on Several Explosives," *Proc., 5th International Symposium on Detonation*, Aug 18-21, 1970, Pasadena, CA, pp. 23–29.

32. Ashaev, V.K., Doronin, G.S., and Levin, A.D., "Detonation Front Structure in Condensed High Explosives," *Fizika Goreniya i Vzryva*, Vol. 24, No. 1, 1988, pp. 95–99, [English trans., *Combustion, Explosion, and Shock Waves*, Vol. 24, No. 1, 1988, pp. 88–92].

33. Dremin, A.N., Shvedov, K.K., and Veretennikov, V.A., "Studying Detonation of Ammonit PZhV-20 and Some Other HE," Explosive Practice, N 52/9, Gosgortekhizdat Publ., Moscow, 1963.

34. Dremin, A.N., and Shvedov, K.K., "Estimation of Chapman-Jouget Pressure and Reaction Time in Detonation Wave of Powerful HE, *Zhurnal Prikladnoi Mekhaniki i Tekhnicheskoi Fiziki*, No. 2, 1964, pp. 154–159.

35. Zaitsev, V.M., Pokhil, P.F., and Shvedov, K.K., "Measurement of Sound Speed in Detonation Products," *Doklady Akademii Nauk SSSR*, Vol. 133, No. 1, 1960, pp. 155–157.

36. Dorokhin, V.V., Zubarev, V.N., Orekin, Yu.K., Panov, N.V., and Shaboldina, N.L., "Motion of Explosion Products Behind a Detonation Wave Front," *Fizika Goreniya i Vzryva*, Vol. 21, No. 4, 1985, pp. 100–104, [English trans., *Combustion, Explosion, and Shock Waves*, Vol. 21, No. 4, 1985, pp. 471–474].

37. Dorokhin, V.V., Zubarev, V.N., Orekin, Yu.K., Panov, N.V., and Shaboldina, N.L., "Continuous Radiographic Recording for Explosion Products Behind a

Detonation Front," *Fizika Goreniya i Vzryva*, Vol. 24, No. 1, 1988, pp. 118–122, [English trans., *Combustion, Explosion, and Shock Waves*, Vol. 24, No. 1, 1988, pp. 109–112].

38. Zubarev, V.N., "Structure of Self-Similar Rarefaction Waves and Expansion Adiabats of Substances," *Fizika Goreniya i Vzryva*, Vol. 20, No. 3, 1984, pp. 66–67, [English trans., *Combustion, Explosion, and Shock Waves*, Vol. 20, No. 3, 1984, pp. 307–308].

39. Sheffield, S.A., Bloomquist, D.D., and Tarver, C.M., "Subnanosecond Measurements of Detonation Fronts in Solid High Explosives," *Journal of Chemical Physics*, Vol. 80, No. 8, 1984, pp. 3831–3844.

40. Utkin, A.V., Kanel, G.I., and Fortov, V.E., "Empirical Macrokinetics of the Decomposition of a Desensitized Hexogen in Shock and Detonation Waves," *Fizika Goreniya i Vzryva*, Vol. 25, No. 5, 1989, pp. 115–122, [English trans., *Combustion, Explosion, and Shock Waves*, Vol. 25, No. 5, 1989, pp. 625–632].

41. Fedorov, A.V., Menshikh, A.V., and Yagodin, N.V., "Detonation Wave Front Structure of Condensed High Explosives," *Proc., New Models and Hydrocodes for Shock Wave Processes in Condensed Matter*, Oxford, UK, 1997, Publ., AWE Hunting BRAE, Aldermaston, UK, 1997, Vol. 2, pp. 830–832.

42. Voskoboynikov, I.M., and Gogulya, M.F., "Shock Front Luminosity in Liquid Near the Detonating Charge Interface," *Khimicheskaya Fizika*, No. 7, 1984, pp. 1036–1041.

43. Evstigneev, A.A., Zhernokletov, M.V., and Zubarev, V.N., "Isentropic Broadening and Equation of State of Trotyl Explosion Products," *Fizika Goreniya i Vzryva*, Vol. 12, No. 5, 1976, pp. 758–763, [English trans., *Combustion, Explosion, and Shock Waves*, Vol. 12, No. 5, 1976, pp. 678–682].

44. Zhernokletov, M.V., Zubarev, V.N., and Telegin, G.S., "Expansion Isentropes of the Explosion Products of Condensed Explosives," *Zhurnal Prikladnoi Mekhaniki i Tekhnicheskoi Fiziki*, Vol. 10, No. 4, 1969, pp. 127–132, [English trans., *Journal of Applied Mechanics and Technical Physics*, Vol. 10, No. 4, 1969, pp. 650-655].

45. Kanel, G.I., Razorenov, S.V., Utkin, A.V., and Fortov, V.E., *Shock-Wave Phenomena in Condensed Media*, Yanus, K. Publ., Moscow, 1996 [*see also*, Kanel, G.I., Razorenov, S.V., and Fortov, V.E., *Shock-Wave Phenomena and the Properties of Condensed Matter*, Springer-Verlag, New York, 2004].

46. Zubarev, V.N., and Evstigneev, A.A., "Equations of State of the Products of Condensed-Explosive Explosions," *Fizika Goreniya i Vzryva*, Vol. 20, No. 6, 1984, pp. 114–126, [English trans., *Combustion, Explosion, and Shock Waves*, Vol. 20, No. 6, 1984, pp. 699–710].

47. Altshuler, L.V., "Use of Shock Waves in High-Pressure Physics," *Uspekhi Fizicheskikh Nauk*, Vol. 85, No. 2, 1965, pp. 197–258, [English trans., *Soviet Physics Uspekhi*, Vol. 8, No. 1, 1965, pp. 52–91].

48. Altshuler, L.V., Doronin, G.S., and Zhuchenko, V.S., "Detonation Regimes and Jouguet Parameters of Condensed Explosives," *Fizika Goreniya i Vzryva*, Vol. 25, No. 2, 1989, pp. 84–103, [English trans., *Combustion, Explosion, and Shock Waves*, Vol. 25, No. 2, 1989, pp. 209–224].

49. Zababakhin, E.I., *Some Problems of the Gasdynamics of Explosions*, RFNC-VNIITF, Snezhinsk, Russia, 1997, [English trans., RFNC-VNIITF, Snezhinsk, Russia, 2001].

50. Azbukina, I.N., Belyaev, A.F., "Estimation of Cutoff Diameters with the Method of Cones," Physics of Explosion. Collected Papers N3., USSR Academy of Sciences Publishing House, Moscow, 1955.

51. Baum, F.A., Derzhavets, A.S., and Duvanova, Zh.M., "Detonation Ability of HE Designed for Operations in Deep Wells," Explosive Practice 63/20., Nedra Publ., Moscow, 1967, pp. 251–259.

52. Ramsay, J.B., "Effect of Confinement on Failure in 95 TATB/5 Kel-F," *Proc., 8th International Symposium on Detonation*, Jul. 15–19, 1985, Albuquerque, NM, pp. 372–379.

53. Belyaev, A.F., and Kurbangalina, R.Kh., "Effect of Initial Temperature on Nitroglycerine and a Trinitrotoluene Cutoff Diameter," *Zhurnal Fizicheskoy Khimii*, Vol. 34, No. 3, 1960, pp. 603–610.

54. Price, D., "Shock Sensitivity, A Property of Many Aspects," *Proc., 5th International Symposium on Detonation*, Aug. 18–21, 1970, Pasadena, CA, pp. 207–217.

55. Apin, A.Ya., and Velina, N.F., "On Cutoff Diameters of Explosive Single Crystal Detonation," *Proc., 2nd All-Union Symposium on Combustion and Detonation,"* Chernogolovka, 1969.

56. Andreev, K.K., and Belyaev, A.F., *Theory of High Explosives*, Oborongiz Publishers, Moscow, 1960.

57. Apin, A.Ya., Bardin, E.P., and Velina, N.F., "Influence of High Explosive Density and Composition on Explosion Impulse," *in* Explosive Practice N 52/9, Gosgortekhizdat Publ., Moscow, 1963, pp. 90–102.

58. Kuznetsov, V.M., and Shatsukevich, A.F., "The Efficiency of Explosives," *Fizika Goreniya i Vzryva*, Vol. 14, No. 2, 1978, pp. 120–125, [English trans., *Combustion, Explosion, and Shock Waves*, Vol. 14, No. 2, 1978, pp. 235–239].

59. Dubnov, L.V., Bakharevich, N.S., and Romanov, A.I., *Industrial High Explosives*, Nedra Publishers, Moscow, 1988.

60. Altshuler, L.V., Kormer, S.B., Brazhnik, M.I., Vladimirov, L.A., Speranskaya, M.P., and Funtikov, A.I., *Zhurnal Eksperimentalnoi i Teoreticheskoi Fiziki*, Vol. 38, 1960, pp. 1061–1073, [Enlish trans., *Soviet Physics JETP*, Vol. 11, No. 4, 1960, pp. 766–775].

61. Smirnov, S.P., Kolganov, E.V., Kulakevich, Ya.S., *et al.*, "Relation of the Launching Action of Mix and Individual HE to Their Composition and Structure," *in* Advanced Methods for Designing and Verification of Rocket-Artillery Arms, RFNC-VNIIEF, Sarov, Russia, 2000, pp. 410–412.

62. Akst, I.B., "Heat of Detonation, the Cylinder Test, and Performance in Munitions," *Proc., 9th International Symposium on Detonation*, Aug 28 – Sep 1, 1989, Portland, OR, pp. 478–488.

Laser Doppler Measuring Systems and Their Use in Shock-Wave Studies

A.V. Fedorov, A.L. Mikhailov, and D.V. Nazarov

As discussed earlier, many of the experimental techniques used in the study of shock wave phenomena involve a recording of free surface velocity, or the velocity on an interface, as a function of time or spatial coordinate. The streak-camera, electro-contact methods, methods employing manganin gages, and methods involving the shorting of current-carrying wires, all, in one way or another involve either a barrier in the path of the sample under study, or an indicator pre-inserted into the sample that leads to a perturbation of the process under study. Capabilities of the non-perturbing capacitive sensor techniques are rather limited, with relatively short gage lengths, and with accuracy assured only for relatively low measured velocities (the measurement of which involves an approximate averaging of the measured displacement over the probe area). In the 1960s, soon after the advent of lasers, the need for the development of more accurate non-perturbing methods of measurement for shock-wave experiments prompted the development of the first interferometric methods for the measurement of velocity. These methods work by recording the Doppler frequency shift of probing monochromatic electromagnetic radiation reflected from the surface of the specimen. A merit of these methods is that they provide a continuous recording of velocity, without imposing any perturbing effects on the process under study. In some of the earliest applications of this technology, measuring relatively low velocities, displacement interferometers were based on the scheme of the two-beam Michelson interferometer with the polished surface under study serving as a mirror in one of the arms of the interferometer [1]. Later on new Doppler systems appeared – Laser Differential Interferometer (LDI) [2,3], and Velocity Interferometer System for Any Reflector (VISAR) [4].

In the late 1970s through the early 1980s, the Fabry-Perot laser interferometer [5] and the optically symmetric ORVIS (Optically Recording Velocity Interferometer System) interferometer [4] came into wide use. The VISAR and ORVIS systems are constructed using an optically symmetric interferometric scheme and differ only in the imaging methods: the VISAR records a change in the intensity of the light signal arriving at the photoelectric multiplier from the central spot of the interference pattern [4], the ORVIS records a shift in fringe positions at the streak camera slit. In the late 1980s through the early 1990s, systems were developed to measure the velocity distribution over the surface, both in frame-by-frame and in continuous multi-channel modes. It should be noted that the area of the probed surface depends primarily on the laser radiation power and the recorder sensitivity. Narrow-band lasers with amplifiers of $\approx 10^4$ W power are currently being developed.

This chapter discusses the principal interferometric schemes, their abilities, merits and demerits, and gives examples of the use of these schemes for the recording of fast processes.

9.1 Theoretical Basis of Interferometric Methods of Velocity Measurement

The relativistic Doppler shift is the basis for all interferometric methods used to measure the velocity of a reflecting surface. Since objects whose velocity is to be measured frequently have diffusively reflecting surfaces, the angles between the target velocity vector and the light beams incident to and reflected from the surface of the target can be arbitrary (Fig. 9.1).

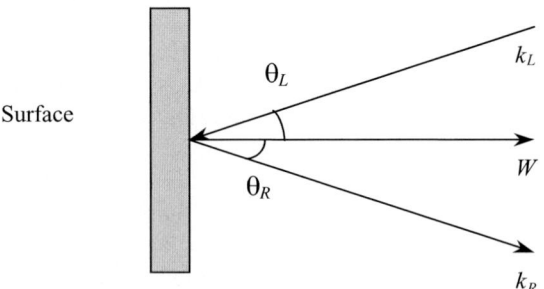

Fig. 9.1. Geometry of incident and reflected light beam

The task is to estimate the frequency of reflected light f_R as a function of target velocity and angles of incidence and reflection for known incident light frequency f_L. The standard relativistic analysis gives:

$$\frac{f_R}{C} = \frac{f_L}{C} \left[1 + (\cos \theta_R) \right] ; \qquad \frac{\Delta f}{f_L} = \frac{\Delta \lambda}{\lambda} = \frac{W}{C} \left[k_L - k_R \right] , \qquad (9.1)$$

where k_L and k_R are wave vectors of the incident and reflected light, C is the speed of light in a vacuum, λ is the probing radiation wavelength, W is the reflecting surface velocity vector, and θ_L and θ_R are the angles of the incident and reflected beams, respectively, relative to the velocity vector.

When $W/C << 1$, the Doppler shifts of wavelength $\Delta\lambda$ are small, typically less than 0.01 Å, and can be registered only by highly sensitive interferometric systems.

Of particular interest are two cases:

if $\theta_L = \theta_R = 0$, then $\Delta f/f_L = 2W/C$ – the case of normal incidence;
if $\theta_L = \theta_R = \theta$, then $\Delta f/f_L = (2W/C)\cos\theta$

The latter case occurs frequently in practice, when the incident laser beam is focused onto the target surface by a lens and a lens is used to collect the reflected light (corrections are required in the analysis of the data obtained).

9.2 Principal Interferometric Systems

9.2.1 Displacement Interferometer

The scheme of the displacement (Michelson) interferometer appears in Fig. 9.2. A reflecting surface of the sample is used as one of the mirrors in the Michelson interferometer measuring arm [1]. The beam is split into two parts by beam splitter (3). On reflection from surface (1) and stationary mirror (2), the beams, which have by this time have passed through the interferometer measuring and support arms, respectively, are mixed again by the beam splitter, and directed by mirror (2) to photomultiplier (6) to form the interference pattern, which will appear as concentric rings. The pattern can be magnified by the adjustment of the arm lengths so that only the central spot will be in the field of vision of the photomultiplier. When the sample surface is set in

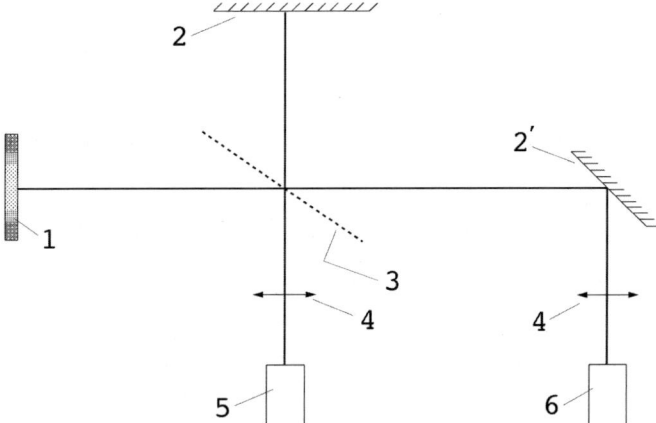

Fig. 9.2. Optical scheme of the displacement interferometer: 1 – sample under study with mirror surface; 2 – immovable mirrors; 3 – beam-splitting plate; 4 – focusing lenses; 5 – laser; 6 – photomultiplier

motion, a running interference pattern develops at the photomultiplier, and at the central spot, the photomultiplier signal will be modulated in amplitude by the Doppler frequency shift.

The surface displacement $S(t)$ is expressed as:

$$S(t) = \frac{\lambda_0}{2} F(t) , \qquad (9.2)$$

where λ_0 is the laser generation wavelength and $F(t)$ is the number of interference pattern central spot intensity changeover periods.

Velocity is determined by differentiation of the above expression. This method provides an improvement in the resolution of surface displacement by 1 to 2 orders of magnitude over other methods. However, its use is restricted to velocities that are ≤ 0.2 mm/µs. At higher velocities, the band passage frequency in the interferometer becomes so high that it is impossible to record it with existing devices. For this reason, the scheme has not found wide application in shock-wave experiments. The capabilities that the scheme offers in terms of high resolution with "slow" displacement velocities, however, make it a valuable tool in experiments aimed at studying the fine details of elastic-plastic material behavior at stresses 3–5 times higher than the elastic limit.

9.2.2 Laser Differential Interferometer

This method uses the interference generated by the interaction of two time-displaced elements of a single probing beam. That is, interference between the segment of the beam that experiences the Doppler shift at time t, and the segment that experienced the Doppler shift earlier, by time τ [2,3]. In this scenario, each fringe corresponds to a certain velocity change. The scheme of the LDI appears in Fig. 9.3.

Since the reference and delayed beams (in delay arm $ABCD$) are Doppler-like shifted in frequency by about the same magnitude (determined by the difference between probed surface velocities at times t and $t + \tau$), their beat frequency is much less than that of the displacement interferometer. When the sample velocity is constant, both beams have identical Doppler shift, and a time-independent signal occurs at the photomultiplier. The surface velocity is determined from:

$$W\left(t - \frac{\tau}{2}\right) = \frac{\lambda_0}{2\tau} F(t) , \qquad (9.3)$$

where τ is the time delay.

The LDI method has not found wide application for the following reasons:

– in strong shock waves, the mirror surface frequently becomes diffusive and the interferometer will not operate;
– the ability of the interferometer to accurately record velocity is dramatically dependent on the inclination of the sample surface during motion.

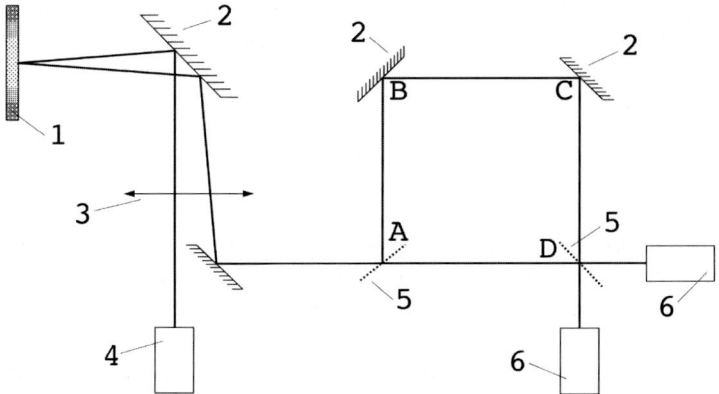

Fig. 9.3. Optical scheme of LDI: 1 – mirror-surface sample to be studied; 2 – immovable mirrors; 3 – focusing lenses; 4 – laser; 5 – beam-splitting plate; 6 – photomultiplier

These limiting factors have been overcome in the interferometric systems known as VISAR, ORVIS, and Fabry-Perot.

9.2.3 Optically Symmetric Interferometers VISAR and ORVIS

The scheme of the optically symmetric interferometer appears in Fig. 9.4. [4]. The VISAR system uses a photomultiplier (1) as the recorder, while the ORVIS system uses a streak camera to obtain a recording of fringe patterns at the streak camera slit (see Fig. 9.4.).

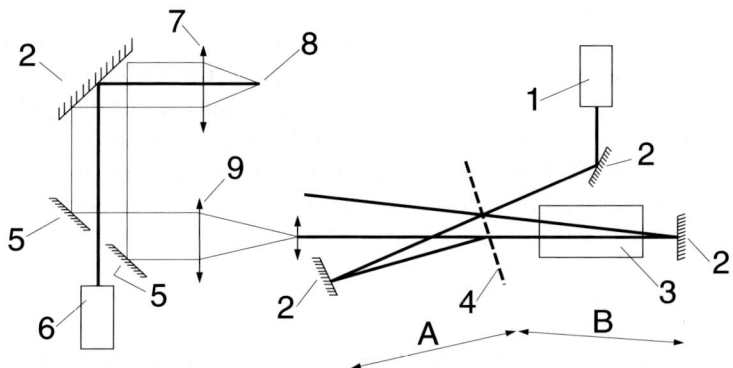

Fig. 9.4. Optical schemes of interferometers VISAR and ORVIS: 1 – recorder; 2 – immovable mirrors; 3 – optical delay line; 4 – beam splitting plate; 5 – apertured mirror; 6 – laser; 7 – lens; 8 – sample to be studied; 9 – telescope

The beam from laser (6) is focused by lens (7) onto the sample surface. The diffusively scattered light is collected by lens (7) and directed by an approximately parallel beam to the interferometer. To ensure a time delay, glass reference (3), with refractive index n, is inserted into one of the arms of the interferometer. It should be noted that the reference material should be optically homogeneous to a small tolerance in order to ensure that any distortions to the light wave that are introduced by it, are no greater than $\sim$0.1 λ. Equal lengths of interferometer arms A and B ensure a high-quality interference pattern using the light scattered by the diffusive surface. The beam delay time in arm B is:

$$\tau = \frac{2h}{C}\left(n - \frac{1}{n}\right),\tag{9.4}$$

where n is the refractive index of the reference material, h is the reference length, and C is speed of light in a vacuum.

For the VISAR system, the surface velocity is determined by the equation:

$$W\left(t - \frac{\tau}{2}\right) = \frac{\lambda_0 F(t)}{2\tau(1 + \delta)}.\tag{9.5}$$

In Eq. (9.5), $F(t)$ is the number of beats on the central interference pattern intensity oscillogram, and δ is the optical dispersion of the reference material, given by:

$$\delta = \frac{(n_0\lambda_0)}{n_0^2 - 1}\left.\frac{dn}{d\lambda}\right|_{\lambda=\lambda_0},\tag{9.6}$$

where n_0 is the reference material refractive index for wavelength λ_0. Typical values of δ for various optical range wavelengths are 0.02–0.04 Å.

A drawback of the VISAR system is that it is hard to differentiate between acceleration and deceleration when interpreting the oscillograms. For this reason, a polarization encoding system is typically used. Before the splitting of the beam by the beam splitter, the beam of light that is reflected from the sample is polarized at an angle of 45° to the horizontal. A quarter-wavelength plate is positioned in one of the interferometer arms, in which a shift in phase of vertically polarized light $\sim$90° relative to the horizontal component occurs. Upon beam rearrangement, the light beam at the interferometer outlet is split into vertical and horizontal polarization by the polarization beam splitter. The intensity of the beats of each component, which are also phase-shifted by 90°, are recorded independently by two photomultipliers. As a result, velocity change is easy to determine from the interference beat phase relationship. An acceleration sign change will be registered by at least one photoreceiver as a turning point in the oscillogram outside the beat extremes. Hemsing [6] used four photomultipliers for a very reliable determination of the acceleration sign, with the polarization plane rotated by 90°, 180°, and 270°, respectively. The velocity of the reflecting surface is determined uniquely at each time, in part

due to control of the intensity of light penetrating the interferometer and polarization encoding. The history of velocity $W(t)$ is determined by processing
three oscillograms corresponding to two interference beat recording channels
and reflected light intensity control. With computer processing, the history of
velocity $W(t)$ is determined to within an error of ± 5 m/s.

In the VISAR system, time resolution depends on the delay time τ (Eq.
(9.4)). In order to achieve better resolution in the velocity, a longer reference
material yielding a lower velocity constant of the system (Eq. 9.5)) must
be used, but this negatively impacts time resolution. To increase the time
resolution of the optically symmetric interferometer (with this leading to a
reduced resolution of the velocity), [7] suggests that the photomultiplier should
be replaced with a photoelectric recorder (PER), whose time resolution is
greater by 1 to 2 orders of magnitude than that of typical photomultiplier-
oscillograph systems. This new system is referred to as ORVIS (Optically
Recording Velocity Interferometer System) [7,8]. Figure 9.5 depicts the beam
run in ORVIS.

While the interfering beams are parallel in the VISAR system, this is not
the case with ORVIS, resulting in the necessity of bringing the rearranging
beams into convergence (through angle α) at the PER slit. For the interfering
beams to intersect at the PER slit, it is necessary to move mirror (1) from
position A to position A' and turn it through an angle $\alpha/2$. When this is done,
a number of parallel equal thickness fringes develop perpendicular to the slit,
with the distance between intensity maxima expressed as:

$$d = \frac{\lambda_0}{2 \sin \alpha} \, , \tag{9.7}$$

where λ_0 is the probing radiation wavelength, and d is the fringe period.

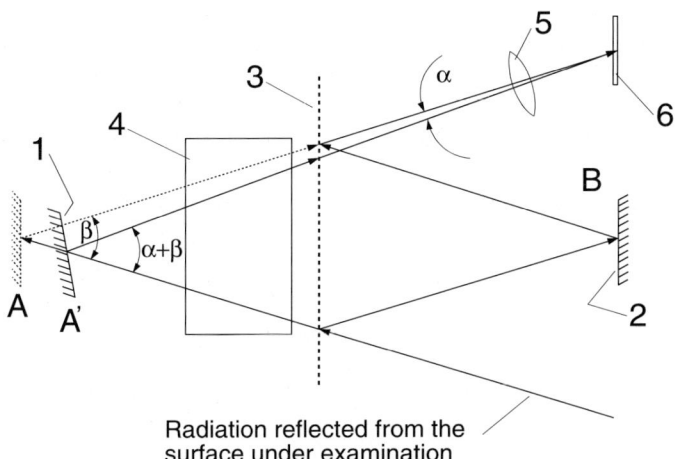

Radiation reflected from the
surface under examination

Fig. 9.5. Optical scheme of beam convergence in interferometer ORVIS: 1 – mirror
in arm A; 2 – mirror in arm B; 3 – beam splitting plate; 4 – glass delay line; 5 –
cylindrical lens; 6 – photorecorder slit

To enhance the interference pattern intensity, the PER slit is placed at the focus of the cylindrical lens. Because of the need for mirror displacement and angular rotation, fringe contrast is somewhat degraded and a correction term, τ_{corr}, is introduced into the time delay, τ, in order to alleviate this. Typically, $\tau_{corr} \ll \tau$ and is neglected in most cases. For ORVIS, unlike for VISAR, one beat period in $F(t)$ corresponds to a fringe displacement through a distance of one period. The fringe center coordinate is related to the beat quantity by:

$$F'(t) = \frac{y(t)}{d} ,$$ (9.8)

where $y(t)$ is the fringe coordinate, and d is the fringe period.

For ORVIS, Eq. (9.5) is modified to:

$$W\left(t - \frac{\tau}{2}\right) = \frac{\lambda_0}{2\tau} \frac{y(t)}{d(1+\delta)} .$$ (9.9)

Thus, when continuously recording velocity using the ORVIS scheme, the Doppler shift reflected from the moving surface is recorded in the form of a change in the interference line intensity maxima on the slit-scan streak camera. The limit in time resolution of the ORVIS system is $\sim 10^{-10}$ s [9]. In [10], the error in velocity measurement is estimated to be ± 30 m/s.

9.2.4 Fabry-Perot Laser Interferometer

In the Fabry-Perot laser interferometer (FPLI), light passing through two parallel partially silvered surfaces is multiply reflected, with a portion of the light being transmitted through the second surface each time the light reaches this surface. The transmitted light consists of multiple beams that can interfere with each other, producing interference rings after focusing by a lens. The optical scheme of the FPLI method appears in Fig. 9.6. The Doppler shift in the frequency of radiation reflected from a moving surface can be recorded (as a function of time), in terms of the change in interference ring diameter using a slit-scan streak camera with continuous velocity measurement [5,11–13], or by using a photographic camera with prompt velocity measurement (see Fig. 9.7) [14]. The FPLI system offers a simple visual method for obtaining results that involves no complicated interpretation of the interferograms. Figure 9.8 presents a typical interferogram for FPLI [15]. The FPLI system is more reliable than other methods, since it records interference ring position, rather than light intensity. This advantage is especially beneficial in shock-wave experiments, where the reflectivity of light coming off a moving surface can change dramatically during the recording.

The recorded velocity is determined as:

$$W = \frac{C\lambda_0}{4L}\left[n + \frac{D_i(t)^2 - D_1^2}{D_2^2 - D_1^2}\right] ,$$ (9.10)

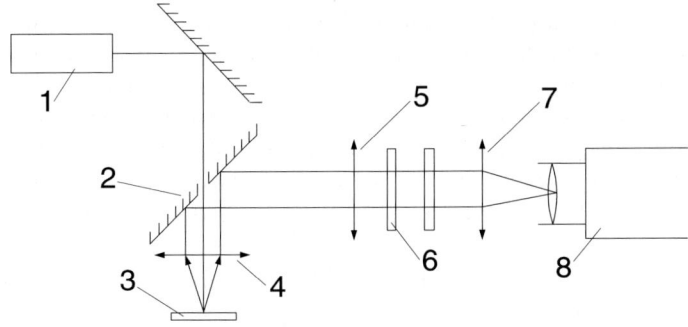

Fig. 9.6. Optical scheme of velocity measurement by FPLI: 1 – laser; 2 – apertured mirror; 3 – moving surface; 4 – short-focus lens; 5 – cylindrical lens; 6 – Fabry-Perot interferometer; 7 – spherical lens; 8 – photorecorder

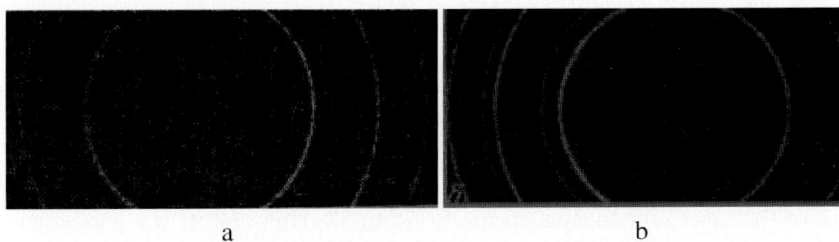

a   b

Fig. 9.7. Prompt velocity spectrograms taken with a photographic camera [16]: **a** – radiation pulse spectrogram (static picture); **b** – spectrogram with Doppler-like mixed component

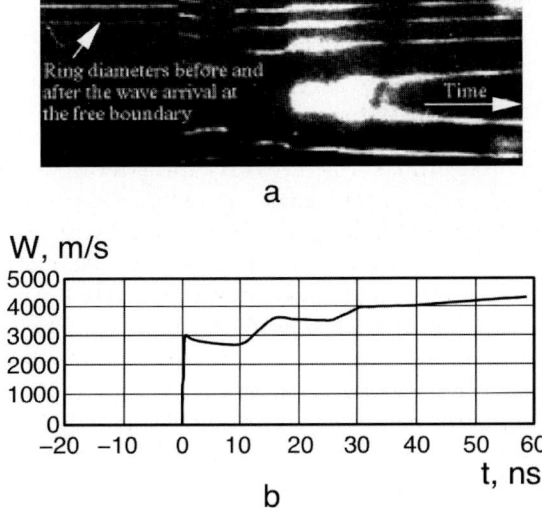

Fig. 9.8. Fabry-Perot interferogram (**a**) and relevant velocity diagram (**b**)

where λ_0 is the laser radiation wavelength, C is the speed of light in a vacuum, L is the distance between the Fabry-Perot mirrors, n is the integer number of the interference ring shift, D_1, and D_2 are diameters of the interference ring corresponding to an immovable target, and $D_i(t)$ is the diameter of the ring located between rings D_1 and D_2.

Similar to the VISAR and ORVIS schemes, the length of the optical delay line in the FPLI scheme is impacted by the length of the optical baseline. With increasing resolution in velocity, the time required for the formation of new interference patterns increases, thereby impairing time resolution.

If the resolution of the recorder is significantly higher than that of the optical scheme, the FPLI time resolution is determined as:

$$\tau_{F-P} = \frac{2L}{C} \left(\frac{\ln\left(1 - \left[\frac{I_p}{I_\infty}\right]^{\frac{1}{2}}\right)}{\ln R} - 1 \right) \qquad (9.11)$$

where τ_{F-P} is the time required for filling the FPLI with light, R is the FPLI mirror reflectivity, and I_P/I_∞ is the FPLI luminous transmissivity.

Of great importance is the choice of FPLI mirror reflectivity. It has to be sufficiently high for the interference pattern contrast to be high and the width of the interference ring profile to be small. However, if the profile width that depends on the interferometer (its spatial resolution) is smaller than that determined by the streak camera characteristics, intensity is lost. Therefore, mirrors of 92–98% reflectivity are typically used. According to [17], the FPLI error in velocity measurement (at $L = 100$ mm, $R = 93\%$) is no more than 5 m/s, with a time resolution of 20 ns.

Owing to its high resolution in velocity measurement and its simplicity and convenience, the system is widely used in shock-wave studies.

9.2.5 Probing Radiation Generator

When recording the reflecting surface velocity using a Doppler interferometric method, the probing radiation generator should be sufficiently powerful and have the needed spectral characteristics. This is why the method was developed soon after the development of lasers (powerful coherent light sources).

The spectral characteristics of the laser are determined by wavelength λ and its spectral width $\delta\lambda$. It is clear that in order to have the capability of recording the Doppler shift $\Delta\lambda$ [see Eq. (9.1)], it is necessary that $\Delta\lambda$ be higher, at least, an order of magnitude higher than $\delta\lambda$. It should be noted that the spectral sensitivity of the recording devices should correspond to the radiation wavelength used in the study. The optical system parameters (reflectivity, transmission and absorption spectrum, optical delay line length) should also correspond to the spectral characteristics. Thus, for the Fabry-Perot interferometer, for example, the spectral width of its device function is determined as:

$$\frac{\delta\lambda}{\lambda} = \frac{\lambda}{L}\frac{1-R}{2\pi\sqrt{R}} , \tag{9.12}$$

where R is the FPLI mirror reflectivity, and L is the distance between the Fabry-Perot mirrors. For the interference pattern maxima to be resolved, $\delta\lambda/\lambda$ should be less than the probing spectral line width.

The required sensitivity of the recorder used depends on the laser power. In the late 1960s, low-power (10^{-2} W) He-Ne lasers were, as a rule, used in the early interferometric systems. The reflecting surface was specular. However, in shock-wave experiments, when strong shock waves approach the probed surface, specularity is compromised and the surface becomes more diffusive. When this occurs, the power of the He-Ne laser is no longer sufficient, even when the probing radiation is focused on the surface to a point and the reflected radiation is directed to the analyzer by a positive lens. Under the assumption of Lambert reflection, the ratio of reflected light flux to light flux incident on the surface can be estimated as:

$$\frac{\phi_r}{\phi_f} = \rho\sin^2\frac{\alpha}{2} , \tag{9.13}$$

where ρ is the diffuse reflectance (equal to 0.4–0.8 for $\lambda = 500$–600 nm for metals, such as steel, copper, or aluminum), and α is the aperture angle of the positive lens. Also impacting the laser power requirement is the fact that light intensity drops along the optical path of the beam because of multiple reflections from boundaries of different media. For a factor of luminous transmission losses of 0.1, aperture angle $\alpha = 10°$, and $\rho = 0.5$, the ratio $\phi_r/\phi_f \sim 3\cdot 10^{-3}$. Using the FPLI method, the illumination of the image that is transmitted from the interferometer is ≈ 0.5 that at the interferometer inlet. Hence, the laser power should be greater than that of the recorder by ~ 3 orders of magnitude. For example, at recorder sensitivity $Q_R = 10^{10}\,\text{J/cm}^2$ and image scanning for $t_R = 10$ µs with speed $V_R = 5$ km/s, under the condition that the illuminated recorded image area is $S = 10$ cm^2, the required probing radiation power will be

$$P = \frac{Q_R S}{t_R}10^3 = 0.1\ W . \tag{9.14}$$

Since the early 1970s, continuous argon lasers with power ~ 5 W have come to be used. That power level is sufficient for recording the velocity of diffusively reflecting surfaces. Velocity field recording requires higher power. In 1982, [16] proposed using a pulsed photodissociation gas iodine laser with wavelength $\lambda = 1.315$ µm and employing radiation transformation into the second harmonic using crystal DKDP or DCDA, $\lambda_{II} = 0.658$ µm. In 1983–1984 a probing radiation generator with power $\sim 10^6$ W, with second harmonic power being up to 100 W, and pulse duration $\tau \approx 5$–25 µs, was used in this scheme with the Fabry-Perot interferometer. The signal that underwent the Doppler shift was recorded onto ordinary film using a high-speed rotating-mirror streak camera [16]. In 1986, [18] used a ruby laser of wavelength

$\lambda = 0.6943$ μm, ~150 W power, and $\tau \approx 200$ μs pulse duration, and a dye laser with a power capacity of ~10^3 W and a pulse duration $\tau \sim 5$ μs. Because of inadequate technological and operational characteristics, the iodine and dye lasers have not found wide application. Current shock-wave studies employ powerful solid-state lasers and laser amplifiers to take simultaneous high-speed (nanosecond) measurements at several points and to enhance the signal-to-noise ratio.

9.3 Features of Interface Velocity Recording

Many shock-wave experiments using laser interferometric techniques include a recording of the time dependence of the velocity of an interface between the material under study and a material of known hydrodynamic properties that is transparent to the probing radiation. The transparent medium may be liquid, ionic crystal, glass, or any other material of appropriate impedance. In [12], McMillan, Goosman, and associates consider qualitatively the effect that the transparent material will have on the recorded velocity. This issue deserves careful consideration since it is apparent that the probing radiation will experience an additional Doppler shift during its transition through the shock front in the window. Moreover, the optical properties of the transparent material in a given pressure range can bring about changes in the frequency and sign of the traversing radiation.

For hypothetical radiation detectors located on the free surface of the window, at the shock front in the window, and on the observed surface, radiation frequency f_i will change in accordance with the velocity of these surfaces relative to one another. For 1D geometry (Fig. 9.9), the recorded frequency is derived from standard relativistic analysis and the requirement of mass continuity at the shock front in the window, and can be expressed as follows:

$$f_4 = f_0 \left[1 + \frac{2W}{C} K \right];$$ \hfill (9.15)

$$K = \frac{n_0 \rho_1 - n_1 \rho_0}{\rho_1 - \rho_0},$$ \hfill (9.16)

where ρ_1 and ρ_0 are the densities of the compressed and non-compressed transparent media, respectively, W is the interface velocity, D is the shock wave velocity in the window material, C is the speed of light in a vacuum, K is a correction factor, and n_1 and n_0 are the refractive indexes of the compressed and non-compressed transparent media, respectively.

Different laws describe the behavior of the refractive index as a function of material density. For low-density materials (including some fluids, e.g., water), the Gladstone-Dale function is appropriate. Under the assumptions of Gladstone-Dale, the difference between the optical path lengths in material

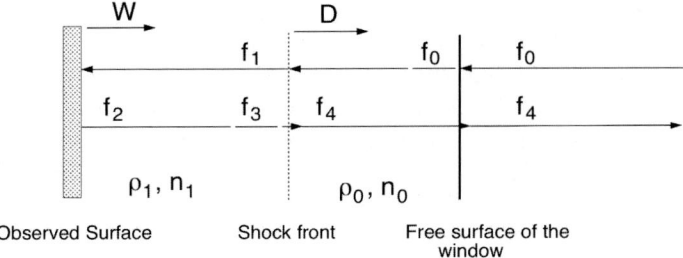

Fig. 9.9. Geometry of beam run through the shock front in the window

versus vacuum, for the same geometric path length, is proportional to the integral material density in terms of the number of atoms per cm^3, irrespective of compression σ within this path length:

$$\sigma = \frac{\rho_1}{\rho_0} = \frac{n_1 - 1}{n_0 - 1} , \qquad (9.17)$$

and $K = 1$.

In fact, the refractive index can vary as a function of material compression [19, 20]. The authors of [19] suggest the following relationship (applicable to a number of transparent materials):

$$n_1 = n_0 + \frac{dn}{d\sigma}(\sigma - 1) . \qquad (9.18)$$

If $dn/d\sigma = n_0 - 1$, then Eq. (9.18) transforms into the Gladstone–Dale function and $K = 1$; otherwise:

$$K = n_0 - \frac{dn}{d\sigma} . \qquad (9.19)$$

Thus the obtained sample-window interface velocity has to be reduced by a factor of K when processing the recorded probing radiation Doppler frequency shift (irrespective of the interferometric scheme).

The method of recording the velocity of the interface between sample and a transparent barrier (window) is widely used to study the detonation wave front mass velocity profile. For this purpose, [21] uses single crystal LiF for the window (Figs. 9.10 and 9.11).

In the interferogram (Fig. 9.11), one can see kinks in the recording of the Doppler shift of the light reflected from the LiF-HE interface, and from the LiF free surface at shock wave arrival (the recording was made to provide a record of mass velocity using the Fabry-Perot method). This is due to the effect of the correction factor K. It is therefore apparent, that in modern studies using interferometric methods that include the use of transparent windows, both shock wave and optical characteristics of the transparent material must be known over the appropriate range of pressure and temperature.

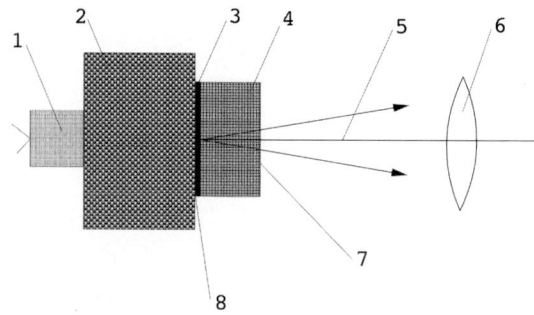

Fig. 9.10. Diagram of setting up experiment [21]: 1 – electric detonator; 2 – HE sample; 3 – Al coating (1 μm); 4 – LiF single crystal; 5 – probing radiation, $\lambda =$ 694.3 nm; 6 – focusing lens; 7 – free boundary of LiF crystal; 8 – interface

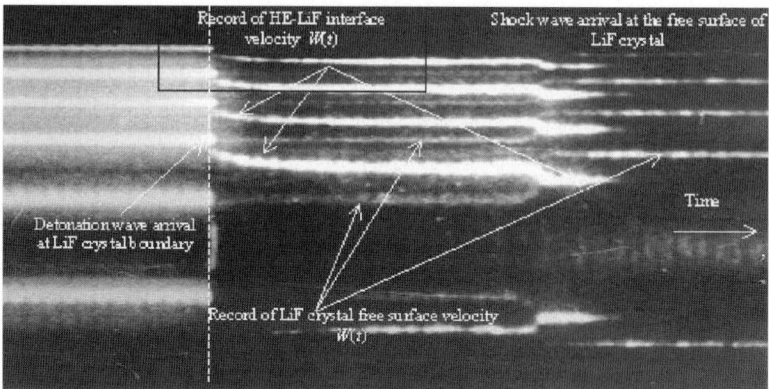

Fig. 9.11. Interferogram of the HE-LiF interface motion [21]

9.4 Multichannel Interferometric Systems

Thanks to the high spatial resolution that is afforded by laser interferometric techniques, it has been possible to develop multi-channel measuring systems and facilities. Optical fibers used for the transport of probing radiation are most effective in this respect. In 1996, D. Goosman and associates developed a multi-channel version of the laser Fabry-Perot velocity interferometer (Fig. 9.12) [22]. It incorporated a 1000-W radiation source with pulse duration of 80-μs. The probing radiation was actuated with a specially developed optical splitter so that the radiation could be provided through five separate optical fibers. Radiation reflected from five distinct points on the surface under examination (Dopler shifted because of surface motion), penetrated (again via five optical fibers) into a focusing system and was then directed to a single Fabry-Perot interferometer common to all five optical fibers. After that the interference patterns were focused on five separate recorders using a

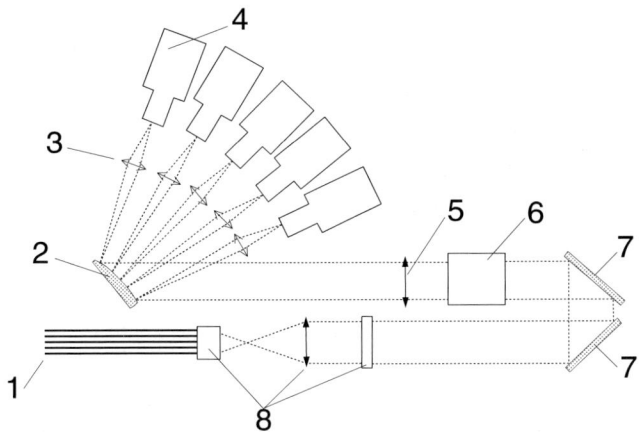

Fig. 9.12. Optical scheme of the five-channel velocimeter for velocity measurement with the Fabry-Perot method: 1 – optical fibers; 2 – pentahedral mirror; 3 – photorecorder lenses; 4 – photorecorders; 5 – spherical lens; 6 – Fabry-Perot interferometer; 7 – rotating mirror; 8 – focusing system

pentahedral mirror and five cylindrical lenses. The authors developed 10- and 20-channel versions of the facility.

In [23], the FPLI system was used to study the shock compressibility of polystyrene. The recording was performed through two channels with incident radiation focused on two points of a stepped target made of the material to be studied (Fig. 9.13). This setup enabled the measurement of both free surface mass velocity u and average (over the step height) shock-wave velocity D in a single experiment, using a single optical scheme and a single interferometer. These experiments find wide application in studies of $D - u$ relationships and EOS model construction for a variety of materials.

In 1991, Hemsing and associates [24] recorded the surface acceleration dynamics of a 25-mm diameter metal plate using a similar method for focusing the probing radiation and a 600-W argon laser radiation amplifier (see Fig. 9.14). The plate velocity field was determined from a recording of the

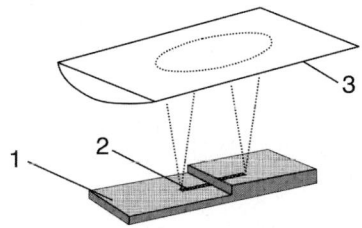

Fig. 9.13. Optical scheme of laser radiation focusing on a stepped surface: 1 – stepped sample; 2 – focused laser spot; 3 – cylindrical lens

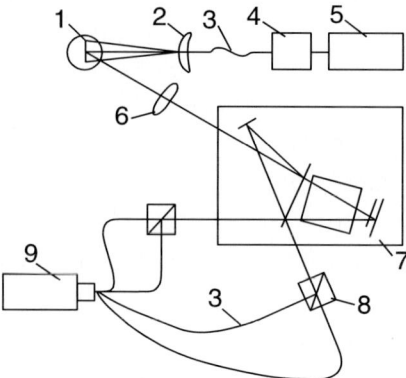

Fig. 9.14. Scheme of experiment [24]: 1 – light image strip; 2 – cylindrical lens; 3 – optical fiber; 4 – light shutter; 5 – laser; 6 – spherical lens; 7 – interferometer; 8 – polarized cubic light splitter; 9 – electron-optical recorder

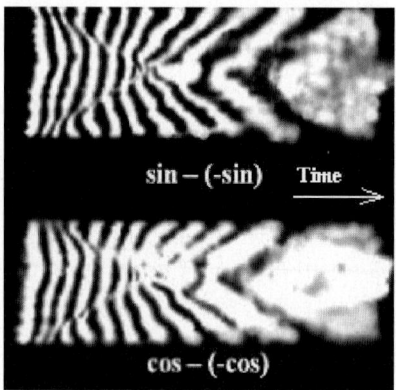

Fig. 9.15. Time scan of vertical (sin) and horizontal (cos) polarization interference spots on photoelectronic recorder slit [24]

interference pattern of a central spot in the interferometric VISAR system, taken with a double-slotted streak camera. The interference pattern of the horizontally polarized image was adjusted to one streak camera slit and that of the vertically polarized image to the other (Fig. 9.15). Velocity time history was recorded for the plate segment (25 mm in diameter) unevenly accelerated to a velocity of 5.5 km/s in 4 μs (Fig. 9.16).

In 1986, Gidon and Behar suggested a method for prompt recording of the surface velocity field using the Fabry-Perot interferometric method [17]. The idea of the method is that a portion of the surface to be examined is exposed to a powerful laser radiation pulse. The reflected light is analyzed with a Fabry-Perot interferometer and the shape and position of the interferometric rings are recorded at different fixed times using a frame-by-frame photographic

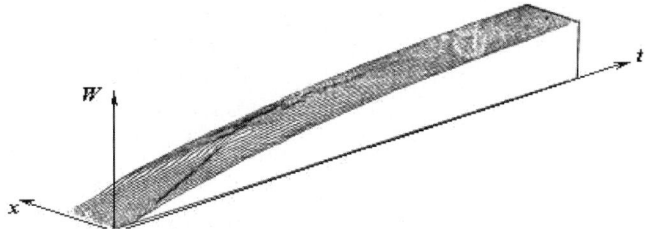

Fig. 9.16. Profile of 25-mm surface segment velocity [24]

method (Fig. 9.17). In this work, the ring coordinate is responsible, both for the point of the surface under study, and for the velocity at that point.

In 1991 Mathews and associates used a 500-W laser amplifier to obtain frame-by-frame images of the velocity field of a 3-mm square surface area accelerated to 3 km/s in 1 μs (see Fig. 9.18). The exposure duration of each frame was 10 ns. The velocity history of the entire exposed surface was constructed from the individual images [25].

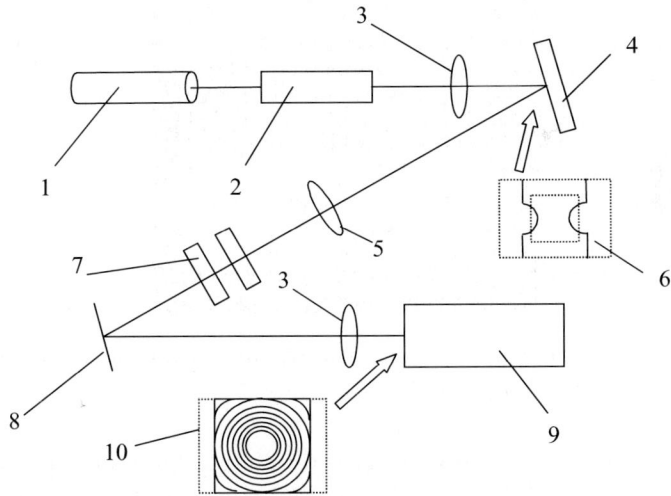

Fig. 9.17. Scheme of Fabry-Perot experiment [17]: 1 – laser; 2 – magnifier; 3 – focusing lens; 4 – surface to be examined; 5 – collective lens; 6 – exposed surface area; 7 – Fabry-Perot interferometer; 8 – rotating mirror; 9 – photoelectronic recorder; 10 – interference pattern

a   b

Fig. 9.18. Velocity profile of the surface points in 10 ns (**a**) and in 200 ns (**b**)

9.5 Using Laser Interferometric Systems in Shock-Wave Studies

Precision and the absence of any perturbing effects on the process under study are reasons for the widespread use of laser interferometric systems in shock wave and high-pressure physics. By 1982, more than 35 types of interferometric systems had been used. Their acceptance may be attributed to the shortage of alternative methods of measurement that would offer similar capabilities with respect to time resolution and velocity measurement accuracy [9]. In the 1990s, shock-wave studies employed more than one hundred types of interferometric systems.

Thanks to their capabilities for the high resolution of time, laser interferometric methods have found wide application in the study of fast physical phenomena, such as occurs with detonation. In 1997, C.M. Tarver and associates used the FPLI method in a study of detonation waves in PETN [26]. The method was applied in three different types of experiments, all driven by the explosion of PETN: (1) experiments requiring the measurement of free surface mass velocity of tantalum disks, (2) experiments requiring the measurement of the mass velocity of an HE-barrier interface (LiF single crystal barrier with ~1000-Å-gold-sprayed along the interface), and (3) experiments requiring the measurement of free surface radial velocity of cylindrical copper shells. These experiments were conducted with a time resolution ≈5 ns. The scheme of the experimental setups is shown in Fig. 9.19. The duration of the chemical reaction in PETN was shown to be shorter than 5 ns. Some PETN charges were pressed to densities close to the crystalline density, and exhibited supracompressed detonation. Good agreement between experimental data and computation was obtained for the chemical reaction delay time and for detonation depth (distance traveled by the initial shock wave pulse until steady detonation) in the pressed PETN. Recording devices of subnanosecond time resolution are needed to register the chemical peak in PETN.

In Russia, the ORVIS system was used for the first time in the work of [10], to record the detonation wave front structure for a number of HEs (Figs. 9.20 and 9.21).

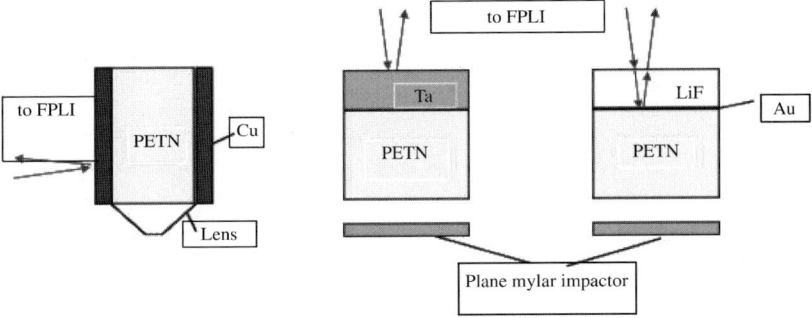

Fig. 9.19. Schemes of setting up experiments [26]

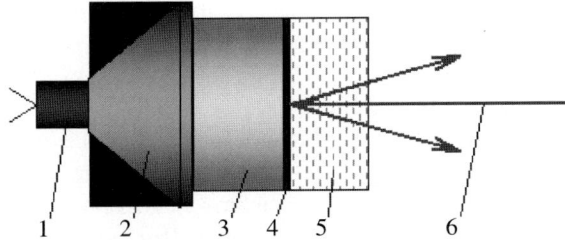

Fig. 9.20. Scheme of setting up the experiment to study the chemical peak in HE:
1 – electric detonator; 2 – plane-wave lens; 3 – HE sample; 4 – Al foil (200 μm); 5 –
transparent window; 6 – probing radiation

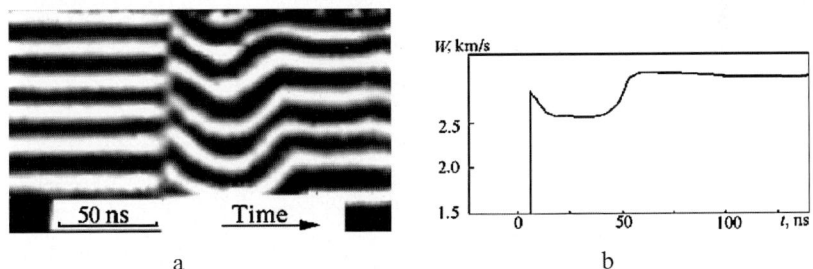

Fig. 9.21. Time-base sweep of the ORVIS interference spot on the photoelectronic
recorder slit (**a**) and relevant velocity plot (**b**) [10]

Setchell [27] also loaded different HE with weak shock waves and recorded
the mass velocity profiles for a variety of HE thicknesses through a quartz
window using the VISAR system. In [28], the impact of thin plates driven
by an electric explosion was used to determine the detonation initiation
threshold. Plate velocities were recorded using the FPLI system. Tantalum
shock compressibility was also recorded for pressures in the range of 190 to
780 GPa (with impact velocities ranging from 4 to 9.7 mm/μs). The capability
afforded by laser interferometric systems for high spatial resolution con-

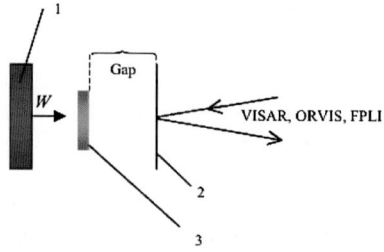

Fig. 9.22. Scheme of the experiment: 1 – Cu impacting plate; 2 – Al foil; 3 – sample

tributed to their wide application in studies of material melting at high pressures, and of the velocity of microparticle ejection from a free surface upon the arrival of a shock wave [29, 30]. The method developed by Asay from Sandia Laboratories [29] is widely used for recording shock-wave induced melting and vaporization of metals. The idea of the method is that a thin (10–100 μm) aluminum foil, "Asay Foil", is placed in the path of metal droplets, vapor, dust, or fine particles coming off a free surface. The foil is separated from the free surface by an evacuated gap (Fig. 9.22). The particle cloud picks up and entrains the foil, which serves as an indicator. The velocity of the indicator foil is recorded using the laser interferometer. Under the assumptions that shock-melted particles are inelastically decelerated on the indicator foil, and the entire melted material mass ejection occurs within a short period of time (initiated by the reflection of the unloading wave from the free surface), the rate of mass accumulation on the foil is:

$$\frac{dM}{dt} = \frac{M_f}{W - W_f} \frac{dW_f}{dt} , \tag{9.20}$$

where M is the specific mass of the sample particles on the foil ($\mu g/cm^2$), M_f is the specific mass of the foil plus sample particles ($\mu g/cm^2$), W is the average velocity of sample particles across the gap (m/s), and W_f is the foil velocity (m/s).

Foil velocity and mass of the particles accumulated onto the foil (from [29]) are plotted against time in Fig. 9.23.

The high accuracy that the laser interferometric techniques offer in the measurement of free surface and interface mass velocity has made them an invaluable tool in support of studies of the elastic-plastic behavior of materials (spall, elastic precursor, etc.), the optical properties of transparent materials, the EOS of materials, and failure theories and phase transformations over a wide range of pressures.

For example, dynamic elastic limits and spall fracture stresses have been measured in structural steel samples with the use of laser interferometry [31]. Planar loading of the samples was accomplished through the impact of aluminum flyer plates (2 mm and 0.4 mm thick) accelerated by explosive devices, or through the detonation of explosive lenses in contact with the samples.

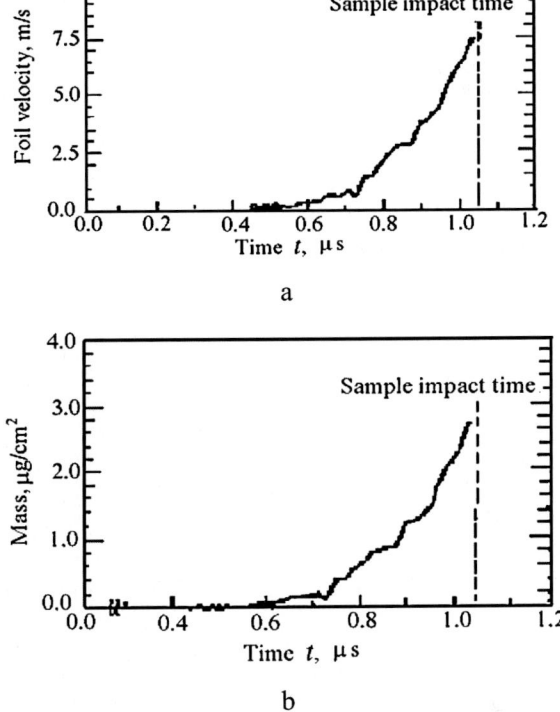

Fig. 9.23. Foil velocity (**a**) and foil mass increment (**b**) versus time [29]

Strength properties were determined from the analysis of the sample's free surface velocity profiles recorded with the aid of a VISAR laser Doppler system.

Chhabildas [32] used two VISAR systems to measure both the normal and shear wave particle velocities of shocked Y-cut quartz. Both longitudinal and transverse velocity profiles are given, along with the corresponding stress levels. The beam of one system was directed at an angle to the surface, that of the other, perpendicular. In [33], the investigators used two LDI systems to record high material strain rates occurring in a target subjected to projectile impact. The VISAR system has been used to study EOS and shock compressibility of zinc chloride water solutions [34]. The VISAR system has been used to study the α-ϵ phase transition that occurs in Armco iron at 13 GPa [35]. Yu. Meshcheryakov and associates studied the spall strength and dynamic strain of a number of materials using the LDI system [36], while Johnson and Barker [37] studied the plastic wave profile in aluminum.

9.6 Optical and Interferometric Techniques for Velocity Measurement

9.6.1 Method of Laser Wave Velocimeter (LWV)

The LWV method (Fig. 9.24) is not interferometric [38]. Indicator barrier (2) is placed in the path of the shock wave immediately following the sample (1). The indicator consists of a set of thin plates made of transparent material (most frequently of organic glass).

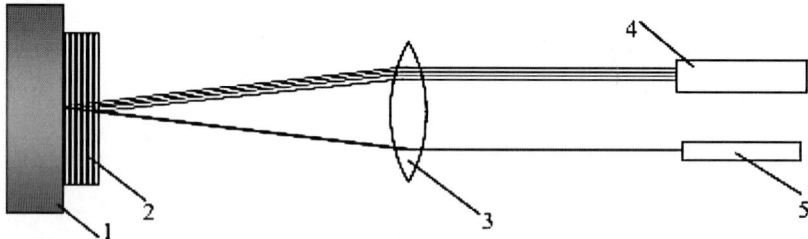

Fig. 9.24. Scheme of the LWV experiment: 1 – sample to be studied; 2 – indicator barrier plate set; 3 – focusing lens; 4 – photoelectric multiplier; 5 – laser

The set of plates is compressed so that the thickness of the gaps between the plates is no greater than a few micrometers and exerts no noticeable influence on the shock process in the set. The plates are probed with the focused laser beam in the direction perpendicular to the base plane. Each surface in the set of plates reflects a small amount of light, which is received by a fast photoelectric multiplier (4). The shock closes the gaps between the plates. Removal of the reflecting surfaces causes a jump-like change in the intensity of light entering the photoreceiver. Plate thicknesses are carefully measured so that from an estimate of the time intervals between the intensity jumps of the light penetrating the photoreceiver, one can determine the shock wave velocity in each plate of the set. The small cross-sectional dimensions of the probing laser beam ensure a high accuracy in the measurement of shock wave velocity in the indicator barrier, although the question of the effect of nonzero gaps and interfaces on shock wave propagation does remain open. The LWV method is used extensively to record detonation wave structure and has exhibited a fairly high efficiency [38].

9.6.2 Radioelectric Method

The basis for the radioelectric velocity measurement method, like that for the laser displacement interferometric method, is the measurement of changes in phase and amplitude of electromagnetic waves reflected from a surface [39]. The only difference is in the frequency of the probing radiation; in this case the wavelengths of the radio-frequency radiation occupy the K band. However, the temporal and spatial resolutions of the radioelectric probing techniques are significantly less than those of the laser systems because of the difference in the electromagnetic wave length (by a factor of $\sim 10^4$). This is the reason for an insufficiently sharp radiation directionality (due to the electromagnetic wave diffraction divergence ($\sim \lambda$) in the limit) and leads to averaging of measured velocity over $\sim$10-cm^{2}area. A merit of the method (a demerit in some cases), is its insensitivity to the perturbing action of material releases from the shock-driven surface under study. It should be noted, that when measuring free surface velocity with microwave methods, the upper limit of the measurement depends on the signal reflected from a layer of plasma (dissociated air) produced following the arrival of the shock wave front at the moving surface. This perturbing effect can be avoided by placing the moving surface in a vacuum.

An advantage of the method is the possibility it affords for the straightforward diagnostics of shock and detonation waves in optically opaque media, for example, in condensed HE. In [39], the investigators use the radioelectric method to study the planar acceleration of copper and aluminum disks propelled by like-diameter pellets of HE (Fig. 9.25). The probing radiowave frequency used in the study was 24 MHz. The principle of recording the change in phase and microwave amplitude is as follows: a part of the permanent wave and a part of the surface-reflected wave are integrated in two electronic adders yielding two independent interference signals. The signals enter two detectors, with at least one of the detectors ensuring a good sensitivity to the surface displacement thanks to phase shifters placed in the permanent port of the radioelectronic interferometer. To ensure a high precision, the interferometer had been calibrated prior to the experiment with a fine adjustment screw (8). An accuracy in the measurement of velocity $\approx$1 km/s (averaged over an area $\sim$10-cm^2) through a 62.5-μm displacement length was estimated to be 3% [39].

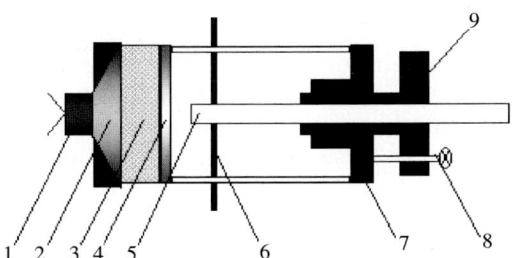

Fig. 9.25. Scheme of experiment [39]: 1 – detonator; 2 – plane wave generator; 3 – HE charge; 4 – plate to be launched; 5 – wave guide; 6 – material sheet absorbing undesirable microwave reflections; 7 – sample holder; 8 – micrometer screw; 9 – guide bushing

References

1. Barker, L.M., and Hollenback, R.E., "Interferometer Technique for Measuring the Dynamic Mechanical Properties of Materials," *Review of Scientific Instruments*, Vol. 36, No. 11, 1965, pp. 1617–1620.
2. Zwick, H.H., and Shepherd, G.G., "Defocusing a Wide-Angle Michelson Interferometer," *Applied Optics*, Vol. 10, No. 11, 1971, pp. 2569–2571.
3. Zlatin, N.A., Mochalov, S.M., Pugachev, G.S., and Bragov, A.M., "Laser Differential Interferometer," *Zhurnal Technicheskoi Fiziki*, Vol. 43, No. 9, 1973, pp. 1961–1964, [English trans., *Soviet Physics Technical Physics*, Vol. 18, No. 9, 1974, pp. 1235–1237].
4. Barker, L.M., and Hollenbach, R.E., "Laser Interferometer for Measuring High Velocities of any Reflecting Surface," *Journal of Applied Physics*, Vol. 43, No. 11, 1972, pp. 4669–4675.
5. Durand, M., "The Use of Fiber Optics to Measure Velocity Means of Laser Doppler Interferometry with Fabry-Perot Interferometry," *Proc., 16th International Congress on High Speed Photography and Photonics*, Aug. 27–31, 1984, Strasbourg, France, SPIE, Vol. 491, Andre, M., and Hugenschmidt, M., eds., pp. 650–656 (in French).
6. Hemsing, W.F., "Velocity Sensing Interferometer (VISAR) Modification," *Review of Scientific Instruments*, Vol. 50, No. 1, 1979, pp. 73–78.
7. Bloomquist, D.D., and Sheffield, S.A., "Optically Recording Velocity Interferometer System (ORVIS) for Subnanosecond Particle Velocity Measurements in Shock Waves," *Proc., 15th International Congress on High Speed Photography and Photonics*, Aug. 21–27, 1982, San Diego, CA, SPIE Vol. 348, Endelman, L.L., ed., pp. 523–528.
8. Sheffield, S.A., and Fisk, G.A., "Particle Velocity Measurements of Laser-Induced Shock Waves Using ORVIS," *Proc., High Speed Photography, Videography, and Photonics*, Aug. 23–25, 1983, San Diego, CA, SPIE Vol. 427, Paisley, D.L., ed., pp. 193–198.
9. Barker, L.M., "Velocity Interferometry for Time-Resolved High-Velocity Measurement," *Proc., High Speed Photography, Videography, and Photonics*, Aug 23–25, 1983, San Diego, CA, SPIE Vol. 427, Paisley, D.L., ed., pp. 116–126.
10. Utkin, A.V., and Kanel, G.I., "Studying Decomposition Kinetics of Trinitrotoluene and Phlegmatized Hexogen in Shock and Detonation Waves," *Proc., 8th All-Union Symposium on Combustion and Detonation*, Detonation and Shock Waves, 1986, Tashkent, pp. 13–16.
11. Durand, M., Laharrague, P., Lalle, P., Le Bihan, A., Morvan, J., and Pujols, H., "Interferometric Laser Technique for Accurate Velocity Measurement in Shock Wave Physics," *Review of Scientific Instruments*, Vol. 48, No. 3, 1977, pp. 275–278.
12. McMillan, C.F., Goosman, D.R., Parker, N.L., Steinmetz, L.L., Chau, H.H., Huen, T., Whipkey, R.K., and Perry, S.J., "Velocimetry of Fast Surfaces Using Fabry-Perot Interferometry," *Review of Scientific Instruments*, Vol. 59, No. 1, 1988, pp. 1–20.
13. Seitz, W.L., and Stacy, H.L., "Fabry-Perot Interferometry Using an Image-Intensified Rotating-Mirror Streak Camera," *Proc., High Speed Photography, Videography, and Photonics*, Aug 23–25, 1983, San Diego, CA, SPIE Vol. 427, Paisley, D.L., ed., pp. 186–192.

14. Kovylov, A.F., Kormer, S.B., Pinegin, A.V., Poklontsev, B.A., and Yushko, K.B., "Laser Doppler Velocimeter," *Pribory i Tekhnika Eksperimenta*, No. 1, 1978, pp. 205–207, [English trans., *Instruments and Experimental Techniques*, Vol. 21, No. 1, Part 2, 1978, pp. 219–220].

15. Fedorov, A.V., Mikhailov, A.L., and Poklontsev, B.A., "Laser Doppler Velocimeter on the Base of High-Power Iodine and Ruby Lasers," *Proc., 22nd International Congress on High-Speed Photography and Photonics*, Oct. 27 – Nov. 1, 1996, Santa Fe, NM, SPIE Vol. 2869, Paisley, D.L., and Frank, A.M., eds., pp. 890–893.

16. Vlasova, G.B., Mikhailov, A.L., Poklontsev, B.A., and Fedorov, A.V., "Iodine-Laser-Based Doppler Meter for Measuring the Velocity of Shock-Accelerated Targets," *Fizika Goreniya i Vzryva*, Vol. 24, No. 1, 1988, pp. 127–130, [English trans., *Combustion, Explosion, and Shock Waves*, Vol. 24, No. 1, 1988, pp. 117–120].

17. Gidon, S., and Behar, G., "Instantaneous Velocity Field Measurements: Application to Shock Wave Studies," *Applied Optics*, Vol. 25, No. 9, 1986, pp. 1429–1433.

18. Fedorov, A.V., and Gerasimov, V.M., "Using Powerful Ruby Lasers in Interferometric Systems," *Proc., International Conference on Applied Optics – 94*, Nov. 1994, St. Petersburg, Russia, p. 353.

19. Kormer, S.B., "Optical Study of the Characteristics of Shock-Compressed Condensed Dielectrics," *Uspekhi Fizicheskikh Nauk*, Vol. 94, 1964, pp. 641–687, [English trans., *Soviet Physics Uspekhi*, Vol. 11, No. 2, 1968, pp. 229–254].

20. Barker, L.M., and Hollenback, R.E., "Shock-Wave Studies of PMMA, Fused Silica, and Sapphire, *Journal of Applied Physics*, Vol. 41, No. 10, 1970, pp. 4208–4226.

21. Fedorov, A.V., Zotov, E.V., Krasovsky, G.B., Menshikh, A.V., and Yagodin, N.B., "Detonation Front in Homogeneous and Heterogeneous High Explosives," Shock Compression of Condensed Matter – 1999, Furnish, M.D., Chhabildas, L.C., and Hixson, R.S., eds., AIP Press, Melville, NY, 2000, pp. 801–804.

22. Goosman, D., Avara, G., Steinmetz, L., Lai, C., and Perry, S., "Manybeam Velocimeter for Fast Surfaces," Proc., 22nd *International Congress on High-Speed Photography and Photonics*, Oct. 27 – Nov. 1, 1996, Santa Fe, NM, SPIE Vol. 2869, Paisley, D.L., and Frank, A.M., eds., pp. 1070–1079.

23. Bernier, H., Durand, M., and Lalle, P., "The Use of a Fabry-Perot Velocimeter to the Study of Shock Loaded Polystyrene," *Proc., High Speed Photography, Videography, and Photonics*, Aug. 23–25, 1983, San Diego, CA, SPIE Vol. 427, Paisley, D.L., ed., pp. 218–223.

24. Hemsing, W.F., Mathews, A.R., Warnes, R.H., George, M.J., and Whittemore, G.R., "VISAR: Line-Imaging Interferometer," Shock Compression of Condensed Matter – 1991, Schmidt, S.C., Dick, R.D., Forbes, J.W., and Tasker, D.G., eds., Elsivier, Amsterdam, 1992, pp. 767–770.

25. Mathews, A.R., Boat, R.M., Hemsing, W.F., Warnes, R.H., and Whittemore, G.R., "Full-Field Fabry-Perot Interferometer," Shock Compression of Condensed Matter – 1991, Schmidt, S.C., Dick, R.D., Forbes, J.W., and Tasker, D.G., eds., Elsivier, Amsterdam, 1992, pp. 759–762.

26. Tarver, C.M., Breithaupt, R.D., and Kury, J.W., "Detonation Waves in Pentaerythritol Tetranitrate," *Journal of Applied Physics*, Vol. 81, No. 11, 1997, pp. 7193–7202.

27. Setchell, R.E., "Velocity Interferometer System for any Reflector (VISAR): Studies of Wave Growth in Granular Explosives," *Proc., High Speed Photography, Videography, and Photonics*, Aug. 23–25, 1983, San Diego, CA, SPIE Vol. 427, Paisley, D.L., ed., pp. 149–154.

28. Froeschner, K.E., Lee, R.S., Chau, H.H., and Weingart, R.C., "Shock Hugoniot Measurements on Ta to 0.78 TPa," Shock Waves in Condensed Matter – 1983, Asay, J.R., Graham, R.A., and Straub, G.K., eds., Elsevier, Amsterdam, 1984, pp. 85–88.

29. Asay, J.R., Mix, L.P., and Perry, F.C., "Ejection of Material from Shocked Surfaces," *Applied Physics Letters*, Vol. 29, No. 5, 1976, pp. 284–287.

30. Kanel, G.I., Baumung, K., Rush, D., Singer, J., Razorenov, S.V., and Utkin, A.V., "Melting of Shock-Compressed Metals in Release," Shock Compression of Condensed matter – 1997, Schmidt, S.C., Dandekar, D.P., and Forbes, J.W., eds., AIP Press, Woodbury, NY, 1998, pp. 155–158.

31. Razorenov, S.V., Kanel, G.I., Anufriev, V.G., and Loskutov, V.F., "Deformation and Failure of Structural Steels in Pulsed Loading," *Problemy Prochnosti*, No. 3, 1992, pp. 42–48, [English trans., *Strength of Materials*, Vol. 24, No. 3, 1992, pp. 270–275].

32. Chhabildas, L.C., Sutherland, H.J., and Asay, J.R., "A Velocity Interferometer Technique to Determine Shear-Wave Particle Velocity in Shock-Loaded Solids," *Journal of Applied Physics*, Vol. 50, No. 8, 1979, pp. 5196–5201.

33. Kim, J.K., and Lee, S.S., "Fast Deformation Velocity Measurement Using Laser Doppler Velocity Interferometer," *Review of Scientific Instruments*, Vol. 53, No. 1, 1982, pp. 65–69.

34. Wise, J.L., Refractive Index and Equation of State of a Shock-Compressed Aqueous Solution of Zinc Chloride," Shock Waves in Condensed Matter – 1983, Asay, J.R., Graham, R.A., and Straub, G.K., eds., Elsevier, Amsterdam, 1984, pp. 317–320.

35. Barker, L.M., and Hollenbach R.E., "Shock Wave Study of the $\alpha \leftrightarrow \varepsilon$ Phase Transition in Iron," *Journal of Applied Physics*, Vol. 45, No. 11, 1974, pp. 4872–4887.

36. Meshcheryakov, Yu.I., Divakov, A.K., and Kudryashov, V.G., "Dynamic Strength with Spalling and Breakthrough," *Fizika Goreniya i Vzryva*, Vol. 24, No. 2, 1988, pp. 126–134, [English trans., *Combustion, Explosion, and Shock Waves*, Vol. 24, No. 2, 1988, pp. 241–248].

37. Johnson, J.N., and Barker, L.M., "Dislocation Dynamics and Steady Plastic Wave Profiles in 6061-T6 Aluminum," *Journal of Applied Physics*, Vol. 40, No. 11, 1969, pp. 4321–4334.

38. Ashaev, V.K., Levin, A.D., and Mironov, O.N., "Measurement of Shock Wave Parameters by an Optical Technique," *Pisma v Zhurnal Tekhnicheskoi Fiziki*, Vol. 6, 1980, pp. 1005–1009, [English trans., *Soviet Technical Physics Letters*, Vol. 6, No. 8, 1980, pp. 433–434].

39. Maron, Y., and Blaugrund, A.E., "Measurements of High Surface Velocities Using K-Band Microwave Interferometry," *Review of Scientific Instruments*, Vol. 51, No. 5, 1980, pp. 666–669.

Index

Printing: Krips bv, Meppel
Binding: Stürtz, Würzburg